Interpreting ECGs

A Practical Approach

Third Edition

3e

Mc Graw Hill Education

Interpreting ECGs

A Practical Approach

Third Edition

3e

Bruce Shade

EMT-P, EMS-I, AAS

Mc Graw Hill Education

INTERPRETING ECGs: A PRACTICAL APPROACH, THIRD EDITION

Published by McGraw-Hill Education, 2 Penn Plaza, New York, NY 10121. Copyright © 2019 by McGraw-Hill Education. All rights reserved. Printed in the United States of America. Previous editions © 2013 and 2007. No part of this publication may be reproduced or distributed in any form or by any means, or stored in a database or retrieval system, without the prior written consent of McGraw-Hill Education, including, but not limited to, in any network or other electronic storage or transmission, or broadcast for distance learning.

Some ancillaries, including electronic and print components, may not be available to customers outside the United States.

This book is printed on acid-free paper.

1 2 3 4 5 6 7 8 9 LMN 21 20 19 18

ISBN 978-1-260-09293-6
MHID 1-260-09293-3

mheducation.com/highered

About the Author

Courtesy Bruce Shade

Bruce Shade is currently employed as the EMS Educator for Cleveland Clinic Hillcrest Hospital in Northeast Ohio. He is also a paramedic instructor at Cuyahoga Community College (Tri-C) in Cleveland. Bruce is also a past Chairperson of the Ohio Emergency Medical Services Board and just recently retired as a part-time firefighter for the City of Willoughby.

Bruce has been involved in emergency services since 1972. He started as a volunteer firefighter/EMT for Granger Township and then served as paramedic, educational supervisor, paramedic training program director, and commissioner for the City of Cleveland's Division of Emergency Medical Service for the next 25 years. During those years, he also worked as a part-time firefighter/paramedic for Willowick Fire Department and the paramedic faculty at Lakeland Community College. For the remainder of his career with Cleveland, he served as an Assistant Public Safety Director. Since retiring, Bruce worked as a Homeland Security Consultant, Operations Director for Community Care Ambulance, and Assistant Safety-Service Director for the City of Elyria, all in Northeast Ohio.

Bruce is past President, Vice President, and Treasurer of the National Association of EMTs and chairperson of the Instructor Coordinator Society. He has served as president of several local associations and chairperson of many committees and task forces. Bruce has authored several EMS textbooks and written many EMS-related articles. He has lectured at local, regional, state, and national EMS conferences.

Dedication

This book is dedicated to my father, Elmer Shade, Jr. He recently passed away at the age of 97. He grew up during the depression, served in France during the Second World War, and worked hard his entire life. He was still mowing 20 acres of property each week at 96 years of age. A lifelong Cleveland sports fan, he had a keen sense of humor and a strong set of values and work ethic. He was known for his ability to tell stories and recall his life experiences. I can say, with great pride, that I acquired many of his traits. My ability to communicate information through textbooks can be directly attributed to what I learned from him.

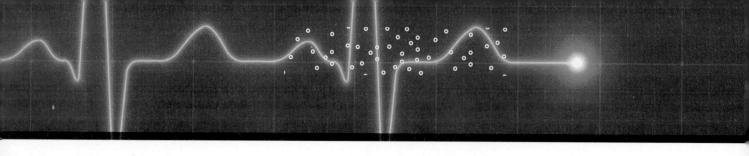

Contents

Preface

This book presents information similar to how an instructor delivers it in the classroom, with lots of illustrations, solid practical content, plentiful reinforcement of material, questions to prompt critical thinking, case presentations, and plentiful practice ECG tracings to promote the application of skills.

One of the first things readers will notice about this text is it is more of a "how-to book" than a "theoretical book." Although there is plenty of detail, the coverage is to the point, telling you and then showing you what you need to know. The breadth of information ranges from simple to complex, but regardless of how advanced the material, the explanations and visuals make the concepts easy to understand. Another aspect of this book is that it truly covers both dysrhythmia and 12-lead analysis and interpretation. It reinforces those core concepts from the beginning to the end using lots of repetition. This book includes plentiful pictures and figures to help readers see what is being discussed in actual use. We have also included coverage of the treatments used to manage the various dysrhythmias and cardiac conditions to give readers a broader perspective and better prepare them for applying what they have learned.

Structure of This Book

This book is divided into five sections:

- **Section 1, Preparatory,** looks at the underlying concepts of the anatomy and electrophysiology of the heart and the electrocardiogram itself.

- **Section 2, The Nine-Step Process,** comprises Chapters 3 through 9 and presents the Nine-Step Process of ECG interpretation. Each chapter provides an in-depth look at one of the steps and introduces the reader to the variances seen with that step.

- **Section 3, Origin and Clinical Aspects of Dysrhythmias,** comprises Chapters 10 through 17 and leads readers through an overview of heart disease and a thorough discussion regarding dysrhythmias. The section covers the origin of dysrhythmias, including the sinus node, the atria, the atrioventricular junction, the ventricles, atrioventricular heart blocks, and pacemakers. And it covers the clinical aspects of each dysrhythmia.

- **Section 4, 12-Lead ECGs,** introduces the concept of 12-lead ECGs in Chapter 18. Then Chapters 19 through 22 cover interpretation and recognition of myocardial ischemia, injury and infarction, bundle branch block and atrial enlargement and ventricular hypertrophy, and a host of other cardiac conditions and their effect on the ECG.

- **Section 5, Review and Assessment,** wraps it all up with the chapter "Putting It All Together" and more practice tracings.

Changes to the Book

Among the changes in this book is that we have retitled it to better reflect its comprehensive nature. While it is still easy to learn to interpret ECGs using this book, its volume and breadth of coverage make it difficult to read from cover to cover in a fast way. The third edition of *Fast & Easy ECGs: A Self-Paced Learning Program* by Bruce Shade is thorough, innovative, and greatly enhanced. We have changed the title to better reflect the comprehensive nature of this book. While we strive to make our approach fast and easy, there are many complicated aspects of learning how to analyze and interpret ECG tracings. For this reason, we cover the material in sufficient depth to provide the reader with everything they need to know in order to be proficient with this important skill.

Whereas the second edition had 22 chapters, this book is expanded and includes 23 chapters. The following chapter is brand new to this edition:

Chapter 10 provides an overview of heart disease, including what it is, the risks for developing it, and its causes and complications. Then we review the common types of heart disease. This chapter is designed to provide the reader with an understanding of how dysrhythmias and cardiac conditions occur. This will make it easier for the reader to understand the characteristics associated with each dysrhythmia and cardiac condition.

In addition to the expanded content, this book has more than 300 figures and close to 400 practice ECG tracings. It also introduces the reader to the treatment modalities for the various dysrhythmias and medical conditions.

We hope this book is beneficial to both students and instructors. Greater understanding of ECG interpretation will lead to better patient care everywhere.

Instructor Resources

Instructors, are you looking for additional resources? Be sure to visit www.mhhe .com/shade3e for answer keys, an Electronic Testbank, and accessible PowerPoint Presentations. Access is for instructors only and requires a user name and password from your McGraw-Hill Learning Technology Representative. To find your McGraw-Hill representative, go to www.mheducation.com and click "Contact," then "Contact a Sales Rep."

Need help? Contact the McGraw-Hill Education Customer Experience Group (CXG). Visit the CXG website at www.mhhe.com/support. Browse our freasked questions (FAQs) and product documentation and/or contact a CXG representative.

Features to Help You Study and Learn

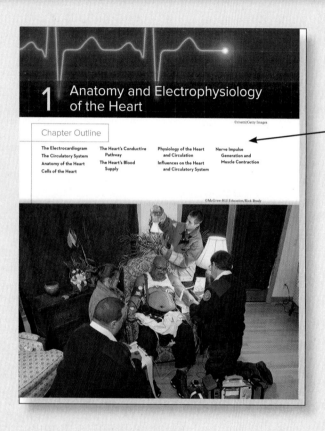

Preview the Chapter Content

Study the Chapter Outline to get an overview of the subjects to be covered in the chapter.

Review the Learning Outcomes to see what you will learn. Note that the Learning Outcomes numbers are keyed to the text and learning assessments.

Learning Outcomes

LO 1.1 Define the term *electrocardiogram*, list its uses, and describe how it works.

LO 1.2 List the components of the circulatory system.

LO 1.3 Describe the anatomy of the heart.

LO 1.4 Identify and contrast the structure and function of the different types of heart cells.

LO 1.5 Identify the structures of the heart's conduction system and describe what each does.

LO 1.6 Identify how the heart receives most of its blood supply.

LO 1.7 Recall how the heart and circulatory system circulates blood throughout the body.

LO 1.8 Describe the influence of the autonomic nervous system on the heart and circulatory system.

LO 1.9 Recall how nerve impulses are generated and muscles contract in the heart.

Case History

Emergency medical services responds to the home of a 65-year-old man complaining of a dull ache in his chest for the past two hours which came on while mowing his lawn. He also complains of a "fluttering" in his chest and "shortness of breath." He has a history of hypertension, elevated cholesterol, and a one-pack-a-day smoking habit.

After introducing themselves, the paramedics begin their assessment, finding the patient's blood pressure to be 160/110, pulse 120 and irregular, respirations 20, and oxygen saturation 92% on room air. The patient is awake and alert, his airway is open, his breathing is slightly labored, and his pulses are strong.

Read the Case History for a real-world scenario that features the type of dysrhythmia covered in the chapter.

Visualize the Content

300 Full-Color Figures show you in detail where each dysrhythmia originates and teaches you step by step how to read the ECGs that demonstrate each dysrhythmia. In addition, algorithms and tables present content visually to help you memorize the most important elements of each type of dysrhythmia and condition.

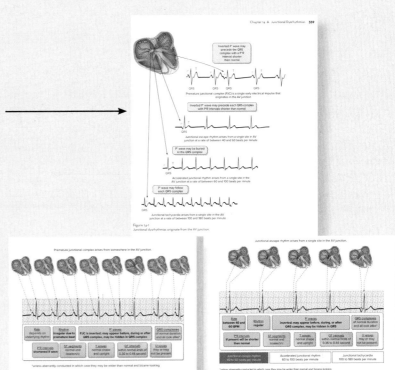

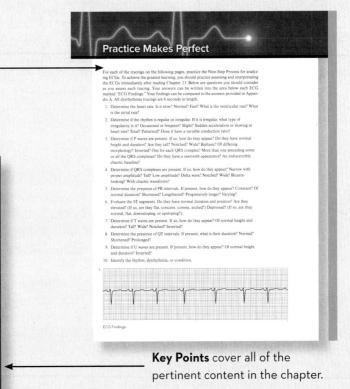

Review the Content

Practice Makes Perfect strips at the end of chapters and sections give you over 400 opportunities to interpret ECG strips using your new knowledge.

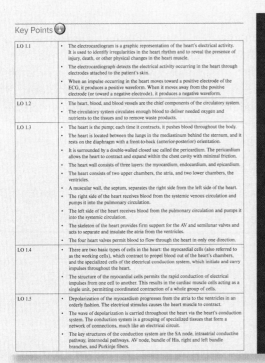

Key Points cover all of the pertinent content in the chapter.

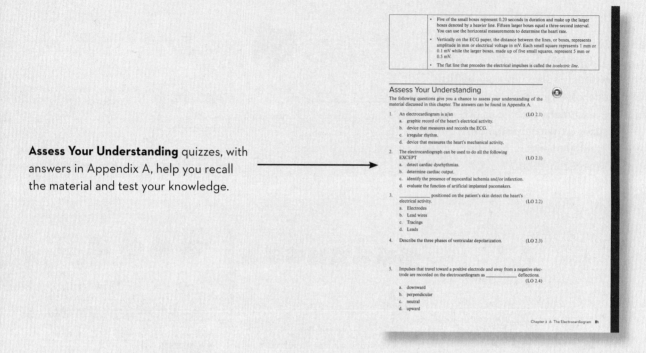

Assess Your Understanding quizzes, with answers in Appendix A, help you recall the material and test your knowledge.

First, I would like to thank Claire Merrick, the editor on the first edition of this textbook. It was her vision for the project that led to the original signing and publishing of *Fast & Easy ECGs*. Next, I would like to thank Melinda Bilecki, the Freelance Product Developer for this edition. Melinda maintained a steady hand to get the chapters rewritten and figures redone despite my many delays. Even with these obstacles, she displayed incredible patience and helped guide completion of the book. Further, her hard work and attention to detail helped ensure the accuracy of the content.

Many thanks go to, Michelle Flomenhoft, the Senior Product Developer, and William Lawrensen, the Executive Portfolio Manager with the Health Professions team at McGraw-Hill. They allowed me to significantly restructure the order of chapters, add more content and practice ECG tracings, and increase the footprint of the textbook. These features make a good book even better. They also convinced me of the need to rename the book to better reflect its comprehensive nature. As hard as it was for me to give up the former title, I recognize the importance of doing so.

Bruce Shade

Interpreting ECGs

A Practical Approach

Third Edition

Mc
Graw
Hill
Education

section 1
Preparatory

1 Anatomy and Electrophysiology of the Heart

©rivetti/Getty Images

Chapter Outline

©McGraw-Hill Education/Rick Brady

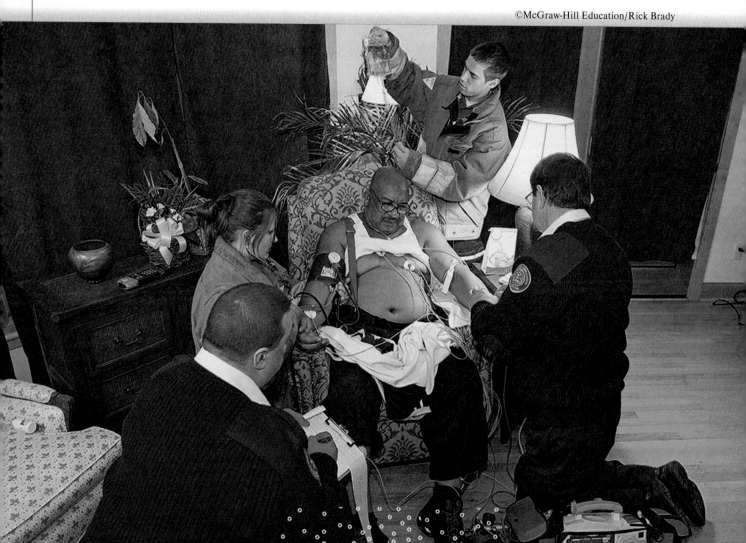

Learning Outcomes

LO 1.1 Define the term *electrocardiogram*, list its uses, and describe how it works.

LO 1.2 List the components of the circulatory system.

LO 1.3 Describe the anatomy of the heart.

LO 1.4 Identify and contrast the structure and function of the different types of heart cells.

LO 1.5 Identify the structures of the heart's conduction system and describe what each does.

LO 1.6 Identify how the heart receives most of its blood supply.

LO 1.7 Recall how the heart and circulatory system circulates blood throughout the body.

LO 1.8 Describe the influence of the autonomic nervous system on the heart and circulatory system.

LO 1.9 Recall how nerve impulses are generated and muscles contract in the heart.

Case History

Emergency medical services responds to the home of a 65-year-old man complaining of a dull ache in his chest for the past two hours which came on while mowing his lawn. He also complains of a "fluttering" in his chest and "shortness of breath." He has a history of hypertension, elevated cholesterol, and a one-pack-a-day smoking habit.

After introducing themselves, the paramedics begin their assessment, finding the patient's blood pressure to be 160/110, pulse 120 and irregular, respirations 20, and oxygen saturation 92% on room air. The patient is awake and alert, his airway is open, his breathing is slightly labored, and his pulses are strong.

The paramedics apply oxygen by nasal cannula and attach the patient to a cardiac monitor by applying electrodes to his chest. The monitor shows a fast, narrow complex rhythm with frequent wide and bizarre-appearing extra complexes. On the basis of what they observe, the paramedics obtain a 12-lead electrocardiogram (ECG) to determine if signs of a heart attack are present. The 12-lead ECG confirms their suspicions. The patient is having a myocardial infarction.

The paramedics then administer aspirin, nitroglycerin, and medication for pain relief to the patient and transport him to the nearest appropriate facility. En route to the hospital, the patient states his pain is less and the paramedics notice that the extra complexes are gone from his heart rhythm.

1.1 The Electrocardiogram

In order for the muscles of the body to contract, they must first be stimulated by electrical impulses generated and conducted by the nervous system. The **electrocardiogram,** often referred to as an ECG or EKG, is a tracing or graphic representation of the heart's electrical activity over time. The device that detects, measures, and records the ECG is called an **electrocardiograph.** The name electrocardiogram is derived of different parts: electro, because it's related to electricity, cardio, a Greek word for heart, and gram, a Greek root meaning "to write."

The ECG provides healthcare professionals with valuable information (Figure 1-1). It is used to identify irregularities in the heart rhythm (called **dysrhythmias**); detect

Figure 1-1
The electrocardiogram provides valuable information in a host of clinical settings.

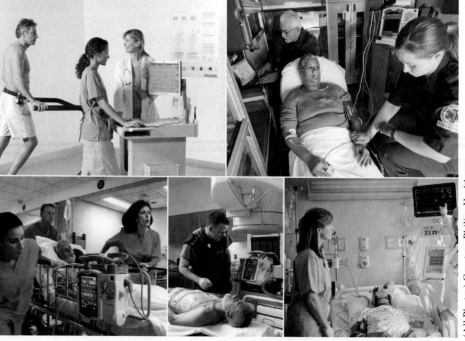

(All Photos) Courtesy Philips Healthcare

electrolyte disturbances and conduction abnormalities; and reveal the presence of, injury, death, or other physical changes in the heart muscle. It is also used as a screening tool for ischemic heart disease during a cardiac stress test. It is occasionally helpful with diagnosing noncardiac conditions such as pulmonary embolism or hypothermia.

The ECG is used in the prehospital, hospital, and other clinical settings as both an assessment and diagnostic tool. It can also provide continuous monitoring of the heart's electrical activity, for instance, during transport to the hospital or in the coronary care unit. The ECG does not, however, tell us how well the heart is pumping. The presence of electrical activity on the cardiac monitor does not guarantee that the heart is contracting or producing a blood pressure. To determine that, we must assess the patient's pulse and blood pressure, as well as perform an appropriate physical examination.

How It Works

In simple terms, the electrocardiograph, or ECG machine, detects the electrical current activity occurring in the heart (Figure 1-2). It does this through electrodes placed on the patient's skin. The ECG electrode must be in good contact with the skin to properly detect the heart's electrical currents. Tips for achieving effective contact will be discussed further in Chapter 2. These impulses, which appear as a series of upward (positive) and downward (negative) deflections (waveforms), are then transferred to the ECG machine and displayed on a screen (called the **oscilloscope** or monitor), or they are printed onto graph paper (often referred to as an ECG tracing or strip).

As the impulse moves toward a positive electrode of the ECG, it produces a positive waveform (upright deflection). Refer to Figure 1-2. In this ECG tracing, all the waveforms (P, QRS, and T) are positive, meaning the impulses are traveling toward

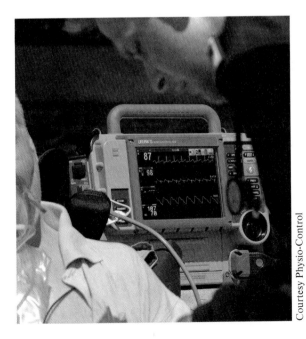

Courtesy Physio-Control

Figure 1-2
The ECG detects electrical activity in the heart.

a positive electrode. When it moves away from a positive electrode or toward a negative electrode, it produces a negative waveform (downward deflection). The sites for the placement of the electrodes vary depending on which area of the heart's activity is being viewed. Different sites provide different views. We discuss this information in more depth in the next chapters.

This book is designed to teach you how to interpret what you see on an ECG. To do this, it is important for you to understand the anatomy and physiology of the circulatory system and the heart. We begin by reviewing the role of the circulatory system and discussing the location and structure of the heart. Then we talk about how the generation and conduction of nerve impulses leads to contraction of the heart chambers, which then pump blood throughout the body. Finally, we discuss the influence of the autonomic nervous system on the heart.

1.2 The Circulatory System

In order to achieve and maintain homeostasis in the body, the circulatory system performs a number of vital functions: It carries nutrients, gases, and wastes to and from the body's cells; it helps fight diseases; and it helps stabilize body temperature and pH. The term perfusion describes the circulatory system's delivery of oxygen and nutrients to the tissues and the removal of waste products from those tissues. Perfusion is necessary for the body's cells to function and survive. The body's cells die if there is insufficient blood supply to meet their needs. The chief elements of the circulatory system are the heart, blood, and blood vessels (Figure 1-3).

The circulatory system includes the pulmonary circulation, a "loop" through the lungs, and the systemic circulation, a "loop" through the rest of the body to provide oxygenated blood to the body's cells. The arteries of the systemic circulation carry oxygenated blood, whereas the veins carry deoxygenated blood. The reverse is true in the pulmonary circulation, where the pulmonary artery carries deoxygenated blood to the lungs and the pulmonary veins carry oxygenated blood back to the heart. The circulatory system of an average adult contains roughly 4.7 to 5.7 L of blood, which consists of plasma that contains red blood cells, white blood cells, and platelets.

1.3 Anatomy of the Heart

The heart is an amazing organ. It is the pump of the circulatory system. Each time it contracts, it pushes blood throughout the body. The typical adult heart beats an average of 75 times a minute, 24 hours a day, 365 days a year, never stopping to take a rest. In an average day it pumps between 7000 and 9000 liters (L) of blood! This circulates enough blood to deliver needed oxygen and nutrients to the tissues and to remove waste products. Depending on the requirements of the body, the heartbeat can either be sped up (during exercise) or slowed down (while resting or sleeping). Try this experiment: count your pulse rate while sitting or lying comfortably reading this book. Then, if you are physically able, go for a brisk walk (or perhaps run) and then recheck your pulse rate. Your heart should be beating faster;

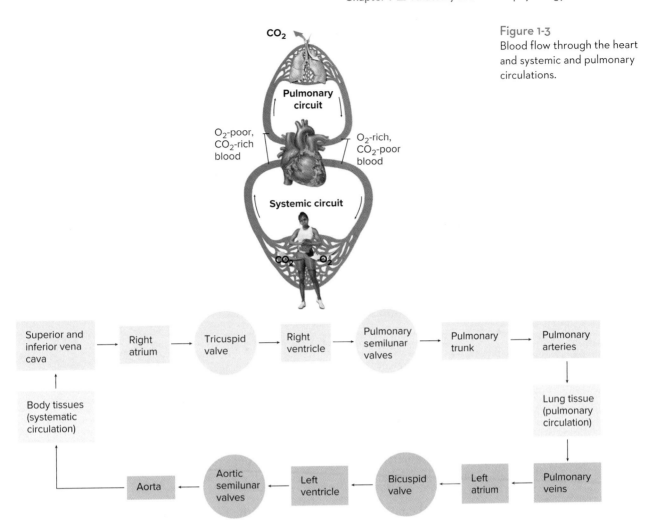

Figure 1-3
Blood flow through the heart
and systemic and pulmonary
circulations.

you may even feel the sensation of it pounding in your chest. Your body increases the heart rate and strength of contractions to circulate more blood (and oxygen and nutrients) to your cells and to remove the waste products that have been produced by those working cells.

Shape and Position of the Heart

Make a fist. Your heart is about the same size as your closed fist (Figure 1-4A). It is shaped like an inverted blunt cone. Its top (called the *base*) is the larger, flat part whereas its inferior end (called the *apex*) tapers to a blunt, rounded point. The heart is located between the lungs in the **mediastinum** behind the sternum (Figure 1-4B). It lies on the diaphragm in front of the trachea, esophagus, and thoracic vertebrae. About two-thirds of the heart is situated in the left side of the chest cavity. Its base is directed posteriorly and slightly superiorly at the level of the second intercostal space. Its **apex** is directed anteriorly and slightly inferiorly at the level of the fifth intercostal space in the left midclavicular line. This gives it a front-to-back (anterior-posterior) orientation. In this position the right ventricle is closer to the front of the left chest whereas the left ventricle is closer to the side of the left chest (Figure 1-4C). This informa-

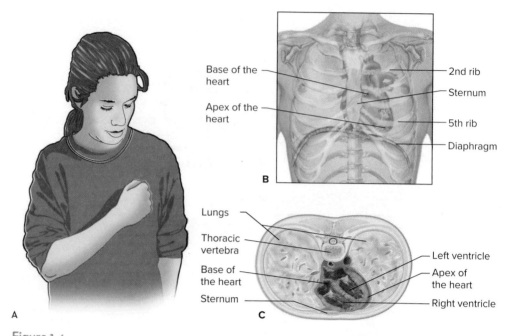

Figure 1-4
(A) The heart is about the size of a closed fist. (B) The position of the heart in the chest.
(C) Cross section of the thorax at the level of the heart.

tion will be particularly useful to you when we discuss placement of the various leads in later chapters.

Knowing the position and orientation of the heart will help you to understand why certain ECG waveforms appear as they do when the electrical impulse moves toward a positive or negative electrode. The location of the various ECG leads permits us to look at the heart from several different directions.

The Pericardial Sac

The heart is surrounded by the pericardial sac (also called the **pericardium**), a double-walled closed sac (Figures 1-5 and 1-6). The tough, fibrous, outer layer is called the *fibrous pericardium* whereas the inner, thin, transparent lining is called the *serous pericardium*. Above the heart, the fibrous pericardium is continuous with the connective tissue coverings of the great vessels, and below, the heart is attached to the surface of the diaphragm. This anchors the heart within the mediastinum. The serous pericardium has two parts: the parietal pericardium, which lines the fibrous pericardium; and the visceral pericardium, which covers the surface of the heart. The pericardial cavity, located between the parietal pericardium and the visceral pericardium, holds a small amount of clear lubricating fluid that allows the heart to contract and expand within the chest cavity with minimal friction.

The accumulation of additional fluid in the pericardial space can restrict the heart's ability to contract. This leads to a condition called *pericardial tamponade*. Pericardial tamponade can be life-threatening.

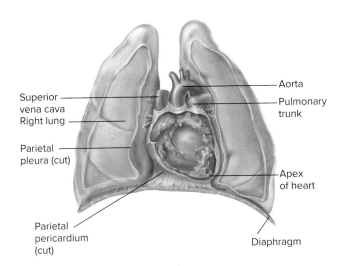

Figure 1-5
The pericardium is the
protective sac that surrounds
the heart.

Superior vena cava

Right lung

Parietal pleura (cut)

Aorta

Pulmonary trunk

Apex of heart

Parietal pericardium (cut)

Diaphragm

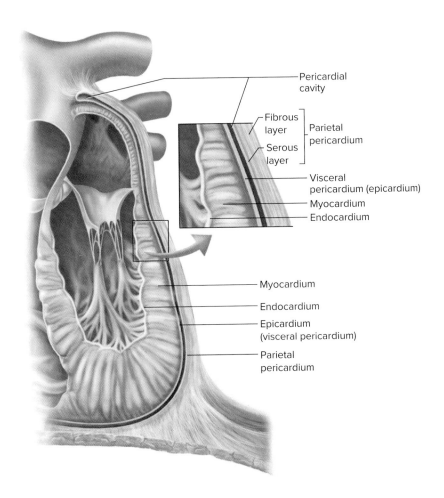

Figure 1-6
This cross section shows the
structure of the heart. The
enlarged section shows that
the wall of the heart has three
distinct layers of tissue: the
endocardium, myocardium,
and epicardium. Also note its
relationship to the
pericardium.

Pericardial cavity

Fibrous layer

Serous layer

Parietal pericardium

Visceral pericardium (epicardium)

Myocardium

Endocardium

Myocardium

Endocardium

Epicardium (visceral pericardium)

Parietal pericardium

The Heart Wall

The heart wall is comprised of three layers (see Figure 1-6). The middle layer, the muscular layer, is called the **myocardium.** *Myo* means muscle whereas *cardia* means heart. It is the thickest of the three layers and is composed of cylindrical cells that look similar to skeletal muscle.

The innermost layer of the heart wall is called the *endocardium.* The endocardium is a serous membrane that lines the four chambers of the heart and its valves. It has a smooth surface and is continuous with the lining of the arteries and veins. It is watertight to prevent leakage of blood into the other layers.

The outermost layer is the epicardium. It is a thin serous membrane that constitutes the smooth outer surface of the heart. The outer layer of the heart wall is called the *epicardium* when someone is referring to the layers of the heart but is called the *visceral pericardium* when someone is referring to the pericardium.

The Internal Heart

The heart is a muscular, hollow organ (Figure 1-7). It has two upper chambers, the **atria,** and two lower chambers, the **ventricles.** You can think of the heart as having two upper pumps (the atria) and two lower pumps (the ventricles). The thin-walled atria serve as low-pressure containers that collect blood from the systemic and pulmonary circulation and deliver it to the ventricles. The larger, more muscular ventricles pump blood to the pulmonary and systemic circulation. The left ventricle is thicker and more muscular because it pumps blood through the larger, higher-pressure systemic circulation. The left ventricle can be thought of as the workhorse of the heart.

Two Functional Pumps

The heart is separated into two functional units by the **septum** (see Figure 1-7). The word *septum* comes from the Latin word *saeptum,* meaning a "dividing wall or enclosure." The interatrial septum is a thin membranous wall that separates the two atria whereas the more muscular wall, the interventricular septum, separates the two ventricles. For this reason the heart is referred to as a *double pump.* The ventricular septum consists of an inferior muscular and superior membranous portion and is extensively innervated with heart cells capable of conducting nerve impulses. The septum also provides strength to the walls of the heart.

Figure 1-7
Internal anatomy of the heart.

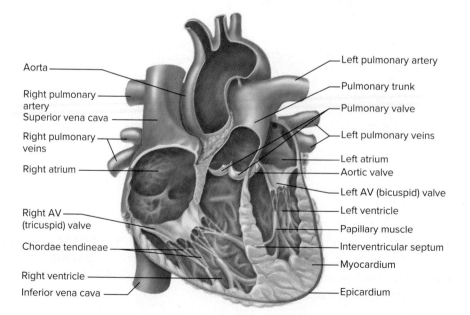

Aorta

Right pulmonary artery
Superior vena cava

Right pulmonary veins

Right atrium

Right AV (tricuspid) valve

Chordae tendineae

Right ventricle
Inferior vena cava

Left pulmonary artery

Pulmonary trunk

Pulmonary valve

Left pulmonary veins

Left atrium
Aortic valve

Left AV (bicuspid) valve

Left ventricle

Papillary muscle

Interventricular septum

Myocardium

Epicardium

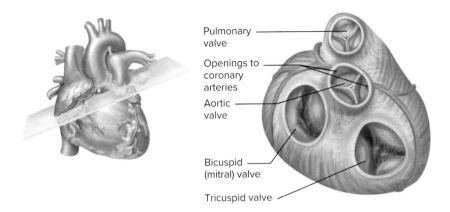

Pulmonary valve

Openings to coronary arteries

Aortic valve

Bicuspid (mitral) valve

Tricuspid valve

Figure 1-8
Superior view of the heart with the atria removed and four heart valves exposed. In the superior view, the top of the heart represents the anterior aspect whereas the bottom of the heart represents the posterior aspect.

Heart Valves

Two atrioventricular (AV) valves are located between the atria and ventricles, whereas two semilunar valves are located between the ventricles and major arteries (Figure 1-8). The two AV valves include the **tricuspid valve,** situated between the right atrium and right ventricle, and the **bicuspid** (or **mitral)** valve, located between the left atrium and left ventricle. These valves prevent blood from flowing backward into the atria when the ventricles contract, resulting in the ejection of blood forward from the ventricles and into the pulmonary and systemic circulation. The mitral valve has two cusps whereas the tricuspid valve has three. The cusps are connected to **papillary muscles** in the floor of the ventricle by thin, strong strings of connective tissue called **chordae tendineae.** These cords prevent the cusps from bulging (prolapsing) backward into the atria during ventricular contraction.

The two semilunar valves include the **pulmonic valve,** found at the base of the pulmonary artery, and the **aortic valve,** situated at the base of the aorta, just as they arise from the right and left ventricles, respectively. They prevent the backward flow of blood after the ventricles have contracted and propelled blood into the pulmonary arteries and aorta. Each valve has three cusps that look somewhat like shirt pockets.

Skeleton of the Heart

Between the atria and ventricles is a plate of fibrous connective tissue called the *skeleton of the heart* (Figure 1-9). This plate forms fibrous rings around the AV and semilunar valves, providing firm support. It also acts to separate the atria from the ventricles, functioning in two important ways.

First, it allows the top and bottom parts of the heart to act as separate, sequential pumps. Second, it electrically insulates the atria from the ventricles. This insulation allows the atria to be depolarized without depolarizing the ventricles and allows the ventricles to be depolarized without depolarizing the atria.

The cardiac muscles are attached to the fibrous connective tissue and arranged in such a way that, when the ventricles contract, they do so in a wringing motion, which shortens the distance between the base and the apex of the heart. This results in the most efficient ejection of blood out of the ventricles.

Figure 1-9
The skeleton of the heart consists of fibrous connective tissue rings that surround the heart valves and separate the atria from the ventricles. Cardiac muscle tissue attaches to the fibrous connective tissue.

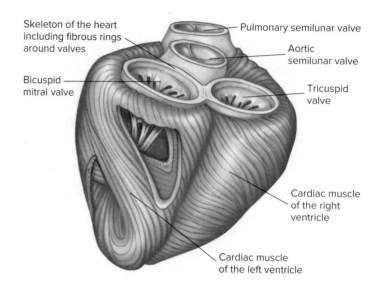

Skeleton of the heart including fibrous rings around valves

Pulmonary semilunar valve

Aortic semilunar valve

Bicuspid mitral valve

Tricuspid valve

Cardiac muscle of the right ventricle

Cardiac muscle of the left ventricle

1.4 Cells of the Heart

There are two basic types of cells in the heart—the myocardial cells (also referred to as the working cells), which contract to propel blood out of the heart's chambers, and the specialized cells of the electrical conduction system, which initiate and carry impulses throughout the heart.

Myocytes

Structurally, **myocytes** (the working cells) are cylindrical branching cells that usually contain only one centrally located nucleus (Figure 1-10). They are enclosed in a plasma membrane called a **sarcolemma.** The individual myocytes are made up of a small latticework of intricate strands composed of two protein filaments referred to as **actin** and **myosin.** These filaments lie side by side and are connected by cross-bridges. These contractile elements permit the muscle fiber to shorten itself and then return to its original length. Their organization gives cardiac muscle a striated (banded) appearance when viewed under a microscope. The process of contraction requires a plentiful supply of energy. This is supplied by the mitochondria, interspersed within the cell.

The myocytes are bound together with adjoining cells both end to end and laterally, forming a branching and anastomosing network of cells called a **syncytium.** The heart consists of two syncytia, the atrial syncytium and the ventricular syncytium. The atrial syncytium consists of the walls of the right and left atria whereas the ventricular syncytium consists of the walls of the right and left ventricles.

Specialized cellular contacts called **intercalated discs** are located where the branches join. The intercalated disks fit together to form electrical connections. **Gap junctions** permit the rapid conduction of electrical impulses from one cell to another. This results in the myocytes acting as a single unit, permitting coordinated contraction of a whole group of cells. This characteristic allows the walls of

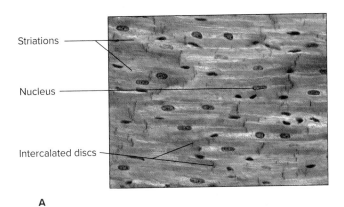

Striations

Nucleus

Intercalated discs

A

Figure 1-10
Myocardial cells. (A) Cells seen through a light microscope. (B) Myocytes and intercalated discs. (C) Actin and myosin filaments.

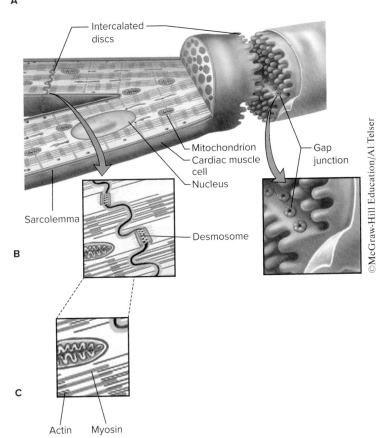

Intercalated discs

Mitochondrion
Cardiac muscle cell
Nucleus

Sarcolemma

Desmosome

Gap junction

B

©McGraw-Hill Education/Al Telser

C

Actin   Myosin

the myocardium to contract almost simultaneously when stimulated by an electrical impulse. The highly coordinated contractions of the heart depend on this characteristic. Specialized structures in the cellular membrane called **desmosomes** hold the myocytes together to prevent them from pulling apart when the heart contracts.

The cells of the heart have four key properties: contractility, automaticity, excitability, and conductivity. The working cells possess the property of **contractility.** This is the ability to contract when stimulated by an electrical impulse (see Figure 1-11). Contraction of the working cells propels blood out of the heart's chambers.

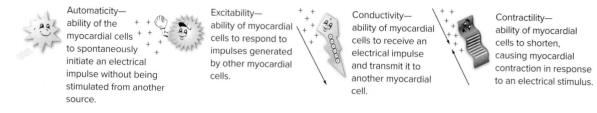

Figure 1-11

The key properties of myocardial cells are automaticity, excitability, conductivity, and contractility.

Pacemaker and Electrical Conducting Cells

In order for the heart muscles to contact, they must first be electrically stimulated. The specialized cells responsible for generating and carrying impulses throughout the heart are called *pacemaker cells* and *electrical conducting cells*. These cells differ from the working cells as they lack myofibrils and cannot contract. Also, these cells contain more gap junctions than do the working cells. This allows them to conduct electrical impulses extremely fast.

1.5 The Heart's Conductive Pathway

The specialized cells responsible for generating and carrying impulses throughout the heart form the **conductive pathway** (Figure 1-12), a grouping of specialized tissues that form a network of connections, much like an electrical circuit. The conductive pathway has two types of cells—the pacemaker cells and the electrical conducting cells. The key structures of the heart's conduction system are the SA node, intraatrial conductive pathway, internodal pathways, AV node, bundle of His, right and left bundle branches, and Purkinje fibers.

SA Node

The electrical event that normally initiates the heartbeat is produced by a group of specialized electrical tissues called the **sinoatrial (SA) node.** The SA node is located high on the posterior wall of the right atrium (see Figure 1-12), just below the opening of the superior vena cava and just under the epicardium. The term *node* means a knot or lump.

It is thought that, once the impulse is generated, it is carried across the interatrial septum and to the left atrium by way of the intraatrial conductive pathway (Bachmann's Bundle) and through the right atrium by way of the anterior, middle, and posterior internodal tracts. The presence of the intraatrial and internodal tracts has not been conclusively proven as they cannot be distinguished structurally from the rest of the atrium. However, the rapid speed at which the impulse moves through the atria strongly supports the belief that these conductive pathways exist.

AV Node

Following their path through the right and left atria, the impulses reach the **atrioventricular (AV) node.** The AV node lies on the floor of the right atrium just medial to the mitral valve and just above the ventricle (see Figure 1-12). It is the only

1. Electrical impulses originate in the SA node and travel across the wall of the atrium (*arrows*) from the SA node to the AV node.

2. Electrical impulses pass through the AV node and along the bundle of His, which extends from the AV node, through the fibrous skeleton, into the interventricular septum.

3. The bundle of His divides into right and left bundle branches, and electrical impulses descend to the apex of each ventricle along the bundle branches.

4. Electrical impulses are carried by the Purkinje fibers from the bundle branches to the ventricular walls.

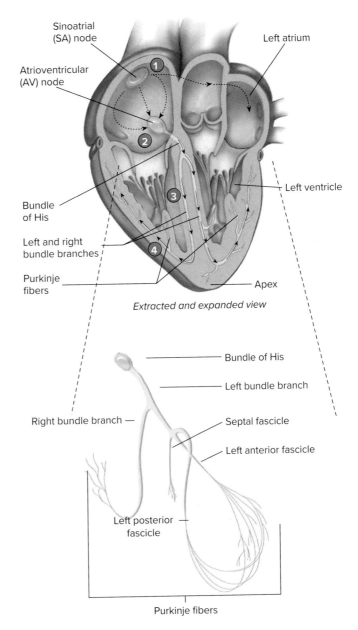

Extracted and expanded view

Figure 1-12
The heart's conduction system generates and carries electrical impulses throughout the heart. The enlarged section shows the His-Purkinje system.

pathway (unless there are accessory pathways as described later in this text) for impulses to move from the atria to the ventricles. Remember, the fibrous skeleton acts as an insulator to prevent electrical impulses from getting to the ventricles by any other route than that described.

The AV node is anatomically less well-defined than the SA node. It is composed of three layers: the upper, middle, and lower, which have unique conduction characteristics.

The AV node, its atrial pathways, and the bundle of His make up the conductive tissue of the AV junction.

Bundle of His and Right and Left Bundle Branches

The bundle of His and right and left bundle branches carry impulses from the AV node to the ventricles. The AV node gives rise to a conducting bundle to the heart, the **bundle of His.** It is also called the *AV bundle* (see Figure 1-12). This bundle passes from the walls of the right atrium through a small opening in the fibrous skeleton to reach the interventricular septum.

In the interventricular septum the bundle divides into the left and right bundle branches. The bundle branches extend beneath the endocardium on either side of the interventricular septum to the apex of each ventricle. The right bundle branch goes to the right ventricle. The left bundle branch goes to the left ventricle.

The left bundle branch divides further into three divisions: the septal fascicle, the anterior fascicle, and the posterior fascicle (see Figure 1-12). The septal fascicle carries the impulse to the interventricular septum in a right-to-left direction. The anterior and posterior fibers spread to their respective sides of the heart.

The bundle of His, bundle branches, and Purkinje fibers are referred to as the His-Purkinje system.

Purkinje Fibers

The right and left bundle branches terminate in the **Purkinje fibers.** The Purkinje fibers consist of countless tiny fibers that spread out widely like the twigs of a tree branch. They extend just underneath the endocardium and terminate in the endocardial cells. These fibers conduct impulses rapidly through the muscle to assist in its depolarization and contraction, resulting in ventricular depolarization starting in the endocardium and proceeding outward to the epicardium.

1.6 The Heart's Blood Supply

The heart constantly pumps blood to the body. Because of this, its oxygen consumption is proportionately greater than that of any other single organ. With the oxygen demand in the myocardial cells being so high, the heart must have its own blood supply. The coronary arteries provide a continuous supply of oxygen and nutrients to the myocardial cells.

The coronary arteries are so named because of how they encircle the heart like a crown. Arising from the aorta just above the aortic valve, the right and left coronary arteries lie on the heart's surface and have many branches that penetrate into all parts of the heart. The terminal branches of the arteries have many interconnections, forming an extensive vascular network—each perfusing a particular portion of the myocardium.

Right Coronary Artery

The **right main coronary artery** originates from the right side of the aorta and passes along the atrioventricular sulcus between the right atrium and right ventricle (Figure 1-13A). It then divides into two branches, the marginal artery and the posterior interventricular artery. The portion of the myocardium supplied by the right coronary artery includes the right atrium, right ventricle, inferior and posterior

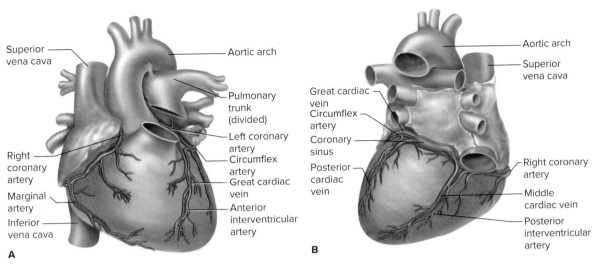

Figure 1-13
The coronary blood vessels supply the heart with oxygen and nutrients and remove waste products. (A) Anterior aspect. (B) Posterior aspect.

walls of the left ventricle, and posterior one-third of the interventricular septum. The right coronary artery also supplies blood to the SA node (in about 50% to 60% of the population), to the AV node (in about 80% to 95% of the population), and to the posterior-inferior fascicle of the left bundle branch.

Left Main Coronary Artery

The **left main coronary artery** originates from the left side of the aorta. It divides into the anterior descending and circumflex branches (Figure 1-13B). The anterior descending artery perfuses the anterior surface and part of the lateral surface of the left ventricle and the anterior two-thirds of the interventricular septum. Branches of the anterior descending artery, the diagonal artery, and the septal perforators help supply blood to the lateral walls of the left ventricle. The circumflex artery supplies the left atrium and the anterolateral, posterolateral, and the posterior wall of the left ventricle.

The anterior descending artery supplies blood to the majority of the right bundle branch. It also supplies blood to both the anterior and posterior fascicles of the left bundle branch. The circumflex artery supplies blood to the SA node (in about 40% to 50% of the population) and the AV node (in about 10% to 15% of the population).

Between the coronary arterioles are anatomic connections called *anastomoses*. These anastomoses allow the development of alternative routes for blood to flow in the event of blockage. This is referred to as *collateral circulation*.

Coronary Veins

Deoxygenated venous blood from the heart drains into five different coronary veins that empty into the right atrium via the coronary sinus. This blood then mixes with the systemic venous return.

If the flow of blood in the coronary arteries is diminished, myocardial ischemia may result. When severe ischemia occurs, necrosis or death of muscle tissue can occur.

1.7 Physiology of the Heart and Circulation

It is now time to talk about how the heart receives and pumps blood throughout the body.

The Cardiac Cycle

The cardiac cycle is the repetitive process of pumping blood that starts with the beginning of cardiac muscle contraction and ends with the start of the next contraction. Pressure changes within the cardiac chambers are produced by myocardial contraction. This causes movement of blood from areas of higher pressure to areas of lower pressure.

There are two phases of the cardiac cycle: diastole and systole (Figure 1-14). Diastole is the relaxation and filling of both the atria and ventricles. However, clinically it is most often used to describe ventricular relaxation and filling. It is during this time that the heart itself receives most of its blood supply from the coronary arteries, as during systole the myocardium compresses the coronary arteries and restricts blood flow. Thus, in addition to the diastolic blood pressure's reflecting the pressure in the heart during the relaxation phase, it also tells us about the perfusion of the heart.

Contraction of the atria and ventricles is referred to as systole. Again, clinically it is most often used to describe ventricular systole. The systolic blood pressure is the pressure within the systemic arteries during ventricular contraction.

Cardiac Output

Contraction of the ventricles normally results in 60 to 100 mL of blood being ejected into both the pulmonary and systemic circulation. The amount of blood ejected from the ventricles is referred to as the stroke volume. It is dependent on preload (stretching force exerted on the ventricular muscle at end diastole), contractile force of the myocardium, and afterload (workload against which the heart must pump).

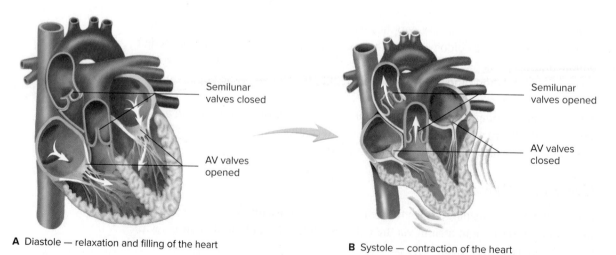

Semilunar
valves closed

AV valves
opened

Semilunar
valves opened

AV valves
closed

A Diastole — relaxation and filling of the heart

B Systole — contraction of the heart

Figure 1-14

The two phases of the cardiac cycle are (A) diastole, the relaxation and filling of both the atria and ventricles and (B) systole, contraction of the atria and ventricles.

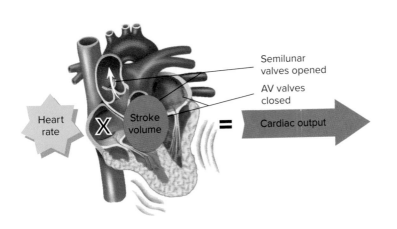

Semilunar
valves opened

AV valves
closed

Heart rate X Stroke volume = Cardiac output

Figure 1-15
Cardiac output is the amount
of blood pumped from the
heart in one minute.

The amount of blood pumped from the heart in one minute is called the cardiac output (Figure 1-15). It is expressed in liters per minute. The cardiac output is the equivalent of the heart rate multiplied by the stroke volume:

$$\text{Heart Rate} \times \text{Stroke Volume} = \text{Cardiac Output}$$

Cardiac output for the right and left ventricles is normally equal. Decreases in either the heart rate or stroke volume result in decreased cardiac output, whereas increases in either result in increased cardiac output. Factors that influence the heart rate, the stroke volume, or both will influence cardiac output and, thus, tissue perfusion.

Blood Pressure

The blood pressure is the force that blood exerts against the walls of the blood vessels as it passes through them. The term blood pressure generally refers to arterial pressure, that is, the pressure in the larger arteries. This pressure causes blood to flow. The blood pressure is the equivalent of the cardiac output multiplied by the peripheral vascular resistance:

$$\text{Cardiac Output} \times \text{Peripheral Vascular Resistance} = \text{Blood Pressure}$$

A decrease in cardiac output or peripheral vascular resistance results in a decrease in the blood pressure, whereas increases in either result in increased blood pressure.

Blood pressure is most commonly measured by using a sphygmomanometer, which historically used the height of a column of mercury to reflect the circulating pressure. Today, blood pressure values are still reported in millimeters of mercury (mm Hg), although modern aneroid and electronic devices do not use mercury.

Blood Flow through the Atria

At the beginning of the cardiac cycle, the right and left atria receive blood from the systemic and pulmonary circulation and deliver it to the ventricles. Deoxygenated blood returned from the body via the superior and inferior vena cava, the largest veins in the body, is delivered to the right atrium. The superior vena cava and the inferior vena cava are sometimes called the great veins. The left atrium receives blood oxygenated in the lungs and returned via the pulmonary veins. As the atria fill with blood, the pressure within the atria rises, forcing the tricuspid and mitral valves open and allowing blood to flow into the right and left ventricles (Figure 1-16). Approximately 70% of the blood coming into the

Figure 1-16
At the beginning of the cardiac cycle, blood flows into the right and left atria, forcing the AV valves open and allowing blood to flow into the right and left ventricles.

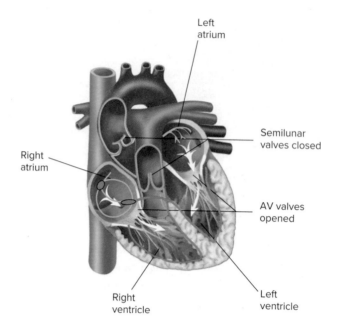

Figure 1-17
Normally, the electrical activity that initiates the heartbeat originates from the SA node.

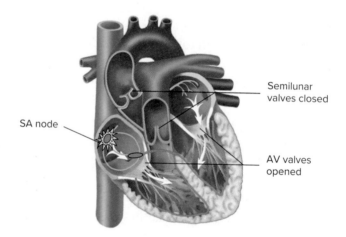

chambers flows passively through the atria and into the ventricles before the atria contract.

Initiation of Impulse in the SA Node

The SA node initiates the electrical event that normally causes the heartbeat (Figure 1-17). The ability of certain myocardial cells to spontaneously discharge electrically (action potential) without the need for external stimulus is called *automaticity.* The heart is unique in this property. All other muscles in the body require stimulation from nerve impulses supplied by the nervous system. If nerves supplying the voluntary muscles are cut, these muscles cease to function; in other words, they are paralyzed. In contrast, if the nerves supplying the heart are severed, the heart will continue to beat. The cells that possess the property of automaticity are called *pacemaker cells.* The SA node serves as the heart's primary pacemaker and has an intrinsic (natural) rate of 60 to 100 beats per minute.

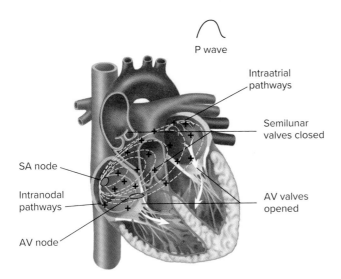

P wave

Intraatrial pathways

Semilunar valves closed

SA node

Intranodal pathways

AV node

AV valves opened

Figure 1-18
From the SA node, the electrical impulse moves in a wavelike manner across the atria. This causes the atria to contract, pushing the remaining atrial blood into the ventricles.

Atrial Depolarization and Contraction

From the SA node, the electrical impulse spreads from cell to cell in a wavelike manner across the atrial muscle via the anterior, middle and posterior internodal pathways and Bachmann's Bundle, depolarizing the right atrium, the interatrial septum, and then the left atrium (Figure 1-18). These conductive pathways do not normally possess the property of automaticity; instead they simply respond to and carry impulses generated by the SA node. This electrical current produces a waveform called the **P wave.** Remember, the structure of the cardiac muscles (which forms the atrial and ventricular syncytium) permits the rapid conduction of electrical impulses from one cell to another, making it possible for impulses to move quickly throughout the heart. The two key properties of the myocardial cells responsible for moving the impulses from cell to cell are excitability and conductivity. Excitability is the ability to respond to an electrical stimulus, whereas conductivity is the ability to transmit an electrical stimulus from cell to cell. Normally, the electrical conducting (nonpacemaker) cells do not possess the property of automaticity but under certain conditions (as will be discussed later in this textbook) can acquire it. This brings about nearly simultaneous contraction of the right and left atria, squeezing the blood in the chambers and forcing it forward into the ventricles. Contraction of the atria pushes the remaining 30% of the blood into the ventricles (remember, up to 70% of the blood flows passively through the atria). The working cells possess the property of contractility. This is the ability to contract when stimulated by an electrical impulse. To contract and pump blood, the cells rely on four key properties: automaticity, excitability, conductivity, and contractility (See Figure 1-11).

Following contraction of the atria, the pressure in the atria and ventricles equalizes, and the tricuspid and mitral valves close. At the same time, the impulse is carried through the atria by the intranodal pathways to the AV node.

Conduction through the AV Node

The internodal pathways carry impulses through the atria to the AV node (Figure 1-19). It is the only pathway (unless there are accessory pathways as described later in this text) for impulses to move from the atria to the ventricles. The impulse travels slowly through the AV node, creating a slight delay that allows the atria to finish pushing any remaining blood into the ventricles. This is referred to as the atrial kick. It also allows the ventricular muscle to stretch to its fullest for peak cardiac output. You can think of

Figure 1-19
The wave of depolarization
slows as it passes through the
AV node. This allows the atrial
contraction to finish filling the
ventricles. This electrical
activity occurs simultaneously
with the P wave.

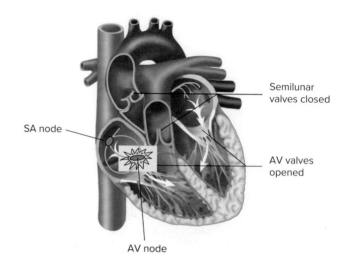

SA node

Semilunar
valves closed

AV valves
opened

AV node

the AV node as being a gatekeeper. The impulse travels through the AV node more slowly because the myocytes are thinner in this area, but, more importantly, they have fewer gap junctions over which the impulse can be conducted.

Like the SA node, the upper and lower AV nodal tissue has pacemaker cells that allow it to initiate the heartbeat if conditions warrant. This nodal tissue has an intrinsic heart rate of 40 to 60 beats per minute. The nodal tissue in the middle lacks automaticity and is slow to depolarize.

On the ECG this electrical stimulation occurs during the P wave. As discussed earlier, the heart skeleton (the plate of fibrous connective tissue) that lies between the atria and the ventricles and includes the heart valves insulates the atrial myocardium from the ventricular myocardium. This allows the electrical stimulus to stimulate the atria as it travels through the upper portion of the heart, permitting the upper chambers of the heart to contract as a unit, pumping blood to the ventricles. Likewise, once the impulse travels through the ventricles, it causes them to contract as a unit, pumping blood to the lungs and to the body.

Conduction through the His-Purkinje System

After slowly moving through the AV node, the impulse then rapidly shoots through the bundle of His, the left and right bundle branches, and the terminal Purkinje fibers (the His-Purkinje system) (Figure 1-20). On the ECG this produces a flat line that follows the P wave. This is because the electrical currents are so small they are not seen on the ECG. This part of the ECG is called the **PR segment.** The His-Purkinje system is responsible for initiating the contraction of the ventricles.

Ventricular Depolarization and Contraction

Depolarization of the whole ventricular myocardium produces contraction of the ventricles and a waveform called the **QRS complex** on the ECG. The sizes of the P wave and the QRS complex are considerably different because of how much thicker the ventricles are compared with the atria. The ventricles, a much bigger mass, produce a bigger waveform. Ventricular depolarization initiates vigorous contraction of the ventricles, causing the ventricular pressure to rise sharply, up to about 120 mm Hg in the left ventricle and 26 mm Hg in the right ventricle (Figure 1-21). Like the SA node and AV junction, the ventricles, including the bundle branches and

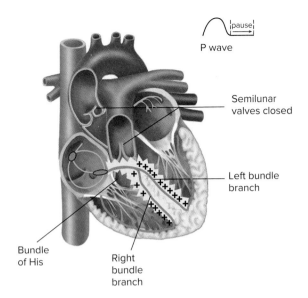

Figure 1-20
After slowly passing through the AV node, depolarization rapidly shoots through and activates the His-Purkinje system. This is represented on the ECG as a flat line called the *PR segment*.

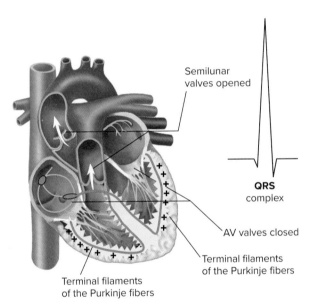

Figure 1-21
The wave of depolarization passing through the whole ventricular myocardium produces contraction of the ventricles and a QRS complex on the ECG.

Purkinje fibers, can also serve as the heart's pacemaker, discharging impulses at a rate of 20 to 40 beats per minute. If the SA node or AV junction fails to initiate a heartbeat or if there is a blockage of conduction through the AV node or bundle of His, the Purkinje fibers will fire as the last resort.

When the ventricular pressure exceeds the pressure in the aorta and pulmonary artery, the tricuspid and mitral valves close completely, and the pulmonic and aortic valves snap open. This allows the blood to be forcefully ejected from the right ventricle into the pulmonary artery, from which it is carried to the lungs. The blood from the left ventricle is ejected into the aorta, from which it is circulated to the tissues of the body.

At the end of the contraction, the ventricles relax, and back pressure in the aorta and pulmonary artery causes the aortic and pulmonic valves to close.

Normal depolarization of the myocardium progresses from the atria to the ventricles in an orderly fashion. The electrical stimulus causes the heart muscle to contract. The ECG shows the electrical stimuli as it travels through the heart.

The heart performs its pumping action over and over in a rhythmic sequence circulating blood through the pulmonary and systemic circulation.

The difference we see in how the atria and ventricles move blood is important as we talk about various cardiac dysrhythmias. Dysrhythmias originating from the atria are generally not life threatening whereas dysrhythmias that rise from the ventricles can be. This is because the ventricles are responsible for pushing blood throughout the body; if they fail, blood circulation throughout the body is compromised.

Atrial and Ventricular Repolarization

Repolarization of the ventricles, which occurs following the conclusion of depolarization, is represented on the ECG by the **ST segment** and the **T wave** (Figure 1-22). Atrial repolarization occurs during ventricular depolarization and is therefore hidden, or obscured, by the QRS complex.

Alternate Pacemaker Sites

Normal depolarization of the myocardium progresses from the atria to the ventricles in an orderly fashion (Figure 1-23). Pacemaker cells in the lower areas of the

Figure 1-22
Repolarization of the ventricles is represented on the ECG by the ST segment and T wave.

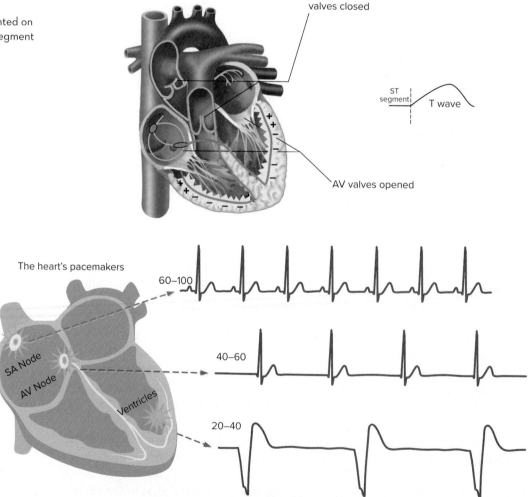

Figure 1-23
Intrinsic rates of pacemaker sites in the heart's conduction system.

heart, such as the junctional tissue and Purkinje fibers, do not normally initiate the heartbeat because they receive impulses from the SA node. However, other sites in the heart can assume control by discharging impulses faster than the SA node. Also, as mentioned above, they can passively take over, either because the SA node has failed or it is generating impulses too slowly. Normally, the farther from the SA node, the slower will be the pacemaker's intrinsic rate.

1.8 Influences on the Heart and Circulatory System

You should be familiar with the terms chronotropic, dromotropic, and inotropic. Chronotropic refers to the heart rate. Dromotropic refers to the rate of impulse conduction through the heart's conduction system. Inotropic refers to the contractility of the heart.

Regulation of the heart rate, speed of electrical conduction, and strength of contraction are influenced by the brain (via the autonomic nervous system), hormones of the endocrine system, and the heart tissue (Figure 1-24). Receptors in the blood vessels, kidneys, brain, and heart constantly monitor the adequacy of the cardiac output. **Baroreceptors** detect changes in pressure, usually within the heart or the main arteries. **Chemoreceptors** have the job of sensing changes in the chemical composition of the blood. If abnormalities are identified, nerve signals are sent to the appropriate target organs, generating the release of hormones or **neurotransmitters** to fix the situation. Once conditions are normal, the receptors stop firing and the signals cease.

1. Baroreceptors are responsible for detecting changes in pressure, usually within the heart or the main arteries. Chemoreceptors sense changes in the chemical composition of the blood. This information is transmitted to the cardioregulatory center in the medulla oblongata.

2. When the blood pressure is elevated, the cardiorespiratory center may activate the parasympathetic nervous system, which acts to slow the heart rate and lower the blood pressure.

3. If the blood pressure is low the cardiorespiratory center will activate the sympathetic nervous system, which acts to increase the heart rate and contractility. This increases cardiac output and raises the blood pressure.

4. The cardioregulatory center causes the release of epinephrine and some norepinephrine from the adrenal medulla into the general circulation. Epinephrine and norepinephrine increase the heart rate and stroke volume.

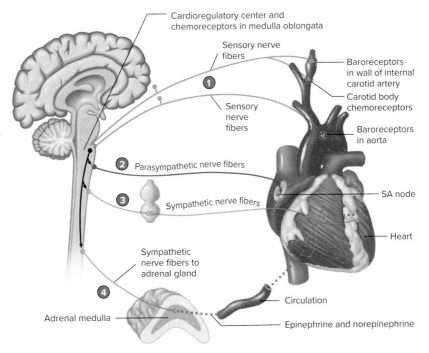

Figure 1-24

Baroreceptor and chemoreceptor reflexes. Sensory (green) nerves carry information from sensory receptors to the medulla oblongata and extend to the heart to regulate its function. Epinephrine and norepinephrine from the adrenal gland also help regulate the heart's action.

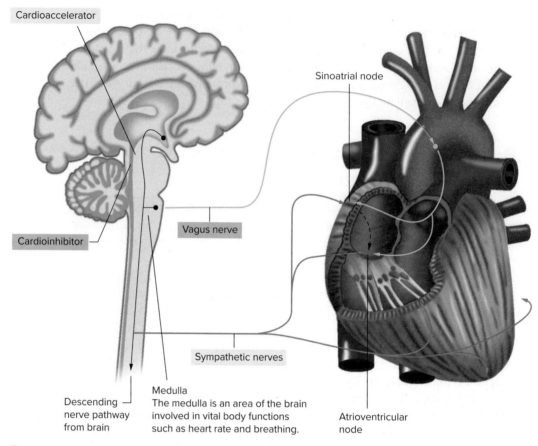

Cardioaccelerator

Sinoatrial node

Cardioinhibitor

Vagus nerve

Sympathetic nerves

Descending nerve pathway from brain

Medulla
The medulla is an area of the brain involved in vital body functions such as heart rate and breathing.

Atrioventricular node

Figure 1-25
Parasympathetic and sympathetic innervation of the heart.

The Autonomic Nervous System

Whereas the myocardial cells can generate electrical impulses on their own, the autonomic nervous system can influence the rate and strength of myocardial contractions (Figure 1-25). This is needed because the cardiovascular system must continually adjust the blood supply to meet demands at the cellular level.

Nervous control of the heart originates from two separate nerve centers located in the medulla oblongata, a part of the brainstem. One of these centers, the cardioaccelerator, is part of the sympathetic nervous system whereas the other, the cardioinhibitor, is part of the parasympathetic nervous system. Both systems exert their effects via neurotransmitters that bind to specific receptors. Impulses from the cardioaccelerator center are transmitted to the heart by way of the sympathetic nerves. Impulses from the cardioinhibitor center are transmitted to the heart by way of the vagus nerve.

The sympathetic and parasympathetic nervous systems work in opposition to one other. The parasympathetic branch dominates during custodial or basal functions whereas the sympathetic system dominates during stress. For example, during physical exertion the sympathetic system causes the heart to beat faster whereas during rest the heart beats slower.

Sympathetic Nervous System

The sympathetic branch of the autonomic nervous system is carried through nerves in the thoracic and lumbar ganglia, causing the release of the neurotransmitter

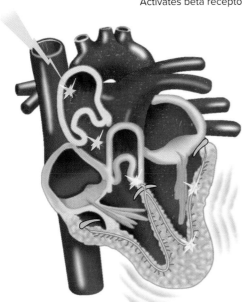

Sympathetic system
Activates beta receptors

Cardioaccelerator
effects

⬆ Rate of pacemaker firing

⬆ Spead of impulse conduction
through heart

⬆ Force of contraction

Coronary vasodilation

Figure 1-26
Activation of the sympathetic
nervous system stimulates the
heart to beat faster and
stronger.

norepinephrine (Figure 1-26). **Epinephrine (adrenalin),** released from the adrenal glands, also functions as a sympathetic branch neurotransmitter. For this reason, the sympathetic branch is called the **adrenergic** system. Stimulation of the sympathetic nervous system produces what we describe as the "fight-or-flight" response—constriction of blood vessels, enhancement of myocardial cell excitability, increased rate of pacemaker firing, increased conduction speed, increased contractility, coronary vasodilation, and a feeling of nervousness.

Sympathetic fibers exert their effect by stimulating special receptors. These receptors are divided into categories depending on their response. The two major categories are alpha and beta receptors. The primary sympathetic receptors in the heart are beta receptors—they affect all areas of the heart.

Parasympathetic Nervous System

The parasympathetic branch of the autonomic nervous system is carried through the vagus nerve and releases the neurotransmitter acetylcholine (Figure 1-27). It is therefore referred to as the **cholinergic** system. The parasympathetic nervous system works just the opposite of the sympathetic branch, causing a slowing of the heart rate and AV conduction. It also controls intestinal activity and affects papillary responses. The areas of the heart most affected by the parasympathetic fibers are the SA node; atria; AV junction; and to a small extent, the ventricles.

The parasympathetic nervous system may also have a modest effect on myocardial contractility by slowing the rate of contraction. In extreme instances, excess stimulation of the parasympathetic nervous system can lead to cardiac arrest.

Increased Myocardial Oxygen Needs

The resting heart extracts most of the oxygen from the coronary blood as it flows through the coronary arteries surrounding the heart muscle. Coronary perfusion occurs during diastole, when the heart is relaxed. Increased oxygen demands of

Figure 1-27
Activation of the parasympathetic nervous system causes the heart to beat slower.

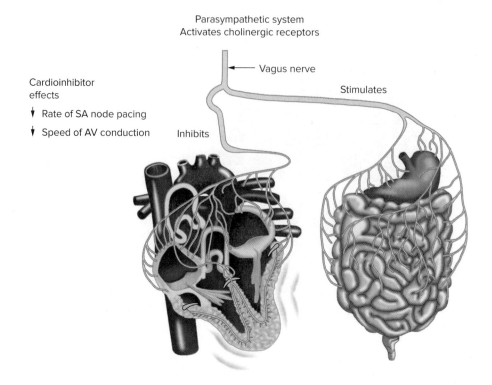

Parasympathetic system
Activates cholinergic receptors

Vagus nerve

Stimulates

Cardioinhibitor effects

↓ Rate of SA node pacing

↓ Speed of AV conduction

Inhibits

the myocardium that result from exercise or emotional stress can only be satisfied by increases in coronary blood flow, mostly through vasodilation.

The need for oxygen in local heart tissue causes the arteries to dilate, which increases blood flow. Stimulation of the sympathetic nervous system also affects coronary blood flow. The coronary blood vessels are innervated by both alpha and beta receptors. Stimulation of the alpha receptors leads to vasoconstriction whereas stimulation of certain beta receptors causes vasodilation.

In situations where coronary blood flow is diminished, the myocardium may become ischemic. If the decreased coronary blood flow persists, the myocardium may become injured or actually die (infarction).

1.9 Nerve Impulse Generation and Muscle Contraction

To better understand how nerve impulses are generated and conducted and how they cause muscles to contract, let's review some basic concepts.

Polarized State

As previously stated, for any muscle to contract, it must first be electrically stimulated. Myocardial cells, like all other cells in the body, are bathed in electrolyte solution. The primary electrolytes responsible for initiating electrical charges are sodium (Na+), potassium (K+), and calcium (Ca++).

In the resting, or **polarized,** state (meaning no electrical activity takes place) the nerve cells have a high concentration of negatively charged ions, proteins, and organelles on the inside of the cell. On the outside, there is a high concentration of

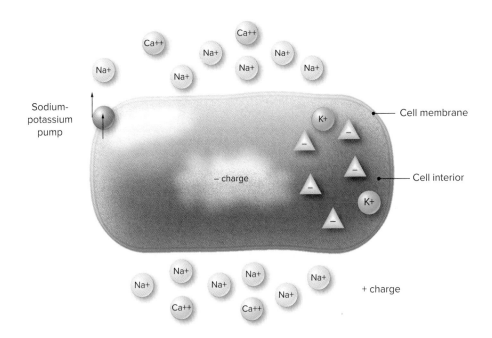

Figure 1-28
In the polarized or resting state, the outside of the cell has a high concentration of positively charged ions while the inside of the cell has a high concentration of negatively charged ions. Under these conditions a negative electrical potential exists across the cell membrane.

positively charged ions (Figure 1-28). This difference in electrical charge between the inside and the outside of the cell creates a **resting membrane potential (RMP).** In the polarized state the RMP is negative, indicating that the cell is ready to fire or discharge. This negative RMP is sustained by the cell membrane's ability to prevent positively charged ions, such as sodium, from entering the cell and the negatively charged ions from leaving the cell.

Even though there are positively charged ions on the inside of the cell, such as potassium (a key ion that facilitates repolarization as discussed below), there are also an abundance of negatively charged ions, proteins, and organelles needed to sustain the cell. This keeps the inside of the cell negatively charged (in relationship to the outside of the cell) during the polarized state.

Depolarization

Impulses are generated and subsequently transmitted when positively charged ions, such as sodium, rapidly move inside the cells, causing the interior to become positively charged (Figure 1-29). This is called **depolarization.** Calcium, another positively charged ion, also enters the cell but more slowly. This rapid change in electrical charge (going from a negative to positive charge) is referred to as the **action potential** of the cell and reflects its ability to depolarize. The action potential is measured in millivolts (mV). Depolarization of myocardial cells causes calcium to be released and come into close proximity with the actin and myosin filaments of the muscle fibers. The filaments then slide together, one upon another, producing a shortening of the muscle fibers and subsequent myocardial contraction.

Different cells in the heart have different action potentials depending on what function they play in the generation and/or conduction of nerve impulses. The way this process occurs differs between the pacemaker cells of the **sinoatrial (SA) node,** the AV node, and the nonpacemaker cells. The positively charged ions move through the cell membrane by way of voltage-gated channels. These channels open and close on the basis of the voltage changes in the cell membrane.

Figure 1-29
Myocyte depolarization begins when an electrical stimulus causes the cell membrane to become permeable to positively charged sodium ions. Sodium rushes into the cell, followed by a slower influx of positively charged calcium ions. This causes the inside of the cell and the cell membrane potential to rapidly become positive.

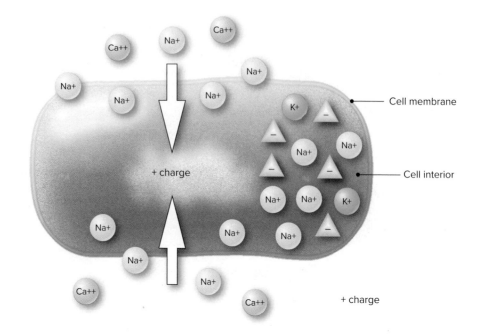

Figure 1-30
As soon as the cell depolarizes, positively charged potassium ions flow out of the cell, initiating a process in which the cell returns to its original, polarized state. This is called repolarization. Aside from the flow of potassium, repolarization involves sodium and calcium being transported from the inside to the outside of the cell by special ion transport systems (i.e., sodium potassium pumps).

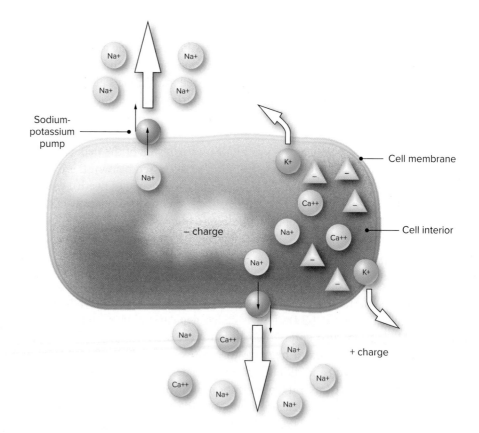

Repolarization

Depolarization is followed by positively charged ions, such as potassium, leaving the cell, causing the positive charge to lower. Then the other positively charged ions, such as sodium, are removed by special transport systems, such as the sodium-potassium pumps, until the electrical potential inside the cell reaches its original negative charge. This is called **repolarization** (Figure 1-30).

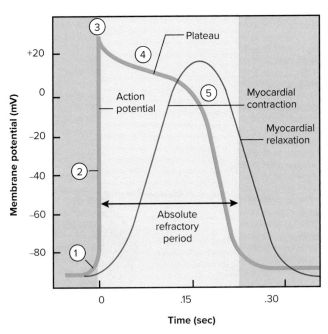

1. Voltage-gated sodium channels open.

2. Na$^+$ inflow depolarizes the membrane and triggers the opening of still more Na$^+$ channels, creating a positive feedback cycle and a rapidly rising membrane voltage.

3. Na$^+$ channels close when the cell depolarizes, and the voltage peaks at nearly +30 mV.

4. Ca^{2+} entering through slow calcium channels prolongs depolarization of membrane, creating a plateau. Plateau falls slightly because of some K$^+$ leakage, but most K$^+$ channels remain closed until end of plateau.

5. Ca^{2+} channels close and Ca^{2+} is transported out of the cell. K$^+$ channels open, and rapid K$^+$ outflow returns the membrane to its resting potential.

Figure 1-31
During the absolute refractory period, no stimulus will depolarize the cell. This figure represents the absolute and relative refractory periods to the movement of electrolytes during depolarization and repolarization.

Refractory Periods

Following depolarization and during the **absolute refractory period** (Figure 1-31), no stimulus, no matter how strong, will depolarize the cell. This prevents spasm-producing (tetanic) contractions in the cardiac muscle. During the later phase of repolarization, the **relative refractory period,** a sufficiently strong stimulus *will* depolarize the myocardium.

Now that we provided a general overview of how nerve impulses are generated with subsequent muscle contraction, let's apply those principles to how the heart's pacemaker cells initiate and the working cells respond to electrical impulses.

Impulse Generation of the SA Node

Through the property of automaticity, the heart's pacemaker, the SA node, fires spontaneously over and over again, normally at a rate of between 60 and 100 times per minute. This occurs because the cells of the SA node have an unstable RMP. Think of the pacemaker cells being in a repetitive state of spontaneous depolarization. In other words, once they depolarize and repolarize, they begin to depolarize again. The action potential of the pacemaker cells consists of three phases, described as follows.

Phase 4

Although the resting potential of the pacemaker cells drops to a minimum negative potential of around −55 to −60 mV between discharges, it is maintained there just for a moment. Then ion channels open, allowing the slow, inward (depolarizing) movement of Na+ currents (called "funny" currents). These currents cause the membrane potential to drift upward (becoming less negatively charged). Once the cell reaches about −50 mV "T," or "transient," Ca++ channels open and Ca++ enters, further neutralizing the intracellular negativity (Figure 1-32). This occurs

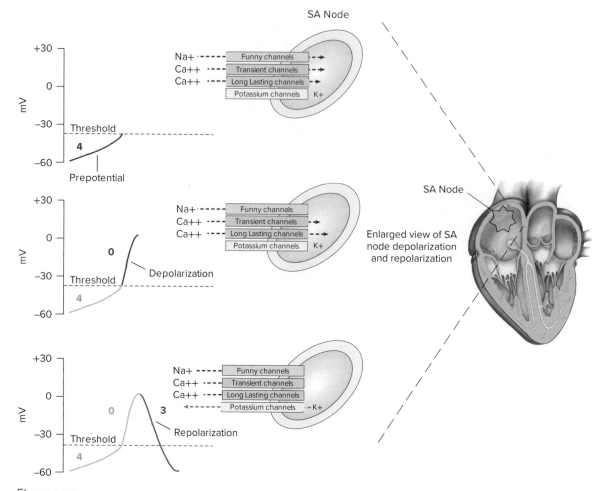

Figure 1-32

SA nodal action potentials occur in three phases. Phase 4 is the spontaneous depolarization that triggers the action potential once the membrane potential reaches threshold between –40 and –30 mV. Phase 0 is the depolarization phase of the action potential. This is followed by phase 3 repolarization. Once the cell is completely repolarized at about –55 to –60 mV, the cycle repeats spontaneously.

without a corresponding outflow of K+ (which would otherwise offset the increase in positively charged ions inside the cell). This initial phase of the action potential is called the *prepotential*. Once the pacemaker potential reaches a threshold of –40 mV, the "L," or "long-lasting," voltage-gated Ca++ channels open and Ca++ slowly flows into the cells from the extracellular fluid.

Phase 0

The increased movement of Ca++ through the L-type Ca++ channels produces the depolarizing phase of the action potential, which peaks slightly above 0 mV. Because the movement of Ca++ through the channels into the cell is not rapid, the rate of depolarization (as seen in the slope of Phase 0) is much slower than that found in other cardiac cells. Also, during this phase the "funny" currents and Ca++ ion movement through the T-type channels decline as the channels close.

Phase 3

Repolarization begins when voltage-gated K+ channels open and K+ ions leave the cell. Also, the L-type Ca++ channels close. This makes the inside of the cell increasingly more negative, bringing about repolarization of the cells. When

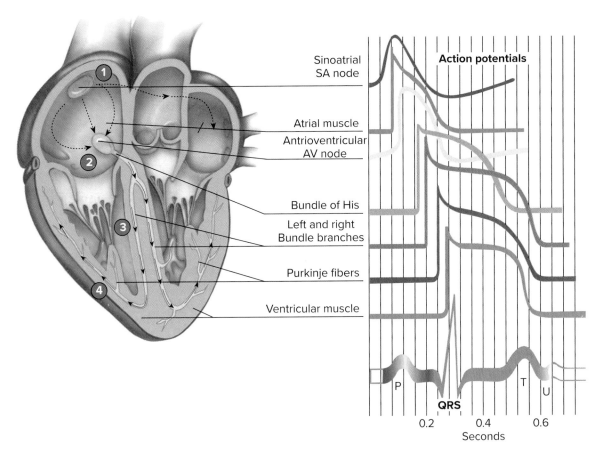

Figure 1-33
This figure shows the action potentials of the heart's pacemaker sites and myocytes from atria and ventricles. Because the SA node depolarizes first, it is the pacemaker of the heart.

repolarization is complete, the K+ channels close. Polarity is restored to the resting state with the ionic concentrations reversed.

The pacemaker potential starts over on its way to producing the next heartbeat. Each depolarization of the SA node initiates one heartbeat. When the SA node fires, it excites the other components in the conduction system which then carry the impulse to the myocytes (the working cells). Normally, the SA node spontaneously depolarizes faster than any other pacemaker site, making it the primary pacemaker for the heart (Figure 1-33).

Depolarization and Repolarization of the Myocytes

In the polarized state, the nonpacemaker myocytes have a very negative resting membrane potential of around -80 to -90 mV. This membrane potential remains stable (the same) until the cell is electrically stimulated, typically via an electric current from an adjacent cell. This electrical stimulus initiates a sequence of events that involve the inflow and outflow of multiple cations and anions that together produce the action potential of the cell and propagate electrical stimulation of the cells that lie next to it. As such, electrical stimulation is conducted from one cell to all the cells that are next to it and then to all the cells of the heart.

The action potential of the myocytes consists of five phases; described as follows.

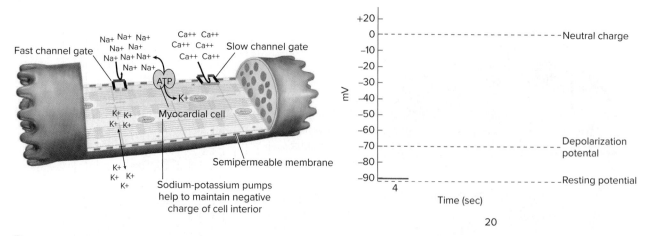

Figure 1-34
Myocyte in polarized state with active transport maintaining a difference in electrical charges between the inside and outside of the cardiac cells.

Phase 4

Phase 4 has a role as the termination of the cardiac electrophysiologic cycle, not just the beginning of it. This is the phase that the cell remains in until it is stimulated by an external electrical stimulus. It is the resting membrane potential. This phase of the action potential is associated with atrial and ventricular diastole. During Phase 4 all the fast Na+ and slow Ca++ channels are closed, and special ion-transporting mechanisms within the cell membrane, called *sodium-potassium pumps,* help to restore the original ionic concentrations (and thus maintain a difference in electrical charges between the inside and outside of the cardiac cells) by continually removing Na+ from inside the cell. This process is referred to as *active transport* and requires the expenditure of energy. Further, the K+ channels are open allowing positive K+ ions to enter and leave the cell (Figure 1-34).

Phase 0

Depolarization begins once the myocyte receives an impulse from a neighboring cell. This is called *Phase 0* (Figure 1-35), or the *rapid depolarization phase.* Na+ quickly enters the cells through voltage-gated channels called *fast Na+ channels* (which are stimulated to open by the electrical impulse). With all this Na+ coming into the cell, it becomes more positively charged. The charge overshoots neutral, rising to about +30 mV. When a certain level is reached, depolarization of the entire cell occurs. The rising phase of the action potential is very brief. Depolarization also causes voltage-gated channels called *slow Ca++ channels* to open and allow some Ca++ to enter the cell. Compared with Na+ fast channels, the Ca++ slow channels open and close slowly. Also, during this phase the K+ channels close, thereby decreasing the movement of K+ across the cell membrane (causing it to remain inside the cell). Think of the movement of Na+ into the cells at the beginning of depolarization as being like a rush of post-holiday shoppers through the doors of the department store when they first open—it is a mass influx.

Phase 1

Phase 1, also referred to as *early repolarization* (Figure 1-36), occurs next. During this phase voltage-gated Na+ channels close, and a small number of the K+

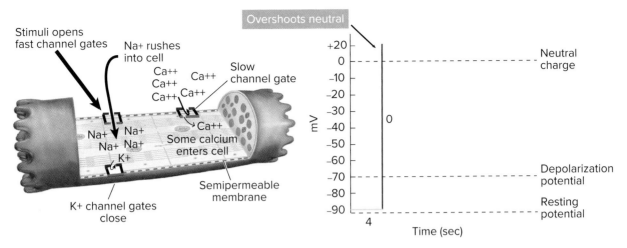

Figure 1-35
Myocyte during Phase 0.

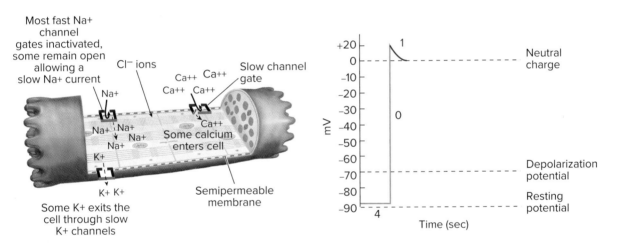

Figure 1-36
Myocyte during Phase 1.

channels open. This stops the fast inflow of Na+ and allows some K+ to move out of the cell. Also, chloride (Cl–) ions enter the cell. These ion movements lower the positive charge inside the cell somewhat.

Phase 2

Next, during Phase 2, or the *plateau phase* (Figure 1-37), Ca++ enters the cell through voltage-gated L-type Ca++ channels, prolonging the depolarization (thereby creating a plateau). This movement of Ca++ into the cell counteracts the potential change caused by the movement of K+ out of the cell through K+ channels. Ca++ then reacts with myosin and actin causing the cell to contract. This prolonged plateau phase distinguishes cardiac action potentials from the much shorter action potentials found in nerves and skeletal muscle.

As one cell depolarizes, the tendency of the adjacent cell to depolarize increases. This leads to a series of depolarizations down the cells in the heart that are capable of generating, responding to, and/or conducting nerve impulses.

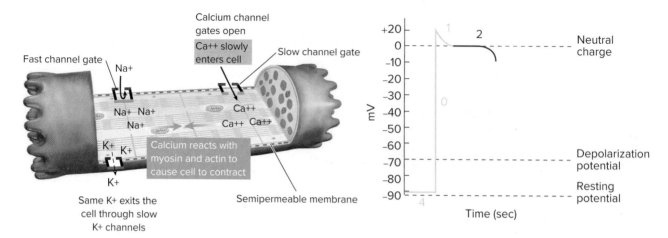

Figure 1-37
Myocyte during Phase 2.

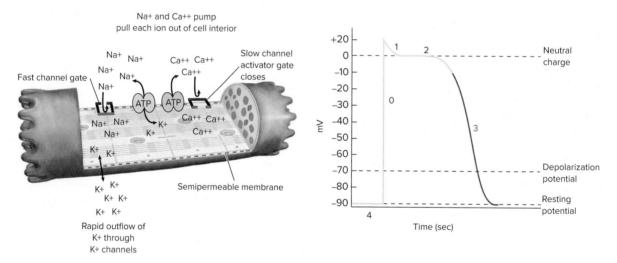

Figure 1-38
Myocyte during Phase 3.

Phase 3

Next, Phase 3, or *rapid final repolarization* (Figure 1-38), results when Ca++ channels close and many K+ channels open. This allows a rapid outflow of K+, causing the cell interior to become more negatively charged. During the next phase (Phase 4) K+ is allowed to reenter the cell interior.

Also, during Phase 3, Ca++ and Na+ are pulled out of the cell interior by Na+ and Ca++ pumps. This helps achieve and maintain the very negative resting membrane potential of the myocytes.

Remember we what said earlier. During what is called the *absolute refractory period*, no stimulus (no matter how strong) will depolarize the cell. This helps assure the rhythmicity of the heartbeat. The absolute refractory period includes Phases 0, 1, 2, and part of Phase 3. It assures that after contraction, relaxation is nearly complete before another action potential can be initiated.

Key Points

LO 1.1	• The electrocardiogram is a graphic representation of the heart's electrical activity. It is used to identify irregularities in the heart rhythm and to reveal the presence of injury, death, or other physical changes in the heart muscle.
	• The electrocardiograph detects the electrical activity occurring in the heart through electrodes attached to the patient's skin.
	• When an impulse occurring in the heart moves toward a positive electrode of the ECG, it produces a positive waveform. When it moves away from the positive electrode (or toward a negative electrode), it produces a negative waveform.
LO 1.2	• The heart, blood, and blood vessels are the chief components of the circulatory system.
	• The circulatory system circulates enough blood to deliver needed oxygen and nutrients to the tissues and to remove waste products.
LO 1.3	• The heart is the pump; each time it contracts, it pushes blood throughout the body.
	• The heart is located between the lungs in the mediastinum behind the sternum, and it rests on the diaphragm with a front-to-back (anterior-posterior) orientation.
	• It is surrounded by a double-walled closed sac called the pericardium. The pericardium allows the heart to contract and expand within the chest cavity with minimal friction.
	• The heart wall consists of three layers: the myocardium, endocardium, and epicardium.
	• The heart consists of two upper chambers, the atria, and two lower chambers, the ventricles.
	• A muscular wall, the septum, separates the right side from the left side of the heart.
	• The right side of the heart receives blood from the systemic venous circulation and pumps it into the pulmonary circulation.
	• The left side of the heart receives blood from the pulmonary circulation and pumps it into the systemic circulation.
	• The skeleton of the heart provides firm support for the AV and semilunar valves and acts to separate and insulate the atria from the ventricles.
	• The four heart valves permit blood to flow through the heart in only one direction.
LO 1.4	• There are two basic types of cells in the heart: the myocardial cells (also referred to as the working cells), which contract to propel blood out of the heart's chambers, and the specialized cells of the electrical conduction system, which initiate and carry impulses throughout the heart.
	• The structure of the myocardial cells permits the rapid conduction of electrical impulses from one cell to another. This results in the cardiac muscle cells acting as a single unit, permitting coordinated contraction of a whole group of cells.
LO 1.5	• Depolarization of the myocardium progresses from the atria to the ventricles in an orderly fashion. The electrical stimulus causes the heart muscle to contract.
	• The wave of depolarization is carried throughout the heart via the heart's conduction system. The conduction system is a grouping of specialized tissues that form a network of connections, much like an electrical circuit.
	• The key structures of the conduction system are the SA node, intraatrial conductive pathway, internodal pathways, AV node, bundle of His, right and left bundle branches, and Purkinje fibers.

LO 1.6	• The oxygen and nutrient demand of the heart is extremely high. The coronary arteries deliver needed blood supply to the myocardial cells.
	• The coronary arteries return deoxygenated venous blood from the heart to the right atrium via the coronary sinus.
LO 1.7	• The cardiac cycle refers to all or any of the events related to the flow of blood that occur from the beginning of one heartbeat to the beginning of the next.
	• Cardiac output is the amount of blood pumped from the heart in one minute. The cardiac output is equal to the heart rate multiplied by the stroke volume.
	• Blood pressure is the force exerted by circulating blood on the walls of the blood vessels.
	• The coronary arteries perfuse the myocardium during diastole.
	• Automaticity, excitability, conductivity, and contractility are the four key properties of myocardial cells. These properties allow the myocardial cells to generate impulses, respond to them, and conduct them throughout the heart, resulting in contraction of the muscle.
	• The electrical impulse that normally initiates the heartbeat arises from the SA node. From there it travels through the atria, generating a positive waveform (the P wave), which shows on the ECG, leading to contraction of the atria. The atrial muscle contraction pushes the remaining blood from the atria into the ventricles.
	• The impulse is slowed as it passes from the atria to the ventricles through the AV node. The AV node is normally the only pathway for impulses to move from the atria to the ventricles. The fibrous skeleton acts as an insulator to prevent electrical impulses from getting to the ventricles by any means other than the AV node.
	• The bundle of His and the right and left branches carry impulses from the AV node to the Purkinje fibers.
	• The Purkinje fibers extend to just underneath the endocardium and terminate in the endocardial cells.
	• Collectively, the right and left bundle branches and Purkinje fibers are referred to as the His-Purkinje system.
	• On the ECG, the impulse traveling through the His-Purkinje system is seen as a flat line following the P wave.
	• The QRS complex is generated and the ventricles contract as a result of the electrical impulse stimulating the ventricles. Ventricular muscle contraction causes the blood to be pushed into the pulmonary and systemic circulation.
	• The ST segment and T wave represent repolarization of the ventricles, which takes place following the conclusion of depolarization. Atrial repolarization occurs but is hidden by the QRS complex.
	• Other sites in the heart can assume control by discharging impulses faster than the SA node. They can also passively take over, either because the SA node has failed or because it is generating impulses too slowly.
	• The intrinsic rate of the SA node is 60 to 100 beats per minute. The intrinsic rate of the AV node is 40 to 60 beats per minute, and the intrinsic rate of the Purkinje fibers is 20 to 40 beats per minute.

LO 1.8	• The autonomic nervous system can influence the rate and strength of myocardial contractions.
	• The two divisions of the autonomic nervous system are the sympathetic and the parasympathetic nervous systems.
	• In the heart, stimulation of the sympathetic nervous system produces enhancement of myocardial cell excitability, increased rate of pacemaker firing, increased conduction speed, increased contractility, and coronary vasodilation.
	• Stimulation of the parasympathetic nervous system slows the heart rate and AV conduction.
	• Increased myocardial demands can only be satisfied by increases in coronary blood flow, mostly through vasodilation.
LO 1.9	• Nerve impulses stimulate cardiac muscles to contract. Without electrical stimulation the muscles will not contract.
	• Sodium, calcium, and potassium are the key electrolytes responsible for initiating electrical activity.
	• In the polarized state, there is a high concentration of sodium (Na+) and calcium (Ca++) on the outside of the cell.
	• Depolarization of the cells occurs when positive electrolytes move from outside to inside the cell. This causes the inside of the cell to become more positively charged.
	• Depolarization of myocardial cells causes calcium to be released and come into close proximity with the actin and myosin filaments of the muscle fibers. The filaments then slide together, one upon another, producing a shortening of the muscle fibers and subsequent myocardial contraction.
	• Repolarization of cells occurs when positively charged ions, such as potassium, leave the cell and sodium and calcium are transported out of the cell. Repolarization is the process of restoring the cell to its pre-depolarization state.

Assess Your Understanding

The following questions give you a chance to assess your understanding of the material discussed in this chapter. The answers can be found in Appendix A.

1. The electrocardiogram measures   (LO 1.1)
   a. pulse strength.
   b. blood pressure.
   c. cardiac output.
   d. the heart's electrical activity.

2. The electrocardiograph can do all of the following EXCEPT   (LO 1.1)
   a. identify irregularities in the heart rhythm.
   b. determine perfusion status of the patient.
   c. reveal the presence of, injury of, death of, or other physical changes in the heart muscle.
   d. diagnose noncardiac diseases such as pulmonary embolism or hypothermia.

3. Components of the circulatory system include   (LO 1.2)

   a. the pulmonary circulation that carries blood to and from the systemic circulation.

   b. blood that consists mostly of blood cells and little plasma.

   c. the systemic circulation that carries blood to and from the lungs where it is oxygenated.

   d. roughly 4.7 to 5.7 L of blood.

4. The heart acts as the _____ of the circulatory system.   (LO 1.3)

   a. pipeline

   b. pump

   c. transportation medium

   d. nervous control center

5. The heart   (LO 1.3)

   a. is located above the lungs.

   b. has a side-to-side orientation.

   c. lies on the diaphragm in front of the trachea, esophagus, and thoracic vertebrae.

   d. lies mostly in the right side of the chest.

6. How would you describe the shape of the heart?   (LO 1.3)

7. Which statement regarding the pericardium is accurate?   (LO 1.3)

   a. It is a double-walled, open sac.

   b. The tough, fibrous, outer layer of the pericardium is called the *serous pericardium.*

   c. The parietal pericardium covers the surface of the heart.

   d. The pericardial cavity holds a small amount of clear lubricating fluid.

8. The myocardium is   (LO 1.3)

   a. the thickest of the three layers of the heart.

   b. smooth and is continuous with the lining of the arteries and veins.

   c. watertight to prevent leakage of blood into the other layers of the heart.

   d. the outermost layer of the heart.

9. The upper chambers of the heart are called the   (LO 1.3)

   a. ventricles.

   b. atria.

   c. vena cava.

   d. venules.

10. Functions of the skeleton of the heart include   (LO 1.3)

   a. allowing the top and bottom parts of the heart to act as separate pumps.

   b. enhancing conduction from the atria to the ventricles.

   c. contracting to allow the atria to finish filling ventricles.

   d. allowing the heart to be depolarized all at once.

11. Describe the difference between the heart's working cells and the specialized cells of the electrical conduction system. (LO 1.4)

12. List the six main components of the heart's conduction system. (LO 1.5)

13. The electrical impulse that normally initiates the heartbeat arises from the (LO 1.7)
    a. SA node.
    b. AV node.
    c. bundle of His.
    d. Purkinje fibers.

14. Depolarization of the heart progresses from the _____ to the _____ in an orderly fashion. (LO 1.7)
    a. atria, ventricles
    b. ventricles, atria
    c. bundle of His, Purkinje fibers
    d. ventricles, bundle branches

15. The myocardium receives its blood supply via the (LO 1.6)
    a. vena cava.
    b. coronary arteries.
    c. blood it pumps to the systemic and pulmonary circulation.
    d. pulmonary arteries.

16. List the coronary blood vessels that supply the right ventricle. (LO 1.6)

17. List the coronary blood vessels that supply the left ventricle. (LO 1.6)

18. Cardiac output is equal to (LO 1.7)
    a. afterload multiplied by preload.
    b. contractility multiplied by the stroke volume.
    c. heart rate divided by the contractility.
    d. stroke volume multiplied by the heart rate.

19. Automaticity, a property of the heart, is the ability to (LO 1.7)
    a. carry an impulse.
    b. respond to an electrical stimulus.
    c. contract when stimulated by an electrical impulse.
    d. produce an electrical impulse without the need for outside nerve stimulation.

20. The impulse traveling through the His-Purkinje system generates a _____ on the ECG. (LO 1.7)
    a. P wave
    b. T wave
    c. flat line
    d. QRS complex

21. The QRS complex represents (LO 1.7)
    a. atrial depolarization.
    b. ventricular depolarization.
    c. delay of the impulse as it travels through the AV node.
    d. ventricular repolarization.

22. The T wave represents (LO 1.7)
    a. atrial depolarization.
    b. AV node depolarization.
    c. ventricular repolarization.
    d. initiation of the electrical stimulus.

23. How soon can the cell respond to its next electrical stimulus?

24. Which of the following is true regarding alternate pacemaker sites of the heart? (LO 1.7)
    a. Unless the SA node fails, no other site can assume control of the heartbeat.
    b. An alternate pacemaker site can initiate the heartbeat if the SA node fails to do so.
    c. The farther it is from the SA node, the faster the pacemaker's intrinsic rate is.
    d. Impulse initiation from an alternate pacemaker site will produce an "extra" P wave.

25. The _____ has an intrinsic rate of 40 to 60 beats per minute. (LO 1.7)
    a. SA node
    b. AV node
    c. Purkinje fibers
    d. bundle of His

26. The sympathetic branch of the autonomic nervous system (LO 1.8)
    a. is mediated by the neurotransmitter acetylcholine.
    b. is often called the *cholinergic system.*
    c. produces the "fight-or-flight" response.
    d. works the same as the parasympathetic branch.

27. The following are true regarding the parasympathetic nervous system EXCEPT (LO 1.8)
    a. it is mediated through the vagus nerve.
    b. its fibers mostly affect the SA and AV nodes of the heart.
    c. it is sometimes referred to as the adrenergic system.
    d. it works just the opposite of the sympathetic branch.

28. _____ stimulate muscle cells to contract. (LO 1.9)
    a. Changes in pressure
    b. Nerve impulses
    c. Outside influences
    d. Polarization of the nerve cells

29. Which of the following is true? (LO 1.9)
    a. During the polarized state, the inside of the cell is positively charged.
    b. Depolarization occurs when there is a rapid influx of positively charged ions from outside to inside the cell.
    c. Repolarization begins with the removal of sodium from the interior of the cell.
    d. During depolarization, sodium enters the cells through the slow channels.

30. Why are the arteries that supply the heart muscle called the coronary arteries? (LO 1.6)

Referring to the scenario at the beginning of this chapter, answer the following questions.

31. If blood flow in the coronary arteries is diminished, what will likely occur? (LO 1.8)
    a. Myocardial ischemia
    b. Hypertension
    c. Increased cardiac output
    d. Decreased diastole

32. What is the intrinsic rate of the SA node?

33. The patient's initial heart rate is fast because of (LO 1.8)
    a. the effects of nicotine.
    b. sympathetic nervous system stimulation.
    c. hypertension.
    d. parasympathetic nervous system stimulation.

34. Which of the following will likely occur if the patient experiences severe myocardial ischemia and it is not corrected? (LO 1.6)
    a. Blood flow through the coronary veins increases.
    b. The heart extracts more oxygen from the blood it pumps through its chambers.
    c. The heart beats erratically.
    d. The affected myocardial tissue becomes necrotic.

2 The Electrocardiogram

Chapter Outline

The Electrocardiogram
and ECG Machines

ECG Leadwires and
Electrodes

Heart's Normal Electrical
Activity

ECG Leads

Planes of the Heart and
Lead Placement

Displays and Printouts

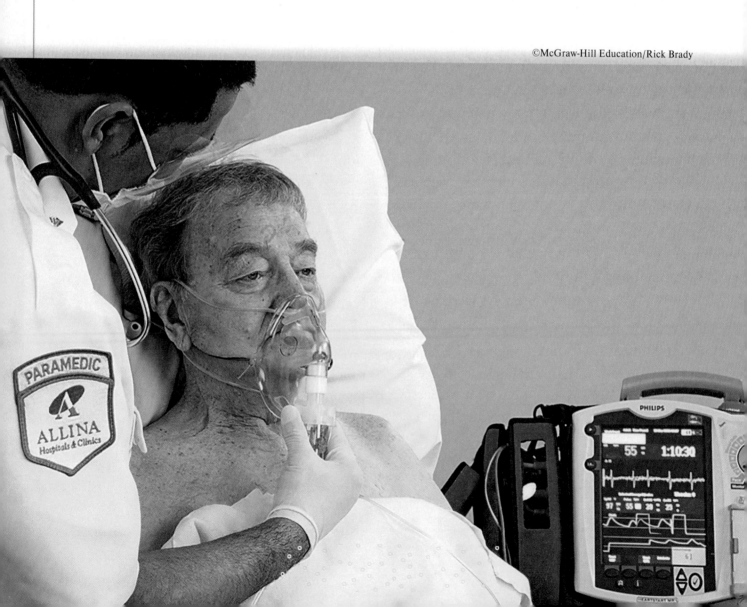

Learning Outcomes

LO 2.1   List the uses for the electrocardiograph.

LO 2.2   Recall the purpose and proper use of ECG lead wires and electrodes.

LO 2.3   Describe the three phases of ventricular depolarization.

LO 2.4   Define the term ECG lead, differentiate between bipolar and unipolar leads, and recall the direction a waveform takes when the electrical current is moving toward a positive electrode.

LO 2.5   Differentiate between the frontal plane and the horizontal plane of the heart, and identify which leads are associated with each plane, the location for proper electrode placement for each lead, and the direction waveforms deflect with each lead.

LO 2.6   Recall what the horizontal and vertical lines on ECG paper represent and list the values of the boxes they create.

Case History

It is a busy morning in the cardiac rehabilitation unit where patients, after undergoing cardiac bypass surgery, come to exercise under observation of trained therapists. A 72-year-old man is on the stationary bike when he begins to feel lightheaded and then passes out, falling off the bike onto the floor. A therapist rushes to his side to find him slowly regaining consciousness.

The therapist is able to print out a continuous recording of his heart rhythm before, during, and after the incident because the patient was on a cardiac monitor during his exercise routine.

An Emergency Medical Service (EMS) ambulance is summoned to transport the patient to the emergency department, and the therapist provides the paramedics with the electrocardiogram (ECG) printout, which shows that the patient was in a slightly fast but otherwise normal heart rhythm before exercise. The rhythm became very slow with pauses up to 10 seconds long during exercise and then returned to normal after he fainted.

2.1 The Electrocardiogram and ECG Machines

As discussed in Chapter 1, the ECG is a tracing of the heart's electrical activity. The device that measures and records the ECG is called an electrocardiograph, although you will probably hear it called the "ECG," "EKG," "monitor," or "12 lead." To avoid confusion, we will use ECG machine when referring to the electrocardiograph.

While many before him contributed to the understanding of heart's electrical activity and how to record it, Dutch physiologist Willem Einthoven, MD, PhD, is credited with inventing the first ECG machine in 1903. Initially, it was a cumbersome and costly device, taking five technicians to operate. During the procedure, patients sat with both arms and the left leg in separate buckets of saline solution (Figure 2-1). These buckets acted as electrodes to conduct the current from the skin's surface to the filament of the device. The three points of electrode contact on the limbs produced what is known as **Einthoven's Triangle,** a principle still used in modern day ECG recording. The EKG was so named because Einthoven had the earliest ECG machines manufactured in Germany, and "kardio" is German for heart.

ECG machines have many uses in both the hospital and out-of-hospital settings. Commonly, they are employed to provide continuous monitoring of the heart rhythm. This is done to check for abnormalities in the heart rhythm called dysrhythmias. Dysrhythmias may be seen as a significant increase or decrease in the heart rate, impulses that arise outside the heart's normal pacemaker site (the SA node), and/or delays or blockage in impulse conduction through the atrioventricular (AV) junction.

Figure 2-1
Picture of the first ECG machine.
Source: Cambridge Scientific Instrument Company

PHOTOGRAPH OF A COMPLETE ELECTROCARDIOGRAPH, SHOWING THE MANNER IN WHICH THE ELECTRODES ARE ATTACHED TO THE PATIENT, IN THIS CASE THE HANDS AND ONE FOOT BEING IMMERSED IN JARS OF SALT SOLUTION

Other uses of the ECG machine are to identify developing or existing cardiac and noncardiac conditions such as myocardial ischemia and/or infarction, inflammation, enlargement and/or hypertrophy of the heart muscle, changes in or blockage of the heart's conduction system, accumulation of fluid in the pericardial sac, or electrical effects of medications and electrolytes and to evaluate the function of artificial implanted pacemakers (devices used to initiate or control the heart rate). The ECG machine is also commonly used to conduct assessments such as stress testing and to obtain a baseline recording before, during, and after medical procedures.

A variety of ECG machines are available for use (Figure 2-2). The type used to provide continuous cardiac monitoring is often referred to as an ECG monitor and is equipped with a display screen (oscilloscope) to view the rhythm. Depending on their use, these devices may view one lead or up to twelve leads at a time. They may also have the capability to switch between several different leads (Figure 2-2A). This type of ECG machine is often equipped with a defibrillator (a device used to deliver an electrical shock to the heart) and a pacemaker (a device used to stimulate the heart to beat at a desired rate) and may have the capability to obtain and record 12-lead ECGs. Some ECG machines are also equipped with an automatic blood pressure device, pulse oximetry sensor, and waveform capnography (end-tidal carbon dioxide/$ETCO_2$, also

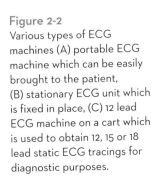

Figure 2-2
Various types of ECG machines (A) portable ECG machine which can be easily brought to the patient, (B) stationary ECG unit which is fixed in place, (C) 12 lead ECG machine on a cart which is used to obtain 12, 15 or 18 lead static ECG tracings for diagnostic purposes.

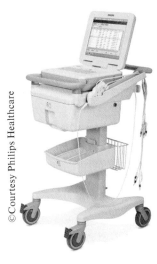

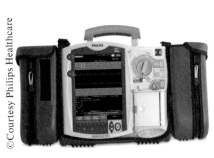

©Courtesy Philips Healthcare

A. Portable

©Courtesy Philips Healthcare

C. 12 lead on cart

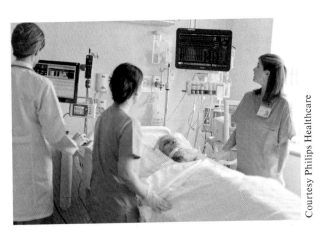

Courtesy Philips Healthcare

B. Stationary

called PetCO$_2$) and/or carbon monoxide detection. Some ECG monitors are portable and used in the prehospital care setting or in the clinical environment where patients are transported from one point to another or are placed on movable cabinets used to store emergency supplies (often called "crash carts") and located in or near treatment areas throughout the hospital. Stationary ECG machines are fixed in place, such as those located in emergency departments, coronary care and intensive care units, specialty areas and regular nursing floors. These ECG machines usually consist of the display monitor, cabling, lead wires and telemetry which continually transmits the patient's ECG and other vital signs to a central monitoring location (Figure 2-2B).

The type of ECG machine used exclusively to perform 12-lead ECGs (Figure 2-2C) in the clinical setting may or may not include a display screen or the features described earlier. It is typically kept on a cart so it can be moved to the location where it is needed. Through the recording of 12 different views of the heart, this machine is used to identify the presence of both cardiac and noncardiac conditions and evidence of injury to the heart. It can also be used to identify cardiac dysrhythmias, although this occurs from the static recording and not from continuous monitoring. Additionally, some ECG machines are equipped to record as many as 15 to 18 leads, giving the clinician the ability to view the right lateral and posterior aspects of the heart.

Many of today's ECG devices are equipped to provide a computerized interpretation of the patient's cardiac rhythm and to analyze normal or abnormal ECG findings. To do this, the device records the patient's ECG tracing and examines it against an internal set of criteria and rules of logic. The result is a computerized ECG analysis or "interpretive" 12-lead ECG that is printed on the ECG report.

To more easily understand the ECG machine, we will first talk about how it obtains the information (the input). Then, we will discuss how it displays the information (the output). Learning about ECGs can be intimidating. For this reason, we discuss how to interpret the ECG rate, rhythm, waveforms, complexes, segments, and intervals in the next several chapters and then learn how to analyze the other, more complex, aspects of the ECG.

2.2 ECG Lead Wires and Electrodes

The heart's electrical activity is detected by electrodes placed on the patient's skin and then transferred to the ECG machine by wires or cables that are connected to these electrodes. The ECG machine then displays the electrical activity on the oscilloscope, or it is printed onto graph paper.

Lead Wires

There may be 3, 4, or 5 lead wires for monitoring purposes and up to 10 lead wires for 12-lead ECGs. Some lead wire sets have a primary cable that includes four standard leads and a connector to which a cable having an additional six lead wires can be attached. This allows the ECG machine to be used either for monitoring purposes or to obtain 12-lead ECGs.

The lead wires each have a clip, snap, or pin-type connector on the distal end. This allows attachment to the metal snap or tab on the ECG electrode. The lead wires then feed into a connector or into a cable that attaches to the ECG machine (Figure 2-3).

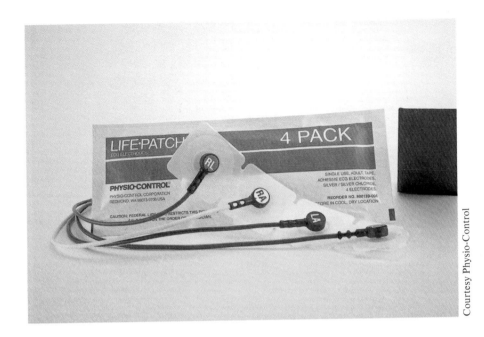

Courtesy Physio-Control

Figure 2-3
Lead wires with ECG electrodes.

The lead wires are color-coded to help you remember where each electrode is placed. However, do not rely solely on the color of the wires for proper placement of the electrodes because the colors vary between two different standard agencies: the American Heart Association (AHA) and the International Electrotechnical Commission (IEC). Instead, use the lettering commonly located on the top of the lead wire connector for each lead. For example, LL stands for left leg, LA stands for left arm, RA stands for right arm, and RL stands for right leg. The leads that are placed across the chest are labeled V_1, V_2, V_3, V_4, V_5, and V_6. Those that are used with an ECG machine capable of viewing 15 or 18 leads and placed across the right side of the chest are labeled V_4R, V_5R, and V_6R while the ones placed across the posterior of the chest are labeled V_7, V_8, and V_9.

Although the lead wires are encased in a plastic or rubberized covering, they are still somewhat fragile, so handle them with care. Bending, stressing, and straining of lead wires during repeated use can cause structural failure, which can result in artifact or prevent transmission of signals.

Another problem with lead wires is that they can easily become tangled with each other or with cables for other sensors, especially during extension and storage. Detangling of the lead wires consumes time, may delay urgently needed medical procedures, and can cause premature failure. For this reason, try to keep the lead wires separated from each other as you put them away or you pull them out (Figure 2-4).

Last, since the lead wires are typically not disposed of following each application, they can become soiled during use by contact with contaminants such as blood, other bodily fluids, and medicinal preparations. For this reason, they should be properly cleaned/decontaminated after each use (be sure to follow the manufacturer's recommendations).

Electrodes

In the past, electrodes were flat, metal plates held in place with straps that went around the arms and legs and suction cups applied with conductive paste. ECG electrodes in use today are typically disposable and consist of a wet or dry

Figure 2-4
Untangling lead wires for proper storing.

Courtesy Bruce Shade

electrolyte gel (which acts to assure good signal pick-up), a metal snap or tab (where the ECG lead wire is attached), and a self-adhesive pad that holds the electrode to the skin. The self-adhesive pad may consist of foam, paper, cloth, or other such material. Electrodes come in a variety of shapes and sizes and offer a number of different features. Some have pull tabs for easier placement and removal; others can be used in diaphoretic patients or those with hairy chests. Some are latex free to eliminate the likelihood of an allergic reaction whereas others have a radiotranslucent snap to prevent their being seen on x-rays or magnetic resonance imaging (MRI) taken of the patient. Some ECG electrodes can be repositioned and still effectively adhere to the skin. Electrodes are available for adults as well as pediatric patients.

Preparing the Electrode Sites

To get the best possible signal (noise-free recording), there must be good contact between the electrode and the patient's skin (Figure 2-5). Otherwise, the ECG tracing may include artifact (wandering baselines, small complexes, fuzzy tracings, etc.), making analyzing and interpreting the ECG tracing difficult.

Before the electrode is placed, the skin site should be cleaned to remove dead skin cells and oils (Figure 2-5A). Gently abrading the skin by using the small abrasive scrub pad found on the peel-away portion of the electrode or briskly rubbing the site with a piece of dry rough material such as a towel or a piece of gauze can adequately prepare the skin. Use soap and water to remove oil or dirt from the skin. If the surface is wet or moist, use a towel or gauze dressing to dry it. Also, if dense hair is present at the area where the electrodes are to be placed, clip or shave it close to the surface.

Some references suggest the use of alcohol to clean the electrode site. Others indicate that the use of alcohol actually increases the impedance of the chest, thus

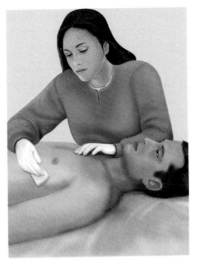

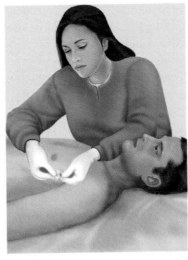

Figure 2-5
Technique for placing
electrodes and attaching
lead wires (A) prepare the site
by gently abrading the skin
surface where the electrode is
to be placed, (B) peel the
protective backing from the
ECG electrode, (C) apply
electrode to correct anatomic
location on skin.

A. Prepare site   **B.** Prepare the electrode

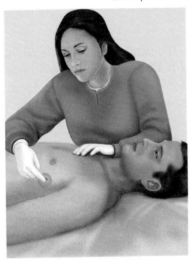

C. Place electrode

interfering with detection of the electrical activity. Follow local protocol and procedure guidelines.

Another hint for assuring a noise-free recording is to eliminate any muscle tension. Place the patient's arms and legs in a comfortable position in which the extremities are resting on a supportive surface. Any self-support of limbs by the patient may introduce fine muscle artifact even though the patient appears not to be moving. If artifact is still present and the patient tolerates it, try laying him or her in a flat position. This often eliminates artifact and allows you to acquire a good tracing.

Placing the Electrodes

When using snap-on lead wires, you should attach the electrode to the lead wire before placing the electrodes onto the patient's skin. This is more comfortable for the patient and preserves the integrity of the electrode gel. When using clip-on type lead wires, apply the gel to the metal snap of the electrode after the electrode has been placed on the skin. Whichever way you use, make sure the leads are connected tightly to the electrodes. Avoid removing the electrodes from their sealed protective envelope before use as the conductive gel may dry. This can decrease the signal detection.

The electrodes can be easily peeled away from the protective backing to expose the gel disc and the sticky surface of the adhesive pad (Figure 2-5B). Make sure the gel is moist (unless it is the dry gel type). If the electrode has become dry, or if it is past its expiration date, discard it and use a new one as a dry electrode decreases electrical contact.

Then, place the electrodes on the patient in the proper positions. Look for a flat surface. While the electrode is flexible enough to keep contact with a surface that is slightly irregular, a flat surface is always preferred. Select sites over soft tissues. Avoid areas where the large bones are near the skin surface as bones are not good conductors and can interfere with detection of the signal. Also, avoid areas where there are thick muscles or skin folds. Those areas can produce ECG artifacts.

Place each electrode onto the patient and, using a circular motion, smooth down the adhesive area (Figure 2-5C). Avoid applying pressure on the gel disc itself as this could result in a decrease in conductivity and adherence. The electrode should lie flat. Repeat the procedure for each electrode. One last point: the lead wires should be positioned so that they do not tug on the electrodes and lift the gel pad away from skin. This will reduce their ability to detect the heart's electrical signals.

2.3 Heart's Normal Electrical Activity

Before we go any further, let's briefly review how the heartbeat originates and is conducted. Each heartbeat arises as an electrical impulse from the SA node. The impulse then spreads across the atria, depolarizing the tissue and causing them both to contract. The initial wave of depolarization spreads anteriorly through the right atrium and toward the AV node. It then travels posteriorly and toward the left atrium. This is seen on the ECG as a P wave. The impulse then activates the AV node, which is normally the only electrical connection between the atria and the ventricles. There, the impulse is delayed slightly, allowing the atria to finish contracting and pushing any remaining blood from their chambers into the ventricles (referred to as the *atrial kick*). The impulse then spreads through both ventricles via the bundle of His, right and left bundle branches, and the Purkinje fibers (seen as the QRS complex), causing a synchronized contraction of the primary pumping chambers of the heart and, thus, the pulse.

We can divide ventricular depolarization into three main phases. In the first phase, ventricular activation begins in the septum as it is depolarized from left to right. Early depolarization of the right ventricle also occurs. In the second phase, the right and left ventricular apex are depolarized, and the depolarization of the right ventricle is completed. In the third phase, the remainder of the left ventricle is depolarized toward the lateral wall. Because the right ventricle has already been depolarized at this point, the left ventricle will depolarize unopposed by the right, so large voltages will be displayed on the ECG at this time. As we describe the view of each lead as follows, we will review how the electrical events occurring in the heart will be displayed.

2.4 ECG Leads

As mentioned earlier, the ECG machine shows the changes in voltage, detectable during the time course of the heartbeat via ECG electrodes placed on the skin. Positioning ECG electrodes in specific positions on the patient's body gives us different views of the heart. These views can be thought of as ECG leads. To help understand the concept of ECG leads, think about taking pictures of a house. From the front you get one picture and from the side of the house you get another picture. If there is damage on the back of the house, you should be able to see it in pictures taken at the rear of the house. The different ECG leads work in the same way; each provides a view that the other leads do not. The location or sites where the electrodes are placed vary depending on which view of the heart's activity is being assessed (Figure 2-6).

You might ask, "Why do I need to see different views of the heart?" If all we are doing is identifying the presence or emergence of dysrhythmias, one lead can typically provide us the information we need. Instead, let's suppose your patient is complaining of chest pain. You are attempting to determine if it is due to a serious condition. The presence of certain findings in two or more contiguous (anatomically next to each other) leads can indicate the presence of myocardial infarction (as well as ischemia and injury). In other words, multiple views are necessary to determine this condition. Further, depending on which lead we see the findings in, we can determine the location of the myocardial infarction.

In addition to identifying the presence and location of myocardial infarction, we can use the different ECG leads to identify physical changes in the heart muscle, conduction defects, electrical effects of medications and electrolytes, and to evaluate pacemaker function.

As we discuss ECG leads further, we review the concepts of bipolar and unipolar leads and how ECG leads view the frontal and horizontal planes of the heart.

Figure 2-6
Think of each lead as being a different view or picture the ECG machine has of the heart. This figure shows the view of each limb lead. Modern ECG machines offer many other views of the heart as described in the following pages.

Bipolar/Unipolar Leads

It is commonly stated that ECG leads are either bipolar or unipolar. A bipolar lead records the flow of the electrical impulse between the two selected electrodes; one represents a positive pole while the other represents a negative pole. Between the two poles (the positive and negative electrodes) lies an imaginary line representing its view of the direction the electrical current is moving through the heart.

The bipolar leads I, II, and III form what is known as Einthoven's triangle (Figure 2-7), which is an electrically equilateral triangle based on these three limb leads' positions relative to one another. These leads intersect at angles of 60 degrees. The axis of lead I extends from shoulder to shoulder with the right-arm electrode being the negative electrode and the positive electrode being the left-arm electrode. The axis of lead II extends from the negative right-arm electrode to the positive left-leg electrode. The axis of lead III extends from the negative left-arm electrode to the positive left-leg electrode. This information will be helpful later in the book as we discuss determining electrical axis.

Bipolar leads have a third (and often a fourth) electrode called a *ground*. The ground is used to help prevent electrical interference from appearing on the ECG and has zero electrical potential when compared with the positive and negative electrodes.

Settings on most ECG machines allow electrodes to be made either positive or negative, depending on the required lead. A lead records the electrical signals of the heart from a particular combination of recording electrodes that are placed at specific points on the patient's body. The direction the electrical current takes, toward a positive or negative electrode, determines the direction a waveform points on an

Figure 2-7
A bipolar lead requires a positive and a negative electrode. Three bipolar leads—I, II, and III—form Einthoven's triangle.

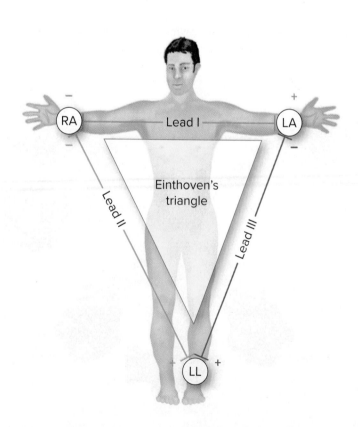

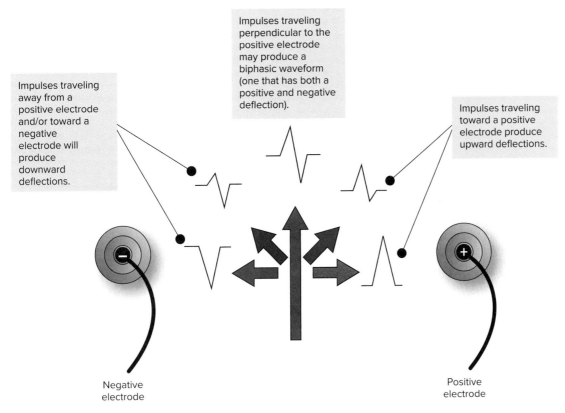

Impulses traveling perpendicular to the positive electrode may produce a biphasic waveform (one that has both a positive and negative deflection).

Impulses traveling away from a positive electrode and/or toward a negative electrode will produce downward deflections.

Impulses traveling toward a positive electrode produce upward deflections.

Negative electrode

Positive electrode

Figure 2-8
The direction the electrical impulse takes toward or away from a positive electrode causes the waveform to deflect either upward or downward.

ECG (Figure 2-8). An electrical current traveling more toward a positive electrode produces a waveform that deflects upward. A current traveling away from a positive electrode or toward a negative electrode produces a waveform that deflects downward.

A biphasic waveform (one that has both a positive and negative deflection) is recorded when the impulse travels perpendicular to where the electrode is positioned.

Given that the left ventricle is the larger muscle mass of the heart, the movement of electrical energy through it produces the largest waveforms. This is discussed in greater detail in Chapter 19.

Unipolar leads are said to use only one positive electrode and a central terminal (also called a reference point) calculated by the ECG machine. The ECG machine averages the input from multiple recording electrodes to create this central terminal which serves as the negative electrode. The central terminal (with zero electrical potential) lies in the center of the heart's electrical field located left of the interventricular septum and below the AV junction (Figure 2-9C). The nine standard leads that employ a positive electrode and a central terminal are termed "V leads."

Note: Because the V leads use a positive pole and a negative pole that are created by the ECG machine, some references refer to them as bipolar leads.

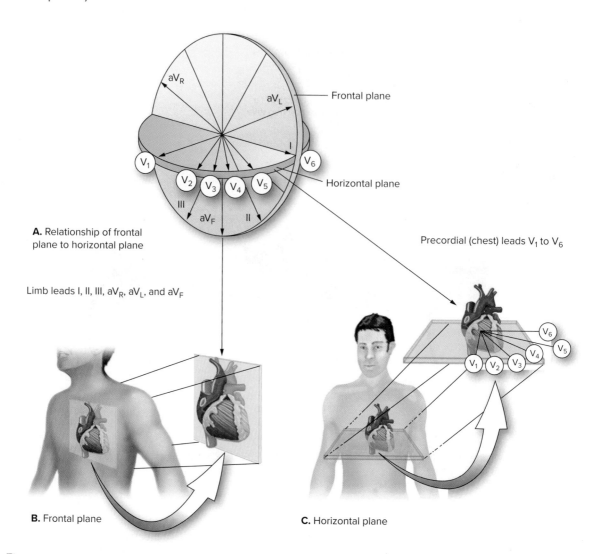

A. Relationship of frontal plane to horizontal plane

Limb leads I, II, III, aV$_R$, aV$_L$, and aV$_F$

Precordial (chest) leads V$_1$ to V$_6$

B. Frontal plane

C. Horizontal plane

Figure 2-9
ECG leads provide a cross sectional view of the heart (A) Graphic representation of the frontal and horizontal planes and the ECG leads that view along each plane, (B) Limb leads view the inferior, superior and lateral aspects of the heart, (C) Precordial leads view anterior, lateral, and posterior aspects of the heart.

2.5 Planes of the Heart and Lead Placement

Electrodes can be placed at specific spots on the patient's extremities and chest wall to view the heart's electrical activity from two distinct planes (Figure 2-9): the frontal and horizontal. These planes provide a cross-sectional view of the heart.

Frontal Plane

The frontal plane is a vertical (front and back, also referred to as an anterior and posterior) cut through the middle of the heart. The leads arranged on the frontal plane view the inferior (below), superior (above), and lateral (side) aspects of the heart. The leads that view the heart along the frontal plane are also referred to as *limb leads*.

The limb leads includes four electrodes; they are placed on the arms and legs or on the upper torso at least 10 centimeters (cm) from the heart (Figure 2-10). When the electrodes are positioned on the limbs, they can be located far down on the limbs or close to

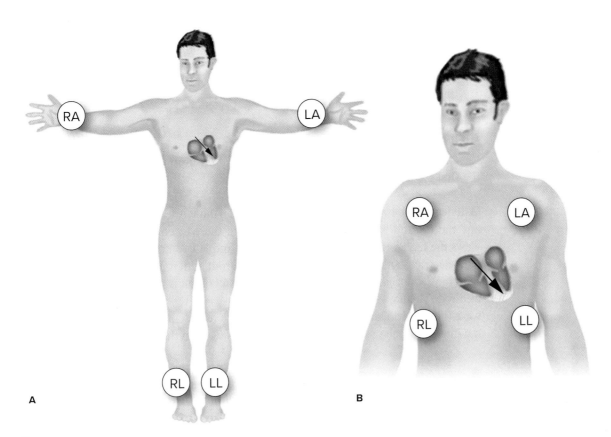

Figure 2-10
Position of limb leads. (A) Location of electrodes applied to the extremities and (B) location of electrodes applied to the chest.

the hips and shoulders, but they must be even (right vs. left). Placement of the electrodes on the upper torso is often preferred over placement on the arms and legs during continuous monitoring as it maximizes patient comfort and minimizes muscle artifact.

Proper positioning of the electrodes for the limb leads is as follows:

- The right arm electrode, labeled RA, is positioned anywhere on the right arm or below the right clavicle in the midclavicular line (middle of the clavicle). However, when placing the electrodes on the arms, it is best to position them on the medial aspect of the lower arms.
- The left arm electrode, labeled LA, is positioned anywhere on the left arm or below the left clavicle in the midclavicular line.
- The left leg electrode, labeled LL, is positioned anywhere on the left leg or left midclavicular line, below the last palpable rib.
- In four-lead wire systems, the right leg electrode, labeled RL, is positioned on the right leg or right midclavicular line, below the last palpable rib.

Horizontal Plane

The horizontal plane is a transverse (top and bottom, also referred to as a *superior* and *inferior*) cut through the middle of the heart. The leads arranged on this plane provide us with anterior, lateral, and posterior (back) views. These leads are called the *precordial leads* (also referred to as *V leads* or *chest leads*). There are six primary precordial leads. As mentioned earlier, depending on the capability of the ECG machine as well as the placement of the ECG electrodes, up to six more leads can be obtained. The precordial leads are placed horizontally across the chest.

Limb Leads

The limb leads include leads I, II, III, aV_R, aV_L, and aV_F.

Leads I, II, and III

Leads I, II, and III are referred to as the *standard* limb leads (based on the use of the first ECG). All three are bipolar leads.

With lead I, the LA lead is the positive electrode, the RA lead is the negative electrode, and the LL and RL leads are the ground and there to complete the circuit. Lead I views the high lateral surface of the left atria and left ventricle and detects the dominant electrical activity of the heart as it moves from right to left. The waveforms in lead I are mostly upright because the wave of depolarization is moving toward the positive electrode (Figure 2-11). This includes the P wave, the R wave, and the T wave. The Q wave (if seen) will take an opposite direction as it represents the brief depolarization of the interventricular septum, which is moving away from the positive lead.

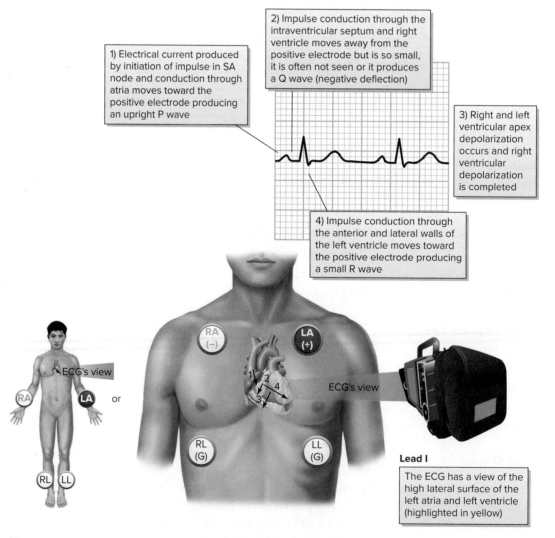

1) Electrical current produced by initiation of impulse in SA node and conduction through atria moves toward the positive electrode producing an upright P wave

2) Impulse conduction through the intraventricular septum and right ventricle moves away from the positive electrode but is so small, it is often not seen or it produces a Q wave (negative deflection)

3) Right and left ventricular apex depolarization occurs and right ventricular depolarization is completed

4) Impulse conduction through the anterior and lateral walls of the left ventricle moves toward the positive electrode producing a small R wave

ECG's view

ECG's view

RA (−)
LA (+)
RL (G)
LL (G)

RA
LA
or
RL LL

Lead I

The ECG has a view of the high lateral surface of the left atria and left ventricle (highlighted in yellow)

Figure 2-11
Electrode placement, ECG's view of the heart, and waveform direction in lead I.

Lead II is a commonly used lead for providing continuous ECG monitoring for emerging or existing cardiac dysrhythmias. With lead II, the LL lead is the positive electrode, the RA lead is the negative electrode, and the LA and RL leads are ground and there to complete the circuit. Lead II views the inferior surface of the right and left ventricles. It detects the dominant electrical activity of the heart as it moves in a right superior to a left inferior direction. The waveforms in lead II are mostly upright because the wave of depolarization is moving toward the positive electrode (Figure 2-12). This includes the P wave, R wave, and T wave. The Q wave (if seen) will take an opposite direction, as it represents the brief depolarization of the interventricular septum, which is moving away from the positive lead.

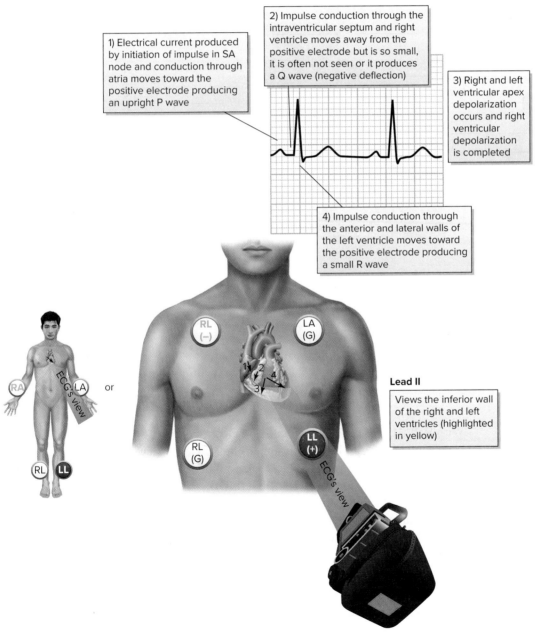

1) Electrical current produced by initiation of impulse in SA node and conduction through atria moves toward the positive electrode producing an upright P wave

2) Impulse conduction through the intraventricular septum and right ventricle moves away from the positive electrode but is so small, it is often not seen or it produces a Q wave (negative deflection)

3) Right and left ventricular apex depolarization occurs and right ventricular depolarization is completed

4) Impulse conduction through the anterior and lateral walls of the left ventricle moves toward the positive electrode producing a small R wave

RL (−)

LA (G)

RL (G)

LL (+)

Lead II

Views the inferior wall of the right and left ventricles (highlighted in yellow)

ECG's view

RA   LA

or

RL   LL

ECG's view

Figure 2-12
Electrode placement, view of the heart, and waveform direction in lead II.

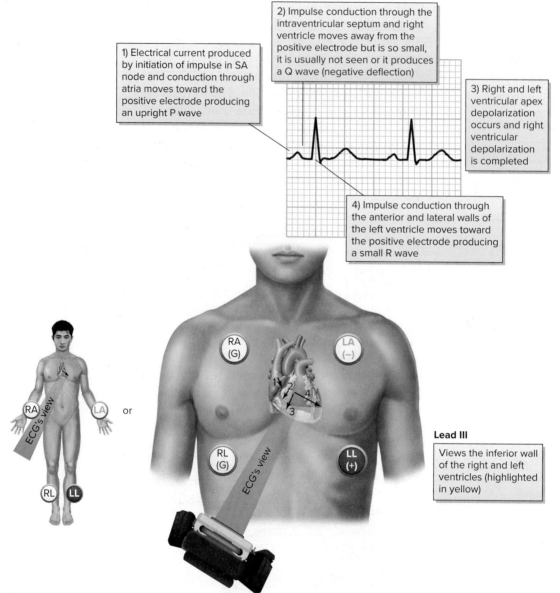

1) Electrical current produced by initiation of impulse in SA node and conduction through atria moves toward the positive electrode producing an upright P wave

2) Impulse conduction through the intraventricular septum and right ventricle moves away from the positive electrode but is so small, it is usually not seen or it produces a Q wave (negative deflection)

3) Right and left ventricular apex depolarization occurs and right ventricular depolarization is completed

4) Impulse conduction through the anterior and lateral walls of the left ventricle moves toward the positive electrode producing a small R wave

Lead III

Views the inferior wall of the right and left ventricles (highlighted in yellow)

Figure 2-13
Electrode placement, view of the heart, and waveform direction in lead III.

With lead III, the LL lead is the positive electrode, the LA lead is the negative electrode, and the RA and RL leads are ground. Lead III views the inferior wall of the right and left ventricles and detects the dominant electrical activity of the heart as it moves in a right superior to a left inferior direction. The P and T waves should be upright while the QRS complex may be mostly positive, although the R wave is not as tall as in lead II. Alternatively, the QRS complex may be biphasic as depolarization of the ventricles intersects the negative to positive layout of the ECG electrodes (Figure 2-13). The Q wave is typically not seen in this lead.

Leads aV_R, aV_L, and aV_F

The other three leads that view the frontal plane are the augmented limb leads, aV_R, aV_L, and aV_F ("a" means augmented, "V" means voltage, "R" means right, "L" means left, and "F" means foot). They are derived from the same three electrodes as leads I, II, and III. The ECG waveforms produced by these leads are small, and for this reason, the ECG machine enhances or augments them by 50%. This results in their amplitude being comparable to other leads. The augmented leads are unipolar. As mentioned earlier, each uses a positive electrode on the body surface and a central terminal as its negative pole (which is a combination of inputs from the other limb electrodes). The augmented limb leads fill in the space between the other limb leads and give us more views of the heart without adding more electrodes.

In lead aV$_R$, the RA lead is the positive electrode. This lead views the right upper side of the heart, and can provide specific information about the right ventricle outflow tract and basal part of the septum. Because the heart's electrical activity moves away from the positive electrode, the waveforms take a negative deflection. (Figure 2-14.) This includes the P wave, R wave, and T wave.

In lead aV$_L$, the LA lead serves as the positive electrode. This lead views the high lateral wall of the left ventricle. Because the heart's electrical activity moves toward the positive electrode (or perpendicular to it), the waveforms take a positive (or biphasic) deflection (Figure 2-15). This includes the P wave, R wave, and T

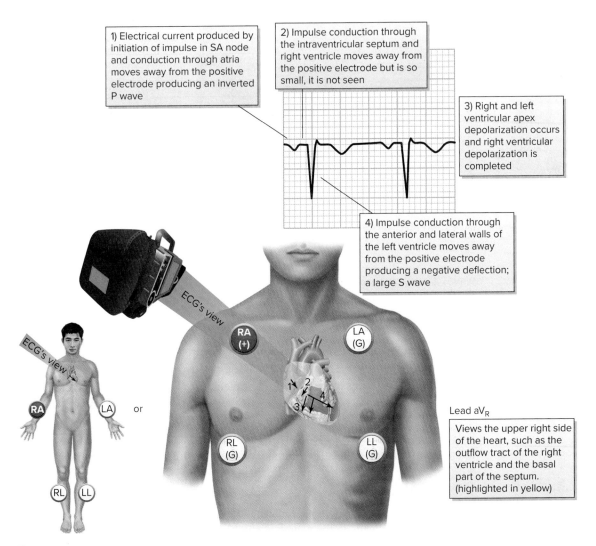

1) Electrical current produced by initiation of impulse in SA node and conduction through atria moves away from the positive electrode producing an inverted P wave

2) Impulse conduction through the intraventricular septum and right ventricle moves away from the positive electrode but is so small, it is not seen

3) Right and left ventricular apex depolarization occurs and right ventricular depolarization is completed

4) Impulse conduction through the anterior and lateral walls of the left ventricle moves away from the positive electrode producing a negative deflection; a large S wave

Lead aV$_R$

Views the upper right side of the heart, such as the outflow tract of the right ventricle and the basal part of the septum. (highlighted in yellow)

Figure 2-14
Electrode placement, view of the heart, and waveform direction in lead aV$_R$.

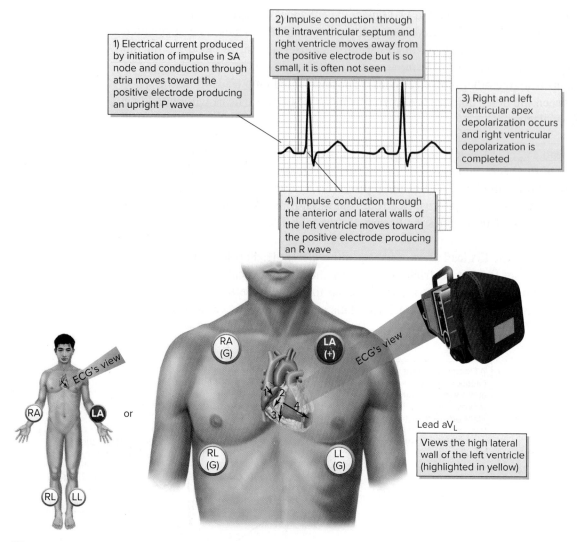

1) Electrical current produced by initiation of impulse in SA node and conduction through atria moves toward the positive electrode producing an upright P wave

2) Impulse conduction through the intraventricular septum and right ventricle moves away from the positive electrode but is so small, it is often not seen

3) Right and left ventricular apex depolarization occurs and right ventricular depolarization is completed

4) Impulse conduction through the anterior and lateral walls of the left ventricle moves toward the positive electrode producing an R wave

Lead aV$_L$
Views the high lateral wall of the left ventricle (highlighted in yellow)

Figure 2-15
Electrode placement, view of the heart, and waveform direction in lead aV$_L$.

wave. The Q wave (if seen) will take an opposite direction because it represents the brief depolarization of the interventricular septum, which is moving away from the positive lead.

Finally in lead aV$_F$, the LL lead is the positive electrode. This lead views the inferior wall of the left ventricle. The waveforms take a positive deflection because the heart's electrical activity moves toward the positive electrode (Figure 2-16). This includes the P wave, R wave, and T wave. The Q wave is typically not seen in this lead.

Lets review what areas of the heart the limb leads view. Leads I and aVL view the lateral left ventricle. Leads II, III, and aVF view the inferior wall of the heart.

Precordial Leads

The precordial leads are unipolar, requiring only a single positive electrode. The opposing pole of those leads is the center of the heart as calculated by the ECG. The six precordial leads are positioned in order across the chest and include V$_1$, V$_2$, V$_3$, V$_4$, V$_5$, and V$_6$. Because of their close proximity to the heart, they do not require augmentation.

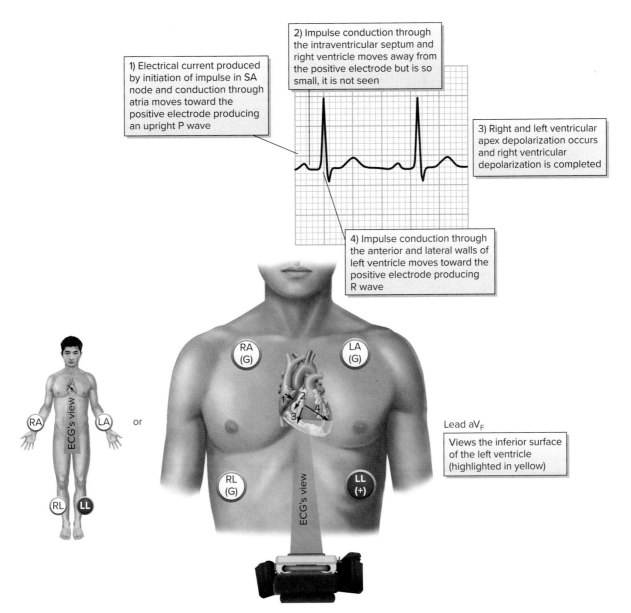

1) Electrical current produced by initiation of impulse in SA node and conduction through atria moves toward the positive electrode producing an upright P wave

2) Impulse conduction through the intraventricular septum and right ventricle moves away from the positive electrode but is so small, it is not seen

3) Right and left ventricular apex depolarization occurs and right ventricular depolarization is completed

4) Impulse conduction through the anterior and lateral walls of left ventricle moves toward the positive electrode producing R wave

Lead aV_F

Views the inferior surface of the left ventricle (highlighted in yellow)

RA (G)   LA (G)

RL (G)   LL (+)

ECG's view

or

RA   LA

RL   LL

Figure 2-16
Electrode placement, view of the heart, and waveform direction in lead aV_F.

With the SA node being above the positive electrode and with the electrical flow moving toward it, the direction of the P waves in the precordial leads is upright. The T waves should also be upright. As discussed earlier, the septum depolarizes from left to right and the ventricles from right to left. For this reason, progressing from leads V1 to V_6, the QRS complexes start out in a downward direction, then go through a transitional zone where they become half upright and half downward, and then become upright (Figure 2-17). If the transition occurs in V_1 or V_2, it is considered early transition, and if it occurs in V_5 or V_6, it is considered late transition. Both early and late transition can indicate the presence of various cardiac conditions.

As you will see from our discussion throughout the text, the leads we discuss cannot view all areas of the heart. Later, as we talk about how to locate ischemia and infarction in the right ventricle and posterior surface of the heart, we will describe ways to accommodate for this.

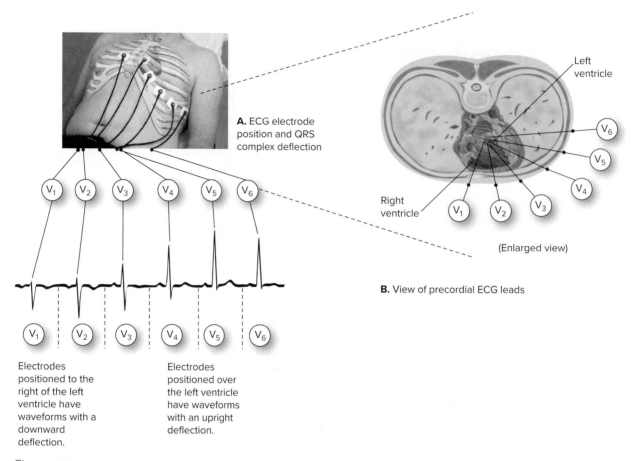

A. ECG electrode position and QRS complex deflection

Left ventricle

Right ventricle

(Enlarged view)

B. View of precordial ECG leads

Electrodes positioned to the right of the left ventricle have waveforms with a downward deflection.

Electrodes positioned over the left ventricle have waveforms with an upright deflection.

Figure 2-17

Precordial leads (A) placement locations and QRS complex deflection for each lead, (B) Sagital view of heart and what area each precordial lead views

Lead V₁

The first precordial lead is V_1. Its electrode is positioned in the fourth intercostal space (between the fourth and fifth ribs) just to the right of the sternum. It faces and is close to the right ventricle. It also has a view of the ventricular septum. Since septal and right ventricular depolarization send the current toward the positive electrode, there is a small initial upright deflection (the R wave). Then, as the current travels toward the left ventricle and away from the positive electrode, it produces a negative deflection, a deep S wave (Figure 2-18).

Lead V_1 is particularly effective at showing the P wave, QRS complex, and the ST segment. It is helpful in identifying ventricular dysrhythmias, ST segment changes, and bundle branch blocks.

Placing the V₁ Electrode   To find the position for the V_1 electrode, gently place your fingers centrally at the base of the patient's throat and move them down until you can feel the top of the sternum (or rib cage). From there, continue moving your fingers down until you feel a bony lump. This is the angle of Louis (also called the *sternal angle*). The angle of Louis is most easily found when the patient is in a supine position, as the surrounding tissue is tighter against the rib cage. From the angle of Louis, move your fingers to the right, and you will feel a gap between the ribs. This gap is the second intercostal space. From this position, move your fingers down across the next rib, and then the next. This space is the fourth intercostal space (Figure 2-19). Where the fourth intercostal space meets the sternum is the location you place the V_1 electrode.

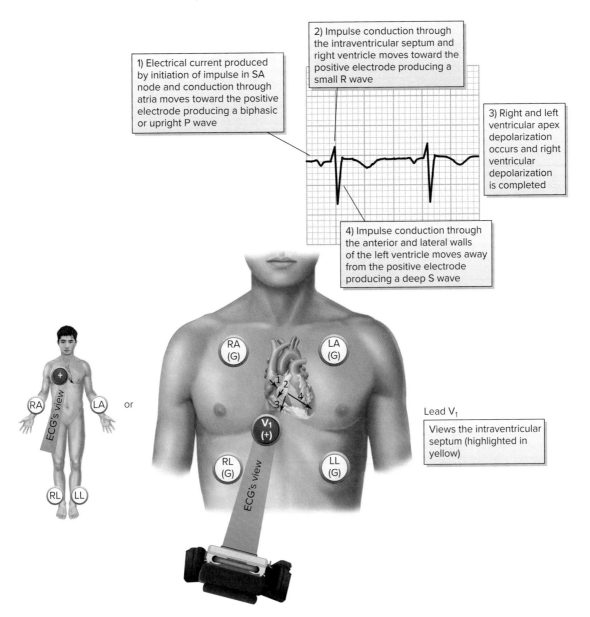

1) Electrical current produced by initiation of impulse in SA node and conduction through atria moves toward the positive electrode producing a biphasic or upright P wave

2) Impulse conduction through the intraventricular septum and right ventricle moves toward the positive electrode producing a small R wave

3) Right and left ventricular apex depolarization occurs and right ventricular depolarization is completed

4) Impulse conduction through the anterior and lateral walls of the left ventricle moves away from the positive electrode producing a deep S wave

Lead V₁
Views the intraventricular septum (highlighted in yellow)

Figure 2-18
Position of positive electrode and direction of electrical current flow, area of heart it views, and direction of waveforms in lead V₁.

Lead V₂

Lead V_2 is positioned in the fourth intercostal space (between the fourth and fifth ribs) just to the left of the sternum. Horizontally, it is at the same level as lead V_1 but on the opposite side the sternum. Just like lead V_1, V_2 faces and is close to the right ventricle. Although it has a view of the right ventricle and anterior wall of the heart, it is more recognized for its view of the ventricular septum (Figure 2-20), particularly when locating myocardial injury. Because it views both the septum and the heart's anterior wall, it is said to have an anteroseptal view. Similar to what we see with lead V_1, in lead V_2 the depolarization initially moves toward the positive electrode, producing a taller R wave (than what was seen in lead V_1). Then, as depolarization moves away from the positive electrode, it produces a less deep S wave.

Placing the V_2 Electrode   There are two quick ways to position the lead V_2 electrode. The easiest is to draw an imaginary line horizontally from the V_1 electrode to the fourth intercostal space on the opposite side of the sternum. Alternatively, you

Figure 2-19
Steps used to place the V₁
electrode.

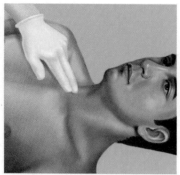

A Locate the base of the patient's throat.

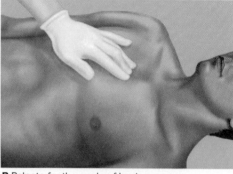

B Palpate for the angle of Louis.

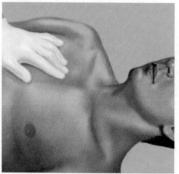

C Move your fingers to the patient's right.

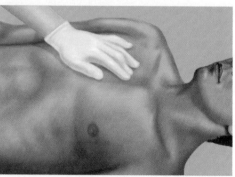

D Locate the second intercostal space (immediately below the second rib).

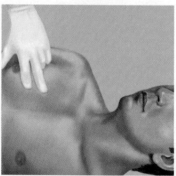

E Move your fingers down two intercostal spaces to the fourth intercostal space.

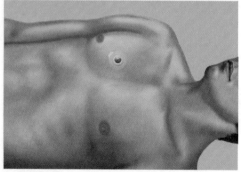

F Lead V₁ positioned in the fourth intercostal space just to the right of the sternum.

can go back to the angle of Louis, slide your finger into the second intercostal space on the left side, and then move down over the next two ribs. You are now at the fourth intercostal space. Where this space meets the sternum is where you place the V_2 electrode (Figure 2-21). Note that leads V_1 and V_2 are essentially in the same location on opposite sides of the chest.

Lead V_3

Lead V_3 is located midway between leads V_2 and V_4. For this reason, you need to locate the position for lead V_4 before you apply the electrode for lead V_3. Lead V_3 views the anterior wall of the left ventricle. Depolarization of the septum and right ventricle produces such a small electrical force that it is not seen whereas the current depolarization of the left ventricle moves perpendicular to the positive electrode, resulting in a biphasic waveform that has an R wave and S wave that are relatively the same amplitude in lead V_3 (Figure 2-22).

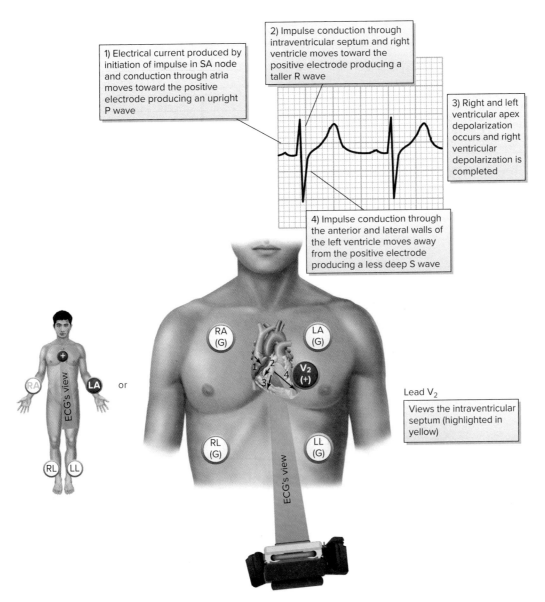

1) Electrical current produced by initiation of impulse in SA node and conduction through atria moves toward the positive electrode producing an upright P wave

2) Impulse conduction through intraventricular septum and right ventricle moves toward the positive electrode producing a taller R wave

3) Right and left ventricular apex depolarization occurs and right ventricular depolarization is completed

4) Impulse conduction through the anterior and lateral walls of the left ventricle moves away from the positive electrode producing a less deep S wave

Lead V₂

Views the intraventricular septum (highlighted in yellow)

Figure 2-20
Position of positive electrode and direction of electrical current flow, area of heart it views, and direction of waveforms in lead V₂.

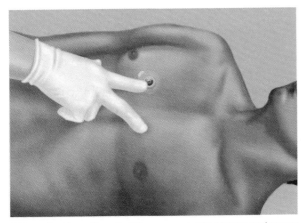

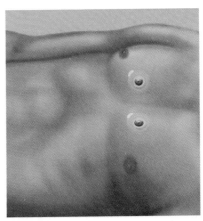

A From the V₁ position, find the corresponding intercostal space on the left side of the sternum.

B Place the V₂ electrode in the fourth space intercostal to the left of the sternum.

Figure 2-21
Steps for placing the V₂ electrode.

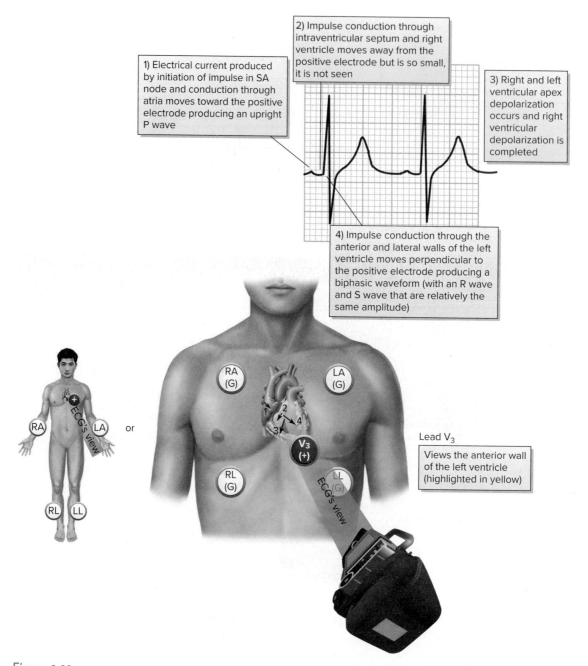

1) Electrical current produced by initiation of impulse in SA node and conduction through atria moves toward the positive electrode producing an upright P wave

2) Impulse conduction through intraventricular septum and right ventricle moves away from the positive electrode but is so small, it is not seen

3) Right and left ventricular apex depolarization occurs and right ventricular depolarization is completed

4) Impulse conduction through the anterior and lateral walls of the left ventricle moves perpendicular to the positive electrode producing a biphasic waveform (with an R wave and S wave that are relatively the same amplitude)

Lead V₃

Views the anterior wall of the left ventricle (highlighted in yellow)

Figure 2-22
Position of positive electrode and direction of electrical current flow, area of heart it views, and direction of waveforms in lead V_3.

Lead V_4

Lead V_4 is placed at the fifth intercostal space (between the fifth and sixth ribs) in the midclavicular line (the imaginary line that extends down from the midpoint of the clavicle). V_4 views the anterior wall of the left ventricle and is close to the heart's apex. Depolarization of the septum and right ventricle produces such a small electrical force that it is not seen while the current depolarization of the left ventricle moves perpendicular to the positive electrode, resulting in a biphasic waveform that has an R wave larger than the S wave in lead V_4 (Figure 2-23). It may also have a mostly positive deflection in this lead.

Placing the V_3 and V_4 Electrodes To locate this spot, position your fingers at the lead V_2 position. Then, slide your fingers down and below the next rib; you are

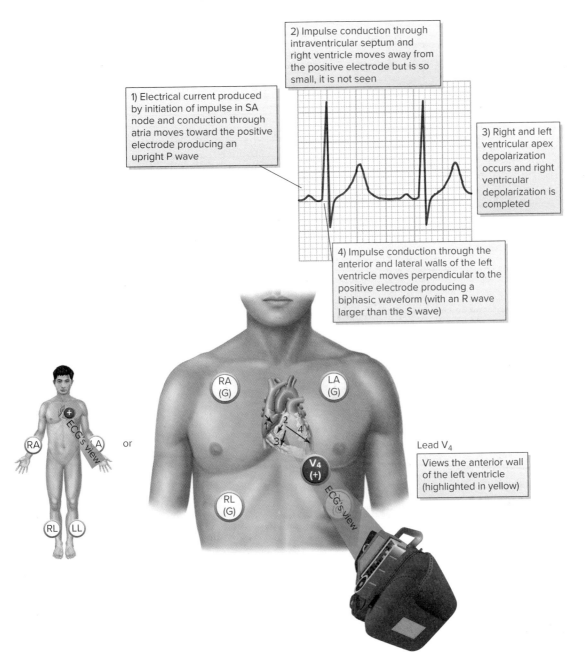

Figure 2-23
Position of positive electrode and direction of electrical current flow, area of heart it views, and direction of waveforms in lead V₄.

now in the fifth intercostal space. Next, look at the chest and identify the left clavicle. Find the middle of the clavicle and draw an imaginary line (called the *midclavicular line*) down to the fifth intercostal space. Position the lead V₄ electrode there. The lead V₃ electrode is then placed in the intercostal space midway between leads V₂ and V₄ (Figure 2-24).

Lead V₅

Lead V₅ is placed in the fifth intercostal space at the anterior axillary line. Horizontally, it is even with V₄ but in the anterior axillary line. Lead V₅ views the lower lateral wall of the left ventricle. As the impulse conducts through the septum and right ventricle, it moves away from the positive lead, producing a small Q wave in lead V₅. Then, as the impulse conducts through the left ventricle, it moves toward the positive electrode, producing a tall R wave (Figure 2-25). Because of its location closest

Midclavicular line | Midclavicle

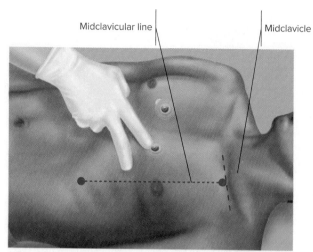

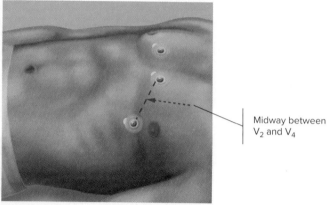

Midway between V_2 and V_4

A From the V_2 position, locate the fifth intercostal space and follow it to the midclavicular line.

B Position the V_4 electrode in the fifth intercostal space at the midclavicular line. Then find the midpoint between the V_2 and V_4 electrodes.

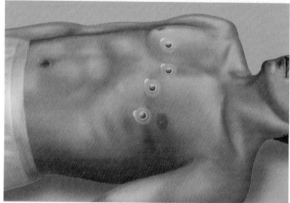

C Lead V_3 is then positioned at the midpoint between V_2 and V_4.

Figure 2-24
Steps for placing leads V_3 and V_4.

to the left ventricle, it will have the tallest R waves of the precordial leads. Then depolarization of the posterobasal right and left ventricular free walls and basal right septal mass produces a small S wave.

Placing the V_5 Electrode To locate the position for the V_5 electrode, while remaining horizontally level with V_4, follow the fifth intercostal space to the left until your fingers are just below the beginning of the axilla (armpit area). This is the anterior axillary line (the imaginary line that runs down from the point midway between the middle of the clavicle and the lateral end of the clavicle; the lateral end of the collarbone is the end closer to the arm). Place the electrode there (Figure 2-26).

Lead V_6

Lead V_6 is located horizontally level with V_4 and V_5 at the midaxillary line (middle of the armpit). Just like V_5, this lead views the lower lateral wall of the left ventricle. As the impulse conducts through the septum and right ventricle, it moves away from the positive lead, producing a small Q wave in lead V_5. Then, as the impulse conducts through the left ventricle, it moves toward the positive electrode, producing a tall R wave (Figure 2-27). Because it is not as close to the left ventricle, the R

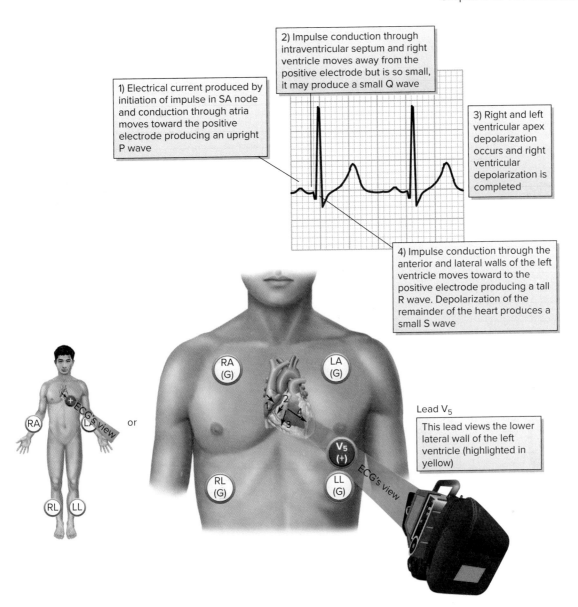

1) Electrical current produced by initiation of impulse in SA node and conduction through atria moves toward the positive electrode producing an upright P wave

2) Impulse conduction through intraventricular septum and right ventricle moves away from the positive electrode but is so small, it may produce a small Q wave

3) Right and left ventricular apex depolarization occurs and right ventricular depolarization is completed

4) Impulse conduction through the anterior and lateral walls of the left ventricle moves toward to the positive electrode producing a tall R wave. Depolarization of the remainder of the heart produces a small S wave

Lead V₅

This lead views the lower lateral wall of the left ventricle (highlighted in yellow)

Figure 2-25
Position of positive electrode and direction of electrical current flow, area of heart it views, and direction of waveforms in lead V₅.

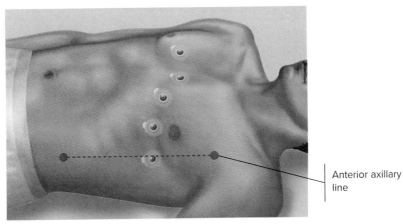

Figure 2-26
Steps for placing the V₅ electrode.

Anterior axillary line

To locate the position for the V₅ electrode, follow the fifth intercostal space across to where it intersects with the anterior axillary line.

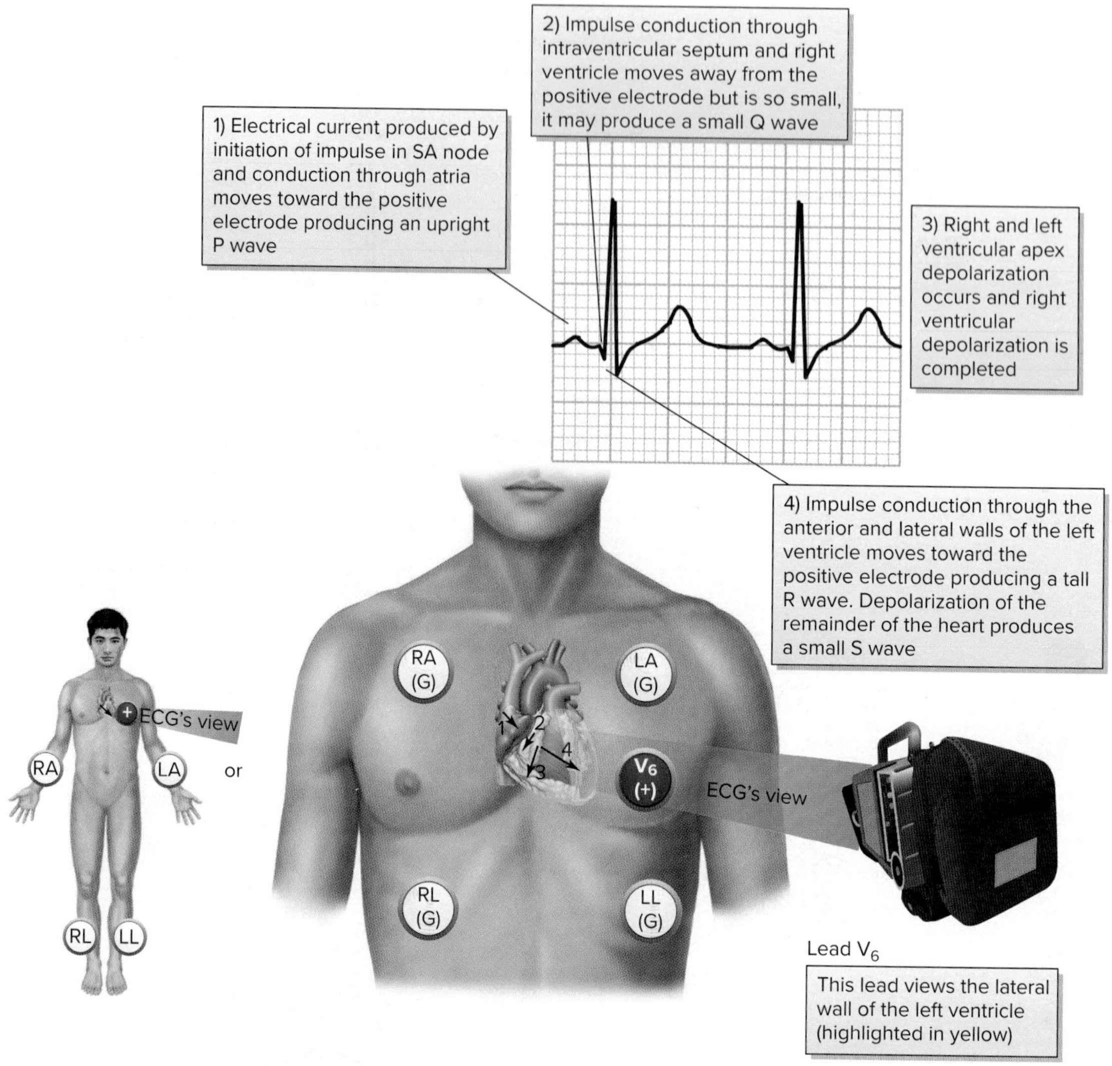

2) Impulse conduction through intraventricular septum and right ventricle moves away from the positive electrode but is so small, it may produce a small Q wave

1) Electrical current produced by initiation of impulse in SA node and conduction through atria moves toward the positive electrode producing an upright P wave

3) Right and left ventricular apex depolarization occurs and right ventricular depolarization is completed

4) Impulse conduction through the anterior and lateral walls of the left ventricle moves toward the positive electrode producing a tall R wave. Depolarization of the remainder of the heart produces a small S wave

Lead V_6

This lead views the lateral wall of the left ventricle (highlighted in yellow)

Figure 2-27
Position of positive electrode and direction of electrical current flow, area of heart it views, and direction of waveforms in lead V_6.

wave in lead V_6 is smaller than that found in Lead V_5. Lastly, depolarization of the posterobasal right and left ventricular free walls and basal right septal mass produce a small S wave.

Placing the V_6 Electrode Follow the line of the fifth intercostal space from the lead V_5 electrode until you are horizontally level with leads 4 and 5 (Figure 2-28) at the midaxillary line (center point of the axilla). This is where you place the lead V_6 electrode.

An important thing to remember is that if you have placed leads V_4 through V_6 correctly, they should line up horizontally.

Each view provides different information. When assessing the 12-lead ECG we look for characteristic normalcy and changes in all leads. Figure 2-29 shows the normal appearance of the 12-lead ECG. Memorizing what view(s) of the heart each lead provides will help you decide which lead to use to gain the information you need to make a proper assessment or diagnosis.

Figure 2-28
Steps for placing lead V_6.

Position leads V_4, V_5, and V_6 in a straight line horizontally

Midaxillary line

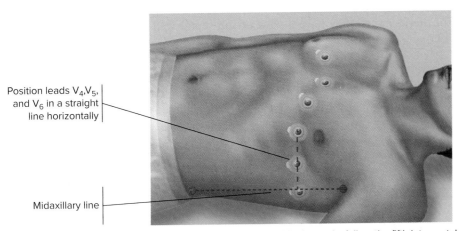

To locate the position for the V_6 electrode, follow the fifth intercostal space across to where it intersects with the mid axillary line.

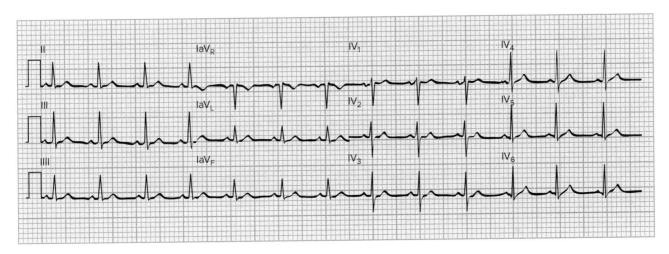

Figure 2-29
Normal appearance of ECG complexes in 12-lead ECG

Putting the Views Together

Now, let's look at the collective view of the heart. This is important as it will help us to identify the location of myocardial ischemia and/or infarction and other cardiac conditions (Figure 2-30).

Lead Groupings and Associated Views

The 12 ECG leads each record the electrical activity of the heart from a different perspective, which also correlates with different anatomical areas of the heart.

In leads II, III, and aV_F, the positive electrode is positioned below the heart, so these leads have a view of the inferior (or diaphragmatic) surface. For this reason, they are called the *inferior leads*.

In leads I, aV_L, V_5, and V_6, the positive electrode is positioned to the left of the heart, so they have a view of the lateral wall. For this reason, they are called the *lateral leads*. Similarly, the positive electrode for leads I and aV_L is located on the left arm or upper torso, these leads are sometimes referred to as the *high lateral leads*. Similarly, due to the positive electrodes for leads V_5 and V_6 being on the patient's chest, they are sometimes referred to as the *low lateral leads*.

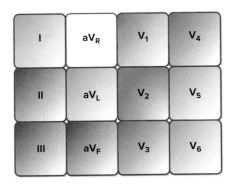

Figure 2-30
Matrix of the ECG leads showing which leads view what area of the heart.

In leads V_1 and V_2, the positive electrode is positioned in front but slightly to the left of the heart, so they have a view of the septal wall of the ventricles and are called the *septal leads*.

In leads V_3 and V_4, the positive electrode is positioned in front of the heart, so they have a view of the anterior wall and are called the *anterior leads*. They also have an apical view of the heart. Lead V_2 can also be considered an anterior lead as it, too, has a view of the anterior wall. The combination of leads V_1, V_2, V_3, and V_4 are referred to as *anteroseptal leads* while the combination of leads V_3, V_4, V_5, and V_6 are referred to as anterolateral. The combination of the inferior leads and the lateral leads is referred to a inferolateral. These combination of ECG groupings will be discussed further in Chapter 19.

Contiguous Leads

Two leads that look at neighboring anatomical areas of the heart are said to be contiguous. As an example, V_4 and V_5 are contiguous (as they are next to each other on the patient's chest), even though V_4 is an anterior lead and V_5 is a lateral lead. The following lists the contiguous leads:

Contiguous Inferior Leads

- Lead II is contiguous with lead III
- Lead III is contiguous with leads II and aV_F
- Lead aV_F is contiguous with lead III

Contiguous Septal, Anterior and Lateral Leads

- V_1 is contiguous with V_2
- V_2 is contiguous with V_1 and V_3
- V_3 is contiguous with V_2 and V_4
- V_4 is contiguous with V_3 and V_5
- V_5 is contiguous with V_4 and V_6
- V_6 is contiguous with V_5 and lead I
- Lead I is contiguous with V_6 and aV_L
- aV_L is contiguous with lead I

The relevance of this is it helps in determining whether an abnormality on the ECG is likely to represent true disease such as acute coronary ischemia or injury or just a false finding.

15- and 18-Lead ECGs

While the 12-lead ECG is useful in the detection of myocardial infarction (MI) of the inferior, anterior, and lateral walls of the left ventricle, it is not as effective in revealing an MI involving the right ventricle and/or the posterior wall of the left ventricle.

Nonstandard ECG leads can be used to increase the detection of right ventricle and posterior wall infarction. These nonstandard leads are commonly identified as *right precordial leads* (V_4R, V_5R, V_6R) and *posterior leads* (V_7, V_8, V_9). By combining these leads, a 15- or 18-lead ECG can be created. Although specialized equipment is available to obtain 15- and 18-lead ECGs, it is not absolutely required. The data can be obtained from these additional leads with the standard 12-lead ECG machine by moving electrodes to specified locations and running the 12-lead a second time.

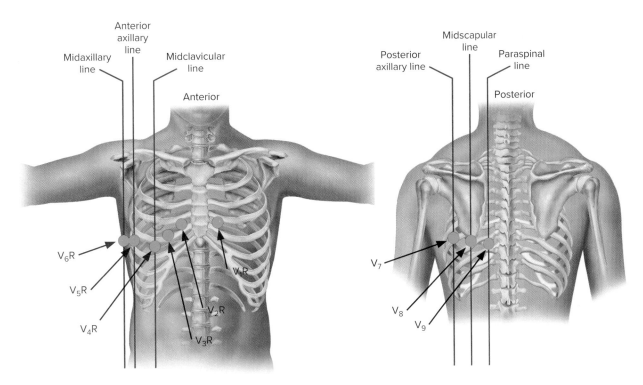

Figure 2-31
15- and 18-lead locations.

To attain the additional leads for a 15-lead ECG, move the lead wires from the precordial electrode positions to the following locations (Figure 2-31):

Right Side Leads

- V_1R is placed in the fourth intercostal space to the left of the patient's sternum (use V_1 lead)
- V_2R is placed in the fourth intercostal space to the right of the patient's sternum (use V_2 lead)
- V_3R is placed in the fifth intercostal space between V_2R and V_4R
- V_4R is placed in the fifth intercostal space even with right midclavicular line (use the V_4 lead)
- V_5R is placed horizontally (straight across) with V_4R even with the right anterior axillary line (use V_5 lead)
- V_6R is placed horizontally with V_5R even with the right midaxillary line (use V_6 lead)

Posterior Leads

- V_7 is placed horizontally with V_6 in the left posterior axillary line (use V_4 lead)
- V_8 is placed horizontally with V_7 in the left midscapular line (use V_5 lead)
- V_9 is placed horizontally with V_8 in the left paraspinal line (use V_6 lead)

Two alternatives to moving all the leads to create the V_R leads is to move just the V_4 lead to the V_4R position or to move just the V_4, V_5, and V_6 to the V_4R, V_5R, and the V_6R positions. Be sure to follow your local protocol.

Then press the 12-lead ECG button to obtain the recording. If the ECG machine does not automatically identify the repositioning of the leads, label the second

12-lead recording with the label of the new leads (V$_4$R, V$_5$R, etc.), so anyone who views the ECG tracing will know you have moved the leads to the other position(s). Also, be sure to have the patient lie as still as possible to avoid artifact in the posterior leads.

Additional leads are recommended for patients in whom changes are seen in the inferior leads or where there are reciprocal changes in leads V$_1$ and V$_2$ (ST segment depression indicating a posterior infarction).

Increasing the detection of right ventricular and posterior wall infarction will lead to more aggressive treatment with reperfusion therapies, as well as avoiding inappropriate treatments that may cause undue harm to the patient. These ECGs can be obtained with a minimal increase in time and cost of care and can result in increasingly positive outcomes.

2.6 Displays and Printouts

The ECG machine translates the electrical impulses generated in the heart into wave-like signals that are recorded on paper or displayed on a monitor as a series of waveforms, intervals, and segments.

ECG rhythms shown on the oscilloscope are called **dynamic ECGs** and represent real-time electrical activity whereas those printed on graph paper are called **ECG tracings** or **static ECGs** (also referred to as *rhythm strips* or *printouts*) and show what has already occurred. Each type has different purposes. It is much easier to analyze a static ECG to determine abnormalities than it is to examine an image moving across a screen. On the other hand, the static ECG tracing is "past news," so it is also necessary to look at the ECG monitor to see what is presently occurring (Figure 2-32). ECG machines are typically capable of producing a tracing, whereas not all are equipped with an oscilloscope.

Figure 2-32
Monitor display vs. printouts.

Another thing to consider is that today's ECG monitors are configured with multiple filters for signal processing. The most common settings are monitor mode and diagnostic mode. The monitor mode ECG display is more filtered than the diagnostic mode. In monitor mode, filtering limits artifact for routine cardiac rhythm monitoring and helps reduce wandering baseline. In diagnostic mode, the filters allow accurate ST segments to be recorded. When we are trying to analyze ST segment changes, a printout must be used as the ECG monitor will not be accurate.

Reading Printouts

Although looking at the tracing on the monitor screen can reveal important information, to more accurately analyze the tracing requires using a printout. Further, to measure the various waveforms, durations, and intervals, you need to see where each begins and ends as well as the amplitude of each waveform.

Special Paper

A special type of paper is used to record the heart's electrical activity. The paper comes in a variety of sizes depending on its use. ECG machines used to provide continuous monitoring may employ narrow paper that prints out a rhythm strip of a single lead whereas ECG machines that produce a 12-lead ECG use wider paper in order to show all the leads.

Regardless of its size, the paper, made of thermally sensitive material, consists of horizontal and vertical lines that form a grid. The narrow lines that run vertically and horizontally intersect to form squares. There are five of these squares between heavier (thicker) lines (that run vertically and horizontally) (Figure 2-33).

Horizontal Lines

The distance between the lines, or boxes, running horizontally represents time or duration. Each small square running horizontally represents 0.04 seconds in duration. Five of these small squares (making up a larger box) represent 0.20 seconds in duration. Five of these larger boxes represent one second. Fifteen larger boxes equal a three-second interval. Horizontal measurements are used to determine the heart rate. You can also use these standardized distances to determine the width or duration of any portion of an ECG complex, segment, or interval.

On the top or bottom (and sometimes both) of the printout, there are often vertical slashes or markings to represent three-second intervals (depending on the manufacturer, there may be shorter or longer time frames). These markings can help you

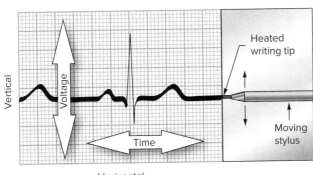

Figure 2-33
ECG paper with stylus-generating waveforms. Vertical lines represent amplitude in electrical voltage in millivolts or millimeters while horizontal lines represent time or duration.

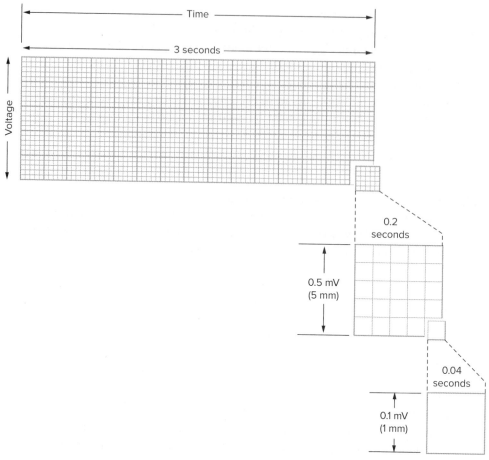

Figure 2-34

From right to left: horizontally each small square represents 0.04 seconds in duration. Five of these small squares (making up a larger box) represent 0.20 seconds in duration. Fifteen of these larger squares represent three seconds. Vertically each small square represents 0.1 millivolt (1 mm) while five of these small squares represent 0.5 mV (5 mm).

quickly determine the approximate heart rate. The ECG recorder is set to a standard speed of 25 mm per second so that tracings made from different machines may be uniformly compared (Figure 2-34). Some ECG machines are equipped with a feature that allows you to speed up the printout to pass 50 mm of paper per second. This effect stretches out the tracings, which can make it easier to diagnose fast rhythms when it is difficult to see the P wave.

If the ECG paper does not have vertical lines to allow you to quickly determine the heart rate, count out 15 boxes and place a vertical mark. Then repeat the step to give you two 3-second sections or a full 6-second section of the tracing.

Vertical Lines

The distance between the lines, or boxes, running vertically represents amplitude in millimeters (mm) or electrical voltage in millivolts (mV). Each small square running vertically represents 1 mm or 0.1 mV. The larger boxes are made up of five small squares and represent 5 mm or 0.5 mV. To determine the amplitude of a wave, segment, or interval, count the number of small boxes from the baseline to its highest or lowest point.

Recording the Tracing

A heated stylus that responds to the electrical stimulus detected through the ECG electrodes moves up and down and produces markings on the paper. When no electrical stimulus is flowing through the electrode (usually representing the polarized state of the cells), or it is too small to detect, the stylus burns a straight line (isoelectric) in the paper. This flat line is also called the **isoelectric line.** We use this line as a baseline or reference point to identify the changing electrical amplitude.

Depending on what you are using the ECG for, you will typically print off either a rhythm strip or a 12-lead tracing. Some ECG machines are capable of analyzing certain aspects of the tracing and including that information on the printout. Even if the ECG machine provides an analysis, you should always take time to analyze the tracing to verify the accuracy of the information.

Key Points

LO 2.1	• The graphic record or tracing is called an *electrocardiogram*. • The ECG machine (electrocardiograph) has many uses in both hospital and out-of-hospital settings.
LO 2.2	• Electrodes placed on the patient's skin detect the heart's electrical activity. The electrical activity is then transferred to the ECG machine by color-coded wires (called *lead wires*) and displayed on a screen (oscilloscope) or printed onto graph paper.
LO 2.3	• Ventricular depolarization can be divided into three main phases. First, ventricular activation begins in the septum as it is depolarized from left to right, along with early depolarization of the right ventricle. In the second phase, the right and left ventricular apex are depolarized, and depolarization of the right ventricle is completed. In the third phase, the remainder of the left ventricle is depolarized toward the lateral wall.
LO 2.4	• Positioning ECG electrodes in specific positions on the patient's body gives us different views of the heart. These views can be thought of as ECG leads. • Bipolar leads record the flow of the electrical impulse between the two selected electrodes including leads I, II, and III. • Unipolar leads use only one positive electrode and a central terminal calculated by the ECG machine and include leads aV_R, aV_L, and aV_F and the precordial leads V_1, V_2, V_3, V_4, V_5, and also include leads aV_R, aV_L, and aV_F and the precordial leads V_1, V_2, V_3, V_4, V_5, and V_6. • The shape of the waveform is described from the perspective of the positive electrode of the selected lead. • Impulses that travel toward a positive electrode are recorded on the ECG as upward deflections. Impulses traveling away from a positive electrode or toward a negative electrode are recorded as downward deflections. • Because the electrodes positioned on the patient's skin detect the heart's electrical activity, placing them in a different location changes the lead or view.

LO 2.5	• The frontal plane is a vertical cut through the middle of the heart. The leads that view this plane of the heart are referred to as the standard limb leads.
	• The limb leads are produced by placing electrodes on the right arm (RA), left arm (LA), left leg (LL), and right leg (RL). The limb leads include leads I, II, and III; augmented voltage right (aV_R); augmented voltage left (aV_L); and augmented voltage foot (aV_F).
	• With lead I, the LA lead is the positive electrode and the RA lead is the negative electrode. In lead II, the LL lead is the positive electrode and the RA lead is the negative electrode. With lead III, the LL lead is the positive electrode and the LA lead is the negative electrode. In leads I, II, and III, the waveforms are positive although the QRS complex can be biphasic in lead III.
	• With lead aV_R, the R_A lead is the positive electrode. In lead aV_L, the LA lead is the positive electrode. With lead aV_F, the LL lead is the positive electrode. In lead aV_R, the waveforms deflect negatively. In leads aV_L and aV_F, the waveforms may be positive or biphasic.
	• The horizontal plane of the heart is a transverse cut through the middle of the heart. The leads that view this plane of the heart are referred to as precordial leads.
	• The precordial leads include leads V_1, V_2, V_3, V_4, V_5, and V_6.
	• With lead V_1, the electrode is placed in the fourth intercostal space just to the right of the sternum. With lead V_2, the electrode is placed in the fourth intercostal space just to the left of the sternum. Lead V_4 is placed at the fifth intercostal space in the midclavicular line. Lead III is then placed midway between leads V_2 and V_4. Lead V_5 is placed in the fifth intercostal space at the anterior axillary line. Horizontally, it is even with V_4 but in the anterior axillary line. Lead V_6 is located horizontally level with V_4 and V_5 at the midaxillary line.
	• With leads II, III, and aV_F, the positive electrode is positioned below the heart so these leads have a view of the inferior surface and are referred to as *inferior leads*. With leads I, aV_L, V_5, and V_6, the positive electrode is positioned to the left of the heart, so they have a view of the lateral wall and are called the *lateral leads*. In leads V_1 and V_2, the positive electrode is positioned in front but slightly to the left of the heart, so they have a view of the septal wall of the ventricles and are called the *septal leads*. In leads V_3 and V_4, the positive electrode is positioned in front of the heart, so they have a view of the anterior wall and are called the *anterior leads*. They also have an apical view of the heart. Lead V_2 can also be considered an anterior lead as it, too, has a view of the anterior wall. The combination of leads V_1, V_2, V_3, and V_4 are referred to as *anteroseptal leads*.
LO 2.6	• The horizontal axis on the ECG tells us about rate and duration, while the vertical axis tells us about electrical voltage or strength of the impulses.
	• ECG rhythms shown on the oscilloscope are called *dynamic ECGs* whereas those printed on graph paper are called *ECG tracings* or *static ECGs*.
	• The ECG paper used to record the heart's electrical activity consists of horizontal and vertical lines that form a grid. The smallest of the squares represents 0.04 seconds in duration.

- Five of the small boxes represent 0.20 seconds in duration and make up the larger boxes denoted by a heavier line. Fifteen larger boxes equal a three-second interval. You can use the horizontal measurements to determine the heart rate.

- Vertically on the ECG paper, the distance between the lines, or boxes, represents amplitude in mm or electrical voltage in mV. Each small square represents 1 mm or 0.1 mV while the larger boxes, made up of five small squares, represent 5 mm or 0.5 mV.

- The flat line that precedes the electrical impulses is called the *isoelectric line*.

Assess Your Understanding

The following questions give you a chance to assess your understanding of the material discussed in this chapter. The answers can be found in Appendix A.

1. An electrocardiogram is a/an   (LO 2.1)
   a. graphic record of the heart's electrical activity.
   b. device that measures and records the ECG.
   c. irregular rhythm.
   d. device that measures the heart's mechanical activity.

2. The electrocardiograph can be used to do all the following EXCEPT   (LO 2.1)
   a. detect cardiac dysrhythmias.
   b. determine cardiac output.
   c. identify the presence of myocardial ischemia and/or infarction.
   d. evaluate the function of artificial implanted pacemakers.

3. _____ positioned on the patient's skin detect the heart's electrical activity.   (LO 2.2)
   a. Electrodes
   b. Lead wires
   c. Tracings
   d. Leads

4. Describe the three phases of ventricular depolarization.   (LO 2.3)

5. Impulses that travel toward a positive electrode and away from a negative electrode are recorded on the electrocardiogram as _____ deflections.   (LO 2.4)
   a. downward
   b. perpendicular
   c. neutral
   d. upward

6. Bipolar leads   (LO 2.4)
    a. record the flow of the electrical impulse between the three selected electrodes.
    b. are used only in the hospital setting.
    c. require two electrodes of opposite polarity.
    d. use two positive electrodes and a reference point calculated by the ECG machine.

7. Define the term "ECG lead."   (LO 2.5)

8. The frontal plane gives us a/an _____ view of the heart's electrical activity.   (LO 2.5)
    a. anterior and posterior
    b. superior and inferior
    c. anterior and superior
    d. lateral and inferior

9. Which leads are referred to as the standard limb leads?   (LO 2.5)

10. The limb leads are obtained by placing electrodes on the   (LO 2.5)
    a. right arm, left arm, left leg, and right leg.
    b. chest and back.
    c. right arm, left leg, and right leg.
    d. left arm and left leg.

11. The lead most commonly used for continuous cardiac monitoring is lead   (LO 2.5)
    a. I.
    b. MCL_1.
    c. V_1.
    d. II.

12. With lead II, the _____ lead wire is the positive electrode. (LO 2.5)
    a. LL
    b. RA
    c. LA
    d. RL

13. List the leads that collectively view the inferior portion of the left ventricle.   (LO 2.5)

14. Describe where lead V_4 is placed.   (LO 2.4)

15. Describe the view of lead V_1. (LO 2.4)

16. Identify where lead VR_4 is positioned. (LO 2.4)

17. An ECG printout (LO 2.6)
    a. is called a "dynamic ECG."
    b. shows the heart's current electrical activity.
    c. provides a more accurate picture of the ST segments than does the oscilloscope.
    d. is harder to analyze than looking at the ECG monitor.

18. Each small square on the ECG paper running horizontally represents _____ seconds in duration. (LO 2.6)
    a. 0.01
    b. 0.04
    c. 0.12
    d. 0.20

19. Thirty larger boxes on the ECG paper represent how many seconds? (LO 2.6)

20. The larger box on the ECG paper, made up of five small squares, represents _____ seconds in duration. (LO 2.6)
    a. 0.12
    b. 0.15
    c. 0.20
    d. 0.32

21. On the top or bottom of the ECG paper, there may be vertical slashes or markings to represent _____ second intervals. (LO 2.6)
    a. 1
    b. 3
    c. 5
    d. 10

22. Each small square on the ECG paper running vertically represents (LO 2.6)
    a. 0.1 mV.
    b. 5 mm.
    c. 10 mm.
    d. 2.0 mV.

23. The reference point used to identify the changing electrical amplitude on the ECG is called the (LO 2.6)
    a. isoelectric line.
    b. ST segment.
    c. PR interval.
    d. QT interval.

Referring to the scenario at the beginning of this chapter, answer the following questions.

24. Examination of the ECG by EMS and healthcare personnel provides the patient's (LO 2.1)
    a. pulse rate.
    b. heart rate.
    c. blood pressure.
    d. cardiac output.

25. The patient fainted because (LO 1.8)
    a. he overexerted himself.
    b. his sympathetic nervous system was overstimulated.
    c. his blood pressure medications are too strong.
    d. his cardiac output dropped too low.

26. Which of the following nerves controls heart rate? (LO 1.10)
    a. Phrenic
    b. Aortic
    c. Vagus
    d. Diaphragmatic

©rivetti/Getty Images

section 2
The Nine-Step Process

3 Analyzing the Electrocardiogram

Chapter Outline

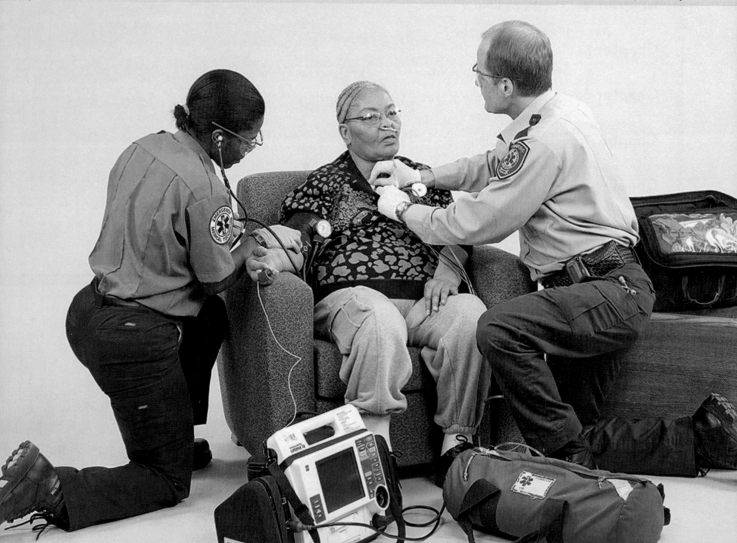

Learning Outcomes

LO 3.1 Identify the characteristics of a "normal" ECG.

LO 3.2 List and describe the elements of the Nine-Step Process.

LO 3.3 Discuss how identifying characteristics of an ECG can help identify dysrhythmias and cardiac conditions.

LO 3.4 Recall how to assess an ECG rhythm on the ECG monitor and ECG tracing.

LO 3.5 Describe what is meant by the expression "calibrating the ECG."

LO 3.6 Define the term *artifact* and give three common examples.

Case History

Your patient is a 67-year-old female whose chief complaint is chest pain. Your assessment of her pulse identifies gross irregularity, prompting you to quickly attach her to your ECG machine. Across the monitor screen, you observe what appears to be a very irregular rhythm that lacks upright P waves but has normal looking QRS complexes. You print out the tracing and show it to your partner who immediately exclaims, "I don't remember seeing anything like that in my ECG book; I don't know what to call that!" You recall what was drilled into your head during your paramedic course: "Use a step-by-step process to analyze your ECG tracings, and don't try to memorize rhythm strips." You then look more closely at the tracing. The rate is approximately 90 beats per minute, the regularity is totally irregular, there is a chaotic-looking baseline in front of each normal-looking QRS complex, and there are no PR intervals. From what you know about ECG dysrhythmias, this has all the characteristics of just one dysrhythmia. You tell your partner, "I bet you lunch this is atrial fibrillation." Later that day, following your free lunch, you call your paramedic instructor to thank her for teaching you how to analyze ECGs using a step-by-step approach.

3.1 Characteristics of the Normal ECG

As discussed earlier, depolarization and repolarization of the atria and ventricles are electrical events. The path of the electrical activity through the heart is displayed in a series of waves and complexes, commonly known as the P wave, QRS complex, and T wave. The ECG detects this electrical activity. However, in order to use the information we must first analyze and interpret it. To analyze an ECG, we can directly observe the rhythm on an ECG monitor screen or print a tracing so that the specific waveforms, segments, and intervals can be measured and examined. We then compare our findings against what is considered the normal ECG, commonly referred to as *sinus rhythm* (Figure 3-1). Sinus rhythm consists of the following:

- Waveforms and intervals that appear at regular intervals at a rate of 60 to 100 beats per minute (in the adult).
- Upright and slightly **asymmetrical** P waves, each followed by a QRS complex of normal upright contour, duration, and configuration.
- A PR interval (PRI) of normal duration that precedes each QRS complex.
- A flat ST segment followed by an upright and slightly asymmetrical T wave.
- A normal QT interval (QTI).
- Sometimes the presence of a U wave.

3.2 Analyzing the Electrocardiogram Using the Nine-Step Process

In this chapter and the next several chapters, we focus on the steps used to analyze the ECG. When looking at an ECG rhythm or tracing, it may appear like just a bunch of scribbles and lines, particularly for those who are new to ECG analysis and interpretation. The key to making sense of what you see is to approach it in a logical and systematic manner. In this book we teach you to use a nine-step process.

The Nine-Step Process assesses the main elements of the ECG tracing. Once learned, it is an effective tool for analyzing ECG tracings regardless of whether for

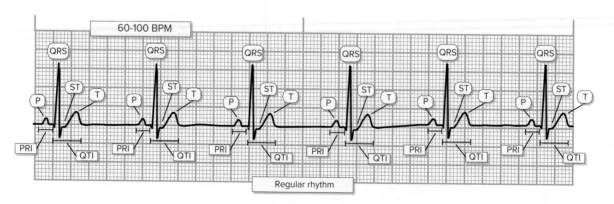

Figure 3-1

Normal sinus rhythm is a repetitive cycle of ECG waveforms, intervals, and segments occurring at a rate of between 60 to 100 beats per minute (BPM).

dysrhythmia recognition or identification of cardiac conditions. It includes the following elements (Figure 3-2):

1. **Heart rate:** Determine the rate. (Is it normal, fast, or slow?)

2. **Regularity:** Determine the regularity. (Is it regular or irregular?)

3. **P waves:** Assess the P waves. (Is there a uniform P wave preceding each QRS complex?)

4. **QRS complexes:** Assess the QRS complexes. (Are the QRS complexes within normal limits? Do they appear normal?)

5. **PR intervals:** Assess the PR intervals. (Are the PR intervals identifiable? Are they within normal limits? Are they constant in duration?)

6. **ST segments:** Assess the ST segment. (Is it a flat line? Is it elevated or depressed?)

7. **T waves:** Assess the T waves. (Are they slightly asymmetrical? Are they of normal height? Is it oriented in the same direction as the preceding QRS complex?)

8. **QT intervals:** Assess the QT intervals. (Are they within normal limits?)

9. **U waves:** Look for U waves. (Are they present?)

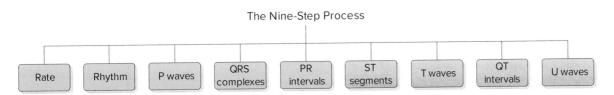

Figure 3-2
The Nine-Step Process.

Step 1: Heart Rate

The heart rate is one of the first characteristics of an ECG tracing that should be assessed (Figure 3-3). The average heart rate in the adult is 60 to 100 beats per minute. A heart rate that is slower or faster is considered abnormal and may indicate significant problems requiring prompt intervention on your part. For this reason, we need to check it as soon as possible. In Chapter 4, we describe a number of methods that can be used to determine the heart rate.

Step 2: Regularity

The regularity of the rhythm is another one of the characteristics of an ECG tracing that should be assessed initially (refer to Figure 3-3). Normally the heart beats in a regular, rhythmic fashion, producing specific waveforms and intervals with each heartbeat. This cycle repeats itself over and over again. The distance between the consecutive P waves should be the same, just as the distance between the consecutive QRS complexes should be the same throughout the tracing. When the distances between the consecutive P waves and/or consecutive QRS complexes differ, the rhythm is irregular. In Chapter 5, we describe several ways to easily identify whether or not the rhythm is regular and describe several types of irregularity that may be seen.

Figure 3-3
Determining the heart rate and regularity are important steps in assessing the ECG. Note: This tracing includes two leads: Lead II is in the top position and lead III is in the second position.

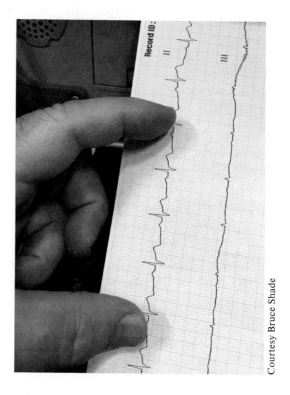

Courtesy Bruce Shade

Characteristics of normal P waves:
Asymmetrical waveform. There is one P wave preceding each QRS complex Its amplitude is 0.5 to 2.5 mm and its duration is 0.06 to 0.10 seconds.

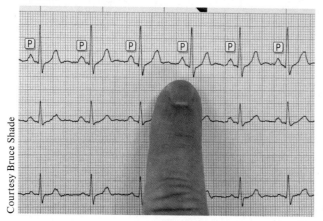

Courtesy Bruce Shade

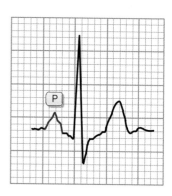

P waves findings in this tracing:
Asymmetrical waveform. There is one P wave preceding each QRS complex. Its amplitude is 2.5 mm and its duration is 0.11 seconds. Further, there appears to be notching of the P wave.

Figure 3-4
Assessing the P waves, the third step, helps to identify if the impulse arose from the SA node and if it traveled through the atria and AV junction in a normal manner.

Step 3: P Waves

Assessing the P waves (Figure 3-4) can tell us if the impulse that initiated the heart beat arose from the sinoatrial (SA) node and if it is traveling through the atria and atrioventricular (AV) junction in a normal fashion. When assessing the waveforms seen on an ECG tracing, we look at the **morphology** of the complexes. The term *morphology* refers to the appearance of the complexes. The morphology may be normal, wide, short, tall, flat, spiked, jagged, or bizarre (meaning its shape is far from normal). It often yields a great deal of information and can help us identify both dysrhythmias as well as cardiac conditions.

Characteristically, the P wave arises from the isoelectric line, appearing as an upright and slightly asymmetrical waveform. There is one P wave preceding each QRS complex. Its amplitude is 0.5 to 2.5 mm and its duration is 0.06 to 0.10 seconds. If the P wave meets all of these characteristics, it has most likely originated from the SA node. P waves that originate from the SA node are referred to as *sinus P waves*.

Step 4: QRS Complexes

Assessing the QRS complexes yields valuable information for both dysrhythmia interpretation and identification of various cardiac conditions (Figure 3-5). In sinus rhythm, the QRS complex follows the P wave and PR segment. It is larger in appearance than the P wave and consists of up to three parts: the Q wave, the R wave, and the S wave.

The waves of the QRS complex are generally narrow and sharply pointed. The Q wave is the first downward deflection from the baseline. Its duration is less than 0.04 seconds (1 mm or one small square) and its amplitude is less than 25% of the R wave in that lead. The Q wave is not always present. The R wave is the first upward deflection after the P wave, and the S wave is the first negative deflection after the R wave that extends below the baseline. The QRS complex is normally 0.06 to 0.10 seconds (one and a half to two and a half small boxes) in duration. The duration is measured from the beginning of the Q wave to the end of the S wave or the beginning of the R wave if the Q is not present. The amplitude of the QRS complex is 5 to 30 mm high (but differs with each lead selected). The QRS complex represents movement of the electrical impulse through the ventricles, causing their contraction.

Characteristics of normal QRS complexes:
There should be one QRS complex following each P wave. The QRS complex is narrow and sharply pointed. It is normally 0.06 to 0.10 seconds in duration and has an amplitude of 5 to 30 mm high (but differs with each lead selected).

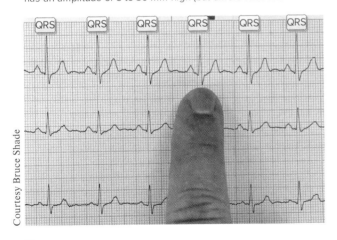

Courtesy Bruce Shade

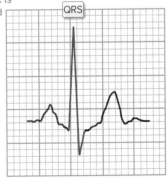

QRS complex findings in this tracing:
There is one QRS complex following each P wave. The QRS complex is narrow at 0.08 seconds in duration and 19 mm in amplitude.

Figure 3-5
Assessing the QRS complexes is the fourth step in assessing the ECG.

Step 5: PR Intervals

Another important step in assessing the ECG is measuring the PR intervals (Figure 3-6). The **PR interval** is the distance from the beginning of the P wave to the

Characteristics of normal PR intervals:
There should be one PR interval preceding each P wave.
Each PR interval is the same at 0.12 to 0.20 seconds in duration.

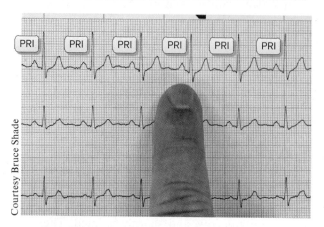

Courtesy Bruce Shade

PR interval findings in this tracing:
There is one PR interval preceding each QRS complex. The duration is the same for each at 0.4 seconds.

Figure 3-6
Assessing the PR intervals is the fifth step in assessing the ECG.

beginning of the Q wave. The PR interval represents depolarization of the heart from the SA node through the atria, AV node, and His-Purkinje system. The first portion of the PR interval is the P wave (discussed earlier), while the second portion is the PR segment. The **PR segment** represents the impulse traveling through the His-Purkinje system and is seen as a flat (isoelectric) line. The PR segment has this appearance because the electrical currents are so small they are not seen on the ECG. The His-Purkinje system is responsible for carrying the impulse to the ventricles.

A normal PR interval indicates the impulse originated from the SA node and traveled through the atria, the AV node, and the His-Purkinje system in a regular, unobstructed course. A normal PR interval is 0.12 to 0.20 seconds in duration; in other words, it extends across three to five small squares. The PR intervals should be the same duration (constant) across the ECG tracing.

With some conditions we may see a PR interval shorter or longer than normal. This tells us that something abnormal is occurring in or around the AV junction. A short PR interval (less than 0.12 seconds in duration) is seen when the electrical impulse arises close to or in the AV junction. A longer PR interval (greater than 0.20 seconds in duration) indicates an even greater delay in conduction of the impulse through the AV junction. It is often associated with a condition called AV heart block. Both shorter and longer PR intervals will be discussed in later chapters.

Step 6: ST Segments

Following the QRS complex is a pause referred to as the ST segment (Figure 3-7). The ST segment starts at the J point (the point or junction at which the QRS complex meets the ST segment) and ends at the beginning of the T wave. The ST segment therefore connects the QRS complex and the T wave. During this period, the ventricles are preparing to repolarize.

The ST segment appears as a flat or isoelectric line that is on the same horizontal line (or close to it) as the PR segment of the preceding complex and has a duration of 0.08 to 0.12 seconds.

Characteristics of normal ST segments:

The ST segment is a flat line that starts at the J point and ends at the beginning of the T wave. It has a duration of 0.08 to 0.12 seconds.

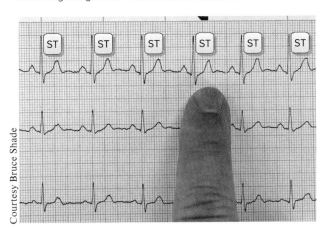

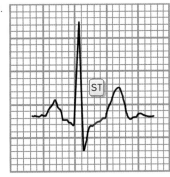

ST segment findings in this tracing:

There is an upward sloping ST segment connecting each QRS complex to the T wave. The duration is 0.08 seconds.

Figure 3-7

Assessing the ST segment, the sixth step, can reveal important information.

Characteristics of normal T waves:

The T wave is larger than the P wave and slightly asymmetrical Normally, the T wave is not more than 5 mm in height in the limb leads or 10 mm in any precordial lead and is oriented in the same direction as the preceding QRS complex.

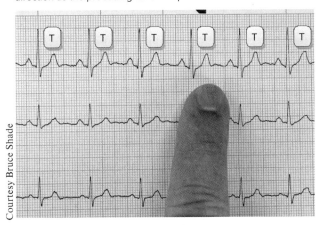

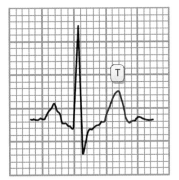

T wave findings in this tracing:

In the middle lead the T wave is the same size as the P wave. However, in the top and bottom leads the T wave is larger than the P waves. It is also biphasic. In the top lead the T wave is 5 mm in height and oriented in the same direction as the preceding QRS complex.

Figure 3-8

Assessing the T waves is the seventh step in assessing the ECG.

Step 7: T Waves

Following the ST segment, we should see the T wave (Figure 3-8). It is larger than the P wave and slightly asymmetrical. The peak of the T wave is closer to the end of the wave than the beginning, and the first half has a more gradual slope than the second half. Normally, the T wave is not more than 5 mm in height in the limb leads or 10 mm in any precordial lead. The T wave is normally oriented in the same direction as the preceding QRS complex. It represents ventricular repolarization (or recovery).

Characteristics of normal QT intervals:
The QT interval is the distance from the onset of the QRS complex until
the end of the T wave and has a normal duration of 0.36 to 0.44 seconds.

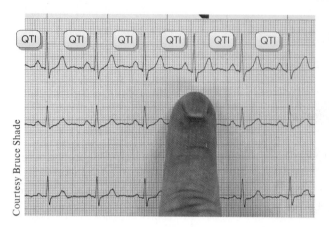

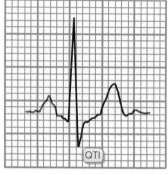

QT interval findings in this tracing:
The QT intervals in this tracing have
a duration of 0.40 seconds.

Figure 3-9
Assessing the QT intervals (QTI), the eighth step, can help identify certain cardiac conditions.

Characteristics of normal U waves:
There may be a small U wave following the T wave.

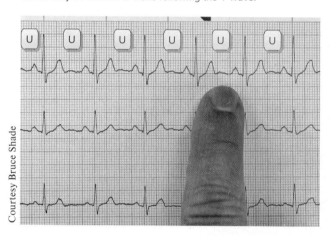

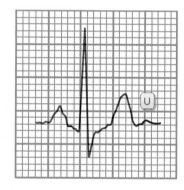

U wave findings in this tracing:
There is a small U wave following
each T wave.

Figure 3-10
Assessing the U waves is the ninth step in analyzing ECG tracings.

Step 8: QT Intervals

Another measurement on the ECG rhythm is the **QT interval** (Figure 3-9). This interval is the distance from the onset of the QRS complex until the end of the T wave and has a normal duration of 0.36 to 0.44 seconds. It measures the time of ventricular depolarization and repolarization. It varies according to age, sex, and heart rate—the faster the heart rate, the shorter the QT interval. The QT interval should not be greater than half the distance between consecutive R waves when the rhythm is regular.

Step 9: U Waves

In about 50% to 75% of ECGs, a small upright (except in lead aV_L) waveform called the **U wave** is seen following the T wave, but before the next P wave (Figure 3-10). Its voltage is so low, however, that the U wave often goes unnoticed.

Flexibility in the Nine-Step Process

Even though we list nine steps, you can sometimes identify dysrhythmias and conditions using just some of these steps. When assessing dysrhythmias, the first five steps (or so) often yield enough information to determine what is occurring on the tracing. Similarly, when evaluating the 12-lead ECG for cardiac conditions, assessment of the ST segment, QRS complex, and T wave usually yields enough information to identify many common conditions. Why then, do you need to memorize them all? Well, the quick answer is, you never know when just a few steps can identify a dysrhythmia or condition and when all nine steps will be needed.

Also, although we list the nine steps of analyzing an ECG tracing in a particular order, there is no requirement to follow that order. For example, determining regularity can be done before identifying the heart rate. Likewise, in some cases the first thing you might notice when looking at an ECG are bizarre-looking QRS complexes suggestive of a dysrhythmia originating from the ventricles. In that case, assessing the QRS complexes may be the first step you use to analyze the tracing. What is important is remembering all the steps and trying to follow the same order so you do not leave out one or two that may help you identify a dysrhythmia or condition.

The information you gather during the Nine-Step Process is referred to as *the characteristics of the tracing.* We compare these findings to what is considered the normal ECG: sinus rhythm.

3.3 Dysrhythmia and Cardiac Condition Characteristics

Each dysrhythmia and cardiac condition has certain characteristics or features. These characteristics are what we use to determine the presence of dysrhythmias and/or cardiac conditions in the ECG tracings.

Over the next three pages, we talk about the characteristics associated with two dysrhythmias. Not to worry, we cover these dysrhythmias more thoroughly in subsequent chapters. Our discussion here is just to make you aware of how characteristics are used in dysrhythmia and cardiac condition analysis and interpretation.

The dysrhythmia sinus bradycardia (Figure 3-11) is a slow rhythm that arises from the SA node and travels through the conduction system in a normal manner. It has the following characteristics:

- Waveforms and intervals that appear at regular intervals at a rate of less than 60 beats per minute (in the adult).
- Upright and slightly asymmetrical P waves, each followed by a QRS complex of normal upright contour, duration, and configuration.
- A PR interval of normal duration that precedes each QRS complex.
- A flat ST segment followed by an upright and slightly asymmetrical T wave.
- A normal QT interval.
- Sometimes the presence of a U wave.

Sinus rhythm

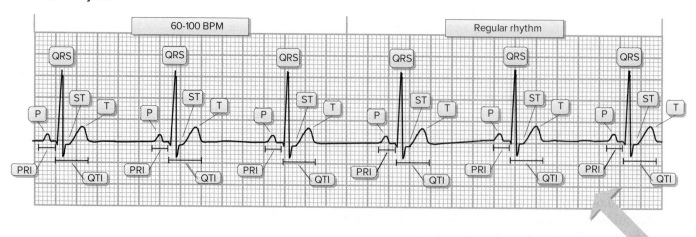

Sinus bradycardia

Compare the characteristics seen to normal sinus rhythm

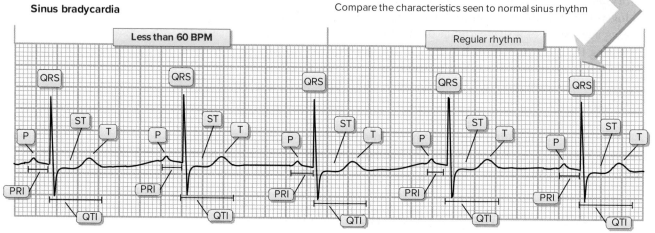

Figure 3-11
When comparing the characteristics of sinus bradycardia to normal sinus rhythm, you can see that the only difference is the heart rate.

You can see that the only difference between sinus bradycardia and sinus rhythm is the heart rate. Thus, the key characteristic of sinus bradycardia is a slow heart rate.

Let's look at another example. Remember our case history at the beginning of this chapter? The dysrhythmia atrial fibrillation (Figure 3-12) occurs when the atria fire chaotically from many different sites. Because the atrial impulses bombard the AV node in such a chaotic manner, not all the impulses get through, producing a totally irregular rhythm. Atrial fibrillation has a number of distinct characteristics including the following:

- Waveforms and intervals appear at totally irregular intervals. The atrial heart rate exceeds 350 beats per minute while the ventricular heart rate depends on how many impulses pass through the AV node.
- There is an absence of P waves; instead there is a chaotic-looking baseline preceding each QRS complex.
- The QRS complexes are of normal upright contour, duration, and configuration.
- There are no PR intervals.

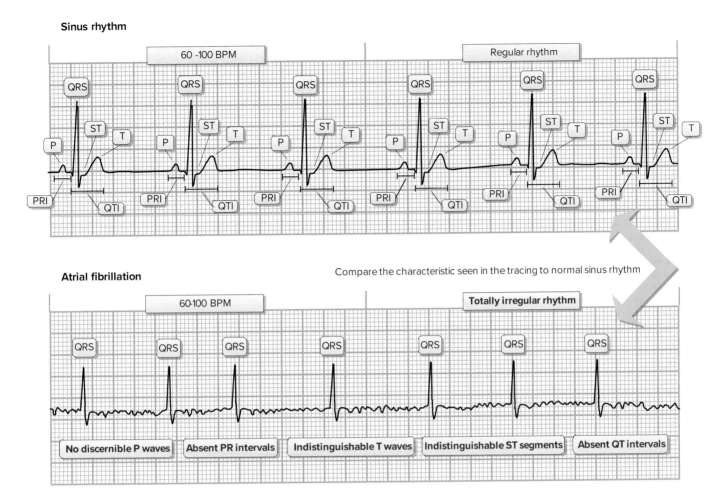

Figure 3-12
Comparing the characteristics of atrial fibrillation to normal sinus rhythm.

- The ST segment and T wave may be indistinguishable or absent.
- The presence of QT intervals depends on whether or not the T waves are identifiable.
- U waves are not seen.

Atrial fibrillation has a number of key characteristics, including a totally irregular rhythm and a chaotic-looking baseline preceding each QRS complex. This is the only dysrhythmia having these characteristics so, when you see them, you know it is likely atrial fibrillation.

If you know the characteristics associated with the various dysrhythmias and conditions, then you can identify what you are seeing. Character recognition is a far better approach for analyzing and interpreting ECG tracings than simply trying to memorize sample ECG tracings you might see in this or any other textbook.

The approach we take in later chapters is to teach you the characteristics of each cardiac dysrhythmia and condition. You can then recall them during your systematic analysis and interpretation of ECG tracings.

3.4 Analyzing the ECG

A quick analysis of an ECG can be done by looking at the tracing as it moves from left to right horizontally across an ECG monitor screen (Figure 3-13). The heart rate; regularity of the rhythm; and morphology of the P waves, QRS complexes, and T waves can all be seen. Many ECG monitor screens also display information such as the approximate heart rate.

Remember, we refer to ECGs seen on the monitor screen as being dynamic; in other words, the tracing demonstrates what is occurring in the heart right now. Seeing what is currently occurring in the heart is the main benefit of observing the ECG tracing on the monitor screen. However, a drawback to the use of monitor screens is the difficulty in analyzing a tracing that is moving across the screen. While it gives us good information, identifying regularity, viewing the waveforms, and measuring intervals and segments is far easier when the tracing is on a piece of paper than when it is moving across a screen.

When analyzing the tracing on the ECG monitor, follow it from left to right, observing the rate, regularity, and presence of P waves and getting a general view of the PR interval, the QRS complexes, and the T waves. If you identify anything abnormal, print a tracing for further analysis.

Then compare your findings against normal sinus rhythm.

Analyzing the Rhythm Strip

A rhythm strip is a printout of just one or a few leads. It is referred to as a static tracing, representing just what occurred in the heart at the time the tracing was

Figure 3-13
Analyzing an ECG tracing using the ECG monitor.

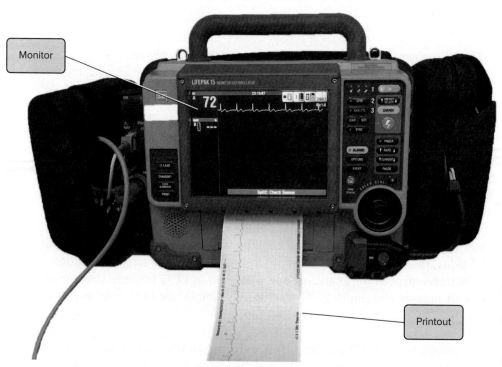

Courtesy Bruce Shade

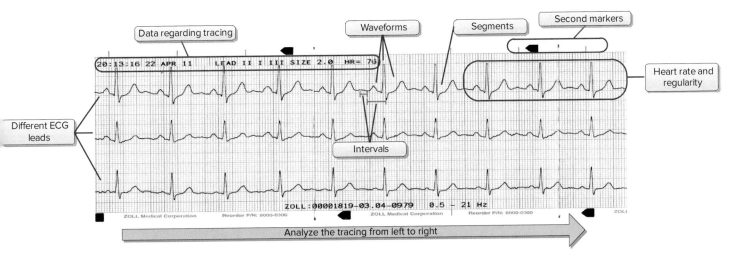

Figure 3-14
Analyzing a rhythm strip.

recorded. The size of the paper varies depending on the manufacturer. The rhythm strip usually gives you enough information to identify existing or emerging dysrhythmias. Sometimes, however, it may be necessary to print out several rhythm strips to capture the necessary information.

To analyze rhythm strips we look at the waveforms, intervals, and durations of each from left to right (Figure 3-14). The benefit to analyzing a static tracing is that we can measure the size of the waveforms, durations, and segments, giving us a much more thorough assessment than what we can gain using the ECG monitor.

We then compare our findings against normal sinus rhythm.

Analyzing the 12-Lead Tracing

The 12-lead tracing is a printout of 12 views of the heart (Figure 3-15). Only a small section of each lead is included, making it difficult to use this type of tracing to identify dysrhythmias. However, some ECG machines print out a rhythm strip along the bottom of the tracing. This rhythm strip is usually one lead but may include up to three different leads. This allows you to more easily find dysrhythmias.

Some devices print standard measurements such as the QRS axis and ECG intervals at the top of the report. The intervals are often reported in milliseconds. The report may also include a 1-mV reference pulse, an annotation of the ECG size setting, the frequency response, and the recorder speed. The 12-lead ECG is usually printed in ×1.0 size in which 1 mV = 10 mm (two large boxes) on the ECG paper. The standard diagnostic frequency response is 0.05 to 150 Hz. Recorder speed is typically 25 mm/sec. Consistency in the areas of frequency response, ECG size, and recorder speed are important when interpreting a 12-lead ECG and comparing reports.

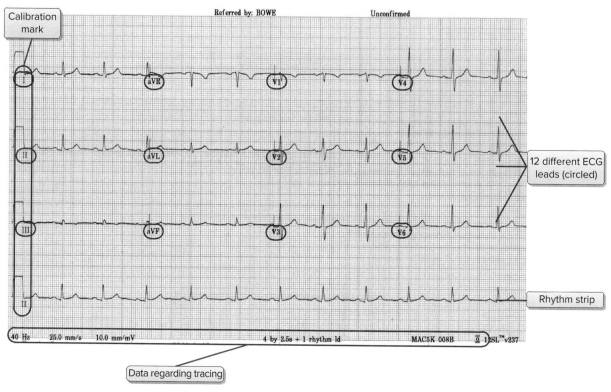

Figure 3-15
Elements of the 12-lead ECG tracing.

There are several ways to analyze the 12-lead tracing. We can look at each lead (from left to right), starting at the top and moving to the bottom in the first column, then in the second column, next in the third column, and then lastly, in the fourth column (Figure 3-16A). Alternatively, depending on what we are looking for, we can analyze groups of ECG leads (Figure 3-16B) to reveal certain conditions such as **axis deviation,** myocardial injury or **infarction, enlargement** or ventricular **hypertrophy,** and **bundle branch** block. The presence of these conditions can be seen in these leads as they view that area of the heart where the condition is occurring.

Again, we then compare our findings against normal sinus rhythm.

Lastly, if a dysrhythmia or abnormality is identified, compare it to your patient assessment findings. This will help identify the significance of the dysrhythmia or abnormality and assist you in making the appropriate treatment decisions.

3.5 Calibrating the ECG

Calibration is the process of determining that the settings on the ECG match the proper standards. If the ECG machine is properly calibrated, the height of the P waves, QRS complexes, and T waves can provide you with certain information.

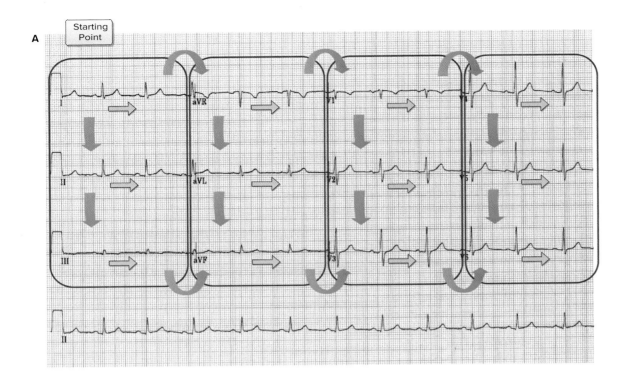

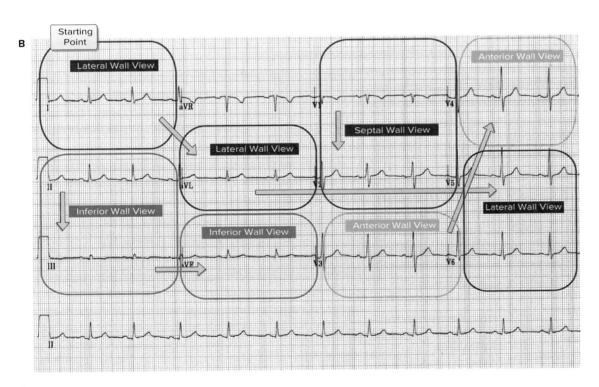

Figure 3-16
Analyzing 12-lead tracings. (A) Tracings can be analyzed from top to bottom. (B) Tracings can be analyzed by looking at groups of ECG leads.

For example, small complexes can indicate conditions such as pericardial effusion whereas tall QRS complexes may point to the presence of left ventricular hypertrophy (both these conditions are discussed in later chapters). To help assure the ECG machine is properly calibrated, you can place a registration or calibration

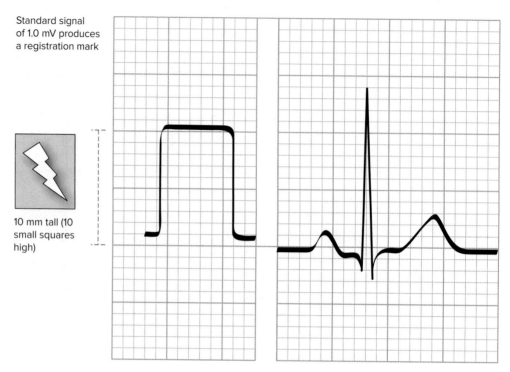

Standard signal of 1.0 mV produces a registration mark

10 mm tall (10 small squares high)

Figure 3-17
Calibration of the ECG.

mark as a reference point on the ECG tracing. This is sometimes called *calibrating the ECG machine.* A standard signal of 1.0 mV will produce a registration point 10 mm tall (10 small squares high). On most ECG machines this mark is created automatically whereas with others it is necessary to manually place this registration point. This registration mark should be included on every ECG tracing (Figure 3-17).

3.6 Artifact

Many factors can interfere with the signal being detected and transferred to the ECG machine. We refer to this as **noise** or **artifact** (Figure 3-18). Noise or artifact has no relationship to the electrical activity of the heart and makes it hard to accurately interpret the ECG. Three common types of artifact are muscle artifact, wandering baseline, and electromagnetic interference (EMI).

Muscle artifact appears as a high frequency, low amplitude, relatively irregular distortion of the baseline and QRS complexes. It is often caused by poorly supported or tense muscles in the arms and legs or shivering. It may be mimicked by loose, improperly placed, dried out, or outdated electrodes; a loose lead wire; poor patient cable connection; worn-out lead wires; a malfunctioning ECG machine; poor skin preparation; or vibration from a moving vehicle.

Ways to prevent or minimize muscle artifact include keeping the electrodes in their unopened package until it is time to use them, avoiding the use of electrodes that are past their expiration date, properly preparing the skin, resting the arms and legs on a supportive surface, ensuring patient comfort and coaching the patient to relax, ensuring good patient/electrode contact, replacing dried out electrodes, securing cable

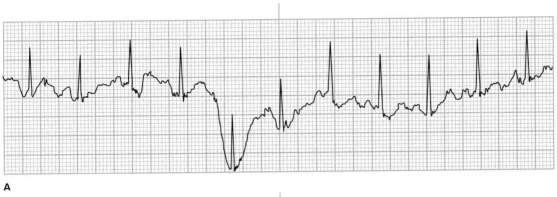

A

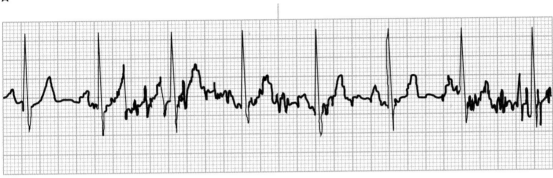

B

Figure 3-18
Artifact is markings on the ECG caused by such things as: (A) the patient moving; (B) shivering, or experiencing muscle tremors; electrodes that are loose or improperly placed; worn-out lead wires; a malfunctioning ECG machine; or the presence of 60-cycle current interference.

connections, and stopping the vehicle while acquiring the ECG. Lastly, lowering the head of the cot or hospital bed from 90 to 45, 30, or 15 degrees or placing the patient into a supine (flat on their back) position is often effective in eliminating artifact.

Wandering baseline is a low-frequency, high-amplitude artifact. It may be observed during deep inhalations and exhalations, patient movement, or the acceleration and deceleration forces of a moving vehicle. Ways to prevent or minimize wandering baseline include asking the patient to breathe quietly while the ECG is being acquired, preventing patient movement, performing good skin preparation, checking all electrodes for good contact, and stopping the vehicle while acquiring the ECG.

Electromagnetic interference originates outside the patient and is caused by interference from electronic sources such as power cords, insufficiently grounded alternating current (AC) powered equipment, and nearby use of portable radios. This is also called alternating current (AC) or 60-cycle interference. Ways to prevent or minimize electromagnetic interference include moving away from a noisy environment, ensuring power cords are not touching or lying near the ECG cable or lead wires, moving away from or unplugging AC-powered equipment that is causing interference, and checking the lead wires and patient cable for observable damage.

The way you can differentiate artifact from waveforms produced by the heart is that there is usually little consistency to the artifact whereas electrical activity produced by the heart usually looks the same (or at least similar) across the rhythm strip. As you will learn throughout this text, some dysrhythmias produce chaotic-looking waveforms (such as atrial and ventricular fibrillation), but even those waveforms generally repeat themselves. In some cases artifact can mimic life-threatening dysrhythmias. For this reason, always compare your rhythm interpretation with the clinical status of the patient. If the patient is awake and alert but the ECG shows a rhythm inconsistent with life, you know you are dealing with artifact.

Key Points 🔑

LO 3.1	• To analyze an ECG, we can directly observe the rhythm on an ECG monitor screen or print a tracing so that the specific waveforms, segments, and intervals can be measured and examined. • We then compare our findings against what is considered the normal ECG, commonly referred to as sinus rhythm.
LO 3.2	• The key to making sense of what you see on any ECG tracing is to approach it in a logical and systematic manner. • The Nine-Step Process assesses the main elements of the ECG tracing and includes the following elements: determine the rate (Is it normal, fast, or slow?); determine the regularity (Is it regular or irregular?); assess the P waves (Is there a uniform P wave preceding each QRS complex?); assess the QRS complexes (Are they within normal limits? Do they appear normal?); assess the PR intervals (Are they identifiable? Are they within normal limits? Are they constant in duration?); assess the ST segments (Is it a flat line? Is it elevated or depressed?); assess the T waves (Are they slightly asymmetrical? Are they of normal height? Is it oriented in the same direction as the preceding QRS complex?); look for U waves (Are they present?); assess the QT intervals (Are they within normal limits?). • The average heart rate in the adult is 60 to 100 beats per minute. A heart rate that is slower or faster is considered abnormal. • Normally the heart beats in a regular, rhythmic fashion, producing specific waveforms and intervals with each heartbeat. This cycle repeats itself over and over again. An irregular rhythm is abnormal. • Throughout the ECG tracing, there should be one normal upright P wave preceding each narrow upright QRS complex. • The PR interval that precedes each QRS complex should be of the same duration and within 0.12 and 0.20 seconds. • After each QRS complex of the ECG tracing, there should be a flat ST segment, followed by an upright and slightly asymmetrical T wave. • The QT intervals in an ECG tracing should be between 0.30 and 0.44 seconds in duration. • U waves may or may not be present in the ECG tracing. • Although following the exact order of the Nine-Step Process is not required, you should try to employ all the steps where indicated.

LO 3.3	• Each dysrhythmia and cardiac condition has characteristics that are unique to it, making it identifiable to anyone who knows for what to look for.
LO 3.4	• We can analyze an ECG on the monitor screen, on a rhythm strip, or on a 12-lead ECG tracing. • ECG rhythms shown on the oscilloscope are called dynamic ECGs whereas those printed on graph paper are called ECG tracings or static ECGs.
LO 3.5	• To help assure the ECG machine is properly calibrated, you can place a registration or calibration mark as a reference point on each ECG tracing.
LO 3.6	• Artifact is markings on the ECG tracing that have no relationship to the electrical activity of the heart. They can be caused by such things as the patient's moving, shivering, or experiencing muscle tremors; electrodes that are loose or improperly placed; worn-out lead wires; a malfunctioning ECG machine; or the presence of 60-cycle current interference.

Assess Your Understanding

The following questions give you a chance to assess your understanding of the material discussed in this chapter. The answers can be found in Appendix A.

1. Using an organized approach to analyzing ECG tracings   (LO 3.2)
   a. takes more time but increases your effectiveness.
   b. is only used with complicated dysrhythmias and cardiac conditions.
   c. helps to ensure that you identify all the characteristics seen in the tracing.
   d. can help identify cardiac output.

2. The elements of the Nine-Step Process include all of the following EXCEPT   (LO 3.2)
   a. heart rate.
   b. P waves.
   c. ST segments.
   d. axis.

3. The normal heart rate in the adult is between _____ and _____ beats per minute.   (LO 3.1)
   a. 60, 75
   b. 70, 90
   c. 60, 100
   d. 80, 100

4. During the _____ initiation of the impulse in the SA node, its movement through the atria and activation of the AV node occurs. (LO 3.2)
   a. P wave
   b. PR interval

    c.   QRS complex

    d.   ST segment

5.   The normal PR interval is _____ seconds in duration.   (LO 3.2)

    a.   0.12 to 0.20

    b.   0.04 to 0.08

    c.   0.36 to 0.44

    d.   0.06 to 0.12

6.   The PR interval represents   (LO 3.2)

    a.   repolarization of the ventricles.

    b.   depolarization of the heart from the SA node through the atria, AV node, and His-Purkinje system.

    c.   depolarization of the atria.

    d.   movement of the electrical impulse through the ventricles.

7.   The Q wave is the first _____ deflection after the PR segment.   (LO 3.2)

    a.   biphasic

    b.   positive

    c.   negative

    d.   none of the above

8.   Characteristics associated with sinus rhythm include   (LO 3.1)

    a.   heart rate of 56 beats per minute.

    b.   an irregular rhythm.

    c.   one upright P wave preceding each QRS complex.

    d.   PR intervals that vary in duration.

9.   The term we use to describe the appearance of waveforms is   (LO 3.2)

    a.   morphology.

    b.   deflection.

    c.   amplitude.

    d.   vector.

10.   Characteristics that differ from those seen with normal sinus can help us to identify: (LO 3.2)

11.   Using the Nine-Step Process   (LO 3.3)

    a.   is required for all ECG tracings.

    b.   requires an extensive understanding of ECGs.

    c.   allows you to identify the characteristics of the tracings you are analyzing.

    d.   is better suited for 12-lead ECGs than it is for dysrhythmia analysis and interpretation.

12. When we analyze an ECG tracing on the ECG monitor,   (LO 3.4)
    a. it shows what is occurring in real time.
    b. it provide us the most accurate picture of what has occurred in the heart.
    c. only the atrial waveforms can be seen.
    d. it is the best way of assessing the regularity of the rhythm.

13. The rhythm strip provides us   (LO 3.4)
    a. all views of the heart.
    b. a good tool for identifying cardiac dysrhythmias.
    c. the information needed to identify cardiac conditions.
    d. a dynamic picture of the heart's electrical activity.

14. A 12-lead ECG tracing   (LO 3.4)
    a. should always be assessed from the left to right and from the top to the bottom of each column.
    b. is the most accurate way of assessing the dynamic electrical activity occurring in the heart.
    c. yields much of the information needed to identify cardiac conditions.
    d. records the heart's electrical activity in four limb leads and six chest leads.

15. You are trying to determine if your patient, who is complaining of chest pain, is experiencing myocardial ischemia. The best way to do this is   (LO 3.4)
    a. view his heart's electrical activity on the ECG monitor screen.
    b. perform a 12-lead ECG.
    c. analyze a recently recorded rhythm strip.
    d. take the patient's blood pressure.

16. A mark serves as a reference point on the ECG tracing from which you can assess the height of the P waves, QRS complexes, and T waves is called: (LO 3.5)

17. Artifact is   (LO 3.6)
    a. a normal part of each ECG tracing.
    b. an indication of decreased cardiac output.
    c. sometimes caused by muscle tremors.
    d. consistently the same across the ECG tracing.

18. Describe what is meant by the term "regular rhythm."   (LO 3.2)

19. Describe the normal morphology of a P wave.   (LO 3.2)

20. What does the T wave represent?   (LO 3.2)

4 Heart Rate

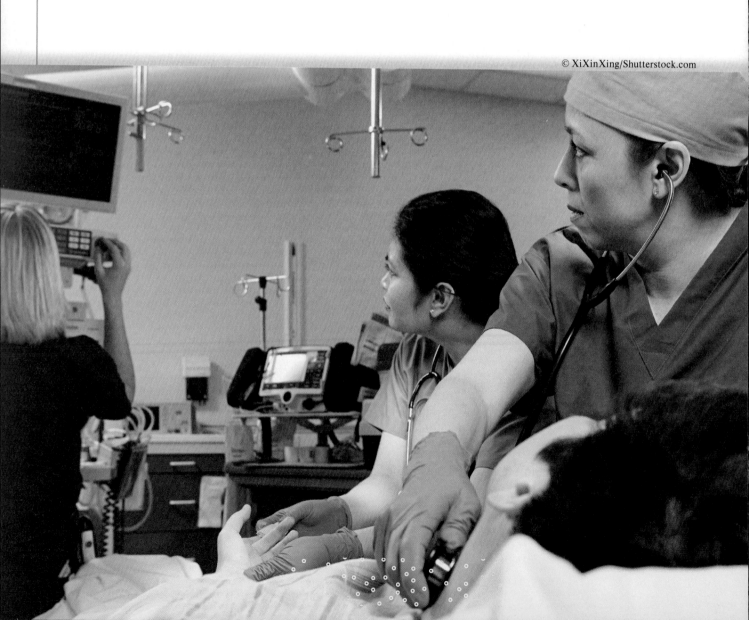

©rivetti/Getty Images

Chapter Outline

Importance of Determining the Heart Rate

Methods of Determining Heart Rate

Counting Both the Atrial and Ventricular Rates

Normal, Slow, and Fast Rates

© XiXinXing/Shutterstock.com

Learning Outcomes

LO 4.1   State why it is important to determine the heart rate in an ECG rhythm.

LO 4.2   Use the various shortcut methods to determine the heart rate.

LO 4.3   Describe how to accurately measure both atrial and ventricular heart rates.

LO 4.4   Define the terms *bradycardia* and *tachycardia*.

Case History

A 36-year-old woman presents to the emergency department complaining of a rapid heartbeat. She was recently diagnosed with hyperthyroidism for which she has begun treatment. She states that, after a heated argument with her husband over finances, she noticed that her heart was "pounding out of her chest." She feels lightheaded but denies chest pain.

The patient is promptly brought back to the major medical room and placed on a cardiac monitor while her vital signs are assessed. Her blood pressure is 105/60, pulse 160, respiratory rate 24, and oxygen saturation 99% on room air. The emergency physician examines the cardiac monitor and observes that the rhythm is fast, with narrow QRS complexes. The heart rate readout on the monitor appears to be malfunctioning, indicating a rate of 320 beats per minute. The physician runs a printout of the rhythm and, using the skills she learned in ECG rhythm recognition, is able to measure the rate using a few rules of thumb. She determines that the patient's heart rate is actually 150 beats per minute.

4.1 Importance of Determining the Heart Rate

One of the first steps of analyzing an ECG is to determine the heart rate (Figure 4-1). The heart rate yields important information. A rhythm that is either slow or fast is considered abnormal. A variety of conditions can produce slow or fast heart rates. Some of these conditions can be quite serious. Further, some slow or fast rhythms can deteriorate into deadly cardiac dysrhythmias (Figure 4-2). For these reasons, each slow or fast rhythm should be properly investigated.

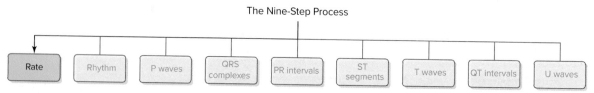

Figure 4-1

One of the first steps to assessing an ECG is to determine the heart rate.

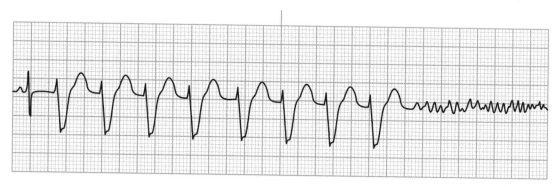

Figure 4-2

Deterioration of an ECG rhythm into ventricular tachycardia and then ventricular fibrillation.

Quick Check of the Heart Rate

Begin by quickly checking the ECG monitor or tracing to see if the rate is slow, normal, or fast. For this initial look, do not worry about calculating the exact heart rate. Instead, identify which group it fits in—slow, normal, or fast (Figure 4-3). Some ECG rhythms may be profoundly slow or fast, in which case you should quickly assess the patient for adequate cardiac output. If cardiac output is compromised, immediate treatment is necessary.

A quick assessment of the heart rate can be done by looking at the space between QRS complexes. More space indicates a slower heart rate while less space indicates a faster heart rate.

You can practice determining whether a heart rate is slow, normal, or fast with the *Section 2 Practice Makes Perfect* exercises beginning on page 217. Answers to these exercises can be found in Appendix A.

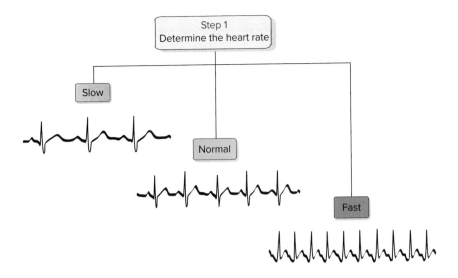

Figure 4-3
The first step in analyzing the heart rate is to determine if the heart rate is slow, normal, or fast.

Many ECG machines (as well as other devices used to assess vital signs) provide a visual display of the patient's heart rate. However, you should not rely on these displays alone. Always confirm the reading by comparing it with the heart rate seen on the ECG tracing and/or by taking a pulse.

The heart rate seen on the ECG tracing should always be compared with the pulse rate (obtained by palpation). The pulse rate should be counted for 15 seconds in regular rhythms and 30 to 60 seconds in irregular rhythms. It is important to palpate a pulse as some impulses may fail to generate contractions of the heart.

Heart Rates Seen with Various Dysrhythmias

Identifying the heart rate of a rhythm can help identify which dysrhythmia we may be facing as certain dysrhythmias are slow or fast whereas others are not. If the rhythm has a normal heart rate, we can automatically exclude those that are characteristically slow or fast. As we discuss this concept further, you will see that excluding a list of dysrhythmias that lacks the features you see in the tracing makes your life a lot easier.

4.2 Methods of Determining Heart Rate

Following the quick assessment, analyze the tracing to determine the approximate or actual heart rate. Rather than printing off a whole minute of ECG tracing and counting the number of QRS complexes, you can use a shortcut method to determine the heart rate. In this text we describe three commonly used methods:

- 6-second interval × 10 method
- 300, 150, 100, 75, 60, 50 method
- 1500 method

The names we've given to the methods for calculating the heart rate may differ from those found in other references. This was done intentionally to make it easier to remember them as each name is descriptive of the way the methods work.

Each method assesses a section of the tracing to approximate the heart rate (Figure 4-4) and has its advantages and disadvantages.

Figure 4-4
Using a shortcut method to
determine the heart rate can
save time.

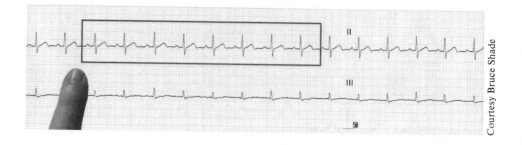

Courtesy Bruce Shade

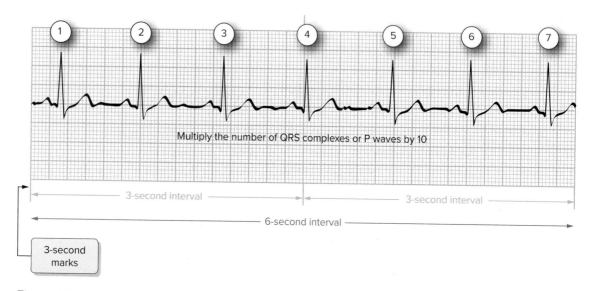

Figure 4-5
To employ the 6-second interval x 10 method, multiply by 10 the number of QRS complexes (for the ventricular rate)
or P waves (for the atrial rate) found in a 6-second portion of the ECG tracing. The rate shown in the ECG tracing
here is 70.

6-Second Interval × 10 Method

The *6-second interval × 10 method* involves multiplying by 10 the number of QRS
complexes found in a 6-second portion of the ECG tracing (Figure 4-5).

Remember that ECG paper typically has markings on the top and/or bottom indicating each 3-second block of time (see Chapter 2). Two of these successive (adjacent to each other) blocks of time equal 6 seconds. When you multiply that amount
of time by 10, it represents 60 seconds or 1 minute.

Not all ECG tracing paper is marked in 3-second blocks of time. For example, some
are measured in 1-second blocks of time. In that case, six of these successive blocks
of time equal 6 seconds.

Therefore to calculate the heart rate

1. Identify two successive 3-second marks.

2. Count the number of QRS complexes in that 6-second section.

3. Multiply that number by 10.

This final number is the approximate heart rate per minute.

Here are some examples of calculating heart rates using this method:

- Eight QRS complexes in the 6-second block—the patient has a heart rate of 80 beats per minute
- Twelve QRS complexes in the 6-second block—the patient has a heart rate of 120 beats per minute
- Four QRS complexes in the 6-second block—the patient has a heart rate of 40 beats per minute.

The 6-second interval $\times$ 10 method is an effective means for assessing the rate of an irregular rhythm. This is because the QRS complexes are measured over 6 seconds. Advantages of the 6-second interval $\times$ 10 method are that it is quick and easy and it does not require any tools or devices. A disadvantage is that it does not calculate the heart rate as accurately as other methods. In other words, the heart rate might actually be 54 beats a minute instead of 50.

Whereas the 6-second interval $\times$ 10 rule is effective for calculating the heart rate in irregular rhythms, there are some dysrhythmias in which the degree of irregularity makes it difficult to find enough consistent tracing to determine an average. For example, you may count eight QRS complexes (and/or P) waves in one 6-second section and 11 QRS complexes (and/or P) waves in another, making it difficult to obtain an accurate heart rate. In these cases, try using a 12-second area of the tracing (4 successive blocks). Count the number of QRS complexes (and/or P waves) and multiply them by 5 (instead of 10).

In addition to determining the ventricular rate, this method can also be used to calculate the atrial rate. Simply multiply by 10 the number of P waves found in a 6-second portion of the ECG tracing.

You can practice determining heart rate using the 6-second interval $\times$ 10 method with the *Section 2 Practice Makes Perfect* exercises on page 217.

300, 150, 100, 75, 60, 50 Method

The *300, 150, 100, 75, 60, 50 method,* also called the *countdown, sequence,* or *triplicate method,* involves locating an R wave on a bold line on the ECG paper (Figure 4-6). Let's refer to that as the "start point." Each bold line to the right of the start point has a value that denotes the heart rate. The first bold line is the 300 line. The bold line after that is the 150 line, the one after that is the 100 line, the next line is 75, the line after that is 60, and the next line is 50. Assume for a minute that you are assessing the heart rate. You have located an R wave that falls on a bold line. If the next R wave falls on the second bold line, the heart rate is 150 beats per minute. Instead, if the next R wave falls on the fourth bold line, then the heart rate is 75 beats per minute.

Therefore, to calculate the heart rate

1. Find an R wave located on a bold line, and then
2. Find the next consecutive R wave. If it is on a bold line, then the heart rate is the value of that line (300, 150, 100, 75, 60, 50).

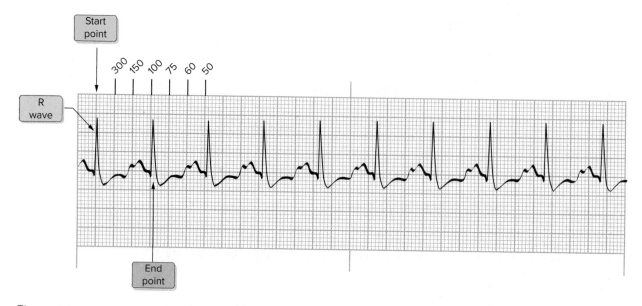

Figure 4-6

The 300, 150, 100, 75, 60, 50 method involves locating an R wave (or P wave) on a bold line on the ECG paper (the start point). Each bold line to the right of the start point has a value that denotes the heart rate. The end point is where the next R wave (or P wave) is located. The rate shown in this ECG tracing is just under 100 beats per minute.

Note that we use the terms *R* and *QRS complex* interchangeably. This is because when you are trying to line up the waveform to a bold line, it is much easier to use the tallest of the three waveforms (of the QRS complex), the R wave. When we say to look for an R wave, we mean look for the most prominent deflection of the QRS complex. In some cases the S wave may be the most prominent.

Advantages of the 300, 150, 100, 75, 60, 50 method are that it is quick; that it is fairly accurate; and that it requires no special tools, calculations, or rulers. A disadvantage is that this method can be used only if the rhythm is regular.

If the heart rate is less than 50 beats per minute, then the next bold line (following the 50 line) has a value of 43, and the line following that has a value of 37. If the heart rate is slower than 37, then you must use the 6-second interval × 10 method to determine the heart rate.

If the second R wave falls between two bold lines, you can more precisely determine the heart rate if you use identified values for each of the thin lines (Figure 4-7). This set of values is not as easy to remember as the 300, 150, 100, 75, 60, 50 method, so having a cheat sheet or rate calculator is a good idea.

Alternatively, when the second R wave does not fall on a bold line, you can approximate the heart rate (Figure 4-8). For example, if it falls between the third and fourth lines, you can say the heart rate is "between 75 and 100 beats per minute."

This method can also be used to calculate the atrial rate. Begin by finding the peak of a P wave that falls on a bold line. Then locate the next consecutive P wave. The values for each bold (and thin) line are the same as previously described.

To determine the heart rate using the 300, 150, 100, 75, 60, 50 method, it is acceptable to move either right or left of the start point. Sometimes you can find

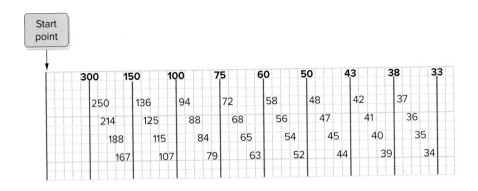

Figure 4-7
Using the identified values for each of the thin lines allows you to more precisely determine the heart rate.

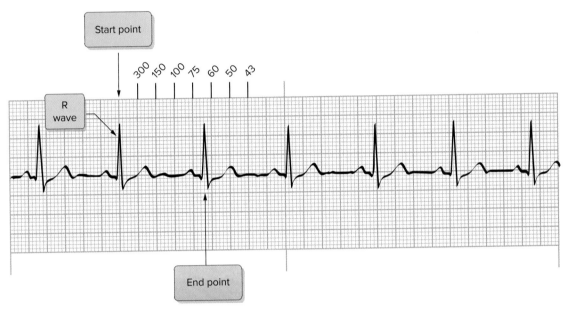

Figure 4-8
When the second R wave does not fall on a bold line, you can approximate the heart rate. The heart rate of this tracing is between 60 and 75 beats per minute.

better alignments of the R waves and the bold lines looking to the left instead of to the right.

An easy way to determine the rate when the R wave does not fall on a bold line is to place a piece of paper over the ECG tracing, then mark it where the first R falls. Next, mark it where the second R wave falls. Slide the paper to the right or left until the first mark lines up with a bold line. Use the second mark to identify the heart rate.

Figure 4-9 shows three examples of calculating the heart rate using the 300, 150, 100, 75, 60, 50 method.

As you will see from the discussion that follows, the two quickest methods for determining the heart rate are the 6-second interval × 10 method and the 300, 150, 100, 75, 60, 50 method.

You can practice determining heart rate using the 300, 150, 100, 75, 60, 50 method with the *Section 2 Practice Makes Perfect* exercises on page 217.

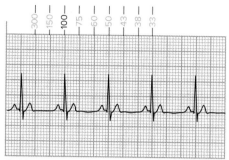

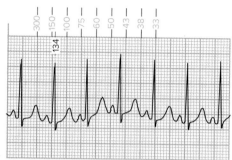

In this example, the first and second R waves fall close to the bold lines making determination of the heart rate easier. This rhythm has a heart rate of 100 BPM.

In this example, the first R wave falls close to the bold line but the second R wave is a box off to the right of the bold line. Here, using the values for the smaller lines (shown earlier), you can see this rhythm has a heart rate of 134 BPM.

1a) Either use calipers or mark two lines on a piece of paper to represent the distance between the first two R waves.

1b) Then align the first leg of the caliper or the drawn line with the bold line to the right of the first R wave. Where the second leg of the caliper or drawn line falls represents where the second R wave would fall.

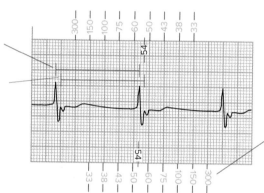

2) Since one of the R waves falls on a bold line you can simply reverse the direction. Go from right to left instead of left to right to align your R waves to the next appropriate line.

In this example, the first R wave does not fall close to a bold line. There are two ways to deal with this problem. Whichever method you use yields a heart rate of 54 BPM.

Figure 4-9
Three examples of calculating the heart rate using the 300, 150, 100, 75, 60, 50 method.

1500 Method

Another way to calculate the heart rate is to use the 1500 method, so named because 1500 small squares on the ECG paper are equal to 1 minute (Figure 4-10). To use this method, count the number of small squares between two consecutive R waves and divide 1500 by that number. For example, if you have 20 small squares (equal to four large boxes) the heart rate is 75 beats per minute ($1500 \div 20 = 75$).

Therefore, to calculate the heart rate

1. Count the number of small squares between two consecutive R waves.
2. Divide 1500 by that number.

This can be used to calculate the atrial rate as well by measuring the number of small squares between the peaks of two consecutive P waves.

Advantages of the 1500 method are that it is the most accurate and requires no special tools or rulers. Disadvantages of this method are that it requires you to do math at a time when things may be extremely hectic and that it cannot be used if the rhythm is irregular.

Following (Figure 4-11) are three examples showing calculating the heart rate using the 1500 method.

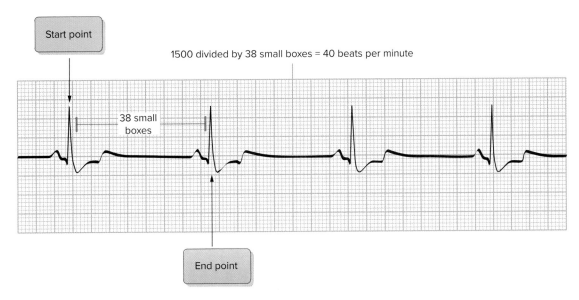

Figure 4-10
To use the 1500 method, count the number of small squares between two consecutive R waves and divide 1500 by that number.

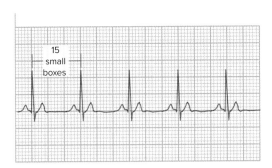

In this example, there are 15 small boxes between consecutive R waves. Dividing 1500 by 15 yields a heart rate of 100 BPM.

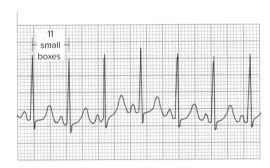

In this example, there are 11 small boxes between consecutive R waves. Dividing 1500 by 11 yields a heart rate of 136 BPM. Comparing that to what we calculated the heart rate to be in Figure 4- (134 BPM) you can see there is a slight difference.

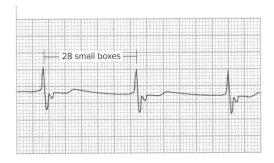

In this example, there are slightly more than 28 small boxes between consecutive R waves. Dividing 1500 by 28 (28.2 to be more accurate) yields a heart rate of 53 BPM. Comparing that to what we calculated the heart rate to be in Figure 4-8 (54 BPM) you can see there is a slight difference.

Figure 4-11
Calculating heart rate using the 1500 method. We are using the same examples show in Figure 4-10 to allow you to see the difference in accuracy between the two methods of calculating heart rate.

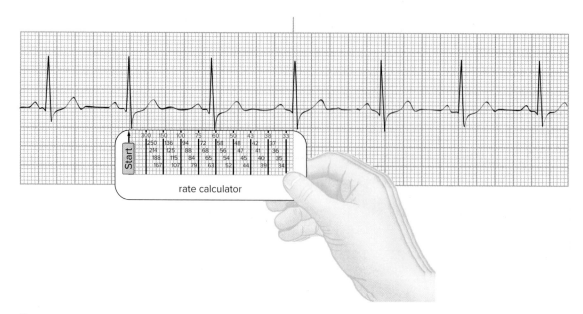

Figure 4-12
To use a heart rate calculator, begin by placing the calculator over the ECG tracing. Position the start mark on an R wave, then find the next consecutive R wave. Where that next R wave lines up is the approximate heart rate.

You can practice determining heart rate using the 1500 method with the *Section 2 Practice Makes Perfect* exercises on page 217.

Rate Calculators

Rate calculators are simple paper or plastic devices about the size of a bookmark. They can be overlaid on the ECG tracing to show the heart rate and, in some cases, the duration of the various waveforms, segments, and intervals. They are easy to use (Figure 4-12) and typically have a start mark at one end and numbers indicating the rate along the bottom. To use a rate calculator, begin by placing the calculator over the ECG tracing. Next, position the start mark on an R wave and then find the next consecutive R wave. The approximate heart rate is where the next R wave lines up.

The advantage of these devices is that they allow you to quickly identify the heart rate. The disadvantages are that you must have it available any time you need to calculate a patient's heart rate and that they cannot be used on irregular rhythms.

4.3 Counting Both the Atrial and Ventricular Rates

We talk about counting the QRS complexes to determine the ventricular rate, but we also count the P waves to determine the atrial rate (Figure 4-13).

In some dysrhythmias, the atrial rate may be less than or greater than the ventricular rate. The rate-calculating methods previously discussed can also be used to identify the atrial rate.

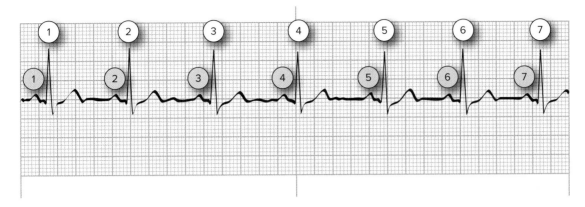

Figure 4-13
When calculating the heart rate, count the number of QRS complexes (for the ventricular rate) and P waves (for the atrial rate). They should be the same. In this 6-second ECG tracing, there are 7 QRS complexes and 7 P waves. Using the 6-second interval × 10 method, the ventricular and atrial rates are 70 beats per minute.

4.4 Normal, Slow, and Fast Rates

In the average adult, the normal heart rate is between 60 and 100 beats per minute. Normal sinus rhythm, as well as many dysrhythmias (to be discussed in later chapters), characteristically have a normal rate.

Rates above 100 beats per minute or below 60 beats per minute are considered abnormal. When we say abnormal, we do not necessarily mean that it is a bad thing but rather that it is outside the normal range.

Slow Rates—Bradycardia

A heart rate less than 60 beats per minute is called **bradycardia** and can occur for many reasons. It may or may not have an adverse affect on cardiac output. In the extreme it can lead to severe reductions in cardiac output and eventually deteriorate into asystole (an absence of heart rhythm).

Fast Rates—Tachycardia

A heart rate greater than 100 beats per minute is called **tachycardia.** Tachycardia has many causes and leads to increased myocardial oxygen consumption, which can have an adverse effect on patients with coronary artery disease and other medical conditions. Extremely fast rates can have an adverse affect on cardiac output. Also, tachycardia that arises from the ventricles may lead to a chaotic quivering of the ventricles called *ventricular fibrillation.*

Stable or Unstable, Narrow or Wide

When cardiac output is not adversely affected by an extremely slow or fast heart rate, it is referred to as stable bradycardia or tachycardia. When cardiac output is compromised and results in the development of signs and symptoms such as hypotension, chest pain, shortness of breath and/or disorientation, etc., it is referred to as unstable bradycardia or tachycardia. Furthermore, for extreme tachycardia, we also categorize it by the width of the QRS complexes, either narrow or wide. The terms stable or unstable narrow complex tachycardia or stable or unstable wide complex tachycardia will be used. This will be discussed in greater depth in subsequent chapters.

Key Points

LO 4.1	• Start assessing the heart rate by quickly checking the ECG monitor or tracing to see if the rate is slow, normal, or fast.
LO 4.2	• The 6-second interval × 10 method of calculating the heart rate involves multiplying by 10 the number of QRS complexes found in a 6-second portion of the ECG tracing. This method is the most effective means for assessing irregular rhythms. • The 300, 150, 100, 75, 60, 50 method for calculating the heart rate involves locating an R wave on a bold line on the ECG paper, then finding the next consecutive R wave and using the 300, 150, 100, 75, 60, 50 values for subsequent bold lines to determine the rate. • To use the 1500 method, count the number of small squares between two consecutive R waves and divide 1500 by that number. • Rate calculators are easy to use and can effectively determine the heart rate.
LO 4.3	• We count the QRS complexes to determine the ventricular rate and count the P waves to determine the atrial rate.
LO 4.4	• Rates above or below 60 to 100 beats per minute in the typical adult are considered abnormal. • A heart rate less than 60 beats per minute is called bradycardia. • A heart rate greater than 100 beats per minute is called tachycardia. • Both extremely slow and fast heart rates can result in decreased cardiac output.

Assess Your Understanding

The following questions give you a chance to assess your understanding of the material discussed in this chapter. The answers can be found in Appendix A.

1. If the QRS complexes are close together the heart rate is   (LO 4.2)
   a. slow.   b. normal.
   c. profoundly slow.   d. fast.

2. Determining the heart rate helps you identify   (LO 4.1)

3. Using a shortcut method for determining the heart rate   (LO 4.2)
   a. can help identify the regularity of an ECG tracing.
   b. helps determine the patient's actual cardiac output.
   c. is faster than counting the complexes found on a whole minute's worth of ECG tracing.
   d. is the most accurate way of identifying the heart rate.

4. The 6-second interval × 10 method   (LO 4.2)
   a. can be used to estimate the heart rate in irregular rhythms.
   b. involves multiplying by 25 the number of QRS complexes found in a 6-second portion of the ECG tracing.
   c. is the most accurate method of determining the heart rate.
   d. can be used only to determine the ventricular rate.

5. The 300, 150, 100, 75, 60, 50 method   (LO 4.2)
   a. can be used to measure the heart rate in irregular rhythms.
   b. involves identifying a T wave that falls on a bold line.
   c. requires the use of a rate calculator to determine the heart rate.
   d. requires you to identify a start point from which to identify where the next R wave falls.

6. To use the 1500 method,   (LO 4.2)
   a. count the number of small squares between two consecutive R waves and divide 1500 by that number.
   b. find an R wave that falls on a bold line and begin counting the small squares.
   c. count the number of QRS complexes in a 6-second section of the ECG tracing.
   d. divide by 1500 the number of QRS complexes in a 3-second section of the ECG tracing.

7. Which one of the following methods for determining the heart rate in a regular rhythm is the most accurate?   (LO 4.2)
   a. 6-second interval $\times$ 10 method
   b. 300, 150, 100, 75, 60, 50 method
   c. 1500 method
   d. Quick assessment method

8. Describe the relationship of the atrial to the ventricular heart rate.   (LO 4.3)

9. The average heart rate in the adult is between _____ beats per minute.   (LO 4.4)
   a. 50 and 75
   b. 70 and 90
   c. 65 and 95
   d. 60 and 100

10. Tachycardia   (LO 4.4)
    a. is a dysrhythmia that has a heart rate greater than 80 beats per minute.
    b. leads to decreased myocardial oxygen consumption.
    c. can be produced by stimulation of the parasympathetic nervous system.
    d. may be brought about by exercise or exertion.

11. If the 6-second interval $\times$ 10 method is used to determine the heart rate, bradycardia is present if there are fewer than how many QRS complexes?   (LO 4.4)

12. What effect can extremely fast tachycardia have on cardiac output?   (LO 4.4)

Referring to the scenario at the beginning of this chapter, answer the following questions.

13. It is unlikely that the patient's heart rate was actually 320 because   (LO 4.4)
    a. she was awake.
    b. her blood pressure was not low.
    c. she was responsive.
    d. All of the above.

14. The most likely cause of the patient's fast heart rate was   (LO 4.4)
    a. anxiety.
    b. hypoxia.
    c. hyperthyroidism.
    d. amphetamine use.

5 Regularity

Chapter Outline

**Importance of Determining
Regularity**

**Methods of Determining
Regularity**

Types of Irregularity

Learning Outcomes

LO 5.1   State why it is important to determine the regularity of an ECG rhythm.

LO 5.2   Describe three methods that can be used to determine atrial and ventricular regularity.

LO 5.3   Identify each of the different types of irregularity.

Case History

An 85-year-old woman presents to her family physician's office for her yearly checkup. She has a history of atherosclerotic heart disease and noninsulin-dependent diabetes. She has been careful to follow her doctor's advice and takes her medications regularly. She denies any new complaints.

While taking her vital signs, the nurse notes that the woman's heart rate is irregular. He examines the patient's medical record to see if this condition has been previously recorded. He finds no evidence that the patient has ever had an irregular heart rhythm and calls the physician into the room to examine the patient. A cardiac monitor is attached to the patient, and an ECG tracing is run. The nurse and physician examine the tracing and observe that the rate is within normal limits but is irregularly irregular. Then, they compare it with a previous tracing, noting that there is a difference between the two.

5.1 Importance of Determining Regularity

An initial step in analyzing an ECG rhythm is to determine if it is regular or irregular (Figure 5-1). This chapter focuses on why it is important to evaluate the regularity of each tracing, details the steps that can be used to determine regularity, and describes the different types of irregularity. Determining the regularity of ECG tracings provides us with significant information. An irregular rhythm is considered abnormal. A variety of conditions can produce irregular rhythms, some of which can be quite serious. Further, some irregular rhythms can deteriorate into deadly cardiac dysrhythmias (Figure 5-2). For these reasons, each irregular rhythm should be properly assessed.

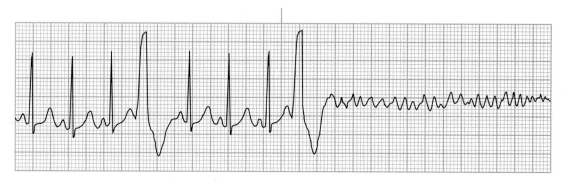

Figure 5-1
Deterioration of an ECG rhythm into ventricular fibrillation.

The Nine-Step Process

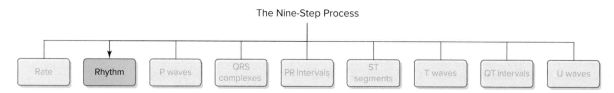

Figure 5-2
Determining the regularity is an initial step in analyzing an ECG tracing.

Quick Check of Regularity

As we discussed in Chapter 2, one of the uses of the ECG is to identify existing or emerging dysrhythmias. We do this by observing the rhythm directly on the ECG monitor screen or by printing an ECG tracing so that the specific waveforms, segments, and intervals can be measured and examined.

As mentioned in Chapter 3, the heart normally beats in a regular, rhythmic fashion, producing a P wave, QRS complex, and T wave with each heartbeat. This cycle repeats itself over and over again. The distance between the consecutive P waves should be the same, just as the distance between the consecutive QRS complexes should be the same throughout the tracing. We call these distances the *P-P interval* and the *R-R interval* (Figure 5-3).

Identifying the regularity of a rhythm can help classify which dysrhythmia we may be facing, as certain dysrhythmias are regular while others are not. (Note that many

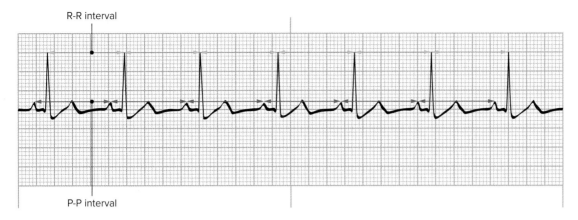

R-R interval

P-P interval

Figure 5-3
The distance between consecutive R waves is called the *R-R interval;* the distance between consecutive P waves is called the *P-P interval.*

dysrhythmias have a regular rhythm; that does not mean they are normal; it just means their rhythm is regular.) If the rhythm is irregular, we can automatically exclude those that are characteristically regular. As we discuss this concept further, you will see that excluding a list of dysrhythmias that lack the features you see in the tracing makes identifying the dysrhythmia easier.

5.2 Methods of Determining Regularity

To assess atrial regularity, we analyze the P-P intervals (Figure 5-4). Again, if the distance remains the same between the P-P intervals, then the atrial rhythm is regular. If the distance differs, the atrial rhythm is irregular. However, rhythms are considered regular even if there is a slight variance in the R-R or P-P interval.

To assess the ventricular regularity, we analyze the R-R intervals. Again, if the distance remains the same between the R-R intervals, then the ventricular rhythm is regular. If the distance differs, the ventricular rhythm is irregular. If an R wave is not present, then you can use either the Q wave or the S wave of consecutive QRS complexes. You should use the uppermost or lowermost (tallest) point of the QRS for consistency in measurement. You should compare the intervals in several cycles. Consistently similar intervals represent a regular rhythm whereas dissimilar intervals indicate an irregular rhythm. A good rule to follow is one small square (0.04 seconds) difference is still considered normal.

The final step is to compare the P-P interval with the R-R interval to determine whether they are the same and appear to be associated with each other. In other words, determine if there is a P wave before every QRS complex and a QRS complex after every P wave.

Several methods can be used to determine the regularity of a rhythm, including marking a paper with a pen (or pencil), employing calipers, and counting the number of small squares between each P-P interval and R-R interval.

In this rhythm, each R-R and P-P interval is 21 small
boxes apart. For this reason it is considered regular.

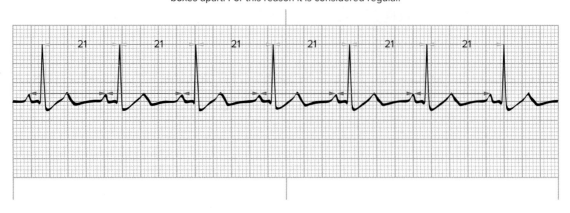

In this rhythm, the number of small boxes differs between some
of the R-R and P-P intervals. For this reason it is considered irregular.

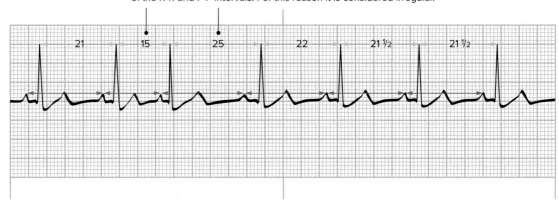

Figure 5-4
The rhythm is regular if the distance between consecutive R waves and P waves is the same. If it changes, the rhythm is irregular.

Paper and Pen Method

Using paper and a pen (or pencil) is simple and quick and requires no special tools (Figure 5-5). To assess the regularity of a rhythm with paper and a pen, do the following:

1. Place the ECG tracing on a flat surface.
2. Place the straight edge of a piece of paper above or over the ECG tracing so that the intervals are still visible.
3. Identify a starting point, the peak of an R wave or P wave, and place a mark on the paper in the corresponding position above it (let's call this mark 1).
4. Find the peak of the next consecutive R wave or P wave, and place a mark on the paper in the corresponding position above it (let's call this mark 2).
5. Move the paper across the ECG tracing, aligning the two marks with succeeding R-R intervals or P-P intervals.

If the two marks line up with subsequent consecutive P or R waves, the rhythm is regular. If the distance between the intervals differs, the rhythm is irregular. You can tell this when mark 2 falls either before or after the next consecutive P wave or R wave.

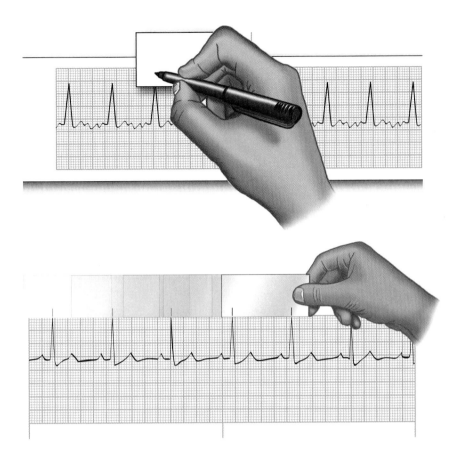

Figure 5-5
To employ the paper and pen method, place marks on a paper above two consecutive P waves or R waves. Then move the paper across the tracing, matching the marks on the paper against subsequent consecutive P waves or R waves.

An additional method similar to the paper marking method is to use the tip of a pen or pencil to determine regularity. *Lay the pen or pencil flat on the ECG strip with the tip on an R or P wave. Slide your fingernail along the pen to the next R or P wave.* Mark the pen in the spot aligned with the second waveform or keep your fingernail in that position. Now, slide the pen along the ECG strip aligning the tip of the pen with subsequent waveforms. If the rhythm is regular, the mark (you made on the pen) or your fingernail will line up with the next waveforms. If the rhythm is irregular, it will not line up.

Caliper Method

The calipers typically used to analyze ECGs are hinged at the top with two legs that extend downward (Figure 5-6). The hinged top allows the legs to be extended away from each other to establish the distance between various waveforms. The legs are pointed at the bottom to allow them to be positioned over even the smallest waveform. To assess the regularity of a rhythm with calipers, do the following:

1. Place the ECG tracing on a flat surface.
2. Place one point of the caliper on a starting point. Most often the starting point is the peak of an R wave or P wave.
3. Open the calipers by pulling the other leg until the point is positioned on the next R wave or P wave.
4. With the calipers open in that position, and keeping the point positioned over the second P wave or R wave, rotate the calipers across to the peak of the next consecutive (the third) P wave or R wave. Alternatively, you can move them so they are lined up on the next consecutive R wave.

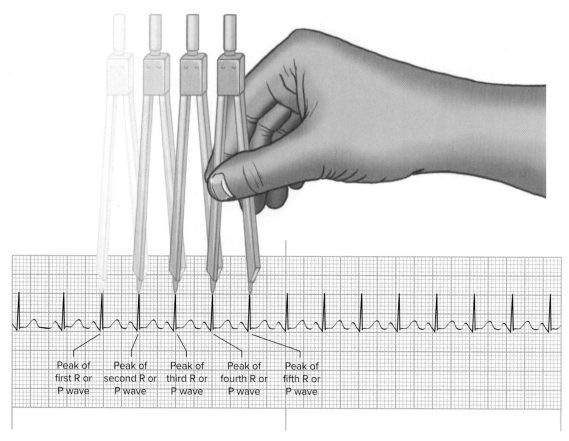

Figure 5-6
The opened calipers are moved across the tracing from the point of one R wave or P wave to the next consecutive R wave or P wave (the R-R or P-P interval).

If the R-R interval or P-P interval is the same, the point will be above the next R wave or P wave. If the R-R interval or P-P interval is different, the point will be situated either before or after the identified waveform. Next, either slide or rotate the calipers across to the next R wave or P wave, then the next, and so on. If the rhythm is regular, the distance of each interval should be the same. If the rhythm is irregular, the distances of the intervals will differ.

Counting the Small Squares Method

Another way to determine the regularity of an ECG is to count the number of small squares on the ECG paper between the peaks of two consecutive R waves (or P waves) and then compare that with the other R-R (or P-P) intervals (Figure 5-7). You can speed up the process by using the larger boxes (made up of five small squares) for all but the last portion.

Using a Rate Calculator

Rate calculators can also be used to determine regularity. Begin by placing the calculator over the ECG tracing. Position the start mark on the peak of an R wave. Then find the next consecutive R wave. Where that next R wave lines up is then noted (this is the heart rate for that ECG). Next, move the calculator over other R-R intervals and compare your findings with what you found in the first R-R interval.

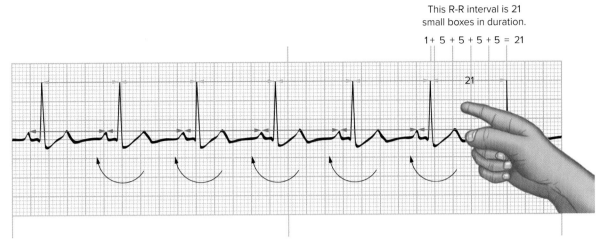

This R-R interval is 21
small boxes in duration.

1+ 5 + 5 + 5 + 5 = 21

For this figure, we started counting from the last R wave and moving to the left because it fell on the bold line making counting of the small squares easier.

Figure 5-7
Count the number of small boxes between consecutive R and/or P waves (each R-R and P-P interval) and then compare the numbers with each other. If they are the same, the rhythm is regular; if they differ, the rhythm is irregular.

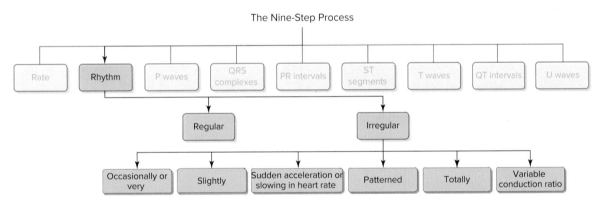

Figure 5-8
Algorithm for regular and irregular rhythms.

5.3 Types of Irregularity

If the rhythm is irregular, we can further analyze it to determine the type of irregularity (Figure 5-8). This can help us identify what dysrhythmia may be present, as certain dysrhythmias have a characteristic type of irregularity. As we discuss each type of irregularity, we also describe, in a general sense, its likely causes. In later chapters, as we discuss the various dysrhythmias, we will identify whether or not an irregular rhythm is one of the characteristics associated with that dysrhythmia and what type of irregularity is present. Note that the names given here for the types of irregularity may differ from those found in other references. This was done intentionally to make them easier to remember.

An irregularity in the rhythm can range from being occasionally irregular to being totally irregular. It is important to have a general sense of the difference as a very irregular rhythm can be a warning that the heart is extremely irritable and may deteriorate to a worse condition.

Occasionally or Very Irregular

When a rhythm is occasionally irregular, it appears mostly regular, but from time to time you see an area where it is irregular (Figure 5-9). The irregularity may be so infrequent that you might not see it at first.

A rhythm may also look very irregular. We need to be concerned with very irregular rhythms as they suggest the heart is extremely irritable. Looking at the ECG tracing in Figure 5-10, you see areas where the distance of the R-R intervals appears to be the same and areas where the distance differs. The areas that are the same are considered the underlying rhythm. We compare the underlying rhythm with the other R-R intervals.

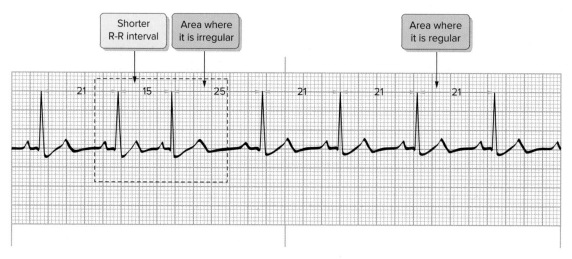

Figure 5-9
An occasionally irregular rhythm appears mostly regular, but from time to time it is irregular.

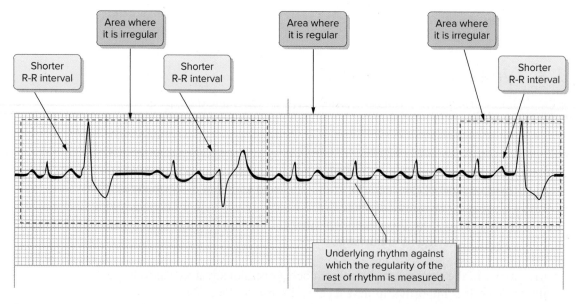

Figure 5-10
A very irregular rhythm has many areas of irregularity.

Normally, the SA node initiates impulses, resulting in a repetitive cycle of P, QRS, and T waveforms.

A premature beat occurs when a site in the a) atria, b) AV junction, or c) ventricles fires before the SA node is able to initiate an impulse.

Following the premature beat, the SA node typically reinitiates impulses in the normal manner.

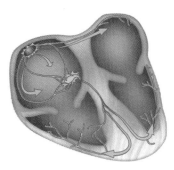
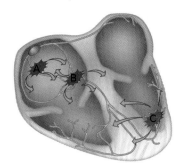
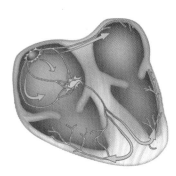

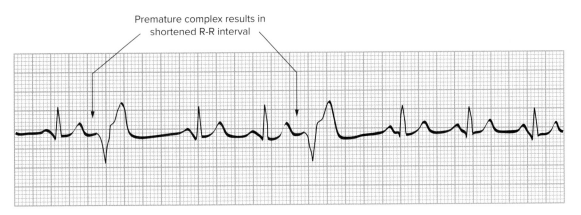

Premature complex results in shortened R-R interval

Figure 5-11
Irregularity can be caused by early beats that arise from the (a) atria, (b) AV tissue, or (c) ventricles.

Occasional or very irregular rhythms occur due to early beats or pauses in or arrest of the impulses arising from the SA node. These are described in the following pages.

Early Beats

Occasionally irregular or very irregular rhythms are often caused by impulses that occur earlier than normal (Figure 5-11). These beats fire before the SA node has a chance to initiate the impulse. These **early beats** (also called *premature beats* or *ectopy*) can arise from any of the cells of the heart, including the atria, AV junctional tissue, or ventricles. The R-R interval between the normal complex and the early beat is shorter than the interval between normal complexes. This makes the rhythm irregular. If early beats occur only a few times a minute, we can describe it as occasionally irregular. If a number are seen, it can be described as very or frequently irregular. Frequent early beats are more likely to progress to very fast atrial, junctional, or ventricular rates (tachycardia). Worse, the rhythm may deteriorate into a quivering of the heart muscle called *ventricular fibrillation*.

Pause or Arrest of the Sinus Node

A rhythm can also appear occasionally or very irregular when the SA node fails to initiate an impulse (Figure 5-12). This is seen as a pause in the ECG rhythm.

Normally, the SA node initiates impulses, resulting in a repetitive cycle of P, QRS, and T waveforms.

When the SA node fails to initiate an impulse, there is a resulting absence of a P wave, QRS complex, and T wave.

Following the skipped beat, the SA node typically reinitiates impulses in the normal manner.

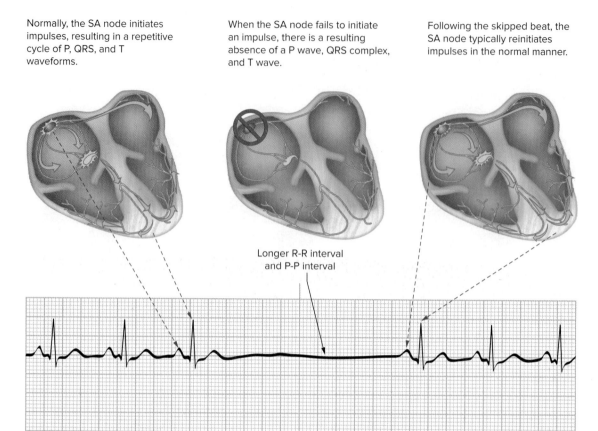

Longer R-R interval
and P-P interval

Figure 5-12
The failure of the sinus node to initiate the heart beat produces an irregular rhythm.

Typically, the rhythm leading up to the pause looks normal. Then there is suddenly an absence of a P wave, QRS complex, and T wave. This creates a gap or pause. Usually with this condition the SA node recovers and fires another impulse. If the SA node fails to fire, then an escape pacemaker from the atria, AV junction, or ventricles initiates an impulse. The prominent characteristic you see is a longer than normal R-R interval that occurs because of the dropped beat. If the SA node frequently fails to fire, there will be many pauses, causing the rhythm to look very or frequently irregular. Frequent pauses in the firing of the sinus node may cause so many dropped beats that the heart rate becomes bradycardic and cardiac output is reduced.

Slightly Irregular

A rhythm may appear to change only slightly as the P-P intervals and R-R intervals vary just a bit (Figure 5-13). It might be so slight that you do not easily see it, but you detect it when you measure the R-R or P-P intervals. The difference between slightly irregular and occasionally irregular rhythms is that slightly irregular rhythms are continuous throughout the ECG tracing; with occasional irregularity, the rhythm is mostly regular but there are areas where it is irregular.

One cause of a slightly irregular rhythm is when initiation of the heartbeat changes from site to site with each beat; instead of arising from the SA node, the pacemaker site shifts to different locations around the atria. As the site changes, so do the P-P and R-R intervals. Again, these differences can be slight.

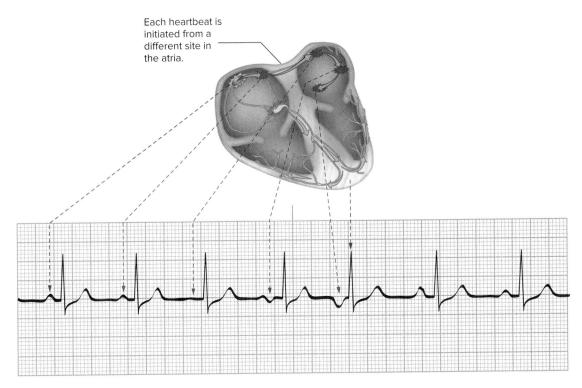

Each heartbeat is initiated from a different site in the atria.

Figure 5-13
A slightly irregular rhythm can occur when, instead of arising from the SA node, the heart's pacemaker shifts from site to site throughout the atria. The number of small boxes for each R-R and P-P interval varies slightly throughout the ECG strip.

Irregularity Caused by Sudden Changes in the Heart Rate

A rhythm may also appear irregular when the heart rate suddenly accelerates (Figure 5-14). We use the term *paroxysmal* to describe a dysrhythmia that has a sudden, rapid onset (and resolution). Because the R-R intervals or P-P intervals go from normal distances to shorter distances, it creates an irregular-looking rhythm. Any of type of tachycardia that arises outside the SA node can occur suddenly. To know this type of irregularity is present, we would have to observe its onset or resolution.

The rhythm can also appear irregular when the heart rate suddenly slows down. Because the R-R intervals or P-P intervals go from normal distances to more space between the respective waveforms, it creates an irregular-looking rhythm. A variety of conditions can lead to an abrupt slowdown of the heart rate. We will describe each in latter chapters. To know this type of irregularity is present, we would have to observe its onset or when the heart began to return to normal.

Irregularly (Totally) Irregular

An irregularly irregular or totally irregular rhythm, also called a *chaotically* or *grossly irregular rhythm*, is one in which there is no consistency to the irregularity. Looking across the tracing, you will find it hard to find an R-R interval that is the same as others (Figure 5-15). This creates the appearance of an irregularly irregular rhythm.

Totally irregular rhythms are typically one condition (which we describe in detail in Chapter 12). Do not worry about knowing the name of the dysrhythmia. Rather, let's

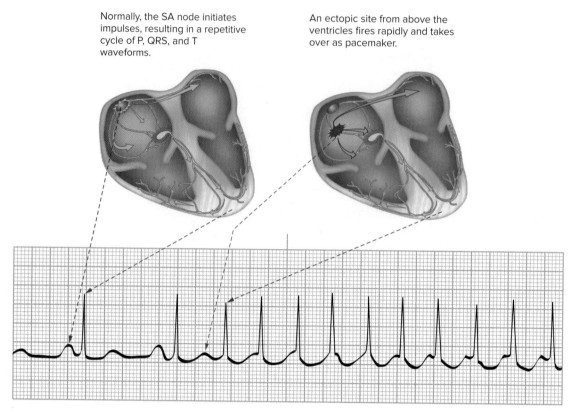

Normally, the SA node initiates impulses, resulting in a repetitive cycle of P, QRS, and T waveforms.

An ectopic site from above the ventricles fires rapidly and takes over as pacemaker.

Figure 5-14
Irregularity can also result from sudden rapid acceleration of the heart rate.

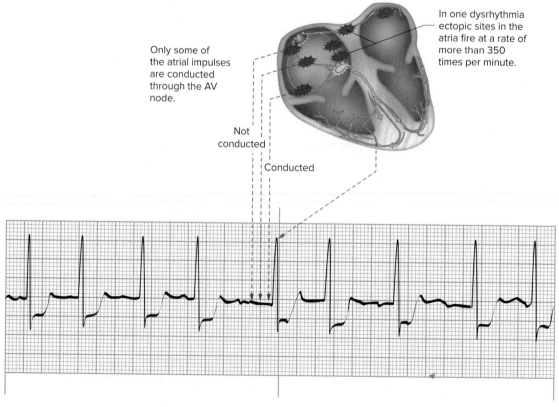

In one dysrhythmia ectopic sites in the atria fire at a rate of more than 350 times per minute.

Only some of the atrial impulses are conducted through the AV node.

Not conducted

Conducted

Figure 5-15
An irregularly irregular rhythm is seen as a complete lack of consistency in the space between the R waves.

just focus on how it produces a totally irregular rhythm. With this dysrhythmia, the atria fire at a rate in excess of 350 beats per minute and from multiple sites. These impulses bombard the AV node so rapidly that it cannot respond to all the impulses, allowing just some to get through to the ventricles. The R-R intervals vary from interval to interval because the impulses are conducted so haphazardly (as many are being generated). This creates the appearance of an irregularly irregular rhythm. Fast rhythms have short R-R intervals, making it difficult to identify the presence of this type of irregularity, so take your time while identifying regularity in fast rates.

Patterned Irregularity

Patterned irregularity, also called *regularly irregular*, is when the irregularity repeats over and over in a cyclic fashion. Early beats occurring at regular intervals, a condition where the heart rate speeds up and slows down in a cyclical way, and one type of conduction blockage that occurs in the AV node, are all characterized as being regularly irregular.

Early beats can be seen in a given pattern, such as those that occur in every other (second) complex (called *bigeminy*), every third complex (*trigeminy*), or every fourth complex (*quadrigeminy*), thus producing a patterned irregularity. These early beats can arise from the atria, AV node, or ventricles (Figure 5-16).

Normally, the SA node initiates impulses, resulting in a repetitive cycle of P, QRS, and T waveforms.

Premature beats occur when a site in the atria, AV node, or ventricles fires before the SA node is able to initiate an impulse.

Following the premature beat, the SA node typically reinitiates impulses in the normal manner.

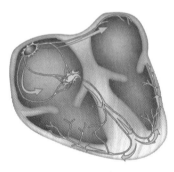
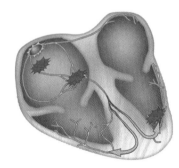
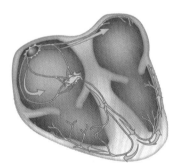

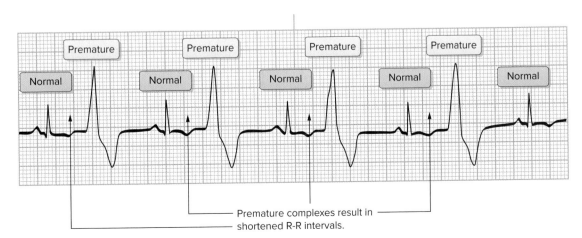

Premature complexes result in shortened R-R intervals.

Figure 5-16
Patterned irregularity can be caused by early beats that occur every other beat, every third beat, or every fourth beat.

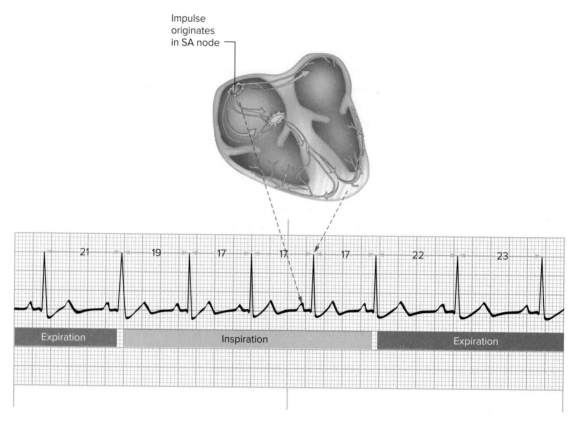

Figure 5-17
Patterned irregularity can be caused by the cyclic speeding up and slowing down of the heart rate.

We may also see patterned irregularity when the rhythm speeds up and slows down in a rhythmic fashion (Figure 5-17). The speeding up and slowing down of the heart rate often corresponds with the respiratory cycle and changes in intrathoracic pressure. The heart rate increases during inspiration and decreases during expiration. As a result of the speeding up and slowing down of the heart, you see a narrowing and then a widening of the R-R intervals. This cycle continually repeats.

Patterned irregularity can also occur because of a blockage in the heart's conduction system when the AV node is weakened and tires more and more with each conducted impulse until finally it is so tired that it fails to conduct the impulse through to the ventricles (Figure 5-18). This results in a dropped ventricular beat. This causes the R-R intervals to be longer wherever there is a dropped ventricular beat. With the dropped ventricular beat, the AV node has a chance to rest. The next impulse conducting through the AV node will be carried through faster, but each subsequent impulse will be delayed until finally another ventricular beat is dropped. This cycle repeats over and over, producing a recognizable pattern.

Irregularity Caused by Varying Conduction Ratios

Normally, each impulse initiated in the SA node is conducted to the ventricles. This results in one P wave preceding each QRS complex. The number of atrial waveforms (P waves) to ventricular waveforms (QRS complexes) is referred to as the *conduction ratio*. It is normally 1-to-1 (in other words there is one P wave for each QRS complex). The conduction ratio should remain constant throughout the tracing. However, if the conduction ratio changes, the R to R intervals will not be the same distance. For example, the ratio might be 1-to-1 but then change to 3-to-1, thus producing an irregular rhythm (Figure 5-19).

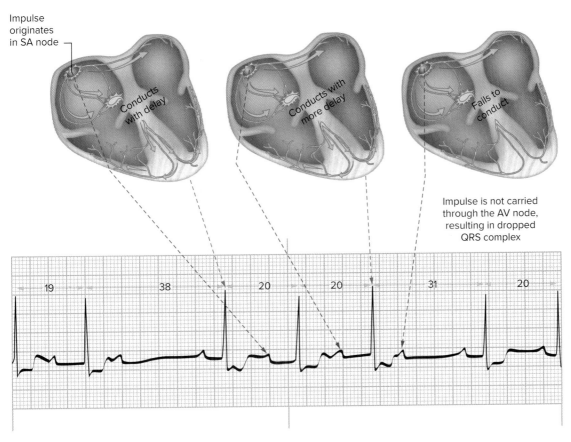

Figure 5-18
Patterned irregularity can be caused by a weakened AV node that delays the conduction of the impulse in a cyclic fashion.

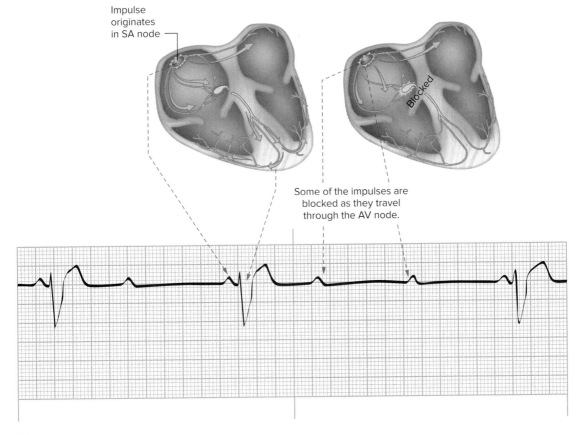

Figure 5-19
Where the atrial to ventricular conduction ratio varies, the distance between the R-R intervals changes, producing an irregular rhythm.

LO 5.1	• An initial step in analyzing an ECG rhythm is determining its regularity. This yields important information.
	• Irregular rhythms are considered abnormal.
	• A variety of conditions can produce irregular rhythms. Some of these conditions can be quite serious, so each irregular rhythm should be properly investigated.
LO 5.2	• The P-P and R-R intervals can be used to determine whether an ECG rhythm is regular or irregular. If the distance of the R-R intervals or P-P intervals is the same, the rhythm is regular. If the distance differs, the rhythm is irregular.
	• Several methods can be used to determine rhythm regularity, including using calipers, marking a paper with a pen, and counting the small squares between each R-R interval.
LO 5.3	• Irregularity may be occasionally irregular or very irregular. It may also be slightly irregular. It can also be patterned or irregularly (totally) irregular.
	• A pacemaker that changes location from site to site can produce a slightly irregular rhythm.
	• A very irregular rhythm is one that has many areas of irregularity. A common cause is frequent early beats.
	• A normal rate that suddenly accelerates to a rapid rate or a normal rate that suddenly slows produces irregularity in the rhythm.
	• A totally irregular rhythm is one in which there is no consistency to the irregularity.
	• Patterned irregularity is where the irregularity repeats over and over in a cyclic fashion.
	• Irregularity can also be seen in dysrhythmias that have a varying atrial-to-ventricular conduction ratio.

Assess Your Understanding

The following questions give you a chance to assess your understanding of the material discussed in this chapter. The answers can be found in Appendix A.

1. Normally, the heart beats in a   (LO 5.1)
   a. slightly irregular manner, producing a series of waveforms with each heartbeat.
   b. regularly irregular, rhythmic fashion, producing a P wave, QRS complex, and T wave with each heartbeat.
   c. regular, rhythmic fashion with periods of sudden heart rate acceleration and slowing.
   d. regular, rhythmic fashion, producing a P wave, QRS complex, and T wave with each heartbeat.

2. Determining the regularity of an ECG tracing (LO 5.2)

   a. is always done first.

   b. should be done only if the health care provider detects an irregular pulse.

   c. is considered an important analysis step.

   d. seldom reveals important information.

3. The distance between two consecutive R waves is called the (LO 5.1)

   a. R-R interval.

   b. RR segment.

   c. PR interval.

   d. ST segment.

4. An irregular rhythm (LO 5.1)

   a. is considered normal.

   b. is caused by only a few conditions.

   c. produces a fast heart rate.

   d. should be properly investigated.

5. Calipers can be used to analyze an ECG by (LO 5.2)

   a. placing one leg on a starting point and extending the second leg four large boxes to the right.

   b. establishing the distance between the first R-R or P-P interval and then adjusting the width of the calipers as you assess each consecutive interval.

   c. placing one leg on a starting point and extending the second leg to the next consecutive waveform being assessed, then comparing this distance across the ECG strip.

   d. moving the calipers across the tracing from each P wave to the next consecutive R wave.

6. The paper and pen method (LO 5.2)

   a. can be used only to measure regular rhythms.

   b. involves identifying an R wave that falls on a bold line.

   c. requires placing a mark over each R wave on the ECG tracing and then using calipers to determine the distance between them.

   d. involves placing two marks on the paper, the first above one of the R waves on the ECG tracing and the second above the next consecutive R wave.

7. To use the counting the small square method, you (LO 5.2)

   a. count the number of small squares between the peaks of two consecutive R waves and then compare that to the other R-R intervals.

   b. find an R wave that falls on a bold line and begin counting small squares.

   c. count the number of small squares in a 6-second section of the ECG tracing.

   d. divide by 1500 the number of QRS complexes in each 6-second section of the ECG tracing.

8. Which of the following methods for determining heart rhythm regularity is the easiest one, for which you can use readily available supplies?   (LO 5.2)

    a. counting small squares method

    b. caliper method

    c. paper and pen method

    d. rate calculator method

9. A/an _____ irregular rhythm is one that appears mostly regular, but from time to time you see an area where it is irregular.   (LO 5.3)

    a. occasionally

    b. patterned

    c. frequently

    d. slightly

10. Early beats   (LO 5.3)

    a. arise from the atria but not the other areas of the heart.

    b. cause the R-R intervals between the normal and early beat to be shorter than normal.

    c. cause the P-P intervals between the normal and early beat to be longer than normal.

    d. cause the rhythm to be chaotically irregular.

11. Failure of the SA node to initiate an impulse causes the   (LO 5.3)

    a. rhythm to be irregularly irregular.

    b. R-R interval between the normal and the beat following the dropped beat to be shorter than normal.

    c. rhythm to have patterned irregularity.

    d. R-R interval that occurs because of the dropped beat to be longer than normal.

12. A/an _____ irregular rhythm is one where the P-P intervals and R-R intervals vary only a bit.   (LO 5.3)

    a. slightly

    b. patterned

    c. occasionally

    d. frequently

13. A patterned irregular rhythm is seen when   (LO 5.3)

    a. the heart rate suddenly accelerates.

    b. there are many early beats.

    c. initiation of the heartbeat changes from site to site with each heartbeat.

    d. early beats occur every other (second) complex, every third complex, or every fourth complex.

14. With an irregular rhythm caused by a variable conduction ratio   (LO 5.3)

    a. the R-R intervals will occasionally be different durations.

    b. the number of atrial impulses preceding each QRS complex varies.

c. there is one P wave preceding each QRS complex.

d. the rhythm is totally irregular.

15. Which method of calculating the heart rate may be used with irregular rhythms?   (LO 5.2)

16. How do you know a heart rhythm is irregular when using either the calipers method or paper and pen method?   (LO 5.2)

Referring to the scenario at the beginning of this chapter, answer the following question.

17. The irregularity with this patient's ECG means there is   (LO 5.3)

a. no pattern to the irregularity.

b. occasional irregularity in the tracing.

c. an area in the tracing where the heart rate suddenly accelerates.

d. a mostly regular rhythm, but from time to time you see an area where it is irregular.

6 P Waves

Chapter Outline

Learning Outcomes

LO 6.1 State why it is important to analyze the P waves of an ECG tracing.

LO 6.2 Describe how to examine the P waves and describe their normal characteristics.

LO 6.3 Describe what is meant by the term *abnormal P waves* and identify the causes of each.

Case History

EMS is dispatched for a 75-year-old man residing in a nursing home complaining of "nonspecific weakness." Upon arrival, the paramedics note that the patient is ashen and diaphoretic, lying on the floor next to his bed. After introducing themselves, the paramedics perform an initial assessment and obtain a history of the events leading up to the patient's collapse. They learn that he apparently felt fine when he went to bed, but after rising he went to the bathroom, began to feel weak, and slumped to the floor.

The patient is now awake and alert but feels lightheaded when he tries to sit up. Vital signs show a BP of 140/90, a pulse of 36, and unlabored respirations of 18. His oxygen saturation is 94% on room air. The cardiac monitor is attached and reveals a slow rhythm with a narrow QRS complex. One paramedic examines the strip more closely while her partner establishes an intravenous line. The rhythm strip reveals the absence of normal P waves and instead has saw-toothed waveforms called *F waves*. A QRS complex occurs after every fourth F wave.

The patient is then transported to the hospital for definitive treatment.

6.1 Importance of Determining the P Waves

As mentioned previously, the heart normally beats in a regular, rhythmic fashion, producing a P wave, PR interval, QRS complex, ST segment, T wave and QT interval, with each heartbeat. There may also be an U wave. This cycle repeats over and over.

The presence or absence of P waves as well as their appearance can help us identify where the impulses that are stimulating the heart to beat are originating. This is another step in analyzing the ECG (Figure 6-1). Obviously, this is important to know. Also, the appearance of the P waves can tell us how well the impulses are being conducted through the atria as well as if there is atrial enlargement.

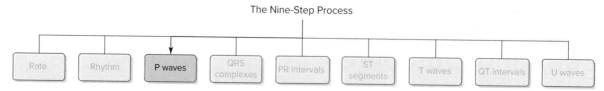

The Nine-Step Process

Rate | Rhythm | **P waves** | QRS complexes | PR intervals | ST segments | T waves | QT intervals | U waves

Figure 6-1
Evaluating the P waves is another important step used to analyze an ECG tracing.

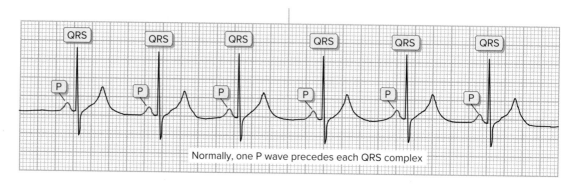

Normally, one P wave precedes each QRS complex

Figure 6-2
When assessing the P waves, look for a normal P wave to precede each QRS complex.

6.2 Examining the P Waves

The initial assessment for P waves can be done quickly; simply determine whether or not they are present. Questions to ask while analyzing the tracing are, "Is there a uniform P wave preceding each QRS complex (Figure 6-2)?" and "Do the P waves appear normal?" (Alternatively, we can ask, "Is every P wave followed by a QRS complex?" or "Are there any P waves without a QRS complex?")

In Chapter 4 we focused on determining the heart rate whereas in Chapter 5 we addressed identifying regularity. In those two steps, the lead (or view) used does not affect how the ECG tracing is analyzed. With the next steps of analyzing an ECG rhythm, however, lead selection does affect the appearance of the waveforms, so you need to be familiar not only with how normal waveforms look but also with how the waveforms look in the different leads. For our discussion in this chapter, we focus primarily on how the P waves appear in lead II. Another point to mention is that our discussion regarding dysrhythmias in this chapter is to give you a sense of the normal P waves and why changes occur. This is not intended to be an in-depth discussion of any dysrhythmia or condition but rather a description of what produces the changes you see in the P waves. Do not worry about learning the names of any of the dysrhythmias at this point.

Normal P Waves

As discussed earlier, the P wave is the first waveform at the start of the cardiac cycle. It begins with its movement away from the baseline and ends on its return to the baseline. Right atrial depolarization produces the initial portion of the P wave whereas left atrial depolarization produces the terminal portion of the P wave. To examine the P waves, we look closely at their characteristics, especially their deflection, location, and morphology.

Characteristics of Normal P Waves	
Location	One P wave precedes each QRS complex
Amplitude	0.5 to 2.5 mm
Duration	0.06 to 0.10 seconds
Morphology	A P wave is usually rounded and upright (or inverted or biphasic in certain leads). It has a more abrupt upslope with a gradual downslope, producing an asymmetrical appearance.

As stated earlier, to examine the P waves (Figure 6-3), first look to see if there is a P wave preceding each QRS complex. Then look at each P wave for uniformity. Does each wave look the same? Then measure the wave's duration and amplitude. Identify where it begins and then where it ends. Count the number of small boxes between the beginning and end. Then look at where the P wave first begins to leave the isoelectric line and then where it peaks. Count the number of small boxes between the baseline and the peak.

If the morphology of a P wave is normal—for example, in lead II it is upright and round but slightly asymmetric—and if it precedes each QRS complex, this electrical impulse most likely originated in the SA node and was carried through the atria and AV node in a normal manner. Normal (regular) sinus rhythm is the rhythm with which we compare all others. All dysrhythmias that originate from the SA node should have normal-looking P waves. However, in faster tachycardias that arise from the SA node, the P wave can sometimes be buried in the T wave of the preceding beat and may have greater amplitude. For the sake of completeness, normal P waves may deflect negatively in lead aVR and appear biphasic in lead V1.

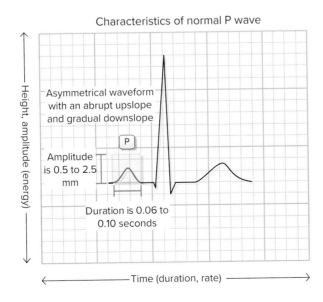

Figure 6-3
P wave characteristics.

6.3 Identifying and Characterizing Abnormal P Waves

Abnormal P waves are those that look different, are inverted, are absent, follow the QRS complex, or are not followed by a QRS complex (Figure 6-4).

A variety of conditions can cause P waves to look different. In some cases, the P wave originates in the SA node, but the atria are altered or damaged, resulting in abnormal conduction of the atrial impulse. In other cases, the P wave appears different because the pacemaker site originates from a site other than the SA node.

Peaked, Notched, or Enlarged Sinus P Waves

Tall (greater than 2.5 mm), narrow, and symmetrically peaked P waves may be seen with increased right atrial pressure and right atrial enlargement (Figure 6-5). This is called **P pulmonale.** We may also see abnormally tall P waves in sinus tachycardia (a faster than normal heart rate that originates from the SA node) and **hypokalemia** (a low level of potassium in the blood). Conversely, waves with decreased amplitude can be seen with **hyperkalemia** (a high level of potassium in the blood).

Notched or wide (prolonged—greater than 0.12 seconds in duration) P waves may be seen in increased left atrial pressure and left atrial enlargement. This is called

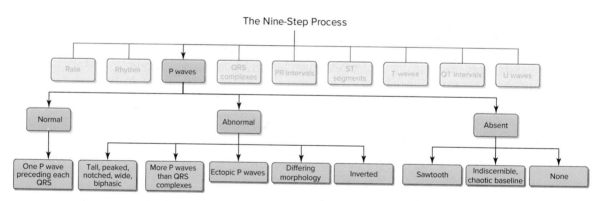

Figure 6-4
Algorithm for normal and abnormal P waves.

Figure 6-5
Types of abnormal atrial waveforms include tall, peaked, notched, wide, or wide biphasic sinus P waves.

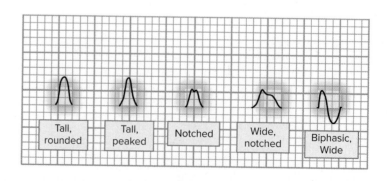

P mitrale. These types of P waves may also be seen when there is a delay or block in the movement of electrical impulses through the interatrial conduction tract between the right and left atria, resulting in one atrium depolarizing before the other.

As mentioned earlier, biphasic P waves may be normal in lead V_1. These waves have an initial positive deflection (reflecting right atrial depolarization) followed by a negative deflection (reflecting left atrial depolarization). However, if atrial enlargement is present, the amplitude or width of the initial or terminal portion of the biphasic P wave will change. Right atrial enlargement leads to an increase in the amplitude of the first part of the P wave. Left atrial enlargement leads to an increase in the amplitude and width of the terminal portion of the P wave.

Atrial P Waves

Some dysrhythmias originate outside the SA node. Those that arise from the atrial tissue or in the internodal pathways are referred to as atrial dysrhythmias. The impulses produce P waves (called *P prime* or *P'*) that look different than the sinus P waves because the impulses arise from outside the SA node.

How the P' wave looks depends on where it originates and the direction the electrical impulse travels through the atria. As a rule, the closer the site of origin is to the SA node, the more it looks like a normal P wave. If the ectopic pacemaker arises from the upper- or middle-right atrium, depolarization occurs in a normal direction—from right to left and then downward. If it is initiated from the upper-right atrium, the P' wave should be upright in lead II, resembling a normal sinus P wave. A P wave initiated from the middle of the right atrium is less positive than one that originates from the upper-right atrium.

If the impulse arises from the lower-right atrium near the AV node or in the left atrium, depolarization occurs in a **retrograde** direction (meaning the impulse conducts upward [backward] through the heart instead of downward)—from left to right and then upward, resulting in the P' wave being inverted in lead II.

P' Waves Seen with Early Beats Arising from the Atria

Early beats that arise from the atria or internodal pathways, before the SA node has a chance to fire, produce P' waves that have a different morphology (appearance) than the other normal beats (Figure 6-6). How they look depends on where they originate (as described previously). The P' waves may be obscured or buried in the T wave of the preceding beat (resulting in a short P'-P' interval as described later in this chapter). This causes the T wave to appear different than those following the other beats. The T wave may look peaked, notched, or larger than normal, suggesting that the P' is buried in the T wave.

You may see an atrial P' wave that is not followed by a QRS complex. This is due to a nonconducted or blocked early beat that arises from the atria (Figure 6-7).

P' Waves Seen with Tachycardia Arising from the Atria

Tachycardia can arise from the atria, resulting in rapid depolarization that overrides the SA node (Figure 6-8). It produces a heart rate of between 150 to 250 beats per

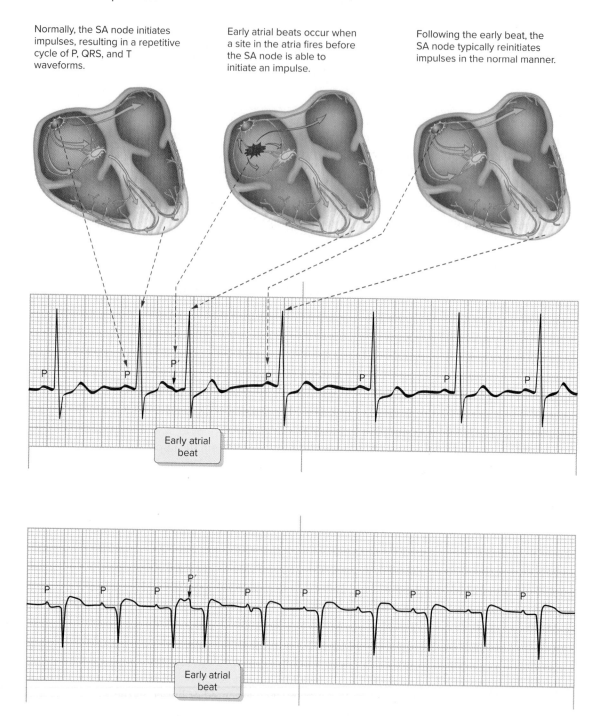

Normally, the SA node initiates impulses, resulting in a repetitive cycle of P, QRS, and T waveforms.

Early atrial beats occur when a site in the atria fires before the SA node is able to initiate an impulse.

Following the early beat, the SA node typically reinitiates impulses in the normal manner.

Early atrial beat

Early atrial beat

Figure 6-6

P′ waves seen with early beats arising from the atria differ from sinus P waves and may be buried in the preceding T wave.

minute. The P wave in tachycardia that arises from the atria has a different morphology than normal P waves. The P′ wave is likely to be buried in the T wave of the preceding beat. For this reason, we refer to tachycardias with normal QRS complexes and no discernable P waves as supraventricular tachycardias (meaning that the tachycardia arises from above the ventricles).

Varying Atrial P Waves

P waves coming from one site will look similar. P′ waves that continuously change in appearance indicate that the impulse is arising from different locations in the

Normally, the SA node initiates
impulses, resulting in a repetitive
cycle of P, QRS, and T
waveforms.

Early atrial beats fire and
produce a P′ wave but the
impulse does not reach
the ventricles.

Following the early beat, the
SA node typically reinitiates
impulses in the normal manner.

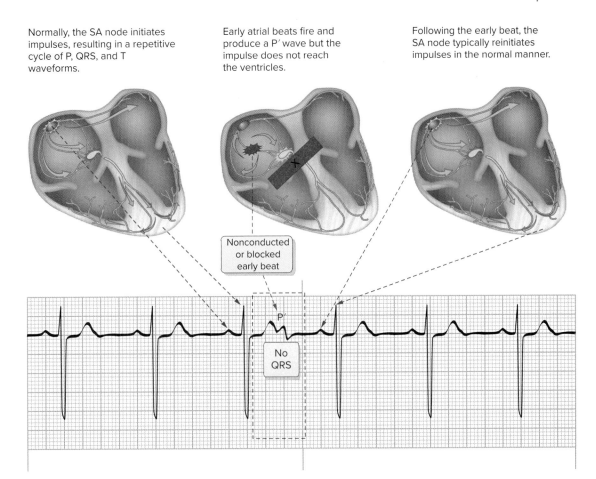

Figure 6-7
With a blocked early atrial beat, there is a P′ wave but no QRS complex.

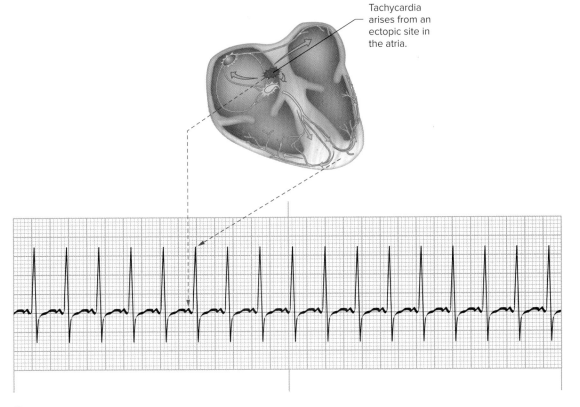

Figure 6-8
If seen, the morphology of the P′ waves associated with tachycardia that arises from the atria is different from normal beats.

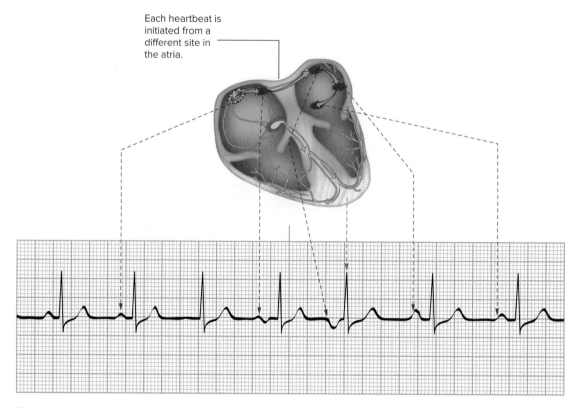

Each heartbeat is initiated from a different site in the atria.

Figure 6-9
Varying P′ waves occur when, instead of arising from the SA node, the heart's pacemaker shifts from site to site throughout the atria.

atria (Figure 6-9). In two dysrhythmias that arise from the atria and are discussed in Chapter 12, the pacemaker site shifts transiently from beat to beat from the SA node to other latent (hidden) pacemaker sites in the atria and AV junction. This produces their most distinguishing characteristic—a change in P′ wave morphology from beat to beat.

Flutter and Fibrillatory Waves

The appearance of the atrial waveforms is very different in two other dysrhythmias that arise from the atria. In one dysrhythmia, the atrial impulses appear as characteristic flutter waveforms whereas in the other there is a characteristic chaotic-looking baseline preceding each QRS complex.

Saw-Toothed Waveforms

In one dysrhythmia that arises from the atria, the atria fire at a rate of between 250 and 350 beats per minute (Figure 6-10). Normal P waves are absent, and instead, we see flutter waves (F waves). They are often described as a saw-toothed pattern.

With this dysrhythmia, the atrial-to-ventricular conduction ratio of impulses is usually 2:1, 3:1, or 4:1. An atrial-to-ventricular conduction ratio of 1:1 is rare because the ventricles are unable to repolarize quickly enough to respond to each of the atrial impulses. The conduction ratios may be constant or variable, meaning that over the course of minutes the conduction ratio changes.

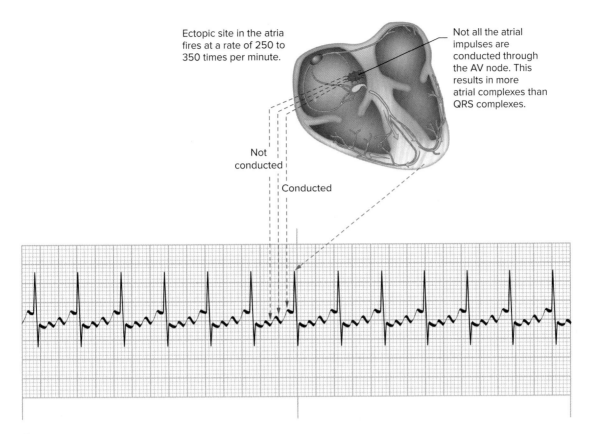

Ectopic site in the atria fires at a rate of 250 to 350 times per minute.

Not all the atrial impulses are conducted through the AV node. This results in more atrial complexes than QRS complexes.

Not conducted

Conducted

Figure 6-10
A rapid atrial rate of between 250 and 350 beats per minute from one site produces a saw-toothed pattern of F waves.

Chaotic-Looking Baseline

In another dysrhythmia that arises from the atria, the atria fire faster than 350 beats per minute (Figure 6-11). This produces a fibrillatory waveform, reflective of what is occurring with the atrial muscle. This is best described as a chaotic-looking baseline with no discernable P waves (in other words there is an absence of identifiable P waves). Instead of P waves, we see uneven baseline off **waves**, waves that occur at a rate of greater than 350 times per minute. In some areas along the baseline you may see what looks like a P wave or two but nothing that really appears the same throughout the tracing.

Inverted and Absent P Waves

In addition to the inverted P′ waves arising from the lower-right atrium near the AV node or in the left atrium, inverted P′ waves can also arise from the AV junction (Figure 6-12). The inverted P′ wave occurs when the electrical impulse travels upward through the AV junction into the atria, causing retrograde (backward) atrial depolarization. It can occur before, during (buried in QRS), or after the QRS complex. We describe those buried in the QRS complex as absent. This may change the morphology of the QRS complex.

P waves are also absent in dysrhythmias that arise from the ventricles. This occurs because the ventricles are below the AV node and do not retrogradedly depolarize the atria.

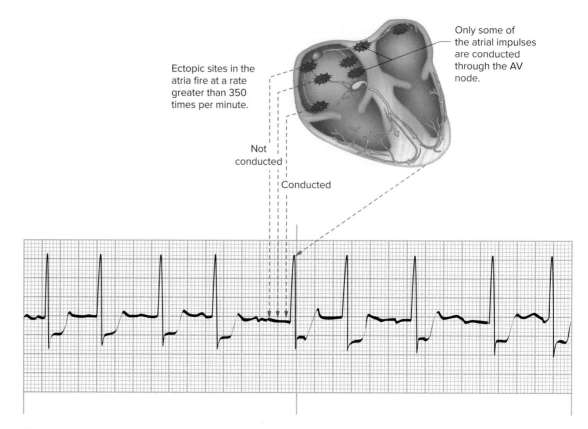

Ectopic sites in the atria fire at a rate greater than 350 times per minute.

Only some of the atrial impulses are conducted through the AV node.

Not conducted

Conducted

Figure 6-11
In one atrial dysrhythmia, the atria fire at a rate in excess of 350 beats per minute from multiple sites, causing a chaotic-looking baseline with no discernable P waves.

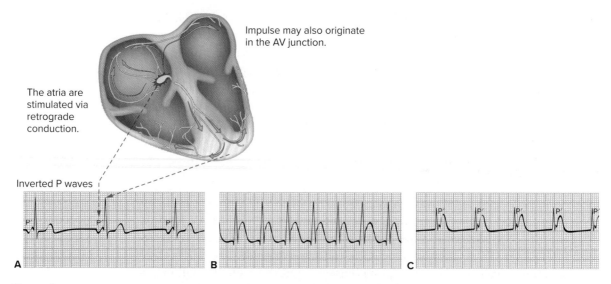

Impulse may also originate in the AV junction.

The atria are stimulated via retrograde conduction.

Inverted P waves

A   B   C

Figure 6-12
Impulses arising from the lower-right atrium near the AV node, in the left atrium, or from the AV junction have inverted P′ waves that can occur before, during (buried in QRS), or after the QRS complex. (A) Inverted P waves preceding the QRS complexes. (B) Absent P waves. (C) P waves that follow the QRS complexes.

More P Waves than QRS Complexes

Normally, there is only one P wave preceding each QRS complex. However, in some dysrhythmias we may see more P waves than QRS complexes. This indicates that the impulse was initiated in the SA node (or other ectopic sites in the atria) but was blocked and did not reach the ventricles. The most common cause for this is a block in the AV node or junction. The term we use to describe this blockage is AV heart block. We can also see this with blocked early beats that arise from the atria as mentioned earlier in this chapter.

P Waves Seen with AV Heart Block

With the most severe cases of AV heart block, only some, or worse yet none, of the electrical impulses initiated by the SA node reach the ventricles (Figure 6-13). This results in the QRS complex being dropped, and there are more P waves than QRS complexes. The P waves should all be normal because the impulses are initiated from the SA node. They just are not all followed by a QRS complex.

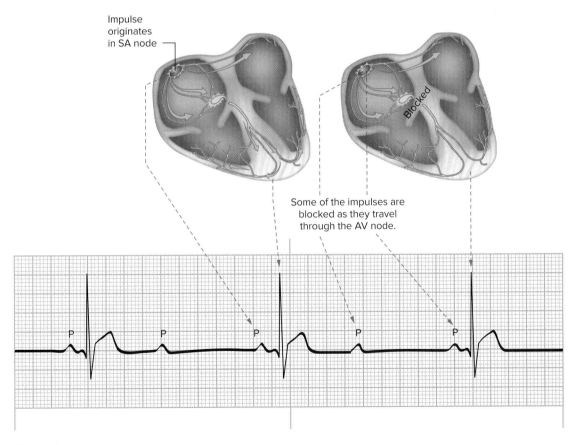

Impulse originates in SA node

Blocked

Some of the impulses are blocked as they travel through the AV node.

Figure 6-13
Impulses are initiated in the SA node but do not all reach the ventricles. This results in more P waves than QRS complexes.

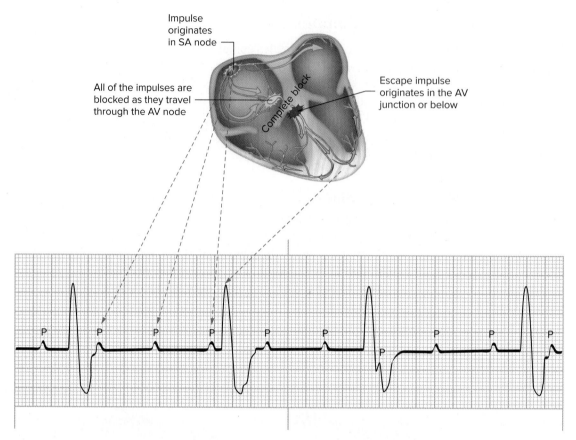

Impulse
originates
in SA node

All of the impulses are
blocked as they travel
through the AV node

Complete block

Escape impulse
originates in the AV
junction or below

Figure 6-14
With the most severe form of AV node blockage, there are more P waves than QRS complexes. Further, the
P waves appear to march through the QRS complexes.

Depending on the degree of AV heart block, the atrial-to-ventricular conduction ratio may be constant, usually 2:1, 3:1, or 4:1 (atrial to ventricular beats); they may change (due to changing conduction ratios or start and stop); or the atrial and ventricular impulses may be unrelated.

In the most severe form of AV heart block, the AV node is completely blocked (Figure 6-14). The atria are stimulated to contract by an impulse that originates in the SA node. For this reason, the atrial rate should be within a normal range. The ventricles are stimulated to contract by an escape pacemaker that arises from somewhere below the AV node, resulting in a slower heart rate. There should be more P waves than QRS complexes because the SA node has an inherently faster rate than the AV junction or ventricles. Also, because the pacemaker sites (SA node and ventricles) are firing independently of each other, the P waves seem to "march through" the QRS complexes.

The P waves associated with blocked early beats that arise from the atria (described earlier in this chapter) can be confused with AV heart block. A key difference between the two is that with early beats, the P′-P interval is shorter whereas, with AV heart block, the P-P interval remains constant. With early atrial beats, the P′ wave also looks different than the other P waves whereas, with AV heart block, the P waves look alike.

Key Points 🌡️

LO 6.1	• Normally, the heart beats in a regular, rhythmic fashion, producing a P wave, PR interval, QRS complex, ST segment, T wave, and QT interval with each heartbeat. This cycle repeats over and over.
	• Another important step used to analyze an ECG rhythm is examining the P waves.
LO 6.2	• The P wave is the first deflection from the baseline at the beginning of the cardiac cycle.
	• A normal P wave (upright and round in lead II) preceding each QRS complex indicates that the electrical impulse likely originated in the SA node and was carried through the atria in a normal manner.
	• The amplitude of a normal P wave is 0.5 to 2.5 mm high, and the duration is 0.06 to 0.10 seconds.
LO 6.3	• P waves that originate in the SA node but conduct abnormally through altered, damaged atria will have an altered morphology.
	• P waves that result from a pacemaker site outside the SA node will have an altered morphology.
	• Tall and symmetrically peaked P waves may be seen with increased right atrial pressure and right atrial enlargement.
	• Notched or wide (enlarged) P waves may be seen with increased left atrial pressure and left atrial enlargement.
	• Also, biphasic P waves having increased amplitude or width of the initial or terminal portion indicate atrial enlargement.
	• Impulses that arise from the atria produce P prime (P′) waves that look different than the sinus P waves.
	• P′ waves that arise from the upper-right atrium should be upright in lead II, resembling a normal sinus P wave.
	• A P′ wave that arises from the middle of the right atrium is less positive than one that originates from the upper-right atrium.
	• A P′ wave that arises from the lower-right atrium near the AV node or in the left atrium results in atrial depolarization occurring in a retrograde direction. This produces an inverted P′ wave in lead II.
	• Early beats can arise from the atria or internodal pathways. The P′ wave in these early beats has a different morphology than normal beats.
	• Tachycardia that arises from the atria produces a heart rate of between 150 and 250 beats per minute. It has P′ waves, which look different than normal beats. These P′ waves are often buried in the T wave of the preceding beat.
	• P waves that continuously change in their appearance indicate that the pacemaker is originating from a different site with each heartbeat.
	• In one dysrhythmia that arises from the atria, the atria fire at a rate of between 250 and 350 beats per minute. Normal P waves are absent, and, instead, flutter waves (waves) are present. These are often described as having a sawtoothed appearance.

- In another dysrhythmia that arises from the atria, there are no discernable P waves. Instead, there is a chaotic-looking baseline off waves preceding the QRS complexes.

- Impulses that arise from the AV junction produce an inverted P′ wave that may immediately precede, or occur during or following, the QRS complex.

- Normally, there is only one P wave preceding each QRS complex. More P waves than QRS complexes indicate that the impulse was initiated in the SA node or atria but was blocked and did not reach the ventricles. The most common causes for this are AV heart block and blocked early beats that arise from the atria.

Assess Your Understanding

The following questions give you a chance to assess your understanding of the material discussed in this chapter. The answers can be found in Appendix A.

1. To analyze the atrial activity of an ECG you should assess the   (LO 6.1)
   a. regularity.
   b. P waves.
   c. PR intervals.
   d. heart rate.

2. The amplitude of the P wave normally does not exceed _____ mm high.   (LO 6.2)
   a. 1.0
   b. 2.5
   c. 3.0
   d. 0.5

3. An upright, round P wave (in lead II) that precedes each QRS complex indicates that the electrical impulse originated in the   (LO 6.1)
   a. AV node and was carried through the atria in a retrograde manner.
   b. ventricles and was carried through the atria in the normal manner.
   c. SA node and was carried through the AV node in a retrograde manner.
   d. SA node and was carried through the atria in a normal manner.

4. The characteristic considered normal (in lead II) is a/an   (LO 6.2)
   a. P wave that follows the QRS complex.
   b. inverted P wave.
   c. peaked or notched P wave.
   d. rounded and upright P wave.

5. Which of the following is best evaluated by choosing the appropriate lead?   (LO 6.2)
   a. heart rate
   b. regularity
   c. P waves
   d. PR interval

6. With increased left atrial pressure and left atrial enlargement, the P wave is normally   (LO 6.3)
   a. inverted.
   b. notched or wide.
   c. pointed.
   d. biphasic with the initial and terminal deflections being the same amplitude and width.

7. Which of the following will produce P waves that look different than sinus P waves?   (LO 6.3)
   a. enlarged or damaged atria
   b. AV heart block
   c. tachycardia arising from the sinus node
   d. all of the above

8. An impulse that arises closer to the SA node   (LO 6.3)
   a. has a very different appearance from a normal P wave.
   b. is inverted.
   c. looks more like a normal P wave.
   d. is biphasic or notched.

9. Early beats that arise from the atria   `(LO 6.3)
   a. occur only in the right atrium.
   b. always have an inverted P wave.
   c. may have P′ waves buried in the T wave of the preceding beat.
   d. have P waves that look the same as P waves that arise from the SA node.

10. With tachycardia that arises from the atria,   (LO 6.3)
    a. the P′ wave is typically notched or widened.
    b. the P′ wave looks different than P waves that arise from the SA node.
    c. the P′ wave is hidden within the QRS complex.
    d. there are more P′ waves than QRS complexes.

11. An atrial pacemaker site that changes from location to location has   (LO 6.3)
    a. P waves that continually change in appearance.
    b. one pacemaker site from which the impulses arise.
    c. inverted P′ waves.
    d. a chaotic baseline preceding each QRS complex.

12. When the atria fire faster than 350 beats per minute, the P waves   (LO 6.3)
    a. are indiscernible; instead, there is a chaotic-looking baseline.
    b. appear in a saw-tooth pattern.
    c. are inverted (in lead II).
    d. are buried in the T wave of the preceding beat.

13. Dysrhythmias that arise from AV junctional tissue have _____ P' waves. (LO 6.3)
    a. upright
    b. inverted
    c. biphasic
    d. peaked

14. With ventricular dysrhythmias, the P' waves (LO 6.3)
    a. change in appearance.
    b. have a saw-tooth appearance.
    c. immediately precede the QRS complexes.
    d. are absent.

15. Which of the following has more P waves than QRS complexes? (LO 6.3)
    a. a dysrhythmia where the pacemaker changes from site to site with each beat
    b. early beats that arise from the atria
    c. AV heart block
    d. dysrhythmias that arise from the AV junction

16. What does a sinus P wave represent? (LO 6.2)

17. What is a P prime (P') wave? (LO 6.3)

18. If a P wave originates from the SA node but the atria are altered or damaged, how might the P wave look? (LO 6.3)

Referring to the scenario at the beginning of this chapter, answer the following questions.

19. From where did the patient's dysrhythmia originate? (LO 6.3)
    a. sinus node
    b. AV junctional tissue
    c. ventricles
    d. atria

20. The firing rate of the F waves is _____ beats per minute. (LO 6.3)
    a. 36
    b. 72
    c. 108
    d. 144

7 QRS Complexes

©rivetti/Getty Images

Courtesy Physio-Control

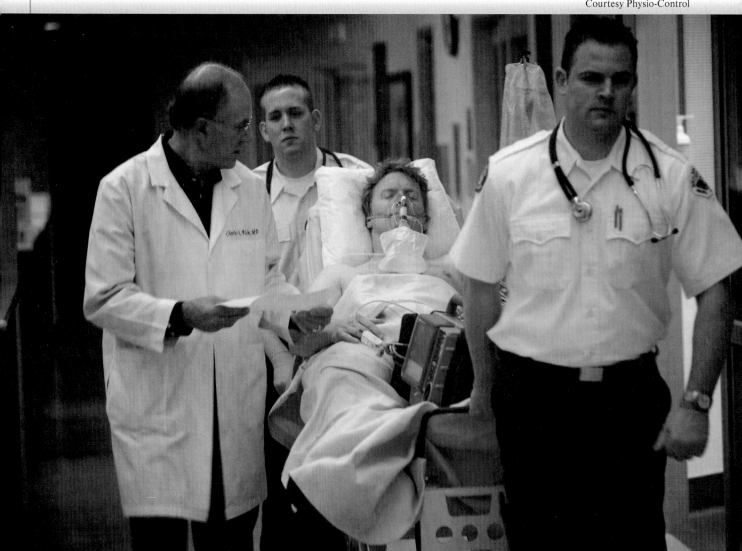

Learning Outcomes

LO 7.1 State why it is important to analyze the QRS complexes of an ECG tracing.

LO 7.2 Describe the characteristics of the normal QRS complex and recall how to measure it.

LO 7.3 Describe the variations in the QRS configuration.

LO 7.4 Recognize that the appearance of the QRS complexes differs in the various ECG leads.

LO 7.5 Recall when normal QRS complexes are seen.

LO 7.6 Describe what is meant when we say "abnormal QRS complexes."

Case History

Emergency medical service (EMS) is called to the home of a 39-year-old man complaining of chest pain. Upon arrival, they find the patient lying on the couch. He is awake and alert. He states that he has a dull ache in his chest. His vital signs are normal, and the ECG leads are quickly attached and show a normal sinus rhythm. The paramedics apply oxygen via nasal cannula to the patient and establish an intravenous line. The patient's chest pain resolves after the application of oxygen and the administration of aspirin and one sublingual nitroglycerin.

En route to the hospital, the paramedics notice the occurrence of early beats. They are wide and bizarre in configuration, without associated P waves, and each complex appears different than the others. The patient continues to be pain free, but the early beats appear to be coming more frequently. The paramedics contact medical control at the receiving hospital and describe the rhythm. The physician approves their request to deliver an antidysrhythmic agent and the early beats cease. The patient is delivered to the hospital without further incident.

7.1 Importance of Examining the QRS Complexes

As part of analyzing an ECG, we examine and measure the specific waveforms, segments, and intervals. In this chapter, we discuss assessment of the QRS complexes (Figure 7-1).

Assessing the QRS complexes is important because it tells us how well the impulses are traveling through the ventricles or if the impulses are originating from the ventricles. This is important to know as delays or interruption of conduction through the ventricles or impulses that arise from the ventricles can lead to serious or even life-threatening conditions.

The initial assessment for QRS complexes can be done quickly. Questions to ask include, "Are the QRS complexes present and within normal limits? Do they appear normal in their configuration? Do they all look alike?" (Figure 7-2.)

7.2 Examining the QRS Complexes

To examine the QRS complexes, we look closely at their characteristics, especially their location, configuration, and deflection (Figure 7-3).

As discussed, the QRS complex is the waveform immediately following the PR interval. Its starting point is where the first wave of the complex starts to move away (sharply or gradually) from the baseline. The QRS complex ends at the point where the last wave of the complex starts to flatten (sharply or gradually) at, above, or

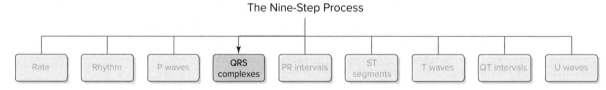

Figure 7-1
Assessing the QRS complexes is another part of The Nine-Step Process algorithm.

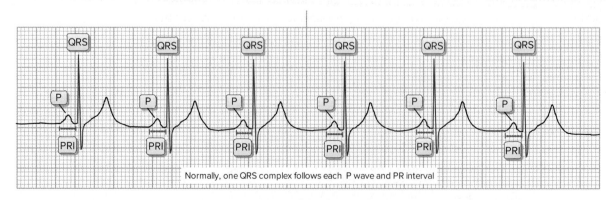

Figure 7-2
When assessing the QRS complexes, look for a normal QRS complex to follow each P wave and PR interval.

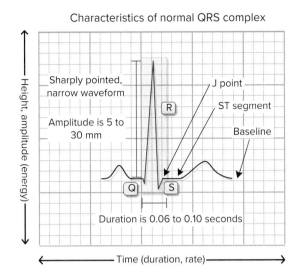

Characteristics of normal QRS complex

Height, amplitude (energy)

Sharply pointed, narrow waveform

Amplitude is 5 to 30 mm

R

Q

S

J point

ST segment

Baseline

Duration is 0.06 to 0.10 seconds

Time (duration, rate)

Figure 7-3
The QRS complex consists of the Q wave, R wave, and S wave. The J point is where the S wave connects to the ST segment. The ST segment is the flat line that connects the S wave with the T wave.

below the baseline. The QRS complex is much bigger than the P wave because depolarization of the ventricles involves considerably larger muscle mass than depolarization of the atria. The QRS complex characteristically looks thinner than the other parts of the ECG because the ventricles depolarize so fast. This minimizes contact time between the stylus and the ECG paper.

Characteristics of Normal QRS Complexes	
Location	A QRS complex follows the PR interval.
Amplitude	5 to 30 mm (but it differs with each lead employed).
Duration	0.06 to 0.10 seconds. The duration is measured from the beginning of the Q wave (or the beginning of the R wave if the Q is not present) to the end of the S wave.
Morphology	The QRS complex includes the Q, R, and S waves.

The elements of the QRS complex are described below:

- *Q wave.* The Q wave is the first negative deflection from the baseline following the P wave. It is always negative. In some cases, the Q wave is absent. The normal duration of the Q wave in the limb leads (when present) is less than 0.04 seconds. The amplitude is normally less than 25% of the amplitude of the R wave in that lead.

- *R wave.* The R wave is the first positive, triangular deflection in the QRS complex. It follows the Q wave (if it is present).

- *S wave.* The S wave is the first negative deflection that *extends below* the baseline in the QRS complex following the R wave.

The point at which the QRS complex meets the ST segment is called the *junction* or *J point.*

The Q wave represents depolarization of the interventricular septum. It is activated from left to right. The R and S waves represent simultaneous depolarization of the right and left ventricles. The QRS complex mostly represents the electrical activity occurring in the left ventricle because of its greater muscle mass.

Generally speaking, men have higher QRS complex amplitude than women, and young people have higher amplitudes than the elderly. The precordial leads have higher amplitude than the limb leads because the electrodes are closer to the heart.

Measuring QRS Complexes

To measure the width (duration) of a QRS complex, first identify the complex with the longest duration and the most distinct beginning and ending (Figure 7-4). Normally, the QRS complex will not extend beyond two and a half small boxes (0.10 seconds in duration). Start by finding the beginning of the QRS complex. This is the point where the first wave of the complex, which will either be the Q wave or the R wave, begins to deviate from the baseline. Then measure to the point where the last wave of the complex transitions into the ST segment. Typically, it is where the S wave or R wave (in the absence of an S wave) begins to level out (flatten) at, above, or below the baseline. This is considered the end of the QRS complex.

To make measuring easier, try to find a QRS complex where the waveform begins on one of the bold or thin lines. Then count the number of small squares (boxes) between the beginning and the end of the complex. If none of the Q or R waves falls squarely on a line, then you can use calipers or the paper and pencil technique. Start by placing the first point at the beginning of the complex; then place the second point at the end of the complex. This establishes the distance from the beginning to the end. Now slide the calipers or paper (with lines you have drawn) horizontally across the tracing so that the first point lines up with a bold or thin line. Now count the number of squares between it and the second point.

Figure 7-4
Measure the QRS complex from where the Q or R wave starts to where the S wave ends (J point).

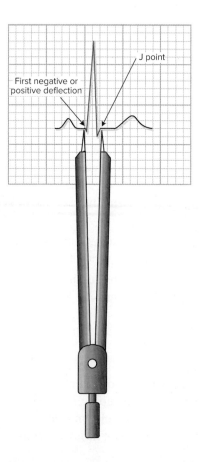

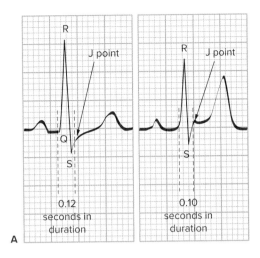

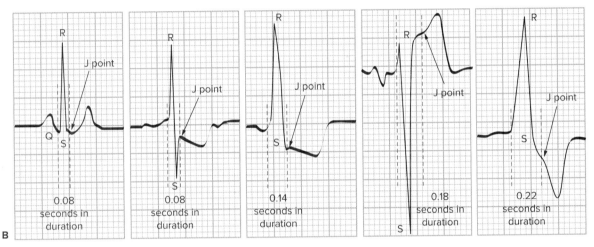

Figure 7-5
Measuring the QRS complex. (A) These two QRS complexes have easy to see J points. (B) These QRS complexes have less-defined transition, making measurement of the QRS complex more challenging.

Determining where the QRS complex ends can be difficult because you don't always see a clear transition with nice, straight lines (Figure 7-5). Instead, particularly when there is depression or elevation of the ST segment, you need to look for a small notch, slope, or other movement that suggests an alteration of electrical flow. Include in your measurement the entire S wave, but don't let it overlap into the ST segment or the T wave. It is helpful to look for the end of the QRS complex in as many leads as possible as you can sometimes see it in one lead but not another. In some ECG tracings, you have to use your best educated guess and common sense to conclude what is the duration of the QRS complex.

7.3 Variations in the QRS Configuration

The QRS complex can take various forms (Figure 7-6). Although we said it consists of positive (upright) deflections called R waves and negative (inverted) deflections called Q and S waves, all three waves are not always seen. Also, the waveforms will look different depending on which lead is used. If the R wave is absent, the complex is called a QS complex. Likewise, if the Q wave is absent, the complex is called an RS complex. Waveforms of normal or greater than normal amplitude are denoted

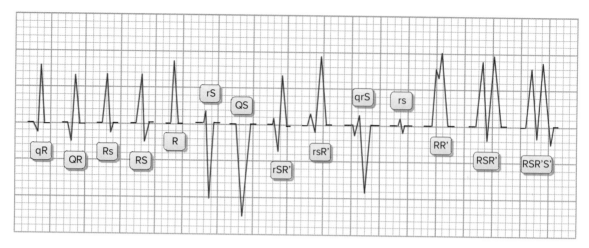

Figure 7-6
Variations in QRS configurations.

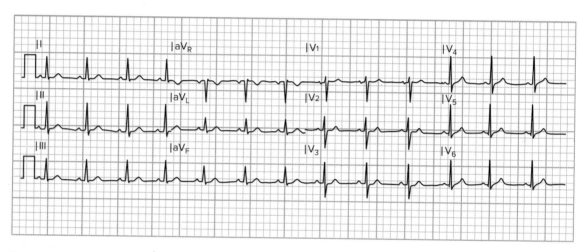

Figure 7-7
Direction of QRS waves in various leads.

with a large case letter, whereas waveforms less than 5 mm in amplitude are denoted with a small case letter (e.g., "q," "r," "s"). Later, when we discuss myocardial infarction, this will become important as it helps us determine whether a Q wave is significant.

Although there is only one Q wave, there can be more than one R wave and S wave in the QRS complex. A positive impulse immediately following the R wave is called R prime (R′). A double positive impulse immediately following the R wave is called double R prime (R″). A negative impulse immediately following the S wave is called S prime (S′). A double negative impulse immediately following the S wave is called double S prime (S″). This discussion becomes more significant in Chapter 19 when we discuss bundle branch block.

7.4 QRS Complexes in Different Leads

Remember, from our discussion in Chapter 2, the appearance of the QRS complex differs depending on which ECG lead is viewed (Figure 7-7). In leads I and II, the deflection of the QRS complex is characteristically positive. In leads III, aV$_L$, and aV$_F$, the QRS complex deflects positively but it may also be biphasic. These five limb leads sometimes deflect negatively but this would be considered unusual. In leads aV$_R$, V$_1$, and V$_2$, the QRS complex mostly deflects negatively. Positively deflecting QRS complexes in V$_1$ and/or V$_2$ would be considered abnormal. In lead V$_3$, the QRS complex is biphasic. In lead V$_4$, the QRS deflects mostly positive or is biphasic. In leads V$_5$ and V$_6$, the QRS complex deflects positively.

7.5 Where We See Normal QRS Complexes

The QRS complexes should appear normal (upright and narrow) if

- The rhythm is initiated from a site above the ventricles—meaning in the SA node, atria, or AV junction.
- Conduction has progressed normally from the bundle of His, through the right and left bundle branches, and through the Purkinje network.
- Normal depolarization of the ventricles has occurred.

Unless there is a conduction delay through the ventricles, normal sinus rhythm and any dysrhythmia initiated by a pacemaker site above the ventricles can be expected to have normal QRS complexes.

Generally, you can look at the QRS complex and get a sense of whether it is narrow or wide.

7.6 Abnormal QRS Complexes

Abnormal QRS complexes are produced by abnormal depolarization of the ventricles. The pacemaker site in these abnormal QRS complexes can be the SA node or an ectopic pacemaker in the atria, AV junction, bundle branches, Purkinje network, or ventricular myocardium.

The onset and end of an abnormal QRS complex are the same as those of a normal QRS complex. The direction can be mostly positive (upright), mostly negative (inverted), or biphasic. Its duration is often greater than 0.10 seconds. The amplitude of waves in the abnormal QRS complex varies from 1 to 2 mm to 20 mm or more. The morphology of the abnormal QRS complex can vary from being only slightly abnormal to extremely wide, notched, or slurred. We refer to these extremely abnormal configurations as bizarre-looking because they deviate so far from normal (Figure 7-8).

A variety of conditions can cause abnormal QRS complexes including

- Ventricular hypertrophy
- Obesity, hyperthyroidism, pleural effusion
- Bundle branch block
- Intraventricular conduction disturbance
- Aberrant ventricular conduction
- Ventricular preexcitation
- An electrical impulse originating in a ventricular ectopic or escape pacemaker
- Ventricular pacing by a cardiac pacemaker

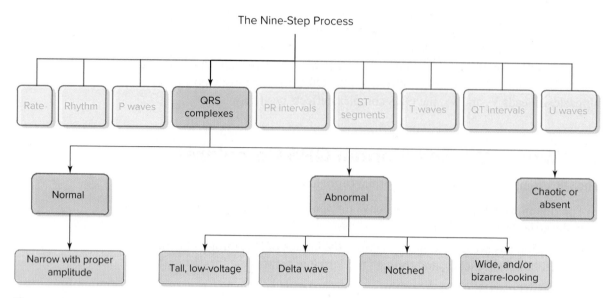

Figure 7-8

Algorithm to evaluate QRS complexes.

As a point of information, waveforms that look alike from beat to beat (such as QRS complexes) are coming from the same location. Occasionally, frequently, or continuously repeating waveforms that look different are coming from a different location(s) or traveling through the conduction systems differently.

Tall and Low-Amplitude QRS Complexes

Very tall QRS complexes are usually caused by **hypertrophy** of one or both ventricles or by an abnormal pacemaker or aberrantly conducted beat (Figure 7-9). Hypertrophied ventricles are larger and, therefore, have more myocardial cells.

Low-voltage or abnormally small QRS complexes may be seen in obese patients, hyperthyroid patients, and patients with pleural effusion.

Wide QRS Complexes of Supraventricular Origin

Any rhythm that originates from above the ventricles (i.e., the SA node or an ectopic pacemaker in the atria or AV junction) is referred to as a supraventricular rhythm. Even though a cardiac rhythm originates from a supraventricular site, the subsequent impulse conduction through the ventricles can be impaired or can occur differently than normal. This can result in wide, bizarre QRS complexes.

Ventricular Conduction Disturbances

One type of abnormal conduction is referred to as a ventricular conduction disturbance. Ventricular conduction disturbances usually occurs as a result of right or left bundle branch block. Bundle branch block takes place when conduction of the electrical impulse is partially or completely blocked in either the right or left bundle branch while conduction continues uninterrupted through the unaffected bundle branch (Figure 7-10). A block in one bundle branch causes the ventricle on that side to be depolarized later than the other. This results

Tall QRS complexes

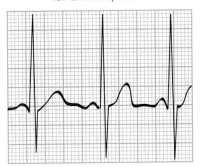

Low-voltage QRS complexes

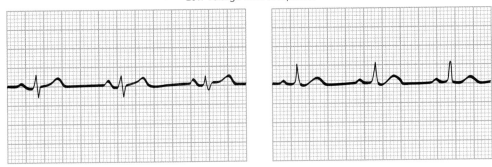

Figure 7-9
Examples of tall QRS complexes and low-voltage QRS complexes.

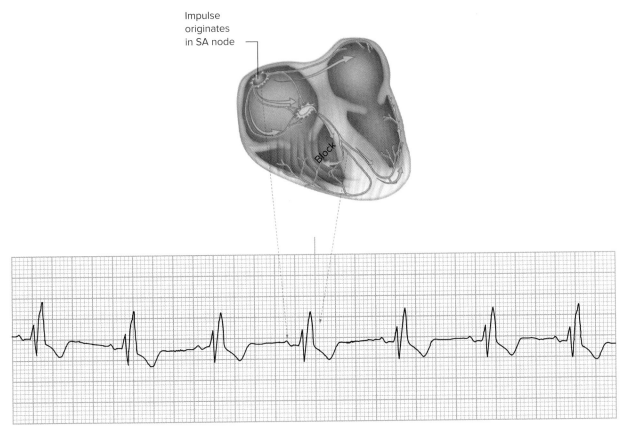

Figure 7-10
In bundle branch block, wide QRS complexes occur because conduction of the electrical impulse is partially or completely blocked in either the right or left bundle branch while conduction continues uninterrupted through the unaffected bundle branch.

in an abnormal QRS complex. In bundle branch blocks, the QRS complex is greater than 0.12 seconds in duration and appears bizarre (abnormal in size and shape).

Bundle branch block may be either partial or complete. As its name implies, partial, or incomplete, bundle branch block results in only part of the conduction being delayed. This results in less of a delay in ventricular depolarization than would be expected in complete bundle branch block. This causes the QRS complex to be greater than 0.10 seconds but not greater than 0.12 seconds in duration. However, the QRS complex has some abnormalities, such as notching, and does not appear completely normal.

Another form of intraventricular conduction disturbance is a conduction abnormality located in the ventricles and not a blockage of the right or left bundle branch (Figure 7-11). Causes include myocardial infarction; fibrosis; electrolyte imbalance, such as hypokalemia and hyperkalemia; and excessive administration of such cardiac drugs such as quinidine, procainamide, and flecainide.

Aberrant Conduction

Aberrant conduction is a brief failure of the right or left bundle branch to normally conduct an electrical impulse (Figure 7-12). It can occur when electrical impulses, such as early beats and tachycardias that arise above the ventricles, reach the bundle branch shortly after it has just conducted an impulse. It is refractory and not

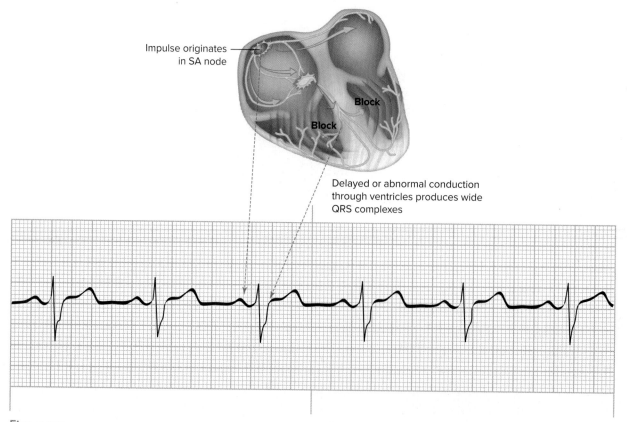

Impulse originates in SA node

Block

Block

Delayed or abnormal conduction through ventricles produces wide QRS complexes

Figure 7-11

The presence of a P wave preceding each QRS complex indicates that the rhythm is arising from the SA node. The wide QRS complexes result from a conduction defect or delayed conduction through the ventricles.

Normally, the SA node initiates impulses, resulting in a repetitive cycle of P, QRS, and T waveforms.

Premature impulse travels down one of the bundle branches before the other.

Following the premature beat, the SA node typically reinitiates impulses in the normal manner.

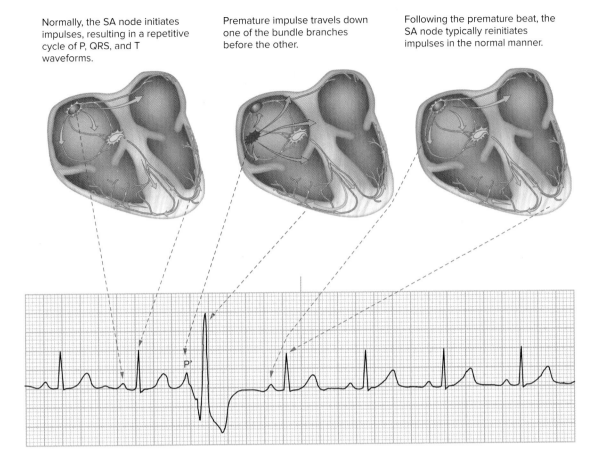

Figure 7-12
In aberrant conduction, the QRS complex appears wider than normal due to the impulse traveling down one unaffected bundle branch while stimulation of the other bundle branch follows.

able to be stimulated. One bundle branch may be less refractory than the other and will conduct the impulse. The refractory bundle branch will then be stimulated following a delay. This produces an abnormal QRS complex. The abnormal QRS complex frequently looks like incomplete or complete bundle branch block and may mimic ventricular dysrhythmias. Aberrant ventricular conduction can occur in many supraventricular dysrhythmias.

Ventricular Preexcitation

Another cause of wider than normal QRS complexes is ventricular preexcitation (Figure 7-13). This is the premature depolarization of the ventricles that occurs when an impulse arises from a site above the ventricles but travels through abnormal accessory conduction pathways to the ventricles. (Ventricular preexcitation will be covered in more detail in Chapter 13.) These abnormal conduction tracts bypass the AV junction, or bundle of His, allowing the electrical impulses to initiate depolarization of the ventricles early. This results in an abnormally wide QRS complex (greater than 0.10 seconds). It also has characteristically abnormal slurring, and sometimes notching, at its onset. This is called the **delta wave**. The abnormal shape and width of the QRS complex is the result of the fusion of the abnormal premature depolarization of the ventricle (which produces the delta wave) with the normal depolarization of the rest of the ventricles.

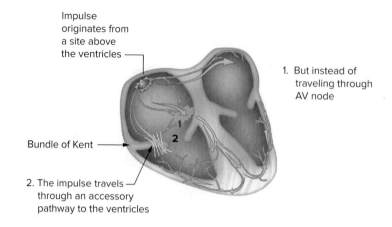

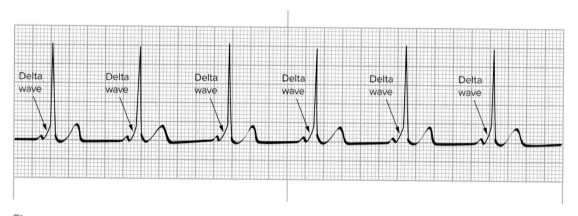

Figure 7-13
Ventricular preexcitation is premature depolarization of the ventricles that occurs when an impulse arises from a site above the ventricles but travels through abnormal accessory conduction pathways to the ventricles instead of traveling through the AV node.

Pacemaker-Induced QRS Complexes

Certain conditions require the insertion of a cardiac pacemaker (Figure 7-14). Cardiac pacemaker–induced QRS complexes are generally 0.12 seconds or greater in width and appear bizarre. Preceding each pacemaker-induced QRS complex is a narrow, often biphasic, deflection called the *pacemaker spike*.

Wide, Bizarre-Looking QRS Complexes of Ventricular Origin

Dysrhythmias that originate from the ventricular tissue are referred to as *ventricular dysrhythmias* (Figure 7-15). The three dysrhythmias we will discuss are early beats, an escape rhythm, and tachycardia, all of which arise from the ventricles. The key characteristic seen with ventricular dysrhythmias is wide (greater than 0.12 seconds in duration) QRS complexes. The QRS complexes are usually bizarre-looking. Beats that arise from the ventricles are not preceded by a P wave (if seen, they are dissociated). The T waves of ventricular beats deflect in an opposite direction of the R waves.

Early Beats That Arise from the Ventricles

Early beats that arise from the ventricles before the SA node has a chance to fire (Figure 7-16) interrupt the normal rhythm. These impulses arise from below the

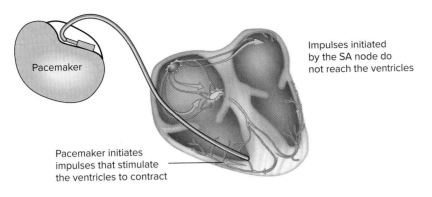

Impulses initiated
by the SA node do
not reach the ventricles

Pacemaker

Pacemaker initiates
impulses that stimulate
the ventricles to contract

Pacemaker
spike

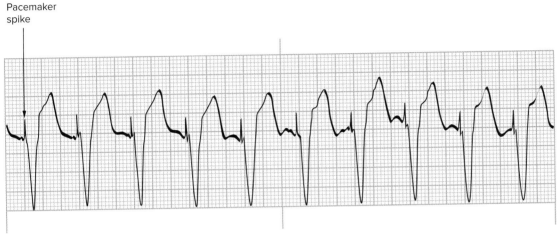

Figure 7-14
Characteristic features of cardiac pacemakers include wide QRS complexes (greater than 0.10 seconds) and a narrow, often biphasic, deflection called the *pacemaker spike*, which precedes each pacemaker-induced QRS complex.

AV junction, producing QRS complexes that look different from those that arise above or at the AV junction. As described, the QRS complexes associated with these early beats are wide, large, and bizarre-looking. A pause, called a **compensatory pause,** follows the early beat. This will be discussed in more detail in Chapter 15.

Escape Rhythm That Arises from the Ventricles

A sustained escape rhythm arises from the ventricles when stimuli from the SA node or AV junction fail to reach the ventricles or their rate falls to less than that of the ventricles (Figure 7-17). The key characteristics of this dysrhythmia are a rate of 20 to 40 beats per minute and QRS complexes that are wide (greater than 0.12 seconds), large, and bizarre-looking. As with PVCs, the T wave typically takes the opposite direction of the R wave. The rhythm arises from one site so it is usually regular. It becomes irregular as the heart dies.

Tachycardia That Arises from the Ventricles

Tachycardia that arises from the ventricles results from rapid depolarization of the ventricles that overrides the SA node (Figure 7-18). It occurs when there are three or more early beats in a row. It may occur in bursts of 6 to 10 early beats or be sustained. In sustained tachycardia that arises from the ventricles, the heart rate is between 100 and 250 beats per minute. The rhythm consists of frequent, wide

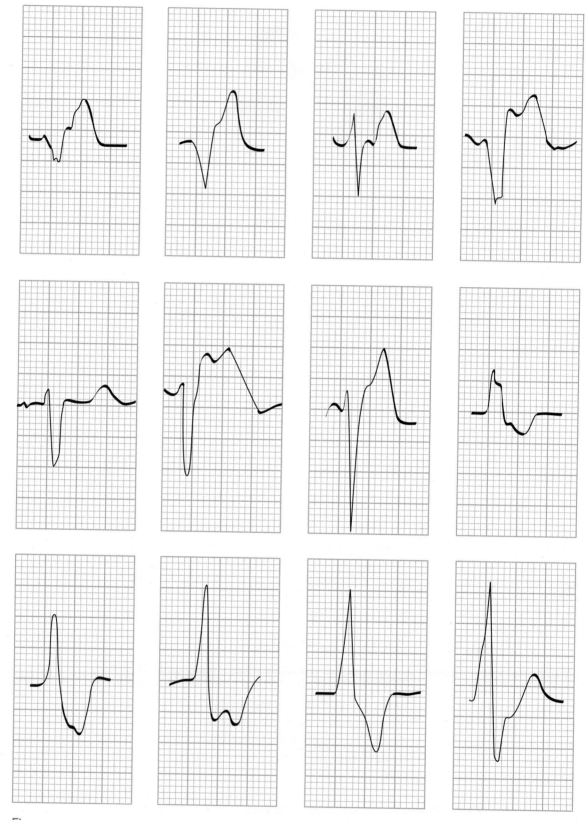

Figure 7-15
Different-looking QRS complexes associated with ventricular beats and/or rhythms.

Normally, the SA node initiates impulses, resulting in a repetitive cycle of P, QRS, and T waveforms.

An early ventricular beat occurs when a site in the ventricles fires before the SA node is able to initiate an impulse.

Following the early beat, the SA node typically reinitiates impulses in the normal manner.

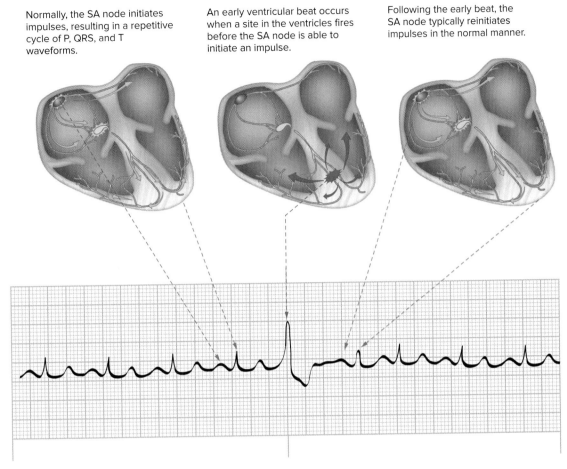

Figure 7-16
QRS complexes associated with early beats that arise from the ventricles are wide, large, and bizarre-looking.

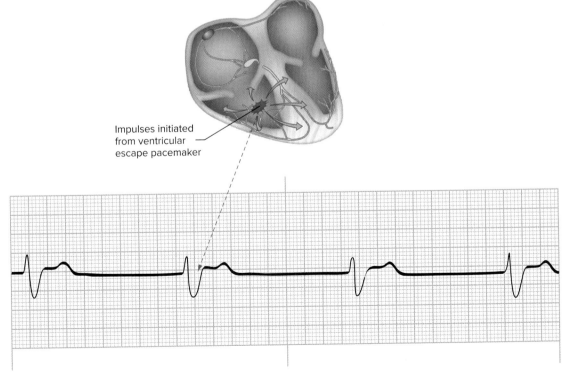

Impulses initiated from ventricular escape pacemaker

Figure 7-17
Wide, bizarre-looking QRS complexes firing at a rate of less than 40 beats per minute and an absence of P waves are seen with an escape rhythm that arises from the ventricles.

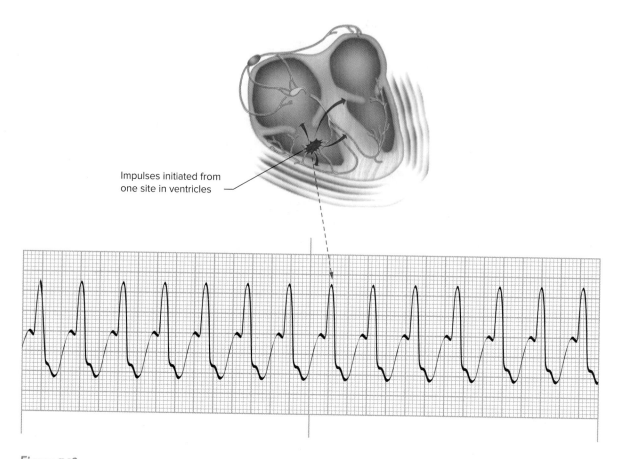

Impulses initiated from one site in ventricles —

Figure 7-18

Wide, bizarre-looking QRS complexes firing at a rate of between 100 and 250 beats per minute and an absence of P waves are seen with tachycardia that arises from the ventricles.

(greater than 0.12 seconds), and bizarre-looking QRS complexes. The QRS complexes occur regularly, and there is no isoelectric line shown between the cardiac cycles. T waves may or may not be present. If seen, they are typically the opposite direction of the R waves.

The QRS complexes associated with tachycardia that arises from one site in the ventricles should all look alike. On the other hand, when the tachycardia arises from more than one site in the ventricles, the shape of the ventricular waveforms can be expected to change (Figure 7-19). In one particular form of tachycardia, the ventricular waveforms appear as a series of QRS complexes that rotate about the baseline (usually gradually) between upright deflections and downward deflections. It is described as a "spindle-shaped rhythm."

AV Heart Block

Another dysrhythmia in which abnormal QRS complexes may occur is in the most severe form of AV heart block (Figure 7-20). Remember, we said that with the most severe form of AV heart block there is complete blockage of the AV node. The atria are stimulated to contract by the impulse that originates from the SA node, whereas the ventricles are stimulated to contract by an escape pacemaker that arises somewhere below the AV node, resulting in a slower heart rate. The location of the ventricular pacemaker site determines the appearance of the QRS complex. Normal

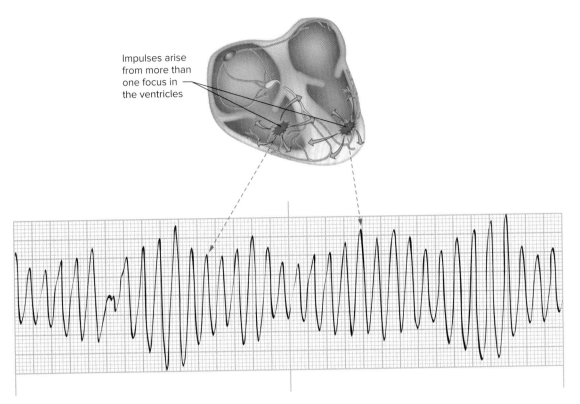

Figure 7-19
Form of tachycardia arising in the ventricles in which the shape of the ventricular waveforms changes between upright and downward deflections. It appears as a spindle-shaped rhythm.

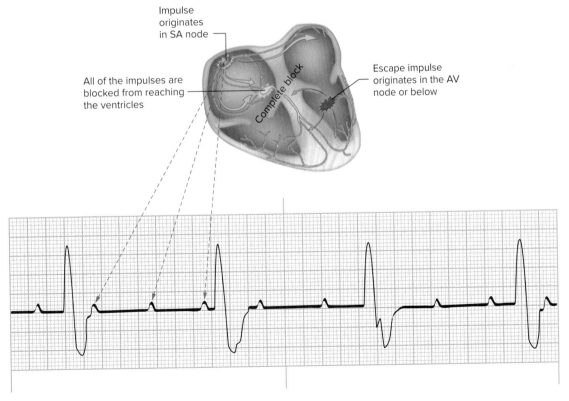

Figure 7-20
With the most severe form of AV heart block, the ventricles are stimulated by an escape pacemaker site that arises from somewhere in the His-Purkinje system. The more distal the pacemaker site, the wider the QRS complex.

(narrow) QRS complexes indicate that the ventricular pacemaker is closer to the AV junction. Wide, large (bizarre-looking) QRS complexes indicate that the ventricles are likely being paced by a ventricular focus.

Absent QRS Complexes

This section describes two dysrhythmias in which the QRS complexes are absent or so bizarre as to indicate that no coordinated depolarization of the ventricles is occurring and therefore the patient is pulseless.

Ventricular Fibrillation

Ventricular fibrillation (VF) is the erratic firing of multiple sites in the ventricles causing the heart muscle to quiver instead of contracting as it normally does (Figure 7-21). On the ECG monitor, it looks like a chaotic wavy line, rising and falling without any logic. There are no discernible waveforms. There is no coordinated contraction of the ventricles so the amount of blood pumped by the heart drops to zero.

Asystole

Asystole is ventricular or cardiac standstill (Figure 7-22). *Asystole* literally means without contraction of the heart (*a* means without and *systole* means heart contraction). It is essentially the absence of any cardiac activity in the ventricles. Asystole appears as a flat (or nearly flat) line on the ECG. It produces complete cessation of cardiac output. Sometimes P waves are seen with asystole. This indicates the presence of electrical activity in the atria. They are typically seen only for a short time.

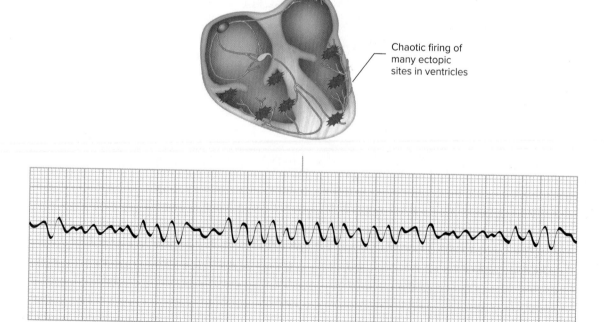

Chaotic firing of many ectopic sites in ventricles

Figure 7-21

Ventricular fibrillation results in a chaotic-looking rhythm with no similarity in the waveforms or the distance between them.

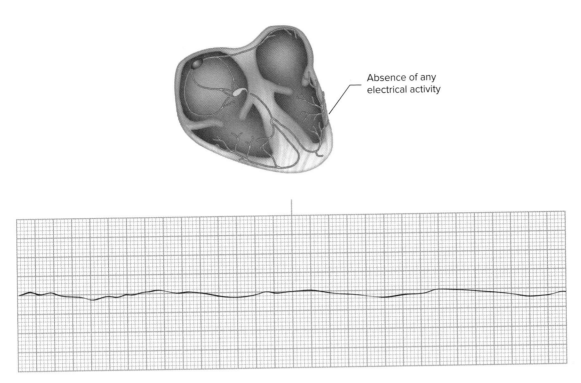

Absence of any electrical activity

Figure 7-22
Asystole is the absence of any cardiac activity in the heart.

Key Points

LO 7.1	• An important step of analyzing an ECG rhythm is examining the QRS complexes. This is important because it tells us how well the impulses are traveling through the ventricles or if the impulses are originating from the ventricles.
	• Delays or interruption of conduction through the ventricles or impulses that arise from the ventricles can lead to serious or even life-threatening conditions.
LO 7.2	• The starting point of the QRS complex is where the first wave of the complex starts to move away (sharply or gradually) from the baseline. It ends at the point at which the last wave of the complex transitions into the ST segment. This is where it begins to flatten (sharply or gradually) at, above, or below the baseline.
	• The QRS complex is larger than the P wave because ventricular depolarization involves a considerably larger muscle mass than atrial depolarization. However, because the ventricles depolarize rapidly; the QRS complex characteristically looks narrower than the other parts of the ECG.
	• A normal QRS complex indicates that the electrical impulse originated at or above the bundle of His and was carried through the ventricles in a normal manner.
	• The Q wave is the first negative deflection from the baseline following the P wave. The R wave is the first positive deflection following the Q wave (the P wave, if the Q wave is absent). The S wave is the first negative deflection that extends below the baseline in the QRS complex following the R wave.

	• The amplitude of a normal QRS complex is 5 to 30 mm high, and the duration is 0.06 to 0.10 seconds. We measure the width of the QRS from the beginning of the Q wave to the end of the S wave. If the Q wave is absent (as it often is), we measure the QRS complex from the beginning of the R wave.
LO 7.3	• While the QRS consists of positive (upright) deflections called R waves and negative (inverted) deflections called Q and S waves, all three waves are not always seen. If the R wave is absent, the complex is called a QS complex. Likewise, if the Q wave is absent, the complex is called an RS complex. Waveforms of normal or greater than normal amplitude are denoted with a large case letter, whereas waveforms less than 5 mm in amplitude are denoted with a small case letter (e.g., "q," "r," "s").
LO 7.4	• The waveforms will look different depending on which lead is used. In leads I, II, V_5, and V_6, the QRS complex characteristically deflects positively. In leads III, aV_L, and aV_F, the QRS often deflects positively but may also be biphasic. In leads aV_R and V_1, the QRS deflects negatively. In leads V_2, V_3, and sometimes V_4, the QRS complex is mostly biphasic.
LO 7.5	• Normal sinus rhythm and dysrhythmias that arise from above the ventricles will usually have normal QRS complexes (unless there is a conduction delay through the ventricles or other type of abnormality as described earlier).
LO 7.6	• Abnormal QRS complexes are produced by abnormal depolarization of the ventricles. • The pacemaker site in abnormal QRS complexes can be the SA node or an ectopic pacemaker in the atria, AV junction, bundle branches, Purkinje network, or ventricular myocardium. • The duration of an abnormal QRS complex is greater than 0.12 seconds while the shape of an abnormal QRS complex varies widely from one that looks almost normal to one that is wide and bizarre-looking and/or slurred and notched. • Very tall QRS complexes are usually caused by hypertrophy of one or both ventricles or by an abnormal pacemaker or aberrantly conducted beat. • Low-voltage or abnormally small QRS complexes may be seen in obese patients, hyperthyroid patients, and patients with pleural effusion. • Wide, bizarre QRS complexes of supraventricular origin are often the result of right or left bundle branch block. Bundle branch block may be either partial or complete. • Aberrant conduction produces an abnormal QRS complex that frequently looks like incomplete or complete bundle branch block and may mimic ventricular dysrhythmias. • Wider than normal QRS complexes may also be due to ventricular preexcitation. This also causes an abnormal slurring, and sometimes notching, at its onset called the delta wave. • Cardiac pacemaker-induced QRS complexes are generally 0.12 seconds or greater in width and appear bizarre-looking. Preceding each pacemaker-induced QRS complex is a pacemaker spike. • Ventricular dysrhythmias are those dysrhythmias that originate from the ventricular tissue. Wide (greater than 0.12 seconds in duration) QRS complexes are the key characteristic seen with ventricular dysrhythmias. Also, the QRS complexes are usually bizarre-looking, and the T waves deflect in an opposite direction to the R waves. • The most severe form of AV heart block is another dysrhythmia in which there may be abnormal QRS complexes. Wide, large (bizarre-looking) QRS complexes indicate that the ventricles are likely being paced by a ventricular focus.
	• Ventricular fibrillation appears on the ECG as a chaotic wavy line, rising and falling without any logic. There are no discernible waveforms. • Asystole appears on the ECG as a flat (or nearly flat) line.

Assess Your Understanding

The following questions give you a chance to assess your understanding of the material discussed in this chapter. The answers can be found in Appendix A.

1. The appearance of the _____ is affected by the lead used. (LO 7.2)

2. In lead II, the deflection of the QRS complex is characteristically   (LO 7.4)
   a. inverted.
   b. positive.
   c. pointed.
   d. biphasic.

3. Analyzing the QRS complex helps identify how the electrical impulse is being carried through the   (LO 7.1)
   a. SA node.
   b. atria.
   c. ventricles.
   d. AV node.

4. The normal duration of the QRS complex is _____ second(s). (LO 7.2)

5. An upright (in lead II), narrow QRS complex indicates that the electrical impulse originated   (LO 7.5)
   a. in the AV node and was carried through the atria in a retrograde manner.
   b. in the ventricles and was carried through the atria in a normal manner.
   c. in the SA node and was carried through the AV node in a retrograde manner.
   d. at or above the AV node and was carried through the ventricles in a normal manner.

6. The R wave is the   (LO 7.2)
   a. first positive deflection in the QRS complex.
   b. second negative deflection that extends below the baseline.
   c. first negative deflection from the baseline following the P wave.
   d. flat line that precedes the Q wave.

7. Which of the following will produce normal QRS complexes?   (LO 7.2)
   a. Altered, damaged, or abnormal ventricles.
   b. Tachycardia that arises from the ventricles.
   c. Early beats that arise from the atria.
   d. An escape pacemaker rhythm that originates from the ventricles.

8. Which is true regarding the amplitude of QRS complexes?   (LO 7.2)
   a. Women have larger amplitudes than men.
   b. The elderly have higher amplitudes than young people.
   c. Precordial leads have higher amplitudes than the limb leads.
   d. Both a and c.

9. Ventricular conduction disturbance (LO 7.6)
   a. is the result of accelerated conduction through the ventricles.
   b. produces tall QRS complexes.
   c. causes the QRS complexes to appear abnormal.
   d. is usually seen in myocardial infarction, fibrosis, and hypertrophy.

10. Low-voltage or abnormally small QRS complexes are seen in (LO 7.6)
    a. hypertrophy.
    b. obese patients.
    c. early beats arising from the ventricles.
    d. slow heart rate that arises from the SA node.

11. The most common cause of ventricular conduction disturbance is (LO 7.6)
    a. hypertrophy.
    b. accessory pathways.
    c. bundle branch block.
    d. a cardiac pacemaker.

12. Bundle branch blocks (LO 7.6)
    a. may be partial or complete.
    b. have a delta wave preceding the QRS complex.
    c. have a pacemaker spike.
    d. have QRS complexes not preceded by a P wave or PR interval.

13. Aberrant conduction (LO 7.6)
    a. occurs when an impulse is late arriving at the bundle branch.
    b. results in narrow, spike-shaped QRS complexes.
    c. mimics supraventricular dysrhythmias.
    d. occurs when the impulse reaches one of the bundle branches while it is still refractory.

14. Early beats that arise from the ventricles (LO 7.6)
    a. are preceded by a P wave.
    b. always have an inverted QRS complex.
    c. produce QRS complexes that look different than those that arise above or at the AV junction.
    d. have narrow, upright QRS complexes.

15. An escape pacemaker rhythm that arises from the ventricles has (LO 7.6)
    a. wide, bizarre-looking QRS complexes.
    b. a P wave that precedes each QRS complex.
    c. T waves that take the same direction as the R wave.
    d. a ventricular rate of 40 to 60 beats per minute.

16. Tachycardia that arises from the ventricles has   (LO 7.6)
    a. a pacemaker site in the SA node.
    b. wide and bizarre-looking QRS complexes.
    c. a fast rate that is always sustained.
    d. narrow QRS complexes.

17. With the most severe form of AV block, the QRS complexes   (LO 7.6)
    a. are always wide and bizarre-looking.
    b. are associated with irregular and frequent early beats.
    c. follow each P wave but at a slower rate.
    d. are slower than the P wave rate because there is complete blockage of the AV node.

18. With ventricular fibrillation,   (LO 7.6)
    a. there is adequate cardiac output.
    b. a chaotic wavy line is seen on the ECG.
    c. there is erratic firing from one site in the ventricles.
    d. there is a flat line on the ECG.

19. Asystole   (LO 7.6)
    a. is the erratic firing of multiple sites in the ventricles.
    b. produces an effective electrical rhythm but no contraction of the heart muscle.
    c. usually produces P waves.
    d. is seen as a flat line on the ECG.

Referring to the scenario at the front of this chapter answer the following questions.

20. Where are the extra QRS complexes originating from?   (LO 7.6)
    a. Ventricles
    b. SA node
    c. AV node
    d. Atria

21. The most likely cause of the patient's extra beats is   (LO 7.6)
    a. anxiety.
    b. pain.
    c. ischemia.
    d. hypertrophy.

8 PR Intervals

©rivetti/Getty Images

©Stockbyte/Stockbyte/Getty Images

Learning Outcomes

LO 8.1 State why it is important to analyze the PR intervals of an ECG tracing.

LO 8.2 Describe the characteristics of the normal PR interval and recall how to examine it.

LO 8.3 Describe what is meant when we say, "PR intervals that are different."

Case History

A 77-year-old woman who had a heart attack three days ago is transferred from the ICU to the medical floor for continued observation. Her heart rhythm is continuously monitored by the telemetry unit. During the patient's morning walk, the nurses receive a call from the telemetry unit informing them that the patient's rhythm has become irregular. After locating the woman, the nurses return her to her bed and obtain the rhythm strips from the telemetry unit. Her attending physician is summoned, and her vital signs obtained and found to be normal except for a regularly irregular pulse.

The patient's physician arrives and examines the rhythm strip with the nurse. He notes that the PR interval increases in length following each P wave until after every third one when the QRS complex is missing entirely. They discuss that this is a common complication after a myocardial infarction. The patient is moved back to the ICU for close observation because of the possibility for further deterioration.

8.1 Importance of Determining the PR Intervals

As part of analyzing an ECG we examine and measure the specific waveforms, segments, and intervals. In this chapter we discuss assessment of the PR intervals (Figure 8-1).

The presence or absence of PR intervals as well as their duration can help us identify if and how well the impulses are traveling through the atrioventricular (AV) junction and bundle of His. This is important to know because delays or interruption of conduction through the AV node and/or junction can lead to serious or even life-threatening conditions.

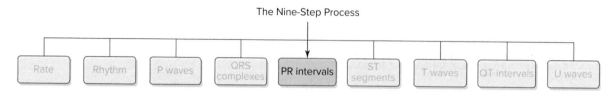

Figure 8-1
Assessing the PR interval is part of the Nine-Step Process algorithm.

The initial assessment for PR intervals waves can be done quickly. Questions to ask include, "Is there a PR interval preceding each QRS complex? Do they appear of normal duration? Are they all the same?"

8.2 Characteristics of Normal PR Intervals

As discussed previously, the PR interval is the distance from the beginning of the P wave to the beginning of the Q wave (or R wave if the Q wave is absent). The PR interval denotes depolarization of the heart from the SA node through the atria, AV node, and His-Purkinje system. A normal PR interval (Figure 8-2) indicates the impulse originated from the SA node (or close to it) and traveled through the atria and AV node in a regular and unobstructed course.

Characteristics of Normal PR Intervals	
Location	Starts at the beginning of the P wave and ends at the beginning of the Q wave (or R wave if the Q wave is absent).
Duration	0.12 to 0.20 seconds.
Morphology	P wave and a flat line.

Figure 8-2
When assessing the PR intervals, look for a normal PR interval to precede each QRS complex.

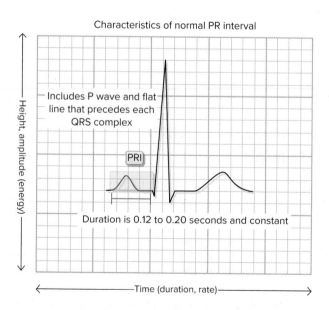

If a P wave precedes each QRS complex and the duration of the PR interval is normal and constant, you can assume that this electrical impulse originated in the SA node and was carried through the atria, AV junction, and His-Purkinje system in a normal manner. As such, we can say that the PR interval is usually normal, with impulses that arise from the SA node. A PR interval within 0.12 to 0.20 seconds can also be present with some dysrhythmias that arise from the atria, provided that the ectopic site is located in the upper atrial wall. As an example, early beats or tachycardia that arises from the atria can have a normal PR interval if the ectopic site from which they arise is high enough in the atria.

Measuring the PR Intervals

First, determine if the PR intervals are identifiable. If they are, measure from where the P wave begins to where the Q wave begins (Figure 8-3). To make measuring the PR interval easier, find a P wave that begins on a vertical line, and then measure to where the Q or R wave begins. Are the waves within normal limits? The PR interval should extend across between 3 and 5 small boxes (0.12 to 0.20 seconds in duration).

Then compare each PR interval to the others. Are they the same (in other words, are they constant?) or do they differ? Last, determine if there a relationship between the P waves and QRS complexes.

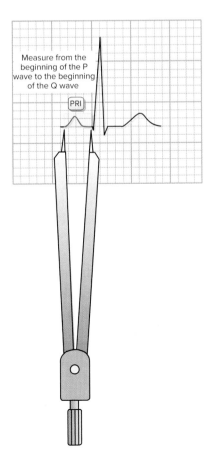

Figure 8-3
Measuring the PR interval.

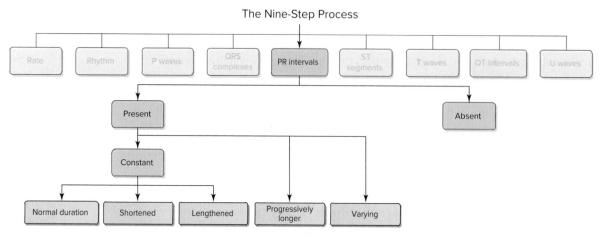

Figure 8-4
Algorithm for normal and abnormal PR intervals.

8.3 PR Intervals That Are Different

PR intervals are considered abnormal if they are shorter than 0.12 seconds, longer than 0.20 seconds, or absent or if they vary in duration (Figure 8-4). A variety of conditions can cause abnormal PR intervals.

Shorter PR Intervals

A shorter than normal PR interval is less than 0.12 seconds in duration. Shorter PR intervals occur when the impulse originates in the lower atria, in the AV junction, or where there is an accessory pathway between the atria and ventricles. The shorter the distance the impulse has to travel, the shorter the duration of the PR interval. If we suspect the P wave is not originating in the SA node, we refer to this shortened PR interval as a P′R (P prime R) interval.

Short P′R Intervals Seen with Dysrhythmias Arising from the Atria

Early beats and tachycardia that arise from the atria or internodal pathways before the SA node has a chance to fire can produce P′R intervals that differ from the underlying rhythm. If the early beat arises from a site in the lower right atrium or in the upper part of the AV junction, the P′R interval should be less than 0.12 seconds in duration (Figure 8-5). Also, with atrial tachycardia the P′ waves may be buried in the T wave of the preceding beat, so it may be hard or even impossible to accurately measure the P′R interval.

If an early beat is initiated in the upper or middle right atrium, the P′R interval is generally normal (0.12 to 0.20 seconds in duration). If an early beat arises from a site high in the left atrium, the PR interval may be prolonged.

Short P′R Intervals Seen with Dysrhythmias Arising from the AV Junction

As mentioned in Chapter 6, when impulses arise from the AV junction, an inverted P′ wave may precede the QRS complex (Figure 8-6). The P′ wave may also occur

Normally, the SA node initiates impulses, resulting in a repetitive cycle of P, QRS, and T waveforms.

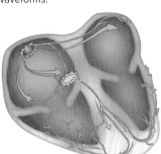

Early atrial beat occurs when a site in the atria fires before the SA node is able to initiate an impulse.

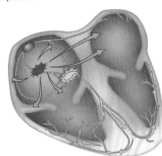

Following the early beat, the SA node typically reinitiates impulses in the normal manner.

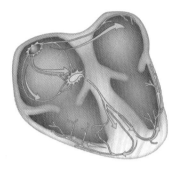

A shorter than normal PR interval occurs when the early atrial beat arises closer to the AV junction.

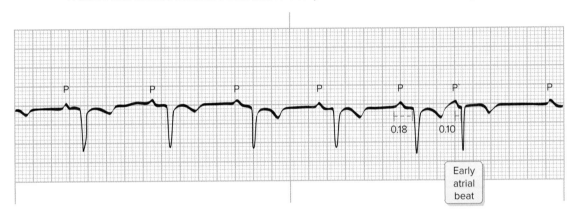

A P`R interval within normal duration occurs when the early atrial beat arises from a site in the upper- or middle-right atrium.

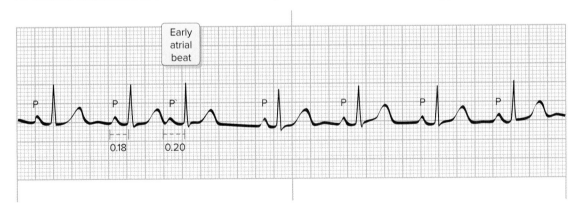

Figure 8-5
Short P′R intervals seen with dysrhythmias that arise from the atria.

during (and be buried in) the QRS complex or follow it. This causes the electrical impulse to travel upward through the AV junction into the atria, causing retrograde atrial depolarization. The P′R interval will be shorter when the P′ wave appears before the QRS complex.

If the P wave is buried in the QRS complex, the P′R interval will be absent. If the P′ wave follows the QRS complex, it is referred to as the RP′ interval and is usually less than 0.20 seconds.

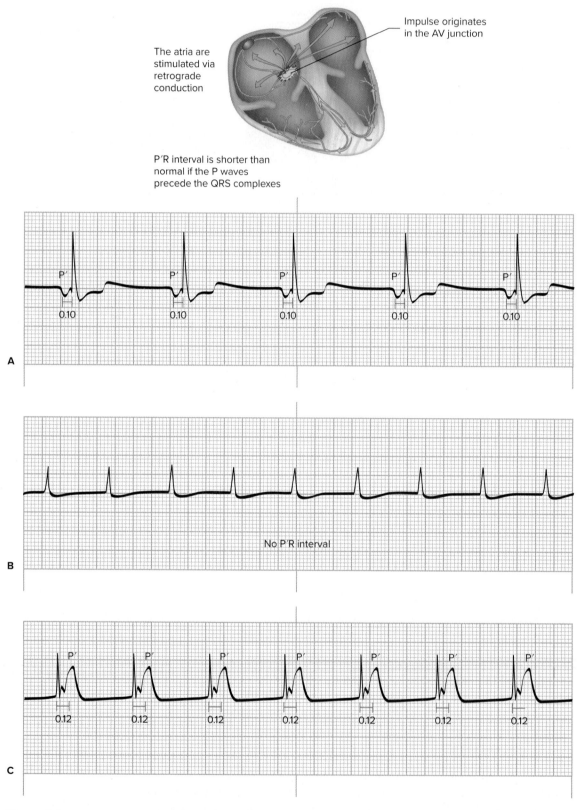

The atria are stimulated via retrograde conduction

Impulse originates in the AV junction

P′R interval is shorter than normal if the P waves precede the QRS complexes

No P′R interval

Figure 8-6

P′R intervals seen with dysrhythmias arising from the AV junctional tissue. (A) Shorter than normal P′R intervals are seen when the P wave precedes the QRS complexes. (B) No P′R intervals are seen when the P waves are buried in the QRS complexes. (C) The RP′ interval is measured when the P wave follows the QRS complex.

As we discussed earlier, because the P′ waves seen with early junctional beats and tachycardia arising from the junctional tissue can be buried in the T wave of the preceding beat, it may be hard to determine the P′R interval.

Short P′R Intervals Seen with Accessory Pathways

Another cause of a shorter than normal PR interval is the presence of accessory pathways between the atria and ventricles. These accessory pathways allow the impulse to bypass the AV node and move to the ventricles more quickly (Figure 8-7). This is referred to as *ventricular preexcitation* and will be discussed in more depth in Chapter 13.

Longer PR Intervals

A longer than normal PR interval is greater than 0.20 seconds in duration. Greater PR intervals occur when the impulse is delayed beyond normal as it passes through the AV junction. The longer it takes the impulse to travel through the AV node, the longer the duration of the PR interval. A variety of conditions can cause the PR interval to be longer than normal.

Blockage of the AV Junction

The most common cause of longer than normal PR intervals is a delay in or blockage in the conduction of the impulse as it passes through the AV junction (Figure 8-8). As mentioned in Chapter 6, we use the term AV heart block to describe these delays or blockage.

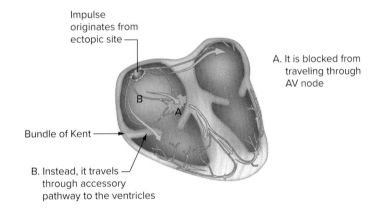

Impulse originates from ectopic site

A. It is blocked from traveling through AV node

B

A

Bundle of Kent

B. Instead, it travels through accessory pathway to the ventricles

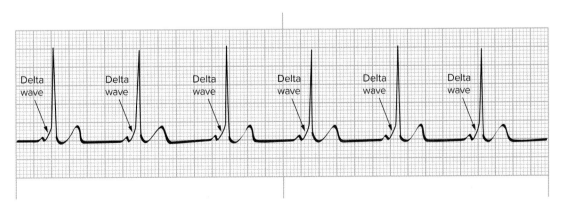

Delta wave

Figure 8-7
A shorter PR interval can occur with the presence of accessory pathways.

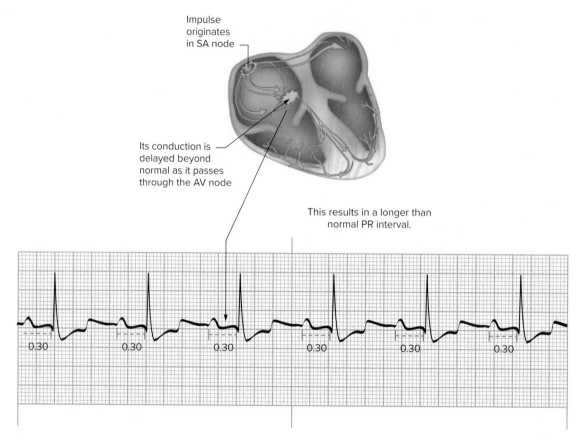

Impulse
originates
in SA node

Its conduction is
delayed beyond
normal as it passes
through the AV node

This results in a longer than
normal PR interval.

0.30   0.30   0.30   0.30   0.30   0.30

Figure 8-8
A PR interval of greater than 0.20 seconds in duration is due to an abnormal delay in conduction through the AV node.

In its milder form, there is a delay in impulse conduction through the AV junction. While there is a delay, the impulse is still able to reach the bundle of His, bundle branches, and the ventricles. The key characteristic of this condition is a PR interval greater than 0.20 seconds that is constant in duration with each beat.

Varying PR Intervals

The PR interval may also be seen to vary or change. This occurs when the pacemaker site in the atria changes from beat to beat and in one form of AV heart block.

Changing Pacemaker Site

When the pacemaker site moves from beat to beat from the sinus node to other latent pacemaker sites in the atria and AV junction (Figure 8-9), the P′ waves appear different and the P′R intervals tend to vary because of the changing pacemaker site.

AV Heart Block

The PR interval may also vary in a milder form of AV heart block in which the weakened AV node fatigues more and more with each conducted impulse until finally it is so tired that it fails to conduct an impulse through to the ventricles. This results in a changing PR interval, one that becomes increasingly longer until there is a dropped QRS complex (Figure 8-10). With the dropped ventricular beat, the AV node has a chance to rest. The next impulse conducting through the AV node is carried through faster, but then each subsequent impulse is delayed until another ventricular beat is dropped.

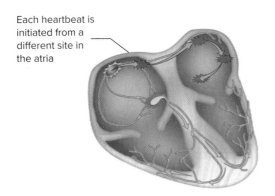

Each heartbeat is
initiated from a
different site in
the atria

P′ waves change in appearance
throughout the tracing

P′R intervals change in
duration throughout the tracing

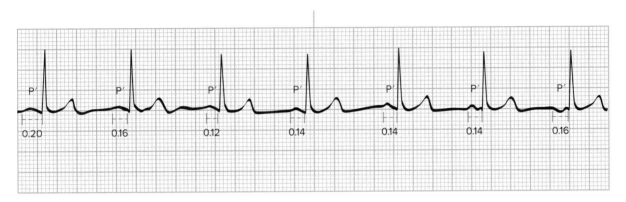

P′ P′ P′ P′ P′ P′ P′

0.20 0.16 0.12 0.14 0.14 0.14 0.16

Figure 8-9
When the pacemaker site continually changes, the duration of the P′R interval varies.

Impulse
originates
in SA node

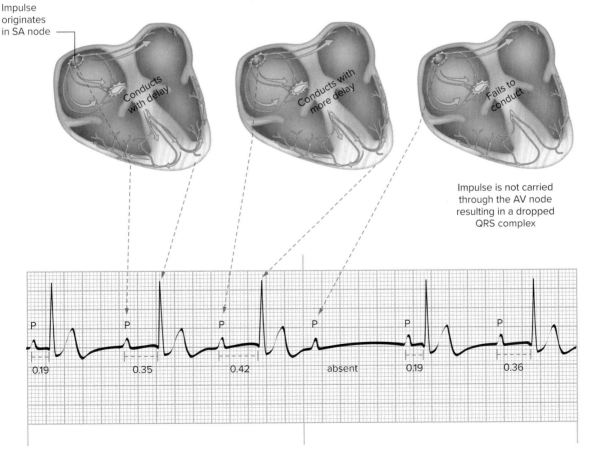

Conducts
with delay

Conducts with
more delay

Fails to
conduct

Impulse is not carried
through the AV node
resulting in a dropped
QRS complex

P P P P P P

0.19 0.35 0.42 absent 0.19 0.36

Figure 8-10
In a milder type of AV heart block, the PR interval becomes increasingly longer until finally a QRS complex is
dropped. The cycle then starts all over.

Absent or Not Measurable PR Intervals

The PR interval will be absent or not measurable in certain dysrhythmias. These dysrhythmias can arise from the atria, AV junction, or ventricles and in the most severe form of AV heart block.

Absent P′R Intervals Seen with Impulses Arising from the Atria

When the atria fire at a rate of between 250 and 350 beats per minute (Figure 8-11) and flutter waves (F waves) are seen instead of the normal P waves, the PR intervals are not measurable.

Also, there are no PR intervals when the atria fire faster than 350 beats per minute (Figure 8-12), producing a chaotic-looking baseline with no discernable P waves.

Ventricular Dysrhythmias

Dysrhythmias that originate in the ventricular tissue are referred to as ventricular dysrhythmias (Figure 8-13). The key characteristics seen with ventricular dysrhythmias are wide, bizarre-looking QRS complexes. Beats that arise from the ventricles are not preceded by a P wave (if seen, the P waves are dissociated), and the PR intervals are not measurable.

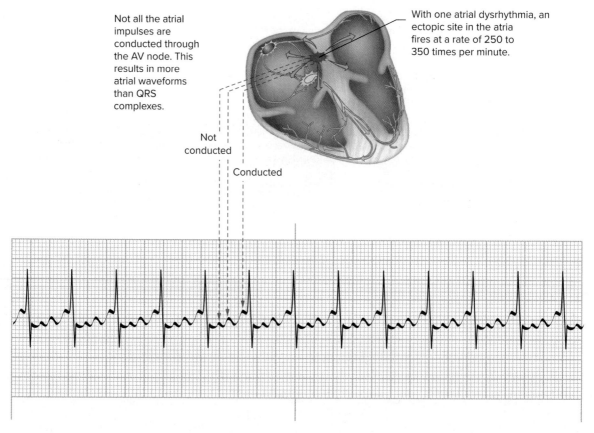

Not all the atrial impulses are conducted through the AV node. This results in more atrial waveforms than QRS complexes.

With one atrial dysrhythmia, an ectopic site in the atria fires at a rate of 250 to 350 times per minute.

Not conducted

Conducted

Figure 8-11
When flutter waves are present, the PR intervals cannot be determined.

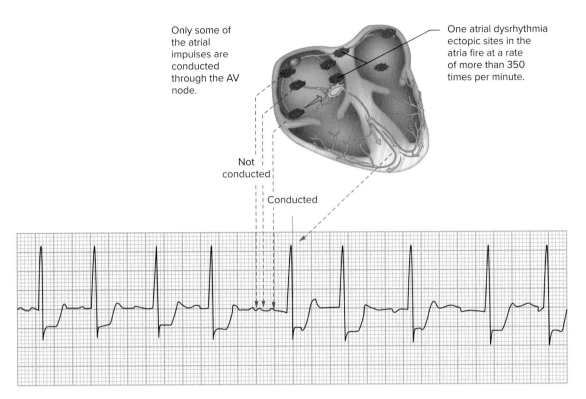

Only some of the atrial impulses are conducted through the AV node.

One atrial dysrhythmia ectopic sites in the atria fire at a rate of more than 350 times per minute.

Not conducted

Conducted

Figure 8-12
With a chaotic atrial baseline, the PR intervals cannot be determined because there are no discernable P waves.

Atrioventricular Heart Block

In the more severe form of AV heart block, there is complete blockage of the AV node (Figure 8-14). The atria are stimulated to contract by the impulse that originates from the SA node. For this reason the atrial rate should be within a normal range. The ventricles are stimulated to contract by an escape pacemaker that arises from somewhere below the AV node, resulting in a slower rate. The pacemaker sites (SA node and ventricles) fire independently of each other, so the P-P intervals and the R-R intervals appear to be disassociated with one another. For this reason, the PR intervals are not measurable. The P waves will march through the QRS complexes.

Constant PR Intervals Seen with More P Waves

AV Heart Block

Normally, there is only one P wave preceding each QRS complex. However, in some forms of AV heart block we may see more P waves than QRS complexes. This indicates that the impulse was initiated in the SA node but was blocked and did not reach the ventricles. The atrial-to-ventricular conduction ratio is usually 2:1, 3:1, or 4:1. PR intervals that are constant (of the same duration) for the conducted beats (Figure 8-15) are characteristic of one type of AV heart block. The PR interval can be normal in duration or prolonged.

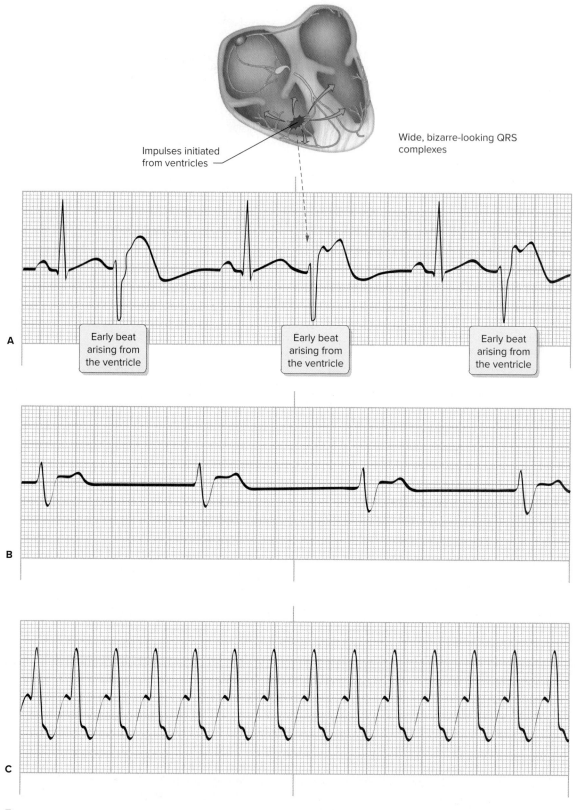

Impulses initiated from ventricles

Wide, bizarre-looking QRS complexes

A   Early beat arising from the ventricle   Early beat arising from the ventricle   Early beat arising from the ventricle

B

C

Figure 8-13
With ventricular dysrhythmias, there is an absence of P waves and no PR intervals. (A) Early beats arising from the ventricles. (B) Escape rhythm arising from the ventricles. (C) Tachycardia arising from the ventricles.

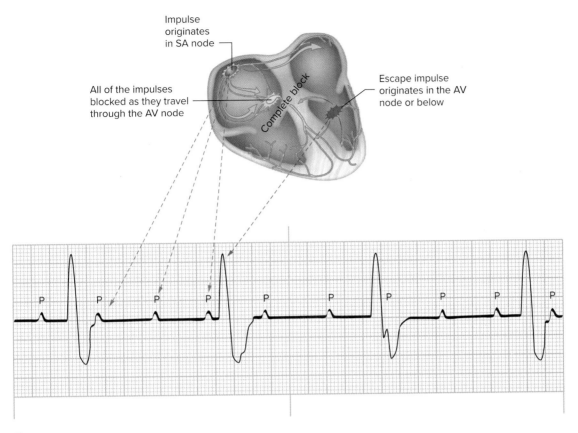

Figure 8-14
With the more severe form of AV heart block, there are no PR intervals as the P waves appear to march through the QRS complexes.

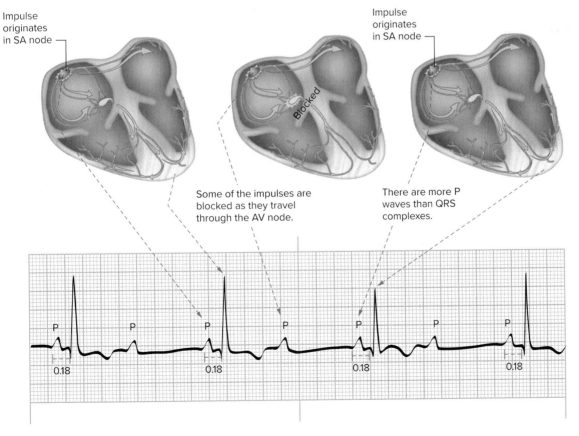

Figure 8-15
With one form of AV heart block, the PR intervals of the conducted beats are constant.

Key Points

LO 8.1	• Another important step in analyzing an ECG rhythm is examining the PR intervals. The presence or absence of PR intervals as well as their duration can help us identify if and how well the impulses are traveling through the atrioventricular (AV) junction and bundle of His.
LO 8.2	• The PR interval is the distance from the beginning of the P wave to the beginning of the Q wave (or R wave if the Q wave is absent).
	• The PR interval denotes depolarization of the heart from the SA node through the atria and AV node.
	• A normal PR interval indicates that the impulse originated in the SA node (or close to it) and traveled through the atria and AV node in a regular and unobstructed course.
	• The duration of the PR interval is normally 0.12 to 0.20 seconds.
LO 8.3	• PR intervals are considered abnormal if they are shorter or longer than normal, absent, or varying in duration.
	• Shorter P'R intervals occur when the impulse originates in the atria close to the AV junction or in the AV junction or when there is an accessory pathway between the atria and ventricles.
	• A longer than normal PR interval is greater than 0.20 seconds in duration. Greater PR intervals occur when the impulse is delayed beyond normal as it passes through the AV junction.
	• When the pacemaker site moves from beat to beat, the P'R intervals will vary.
	• With one type of AV heart block, the PR intervals are progressively longer until a QRS complex is dropped, and then the cycle starts over.
	• PR intervals are not measurable in the presence of flutter waves, a chaotic atrial baseline, or dysrhythmias that originate from the ventricles.
	• In the most severe form of AV heart block, the PR intervals are not measurable.
	• With one form of AV heart block, some of the sinus beats are blocked in the AV node and do not reach the ventricles. This leads to more P waves than QRS complexes. The PR intervals associated with the P waves conducted through to the ventricles are constant.

Assess Your Understanding

The following questions give you a chance to assess your understanding of the material discussed in this chapter. The answers can be found in Appendix A.

1. The PR interval duration _____ from beat to beat. (LO 8.2)
   a. may change
   b. is constant
   c. may be progressively longer
   d. increase and decrease

2. PR interval is the distance from the (LO 8.2)
   a. end of the P wave to the beginning of the Q wave.
   b. beginning of the P wave to the beginning of the Q wave.
   c. beginning of the Q wave to the end of the S wave.
   d. end of the Q wave to the beginning of the R wave.

3. The PR interval denotes depolarization of the heart from the (LO 8.1)
   a. bundle of His through the ventricles.
   b. SA node through the atria, AV node, and His-Purkinje system.
   c. AV node through the bundle of His.
   d. SA node through the atria and AV node.

4. The normal duration of the PR interval is _____ second(s). (LO 8.2)
   a. 0.06 to 0.10
   b. 0.8 to 0.12
   c. 1.0
   d. 0.12 to 0.20

5. How many squares on the ECG paper make up the normal PR interval?
   (LO 8.2)

6. A shorter than normal PR interval is less than _____ second(s) in
   duration. (LO 8.2)
   a. 0.12
   b. 0.08
   c. 0.10
   d. 0.20

7. Which of the following characteristics is considered normal? (LO 8.2)
   a. a P wave that follows the QRS complex
   b. a PR interval that changes continually
   c. a PR interval that is 0.18 seconds in duration
   d. a P'R interval that is 0.10 seconds in duration

8. Which of the following is characteristic of delayed conduction through the
   AV node? (LO 8.3)
   a. varying PR intervals
   b. PR intervals of less than 0.12 seconds
   c. more P waves than QRS complexes
   d. constant PR intervals of greater than 0.20 seconds

9. With a pacemaker site that changes from beat to beat throughout the atria,
   the P'R intervals (LO 8.3)
   a. are constant.
   b. vary.
   c. are longer in duration than 0.20 seconds.
   d. become progressively longer.

10. With AV heart block, the PR intervals may be all of the following EXCEPT (LO 8.3)

    a. constant.

    b. absent.

    c. shorter in duration than 0.12 seconds.

    d. progressively longer.

11. There will be an absence of PR intervals in (LO 8.3)

    a. early beats that arise from the atria.

    b. the most severe form of AV heart block.

    c. beats that arise from the SA node.

    d. a pacemaker site within the atria that changes from beat to beat.

12. If there is an accessory pathway between the atria and ventricles, the PR intervals may (LO 8.3)

    a. continually change in duration.

    b. be longer than normal.

    c. be absent.

    d. be less than 0.12 seconds in duration.

13. How are PR segments commonly described? (LO 8.2)

14. Why would an early beat arising from the atria have a PR interval longer in duration? (LO 8.3)

Referring to the scenario at the beginning of this chapter, answer the following questions.

15. This patient is most likely experiencing (LO 8.3)

    a. early beats arising from the atria.

    b. normal sinus rhythm.

    c. AV heart block.

    d. tachycardia that is arising from the atria.

16. The PR intervals that change in duration are considered (LO 8.3)

    a. normal.

    b. abnormal.

    c. characteristic of specific dysrhythmias.

    d. b and c.

9 ST Segments, T Waves, QT Intervals, and U Waves

©rivetti/Getty Images

Chapter Outline

Analyzing the Specific
  Waveforms, Segments,
  and Intervals

Normal and Abnormal
  ST Segments

Normal and Abnormal
  T Waves

Normal and Abnormal
  QT Intervals

Normal and Abnormal
  U Waves

Courtesy Philips Healthcare

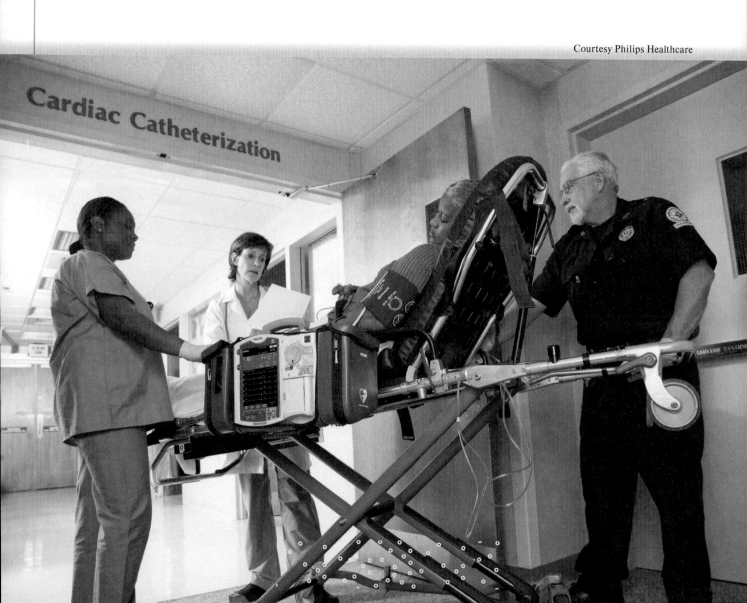

Learning Outcomes

LO 9.1   State why it is important to analyze the ST segments, T waves, QT intervals, and U waves of an ECG tracing.

LO 9.2   Describe the characteristics of normal and abnormal ST segments and recall how to measure them.

LO 9.3   Describe the characteristics of the normal and abnormal T waves and recall how to measure them.

LO 9.4   Identify the characteristics of the normal and abnormal QT intervals and recall how to measure them.

LO 9.5   Describe the characteristics of the normal U wave and list those conditions in which U waves may be seen.

Case History

EMS is called to a local plant for a 53-year-old female who is experiencing shortness of breath. Upon arrival they find the patient sitting on a bench in the locker room. She is awake and able to answer questions. She states she was lifting some heavy boxes when suddenly she couldn't breathe. She denies chest pain but does admit to having pressure "under her breastbone." Except for an elevated blood pressure, her vital signs are normal. The 12-lead ECG leads are quickly attached. It shows a slow sinus rhythm with elevated ST segments and inverted T waves in several leads. The paramedics apply oxygen and establish an intravenous line. The patient's chest discomfort remains despite the application of oxygen and the administration of aspirin and sublingual nitroglycerin.

En route to the hospital the patient continues to experience chest discomfort. The paramedics contact medical control and report their findings. The physician approves their request to deliver an analgesic agent and a second dose of nitroglycerin. The patient is delivered to the hospital without any improvement in her condition.

9.1 Analyzing the Specific Waveforms, Segments, and Intervals

As we said in previous chapters, the heart normally beats in a regular, rhythmic fashion, producing a P wave, a PR interval, a QRS complex, an ST segment, a T wave, a QT interval, and possibly a U wave with each heartbeat. This cycle repeats over and over.

As part of analyzing an ECG, we examine and measure the specific waveforms, segments, and intervals. In this chapter we discuss assessing the ST segments, T waves, QT intervals, and U waves (Figure 9-1). Specific characteristics seen with these elements of the ECG can help identify serious cardiac conditions.

9.2 Normal and Abnormal ST Segments

The ST segment yields important information because it represents the end of ventricular depolarization and the beginning of ventricular repolarization. The ST segment is the isoelectric line that follows the QRS complex and connects it to the T wave (Figure 9-2). Elevation or depression of the ST segment is a hallmark feature of myocardial ischemia and injury.

Figure 9-1
Another important step in assessing the ECG are to assess the ST segment, T wave, QT interval, and U waves (when seen).

Figure 9-2
The ST segment is the line that follows the QRS complex and connects it to the T wave.

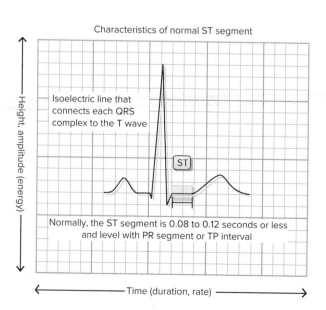

Normal Characteristics

The ST segment begins at the J point. Extending horizontally, it gradually (almost imperceptibly) curves upward to the beginning of the T wave. Under normal circumstances, the ST segment appears as an isoelectric line (neither positive nor negative), although it may vary by 0.5 to 1.0 mm in some leads (precordial). The duration of the normal ST segment is 0.08 to 0.12 seconds or less and dependent on the heart rate. With a faster heart rate, the duration of the ST segment is less compared with longer ST segments that occur with a slower heart rate. The term *ST segment* is used regardless of whether the final wave of the QRS complex is an R or S wave.

Characteristics of a Normal ST Segments	
Location	Follows the QRS complex and connects it to the T wave.
Deflection	Isoelectric line (neither positive or negative). Can vary by 0.5 to 1.0 mm in some precordial leads.

Measuring the ST Segments

When assessing the ST segment, the question to ask is, "What is the deflection?" To determine this, we use the TP segment as the baseline from which to evaluate the amount of displacement from the isoelectric line. The TP segment is the space between the T wave and the P wave of the beat that follows it. Draw a straight line (or use a paper) extending from the end of the T wave to the end of the next T wave (Figure 9-3). Alternatively, some use the PR segment to establish the baseline. To employ the PR segment, draw a straight line (or use a paper) extending from the end of the P wave out past the T wave of the beat that follows it. Then, find the J point. From the J point measure over to the right 0.04 seconds (one small box).

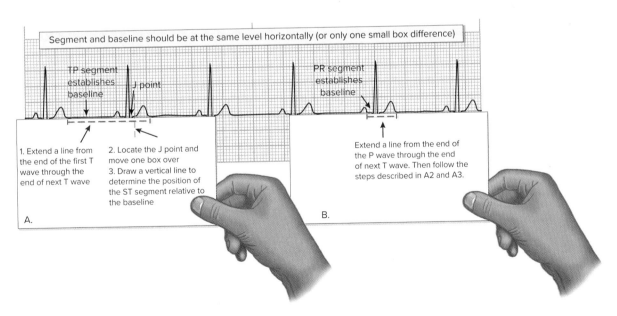

Figure 9-3
Two methods for determining elevation or depression of the ST segment are: (A) use the TP segment or (B) PR segment to determine the baseline.

From there, draw a vertical line up or down (depending on whether the ST segment is depressed or elevated) to where it intersects with the ST segment. Then, measure this distance in the number of small squares or millimeters from the baseline. That is the amount of ST segment elevation or depression that exists in its place. Remember from our discussion in Chapter 8 that the J point is the point that marks the end of the QRS complex and the beginning of the ST segment. It isn't always easy to identify the J point. Sometimes you have to look for that small notch, slope, or other movement that suggests an alteration of electrical flow (see Figure 8-4). Further, it may be difficult to determine the J point in patients experiencing rapid heart rate or hyperkalemia.

The ST segment is considered elevated if it is above the baseline and depressed if it is below it. (Remember, it can vary by 0.5 to 1.0 mm in some leads.)

Abnormal ST Segments

Abnormal ST segments are indicative of abnormal ventricular repolarization. Two types are ST segment elevation and ST segment depression.

ST Segment Elevation

An elevated ST segment is one that is 1 mm or more above the baseline. Its presence may indicate marked myocardial injury and be an early indicator of myocardial infarction. Elevated ST segments can appear as flat, concave (curving inward), convex (curving outward), or arched (Figure 9-4). Convex-shaped ST segment elevation is most often due to myocardial injury and infarction. ST segment elevation on the single-lead ECG pattern is usually not as significant as it is on the 12-lead ECG.

Other causes of elevated ST segments include bundle branch block, pericarditis, ventricular hypertrophy, ventricular fibrosis or aneurysm, hyperkalemia, and hypothermia. These will be discussed in more detail in later chapters.

Although we should be concerned whenever we see ST segment elevation, it is important to note that physiological (also called *normal variant*) ST elevation is quite common. This presents as ST segment changes that are usually limited to leads V_1, V_2, and V_3 and an absence of cardiac symptoms. Further, cardiac ultrasound shows no myocardial wall abnormality, and the changes are fixed and do not evolve like those seen with myocardial infarction.

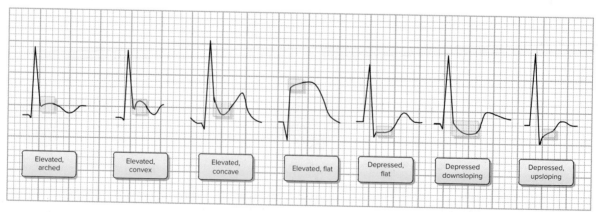

Figure 9-4
Different appearances of ST segment elevation and depression.

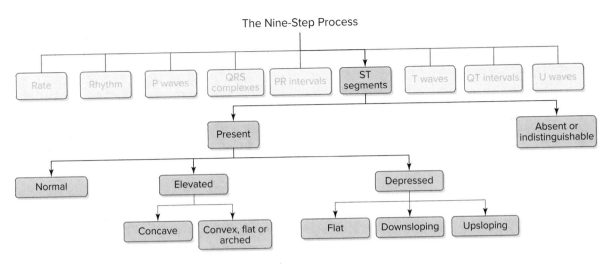

Figure 9-5
ST segment algorithm.

ST Segment Depression

An ST segment is considered depressed when it is more than 1.0 mm below the baseline. A depressed ST segment may indicate myocardial ischemia or reciprocal changes opposite an area of myocardial injury. The depressed ST segment can appear as flat, upsloping, or downsloping (Figure 9-5; also see Figure 9-4).

ST segment depression may also occur with right and left ventricular hypertrophy (strain pattern), right and left bundle branch block, hypokalemia, and hyperventilation, among other conditions. It can also occur with digitalis use.

9.3 Normal and Abnormal T Waves

Next, we should see the T wave. Assessing T waves is important as it represents the completion of ventricular repolarization (recovery). The T wave is a wide, broad waveform. Remember, atrial repolarization occurs during the QRS complex and is not visible on the regular ECG.

Normal Characteristics

The T wave is larger than the P wave and slightly asymmetrical (Figure 9-6). The peak is closer to the end than to the beginning, and the first half has a more gradual slope than the second half. Normally, the T wave is not more than 5 mm in height in the limb leads or 10 mm in any precordial lead. Also, it is normally oriented in the same direction as its associated QRS complex. This is because depolarization begins at the endocardial surface and spreads to the epicardium, whereas repolarization begins at the epicardium and spreads to the endocardium.

T waves are normally positive in leads I, II, and V_2 to V_6 and negative in lead aV_R. They can also be positive in aV_L and aV_F but may be negative if the QRS complex is less than 6 mm in height. In leads III and V_1, the T wave may be positive or negative.

Figure 9-6
The T wave follows the ST
segment and represents
ventricular repolarization.

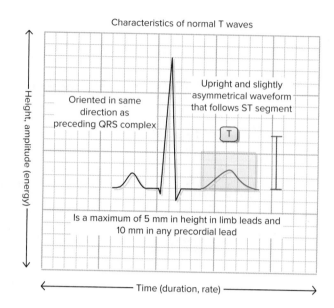

Figure 9-6
The T wave follows the ST segment and represents ventricular repolarization.

Characteristics of a Normal T Wave	
Location	Follows the ST segment.
Configuration	Upright and slightly asymmetrical.
Amplitude	Maximum of 5 mm in height in the limb leads and maximum of 10 mm in any precordial lead.
Deflection	Oriented in the same direction as the preceding QRS complex.

Measuring the T Waves

When evaluating the T wave, the three most important questions to ask are:

1. What is the configuration?
2. What is the amplitude?
3. What is the deflection?

To measure the T wave, identify its starting and ending point, then move to the center of the waveform. Next, count the number of small boxes between the isoelectric line and the top of the waveform (Figure 9-7).

Abnormal T Waves

Abnormal T waves indicate the presence of abnormal ventricular repolarization. Abnormal T waves include those that are tall, peaked or tented, inverted (in leads where they should be upright), biphasic, flat, wide, and heavily notched and those having bumps (Figure 9-8).

Tall or peaked T waves, also known as *tented T waves*, are seen in myocardial ischemia or hyperkalemia (Figure 9-9). Heavily notched or pointed T waves in an adult may indicate pericarditis. Inverted T waves in leads I, II, and V_3 through V_6 may be seen with myocardial ischemia. Deeply inverted T waves are seen with significant cerebral disease such as subarachnoid hemorrhage. Flat T waves are seen with certain

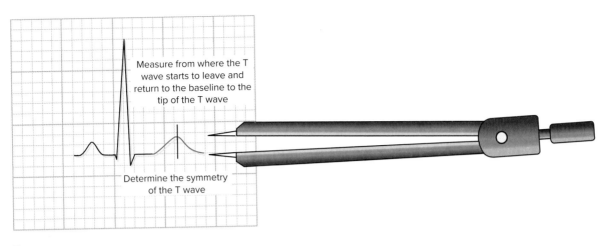

Figure 9-7
Measuring the T wave.

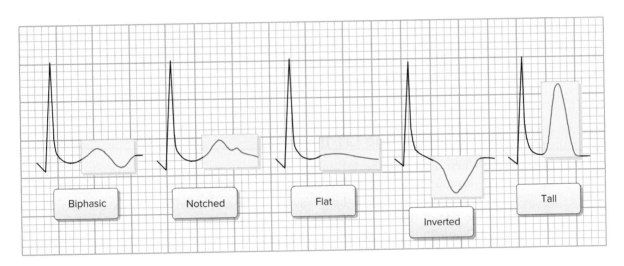

Figure 9-8
Appearance of abnormal T waves.

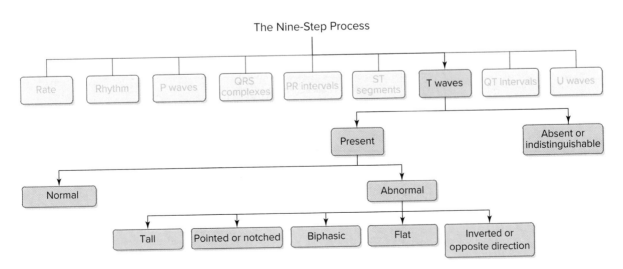

Figure 9-9
T wave algorithm.

electrolyte imbalances. T waves with bumps may be seen when P waves are buried in the T waves. This can occur with sinus tachycardia, atrial tachycardia, early beats that arise from the atria or AV junctional tissue or in certain types of AV heart block.

Also, T waves seen with impulses that arise from the ventricles take a direction opposite of their associated QRS complexes.

9.4 Normal and Abnormal QT Intervals

In addition to the other steps used to analyze the ECG, it is also important to assess the QT intervals. The QT interval represents the time needed for ventricular depolarization and repolarization.

Normal Characteristics

The QT interval extends from the beginning of the QRS complex to the end of the T wave. Its duration varies according to age and sex, but it is usually between 0.36 and 0.44 seconds (Figure 9-10). The QT interval also changes with the heart rate. The QT interval is shorter with faster rates and longer with slower heart rates. A good rule to follow is that the QT interval should not be greater than half the distance between consecutive R waves when the rhythm is regular.

Measuring the QT Interval

When assessing the QT interval, focus on its most important characteristics —location and duration. To measure the QT interval, first locate the isoelectric line immediately preceding the Q wave (or R wave if there is no Q wave). Next, locate the end of the T wave. Count the number of boxes in between these two points (Figure 9-11). Since each small box represents 0.04 seconds, the normal QT interval should extend across at least 9 but not more than 11 small boxes. Preferably, you should record three consecutive measurements of the QT interval. Then, calculate

Figure 9-10
Characteristics of normal QT intervals.

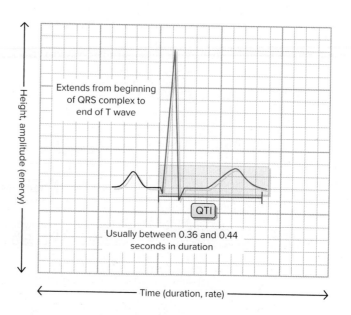

Height, amplitude (eneryy)

Extends from beginning of QRS complex to end of T wave

QTI

Usually between 0.36 and 0.44 seconds in duration

Time (duration, rate)

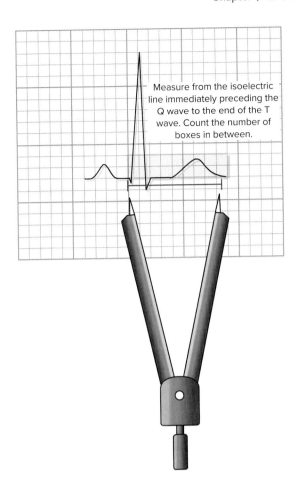

Figure 9-11
Measuring the QT interval.

Measure from the isoelectric line immediately preceding the Q wave to the end of the T wave. Count the number of boxes in between.

the mean. It is best to use lead II to measure the QT interval. If lead II is not suitable, then leads in the sequence of V_5, V_4, V_3, V_2 should be used.

Characteristics of a Normal QT Interval	
Location	Extends from the Q wave (or R wave) to the end of the T wave.
Duration	Usually between 0.36 and 0.44 seconds.

Abnormal QT Intervals

An abnormal QT interval duration can indicate myocardial irregularity. The two types of abnormal QT intervals are prolonged QT intervals and shortened QT intervals (Figure 9-12).

Prolonged QT Intervals

A prolonged QT interval indicates prolonged ventricular repolarization, which means the relative refractory period is longer. In certain conditions, such as myocardial ischemia or infarction, a prolonged QT interval can predispose the patient to life-threatening ventricular dysrhythmias such as torsades de pointes.

Causes of prolonged QT interval include congenital conduction system defect. Called *prolonged QT interval syndrome,* it is present in certain families. Also, drugs

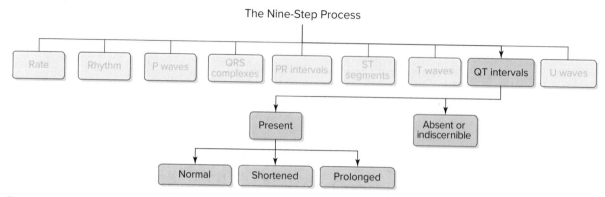

Figure 9-12
QT interval algorithm.

such as haloperidol, methadone, and Class IA antidysrhythmic drugs like amiodarone or sotalol can produce QT interval prolongation.

Shortened QT Intervals

A shortened QT interval can be caused by hypercalcemia or digoxin toxicity.

9.5 Normal and Abnormal U Waves

The U wave is a small wave that follows the T wave. It is not always seen. The origin of U waves is speculative; however, they are thought to represent repolarization of the papillary muscles or Purkinje fibers. U waves are considered normal in younger persons (less than 35–40 years of age) but not so common in the elderly.

Normal Characteristics

When present, the U wave is rounded, has the same polarity as the T wave, and is usually less than one-third its amplitude (Figure 9-13). The U wave will sometimes merge with the T wave. U waves are usually best seen in the right precordial leads, especially V_2 and V_3. The normal U wave is asymmetrical, with the beginning portion being more abrupt than the ending portion (just the opposite of the normal T wave).

Characteristics of a Normal U Wave	
Location	When present, follows the T wave.
Configuration	Upright and slightly asymmetrical.
Amplitude	Usually less than one-third the amplitude of the T wave.
Deflection	Oriented in the same direction as the preceding T wave.

Abnormal U Waves

Prominent U waves are most often seen in hypokalemia but may be present in hypercalcemia, thyrotoxicosis, digitalis toxicity, epinephrine and Class 1A and 3 antidysrhythmics use, mitral valve prolapse, and left ventricular hypertrophy, as well as in congenital long QT syndrome and in the setting of intracranial hemorrhage.

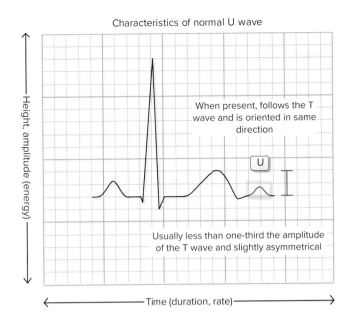

Characteristics of normal U wave

Figure 9-13
Characteristics of U waves.

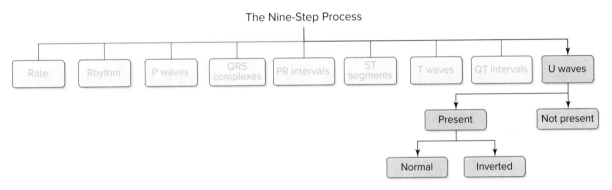

Figure 9-14
U wave algorithm.

An inverted U wave may represent myocardial ischemia or left ventricular volume overload (Figure 9-14).

Key Points

LO 9.1	• Assessing the ST segments, T waves, QT intervals, and U waves are an important part of analyzing an ECG tracing.
LO 9.2	• The ST segment is the isoelectric line that follows the QRS complex and connects it to the T wave. It represents the end of ventricular depolarization and the beginning of ventricular repolarization.
	• ST segment depression or elevation can be evaluated by comparing the ST segment with the TP segment or the PR segment. It is considered elevated if it is above the baseline and depressed if it is below it.
	• Elevation or depression of the ST segment is a hallmark feature of myocardial ischemia and injury.

LO 9.3	• The T wave represents the completion of ventricular repolarization (recovery).
	• The T wave is larger than the P wave and slightly asymmetrical. It is oriented in the same direction as the preceding QRS complex.
	• When evaluating the T wave, focus on its three most important characteristics: configuration, amplitude, and deflection.
	• Abnormal T waves indicate the presence of abnormal ventricular repolarization.
	• Tall or peaked and inverted T waves (in certain leads) may be seen in myocardial ischemia.
LO 9.4	• The QT interval represents the time needed for ventricular depolarization and repolarization.
	• The QT extends from the beginning of the QRS complex to the end of the T wave. Its duration varies according to age, sex, and heart rate, but it is usually between 0.36 and 0.44 seconds.
	• In certain conditions, such as myocardial ischemia or infarction, a prolonged QT interval can predispose the patient to life-threatening ventricular dysrhythmias.
LO 9.5	• When present, the U wave is a small, rounded wave that follows the T wave. It has the same polarity as the T wave.
	• Prominent U waves are most often seen in hypokalemia.

 ## Assess Your Understanding

The following questions give you a chance to assess your understanding of the material discussed in this chapter. The answers can be found in Appendix A.

1. Describe why assessing the ST segments, T waves, and QT intervals is important. (LO 9.1)

2. The ST segment represents (LO 9.2)
   a. atrial repolarization.
   b. ventricular depolarization.
   c. the beginning of ventricular repolarization.
   d. atrial and ventricular repolarization.

3. The ST segment is characteristically (LO 9.2)
   a. neither positive nor negative.
   b. curved downward.
   c. 1 to 2 mm above the isoelectric line.
   d. biphasic.

4. Describe two ways to measure the ST segment. (LO 9.2)

5. An elevated ST segment is _____ the isoelectric line. (LO 9.2)
   a. below
   b. horizontal to
   c. vertical to
   d. above

6. Convex-shaped ST segment elevation is most often due to (LO 9.2)
   a. myocardial injury.
   b. pericarditis.
   c. hyperkalemia.
   d. digitalis use.

7. The T wave represents (LO 9.3)
   a. ventricular depolarization.
   b. atrial repolarization.
   c. ventricular repolarization.
   d. abnormal electrical activity.

8. Which of the following is characteristic of normal T waves? (LO 9.3)
   a. It has a symmetrical shape.
   b. It is larger than the P wave.
   c. It has a deflection that is opposite of its associated QRS complex.
   d. Its beginning has a more abrupt slope than the end.

9. What direction do the positive electrolytes move during repolarization?
   (LO 9.2)

10. The maximum amplitude of T waves in the precordial leads is _____ mm
    in height. (LO 9.3)
    a. 5
    b. 8
    c. 10
    d. 15

11. To determine the amplitude of the T wave, what do you measure? (LO 9.3)

12. Inverted T waves may be seen with (LO 9.3)
    a. hyperkalemia.
    b. pericarditis.
    c. early beats arising from the atria.
    d. myocardial ischemia.

13. What does the QT interval represent? (LO 9.4)

14. The normal QT interval has a duration of _____ to _____
    seconds. (LO 9.4)
    a. 0.24, 0.48
    b. 0.28, 0.44
    c. 0.36, 0.44
    d. 0.44, 0.56

15. To measure the QT interval, we measure from the beginning of the
    _____ to the end of the _____. (LO 9.4)
    a. Q wave, T wave
    b. S wave, T wave

     c. R wave, U wave

     d. P wave, S wave

16. Prolonged QT intervals may be seen with   (LO 9.4)

     a. pericarditis.

     b. digitalis toxicity.

     c. hypercalcemia.

     d. congenital conduction system defect.

17. Which of the following describes U waves?   (LO 9.5)

     a. have sharply pointed waveforms

     b. immediately follow the ST segment

     c. take the same direction as the T wave

     d. always merged with the T wave

18. Prominent U waves are most often seen with   (LO 9.5)

     a. hypokalemia.

     b. bradycardia.

     c. myocardial ischemia.

     d. the use of haloperidol and methadone.

19. Your patient has a QT interval that is 0.48 seconds in duration. This is referred to as a _____ QT interval.   (LO 9.4)

     a. prolonged

     b. normal

     c. shortened

     d. refractory

20. In the presence of myocardial ischemia or infarction, a prolonged QT interval can predispose the patient to   (LO 9.4)

     a. bradycardia.

     b. asystole.

     c. torsades de pointes.

     d. heart failure.

Referring to the case history at the beginning of this chapter, answer the following questions:

21. Elevated ST segments and inverted T waves on the 12-lead ECG suggest the patient is experiencing   (LO 9.2)

     a. pericarditis.

     b. hypercalcemia.

     c. heart failure.

     d. myocardial ischemia and injury.

22. The patient's heart rate can be described as   (LO 9.4)

     a. tachycardic.   b. irregular.

     c. bradycardic.   d. normal.

Practice Makes Perfect

For each of the 13 ECG tracings on the following pages, practice the Nine-Step Process for analyzing ECGs. To achieve the greatest learning, you should practice the step immediately after reading the related chapter (i.e., practice determining the heart rate after reading Chapter 4). Below are questions you should consider as you complete each step. All tracings are six seconds in duration.

1. Determine the heart rate. Is it slow? Normal? Fast? What is the ventricular rate? What is the atrial rate?

2. Determine if the rhythm is regular or irregular. If it is irregular, what type of irregularity is it? Occasional or irregular? Slight? Sudden acceleration or slowing in heart rate? Total? Patterned? Does it have a variable conduction ratio?

3. Determine if P waves are present. If so, how do they appear? Do they have normal height and duration? Are they tall? Notched? Wide? Biphasic? Of differing morphology? Inverted? One for each QRS complex? More than one preceding some or all the QRS complexes? Do they have a sawtooth appearance? An indiscernible chaotic baseline?

4. Determine if QRS complexes are present. If so, how do they appear? Narrow with proper amplitude? Tall? Low amplitude? Delta wave? Notched? Wide? Bizarre? With chaotic waveforms?

5. Determine the presence of PR intervals. If present, how do they appear? Constant? Of normal duration? Shortened? Lengthened? Progressively longer? Varying?

6. Evaluate the ST segments. Do they have normal duration and position? Are they elevated? (If so, are they flat, concave, convex, arched?) Depressed? (If so, are they normal, flat, downsloping, or upsloping?)

7. Determine if T waves are present. If so, how do they appear? Of normal height and duration? Tall? Wide? Notched? Inverted?

8. Determine the presence of QT intervals. If present, what is their duration? Normal? Shortened? Prolonged?

9. Determine if U waves are present. If present, how do they appear? Of normal height and duration? Inverted?

1.

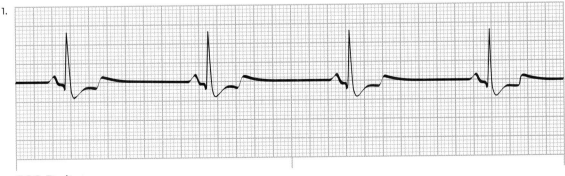

ECG Findings:

2.

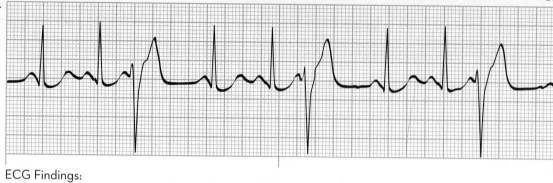

ECG Findings:

3.

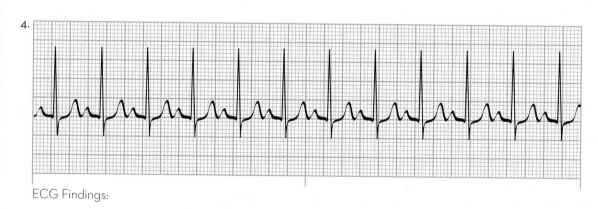

ECG Findings:

4.

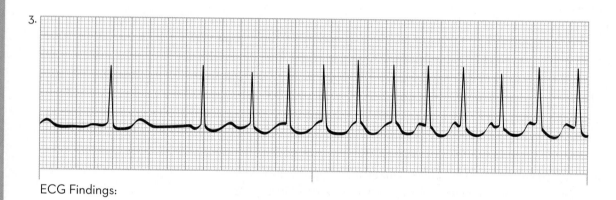

ECG Findings:

5.

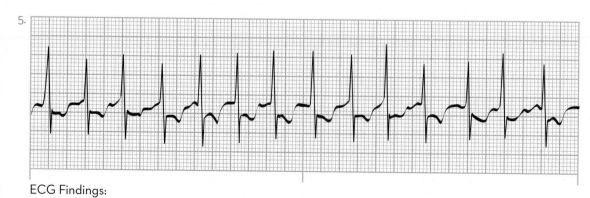

ECG Findings:

6.

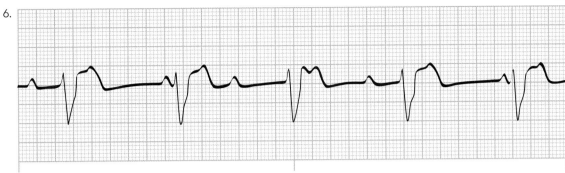

ECG Findings:

7.

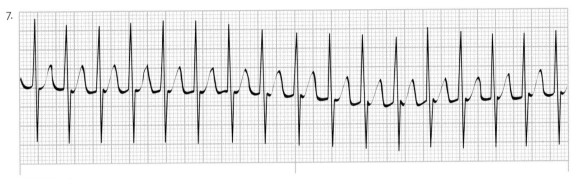

ECG Findings:

8.

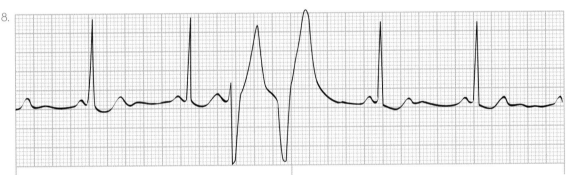

ECG Findings:

9.

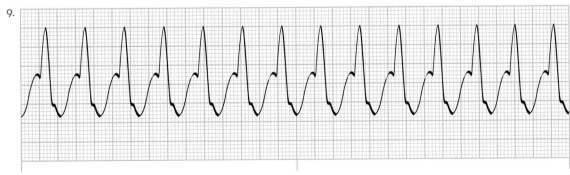

ECG Findings:

10.

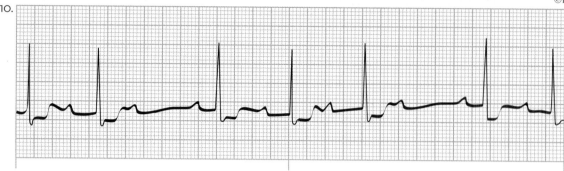

ECG Findings:

11.

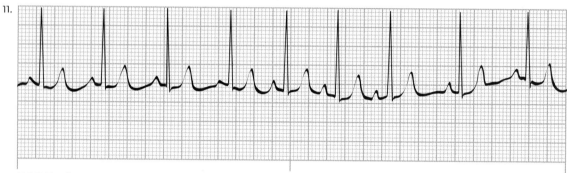

ECG Findings:

12.

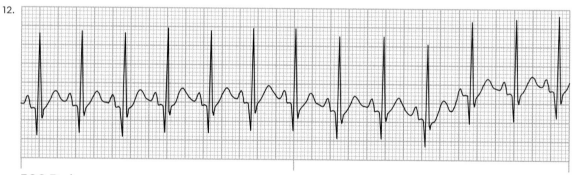

ECG Findings:

13.

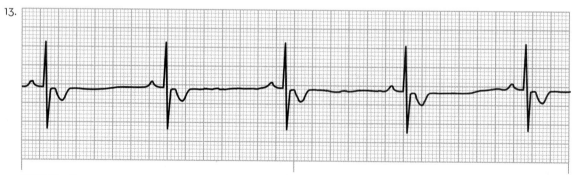

ECG Findings:

10 Heart Disease

Chapter Outline

Defining Heart Disease

Risk Factors of Heart Disease

Complications of Heart Disease

Types of Heart Disease

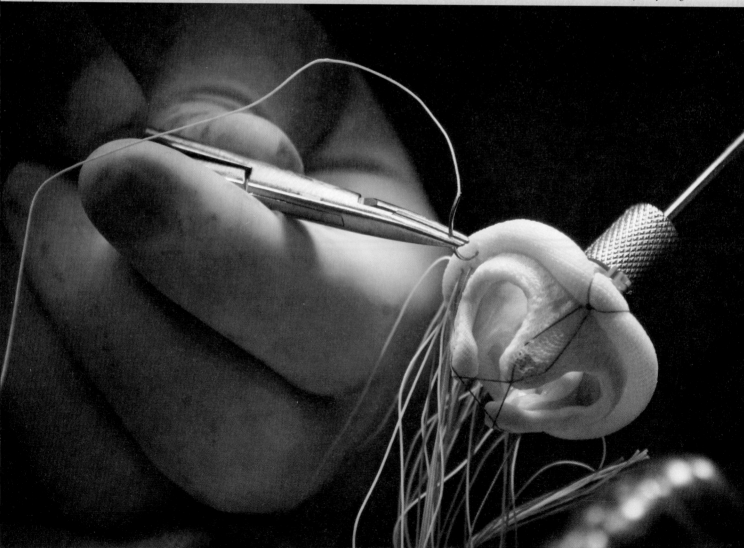

Learning Outcomes

LO 10.1 Define the term heart disease.

LO 10.2 Identify and describe risk factors for heart disease.

LO 10.3 Recall the complications of heart disease.

LO 10.4 Describe the various forms of heart disease.

Case History

On a Thursday morning, at approximately 2:15 A.M. in a small rural town, an EMS unit is dispatched to a residence for an 87-year-old male with difficulty breathing. The EMS unit arrives on scene before the fire engine arrives and goes inside the residence. Upon entering the bedroom, the paramedics find an elderly male, sitting on the edge of the bed, vomiting. He is alert and appears to be in severe respiratory distress. They note that he is diaphoretic, with rapid and labored breathing and intercostal retraction, yet he tries to explain what happened. He tells the paramedics he woke up 15 minutes ago because he could not breathe while lying down. Questioning reveals that he has a history of hypertension and heart failure for which he takes several medications. He also tells the paramedics that he has a "leaking heart valve."

In this chapter, we cover heart diseases that can produce changes on the electrocardiogram (ECG). Learning about these diseases will help you understand how the dysrhythmias and cardiac conditions we discuss later in this book occur.

10.1 Defining Heart Disease

Now that you have learned the anatomy and physiology of the heart, how the ECG machine works and the 9-step process for analyzing the electrocardiogram, it is time to learn about the dysrhythmias and conditions the ECG machine can reveal. It is helpful to have an understanding of the causes and pathophysiology of heart disease, so let's begin with that.

Heart disease is defined any medical condition of the heart or the blood vessels supplying it, or of muscles, valves, or internal electrical pathways, that impairs cardiac functioning. It is the leading killer in the United States and affects an estimated 14 million adults. Heart disease is responsible for more deaths in the United States than the second through seventh leading causes of death combined.

Heart diseases discussed in this chapter include coronary artery disease (CAD), heart muscle diseases, infections, **valvular diseases**, and **congenital heart defects**.

First, let's review the commonly recognized risk factors for heart disease and some of the complications associated with it.

10.2 Risk Factors of Heart Disease

Risk factors that can lead or contribute to heart disease include the following:

Age, Gender, and Family History

As people age, there is a greater risk of damaged and narrowed arteries due to fatty deposits called **plaques** collecting along artery walls and weakened or thickened heart muscle. Also, poor nutrition and exercise habits, high blood pressure, smoking, and diabetes can increase the risk of heart disease. Most people who die of heart disease are 65 or older. At older ages, women who experience myocardial infarction are more likely than men are to die within a few weeks from the myocardial infarction.

Men have a greater risk of heart disease than premenopausal women. However, after menopause, women have about the same risk of heart disease as men.

People are more likely to have coronary artery disease if they have a family history of heart disease. This is particularly true if a parent developed it before age 55 (for a male relative, such as a brother or father) or age 65 (for a female relative, such as a mother or sister).

Smoking and Alcohol Intake

Smokers are at greater risk for myocardial infarction than nonsmokers as nicotine constricts the blood vessels, and carbon monoxide can damage the inner lining, making the blood vessels more susceptible to atherosclerosis. It can also lead to unhealthy cholesterol levels and raise blood pressure. Further, smoking can limit how much oxygen reaches the body's tissues. The risks associated with tobacco use result from both direct consumption of tobacco and exposure to second-hand smoke.

Drinking alcohol excessively can raise blood pressure and increase risk of cardiomyopathy and stroke, cancer, and other diseases. It can also contribute to high **triglycerides**, obesity, alcoholism, suicide, and accidents, as well as produce dysrhythmias. However, on the positive side, moderate alcohol consumption provides a cardioprotective effect.

Poor Diet, Obesity, and Physical Inactivity

Persons having a diet that is rich in fat, trans fat, salt, sugar, and cholesterol and low in fruits, vegetables, and fish are associated with a higher risk for cardiovascular disease. Also, it is widely recognized that excess weight typically worsens other risk factors. A lack of exercise or a sedentary lifestyle is also associated with many forms of heart disease and some of its other risk factors, as well.

High Blood Pressure

Defined as a persistent, abnormally elevated blood pressure, uncontrolled high blood pressure can lead to hardening and thickening of the arteries, which narrows the vessels through which blood flows. High blood pressure is also referred to as hypertension.

High blood pressure is one of two types; primary or secondary. Primary, or essential, high blood pressure is the most common type of high blood pressure. It tends to develop as a person gets older. Secondary high blood pressure is caused by some other medical condition or use of certain medicines. This type usually resolves after the cause is treated or removed.

Causes of high blood pressure include changes, either from genes or from the environment, or changes in the body's normal functions, including changes to kidney fluid and salt balances, the **renin-angiotensin-aldosterone system**, sympathetic nervous system activity, and blood vessel structure and function. An unhealthy lifestyle can cause high blood pressure, including factors such as high dietary sodium intake, excess intake of alcohol, and/or an absence of physical activity.

High Blood Cholesterol Levels

High levels of **cholesterol** in the blood can increase the risk of formation of plaques and **atherosclerosis**.

The human body needs some cholesterol to make hormones, vitamin D, and substances that help digest foods. Normally, the liver manufactures the required amount on its own. However, cholesterol is found in some foods and if one eats too many of those foods, cholesterol levels can rise.

Cholesterol travels through the bloodstream in small packages called **lipoproteins**. These packages are made of fat (lipid) on the inside and proteins on the outside. Two kinds of lipoproteins carry cholesterol throughout the body: low-density lipoproteins (LDL) and high-density lipoproteins (HDL). Having healthy levels of both types of lipoproteins is important.

LDL cholesterol sometimes is called "bad" cholesterol. A high LDL level leads to a buildup of cholesterol in the arteries. HDL cholesterol is sometimes called "good" cholesterol. This is because it carries cholesterol from other parts of the body back to the liver. The liver removes cholesterol from the body in order to maintain appropriate amounts.

Diabetes

Diabetes increases the risk of heart disease. Both diabetes and heart disease share similar risk factors, such as obesity and high blood pressure.

Stress and "Type A" Personalities

Unrelieved stress may damage the arteries and worsen other risk factors for heart disease. Also, "type A" personalities (impatient, aggressive, and/or competitive) are more susceptible to heart disease.

Poor Hygiene

An absence of regular hand washing and a failure to maintain other habits that help prevent viral or bacterial infections put a person at risk for heart infections, especially if they already have an underlying heart condition. Poor dental health can also contribute to heart disease.

10.3 Complications of Heart Disease

Complications of heart disease are the conditions that can result from the different forms of heart disease and include:

Dysrhythmias

Dysrhythmias are irregularities or abnormalities in the heart rhythm and/or heart rate. In a person with a normal, healthy heart, it's unlikely that fatal dysrhythmias will develop in the absence of an outside trigger, such as an electrical shock or use of illicit drugs. However, in a heart that's diseased or deformed (such as one having an area of scarred tissue), the heart's electrical impulses do not properly form or may travel through the heart abnormally, making dysrhythmias more likely to occur. Dysrhythmias are covered in more detail in subsequent chapters.

Angina

Angina is defined as chest pain or discomfort that occurs if an area of the heart muscle does not receive enough oxygen-rich blood. Angina may present as pressure or squeezing in the chest. The pain can also radiate to the shoulders, arms, neck, jaw, or back. Angina pain may even feel like indigestion. Angina isn't a disease but rather it is a symptom of an underlying heart problem. Angina usually is a symptom of coronary heart disease.

Myocardial Infarction

Myocardial infarction occurs when there is a sudden decrease or total cessation of blood flow through a coronary artery to an area of the myocardium. This results in injury or death of that part of the heart muscle. Atherosclerosis is commonly the cause of myocardial infarction. This condition is discussed in more depth later in this chapter and in Chapter 19.

Both angina and myocardial infarction are described in greater detail a little later in this chapter.

Dilation and Hypertrophy

Dilation or enlargement of a chamber (Figure 10-1A) occurs as the result of volume overload—the chamber dilates to accommodate the increased blood volume. The chamber is therefore larger and can hold more blood than normally. Unlike hypertrophy, the muscular wall of the dilated chamber typically does not become thicker. Enlargement is most often seen with certain types of valvular disease. For example, left atrial enlargement can result from mitral valve insufficiency.

Hypertrophy means an increase in muscle mass. It is a condition in which the muscular wall of the ventricle(s) becomes thicker than normal (Figure 10-1B). It results from the ventricle having to pump against increased resistance within the cardiovascular system. Many conditions cause hypertrophy, including systemic hypertension and aortic stenosis.

Heart Failure

Heart failure, also known as congestive heart failure, occurs when the heart is weak and unable to pump enough blood to meet the body's metabolic needs. The diminished

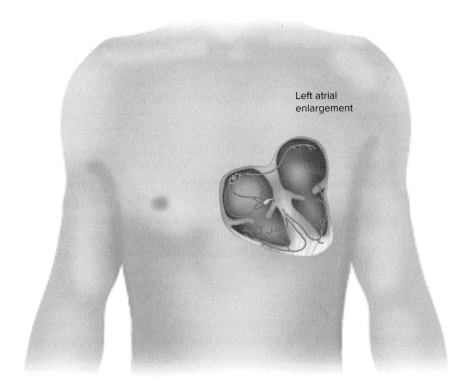

A

Figure 10-1
(A) Enlargement of a heart chamber occurs to accommodate increased blood volume. (B) Hypertrophy is a condition in which the muscular wall of the ventricle becomes thicker.

Left atrial enlargement

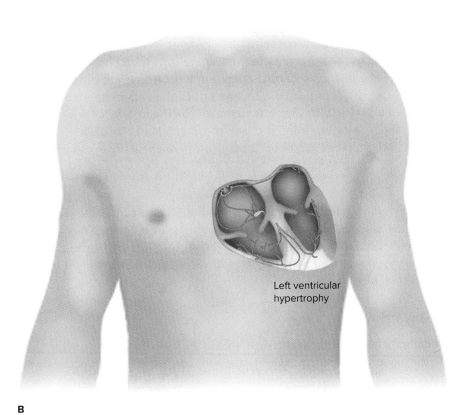

B

Left ventricular hypertrophy

pumping ability leads to blood and fluid backing up into the lungs (called pulmonary edema), and to the buildup of fluid in the abdomen, liver, legs, ankles, and feet (called peripheral edema). Sometimes the heart is dilated and weak; at other times, it can be stiff and thickened.

Heart failure is often the end stage or outcome of many forms of heart disease. It may result from conditions, such as myocardial infarction, that directly affect the heart muscle and impair ventricular function, but it can also be caused by problems such as coronary artery disease, hypertension, drugs that adversely affect the ability of the heart to pump, thyroid disease, heart defects, valvular heart disease, heart infections, and **cardiomyopathy**.

Heart failure is classified as right-sided or left-sided. With left-sided heart failure, the pumping ability of the left ventricle is diminished and fluid backs up into the lungs, causing pulmonary congestion. In its worst form, the lungs fill with large amounts of fluid resulting in the development of pulmonary edema. With right-sided heart failure, the pumping ability of the right ventricle is impaired and fluid backs up into the body. Right-sided heart failure is often caused by left-sided heart failure which makes pumping blood through the pulmonary circulation difficult, causing the right ventricle to weaken. In its weakened state, the right ventricle is unable to pump effectively.

Heart failure can affect both children and adults, but it mostly affects older adults. It is more common in people who are 65 years old or older, African Americans, people who are overweight, and people who have had myocardial infarction. Men have a higher rate of heart failure than women.

Cardiogenic Shock

Cardiogenic shock is a condition in which a suddenly weakened heart cannot pump enough blood to meet the body's needs. It is caused by profound failure of the cardiac muscle, primarily the left ventricle. The most common cause of cardiogenic shock is damage to the heart muscle from severe myocardial infarction. However, not everyone who has a myocardial infarction has cardiogenic shock. In fact, on average, only about 7% of people who have myocardial infarction develop the condition. When a person is in shock, not enough blood and oxygen reach the body's organs. If shock lasts too long, the lack of oxygen starts to damage the body's organs. If shock isn't treated quickly, it can cause permanent organ damage or death.

Stroke

The risk factors that lead to cardiovascular disease also can lead to an ischemic **stroke**, which happens when the arteries to the brain are narrowed, blocked or compressed so that too little blood reaches the brain. Like a heart attack, a stroke is a medical emergency—brain tissue begins to die within just a few minutes of compromised cerebral blood flow.

Aneurysm

Most commonly, an **aneurysm** is a weak spot in the wall of a blood vessel, which can cause the blood vessel to bulge. This is a serious condition that can occur anywhere in the body.

Aneurysms are often caused by atherosclerosis; however, people with certain connective tissue disorders may also be at risk of developing an aneurysm.

Aneurysms commonly occur in the aorta, the large artery that carries blood from the heart to the systemic circulation. They occur in this major artery largely because of the constant volume of blood which is under high pressure that courses through it.

An aortic dissection occurs when the arterial inner wall tears, allowing blood to enter the space between the layers. If the aorta ruptures, hemorrhage occurs and it is often fatal.

Also, the ventricle can develop an aneurysm and potentially rupture. This is usually caused by myocardial infarction.

Peripheral Artery Disease

Atherosclerosis can lead to **peripheral artery disease**. When a person develops peripheral artery disease, the extremities—usually the legs—don't receive enough blood flow. This causes symptoms such as leg pain when walking (claudication).

Pulmonary Embolism

A **pulmonary embolism** is an acute blockage of one of the pulmonary arteries by a blood clot or other foreign matter (Figure 10-2). This leads to obstruction of blood flow to the lung segment supplied by the artery. The larger the artery occluded, the more massive the pulmonary embolus and therefore the larger the effect the embolus has on the lungs and heart. Due to the increased pressure in the pulmonary artery caused by the embolus, the right atrium and ventricle become distended and unable to function properly, leading to right heart failure. This condition is referred to as **acute cor pulmonale**. Massive pulmonary embolism impairs oxygenation of the blood, and death may result. The most common source of the clot is in one of the large pelvic or leg veins. The pain that accompanies a pulmonary embolus is pleuritic, and shortness of breath is often present.

Sudden Cardiac Arrest

Sudden cardiac arrest is the sudden, unexpected loss of heart function, breathing, and consciousness, often caused by a dysrhythmia such as ventricular fibrillation. Sudden cardiac arrest is a true medical emergency. If not immediately treated, it is fatal. The prompt delivery of cardiopulmonary resuscitation (CPR) and

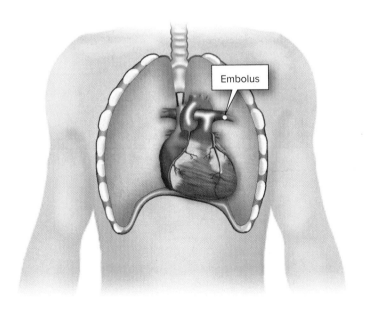

Figure 10-2

A pulmonary embolism is an acute blockage in one of the pulmonary arteries.

defibrillation may restore effective pumping of blood and, thus, be lifesaving for some individuals.

Now that have discussed some of the many risks and complications of heart disease, let's review several types of heart disease.

10.4 Types of Heart Disease

Coronary Artery Disease

Any major interruption of blood flow through the coronary arteries can cause ischemia, injury, and infarction of the heart muscle. This can lead to impairment of myocardial function and sudden death.

Coronary artery disease is an umbrella term for the various diseases that reduce or stop blood flow through the coronary arteries. Atherosclerosis accounts for over 90% of the cases of coronary artery disease.

Atherosclerosis is a gradual, progressive process involving thickening and hardening of the arterial wall. It begins with small deposits of fatty material, particularly cholesterol, invading the intima, the inner lining of the arteries (Figure 10-3). The body

Figure 10-3

Development and progression of coronary artery disease.

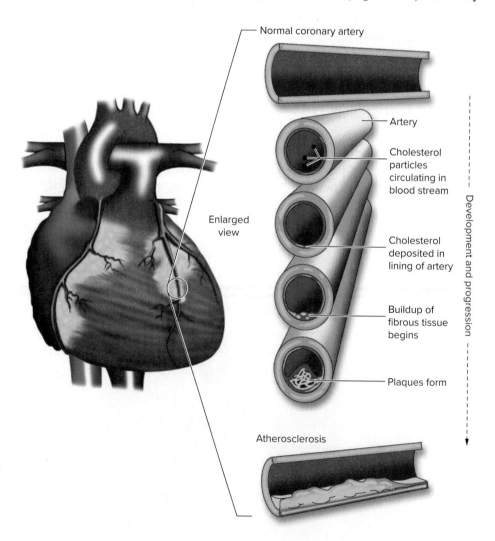

Normal coronary artery

Enlarged view

Artery

Cholesterol particles circulating in blood stream

Cholesterol deposited in lining of artery

Buildup of fibrous tissue begins

Plaques form

Atherosclerosis

Development and progression

has an inflammatory response to protect itself and sends white blood cells called macrophages to engulf the invading cholesterol in the arterial wall. When the macrophages are full of cholesterol, they are called foam cells because of their appearance. This engorgement of the macrophages produces a nonobstructive lesion called a fatty streak, which is situated between the intima and media of the artery. Later, these areas of deposit become invaded by fibrous tissue, including lipoprotein-filled smooth muscle cells and collagen that become calcified and harden into plaque called a **fibrous cap**—hard on the outside but soft and mushy or sticky on the inside. This leads to narrowing of the affected vessels and a reduction of blood flow through them. If narrowing of the artery reaches the stage where blood flow is insufficient to meet the oxygen demands of the myocardium, coronary artery disease is said to exist. If the hard, shell-like outer edge of the cap eventually ruptures, blood components like platelets and small blood clots form a large clot, effectively blocking blood flow through the artery. The heart muscle downstream from the clot then suffers from a lack of blood and becomes damaged or dies.

We use the term **acute coronary syndrome** to denote any group of clinical symptoms compatible with acute myocardial ischemia. Acute myocardial ischemia is chest pain due to insufficient blood supply to the heart muscle that results from coronary artery disease (also called coronary heart disease).

Myocardial Ischemia

Myocardial ischemia is a lack of oxygen and nutrients to the myocardium (Figure 10-4). It typically occurs when the heart has a greater need for oxygen than the narrowed coronary arteries can deliver or when there is a loss of blood supply to the myocardium. Causes of myocardial ischemia include atherosclerosis, **vasospasm**, **thrombosis**, **embolism**, decreased ventricular filling time (such as that produced by tachycardia), and decreased filling pressure in the coronary arteries (such as that caused by severe hypotension or aortic valve disease).

Angina

When ischemia occurs due to an intermittent shortage of oxygen to the myocardium brought on by exertion, emotional stress, or even the stress of cold weather, it produces chest pain, also known as angina or angina pectoris. The ischemia is usually relieved within 2 to 10 minutes of rest or by administration of a medicine such as nitroglycerin that dilates the coronary arteries and allows the needed blood (and oxygen) to flow to the myocardium.

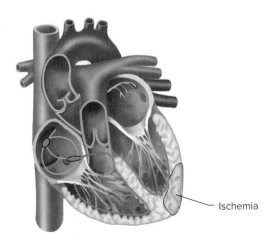

Figure 10-4
Myocardial ischemia is a lack of nutrients and oxygen to the myocardium.

Ischemia

Figure 10-5
Myocardial injury is a degree
of cellular damage beyond
ischemia.

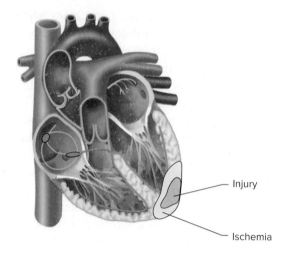

A patient who has angina may remain stable for many years or may go on to experience an infarction. Angina is considered unstable if it is more easily provoked, becomes less predictable, worsens over time and/or medicines such as nitroglycerin do not relieve the pain. Another type is variant **Prinzmetal's angina**. This a rare condition where coronary artery spasm causes the angina. It usually happens between midnight and early morning. Medicine can relieve this type of angina.

Ischemia as the First Phase of Myocardial Infarction

Ischemia may also be the onset of myocardial infarction. As a point of information though, permanent damage can often be avoided if the blood supply is quickly restored to the ischemic tissue.

Myocardial Injury

If the myocardial oxygen demand fails to lower, the coronary artery blockage worsens, or the ischemia progresses untreated, it can lead to **myocardial injury** (Figure 10-5). Myocardial injury reflects a degree of cellular damage beyond ischemia. It occurs if the blood flow is not restored (and ischemia reversed) within a few minutes. The degree of injury depends on if and how quickly the blood supply is restored, as well as on how much myocardium is involved. If blood flow is not restored to the affected area, tissue death will occur. Treatments used to restore blood flow include administration of **fibrinolytic agents**, **coronary angioplasty**, and **coronary artery bypass graft**.

Myocardial Infarction

Myocardial infarction is the death of injured myocardial cells (Figure 10-6). It occurs when there is a sudden decrease or total cessation of blood flow through a coronary artery to an area of the myocardium. Myocardial infarction commonly occurs when the intima of a coronary artery ruptures, exposing the atherosclerotic plaque to the blood within the artery. This initiates the abrupt development of a clot (thrombus). The vessel, already narrowed by the plaque, becomes completely blocked by the thrombus. The area of the heart normally supplied by the blocked artery goes through a characteristic sequence of events described as zones of ischemia, injury, and infarction.

Each zone is associated with characteristic ECG changes. There is ischemic and injured myocardial tissue surrounding the infarcted area so the size of the infarction can increase if coronary perfusion is not restored. If the patient survives the acute myocardial infarction, the infarcted tissue is replaced with scar tissue.

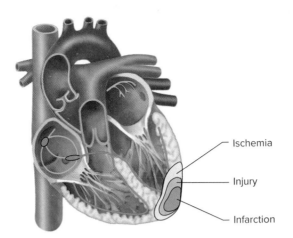

Figure 10-6
Myocardial infarction is the
death of injured myocardial
cells.

- Ischemia
- Injury
- Infarction

When blockage occurs in the left anterior descending artery, it is particularly dangerous because this vessel supplies a much larger portion of the total myocardial mass than the right coronary and left circumflex arteries. Even more serious is obstruction of the left main coronary artery. Significant narrowing of this short vessel causes reduction in blood flow through both the left anterior descending and circumflex arteries. This compromises the blood supply of almost the entire left ventricle. Fortunately, obstruction of the left main coronary artery occurs in only 5% to 10% of patients with symptoms of coronary artery disease.

Cardiomyopathy

Cardiomyopathy is any disease of the heart muscle that makes it harder for the heart to pump blood to the rest of the heart. It may result in heart failure, where the heart is unable to pump as much blood as the body needs.

A common cause of cardiomyopathy is lack of blood flow to the heart muscle, usually because of myocardial infarction. Other causes include excess alcohol consumption, nutritional or hormonal imbalances, viral infection of the heart, connective tissue disorders, excessive iron buildup in the body, some cancer treatments such as chemotherapy and radiation, and certain inherited diseases. There are several types

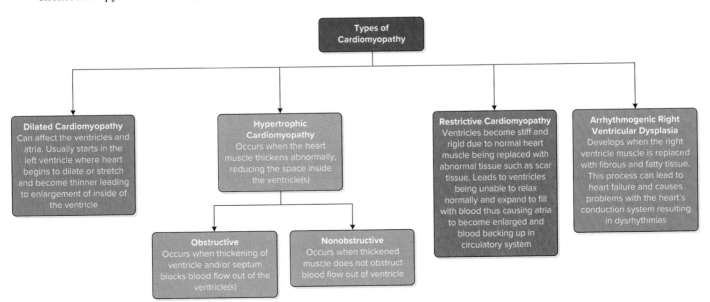

Figure 10-7
Comparison of different types of cardiomyopathy.

including dilated cardiomyopathy, hypertrophic cardiomyopathy, and restrictive car-diomyopathy (Figure 10-7). Arrhythmogenic right ventricular dysplasia (ARVD) is another form of cardiomyopathy.

Dilated Cardiomyopathy

Dilated cardiomyopathy is a disease of the heart muscle fibers. It is the most common type of cardiomyopathy, typically occurring in adults 20 to 60 years of age. Usually, it isn't diagnosed until it is in an advanced stage. Once identified, prognosis is often poor.

Frequently the disease begins in the left ventricle. The heart muscle begins to dilate, meaning it stretches and becomes thinner. Thus, the inside of the chamber enlarges (Figure 10-8). The problem often spreads to the right ventricle and then to the atria, giving the heart a globular shape (Figure 10-9). As the heart chambers dilate, the heart muscle is no longer an effective pump as it lacks effective contractility. As the heart becomes weaker, heart failure can occur. Dilated cardiomyopathy primarily affects systolic function.

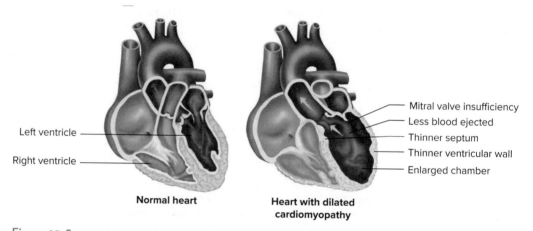

Normal heart

Heart with dilated cardiomyopathy

Left ventricle

Right ventricle

Mitral valve insufficiency

Less blood ejected

Thinner septum

Thinner ventricular wall

Enlarged chamber

Figure 10-8
Anatomy and contraction of normal heart compared to heart with dilated cardiomyopathy where the heart muscle is weakened and the chambers enlarged.

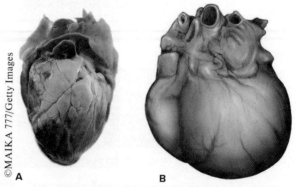

©MAIKA 777/Getty Images

A   B

Figure 10-9
External features of (A) normal heart and (B) heart with dilated cardiomyopathy.

Dilated cardiomyopathy also can lead to heart valve problems, dysrhythmias, and blood clots in the heart.

Hypertrophic Cardiomyopathy

Hypertrophic cardiomyopathy is a primary disease of cardiac muscle and the intraventricular septum of the heart. In this type of cardiomyopathy, the heart's myocytes increase in size, thickening the ventricular walls and making the inside of the ventricle(s) (usually the left) smaller (Figure 10-10). This results in the ventricle(s) holding less blood. The entire ventricle may thicken, or the thickening may happen only at the bottom of the heart. Also, thickening of the ventricle may cause it to become stiff, noncompliant, and unable to relax during ventricular filling. Pressure then rises in the left atria and ventricle, leading to an increase in pulmonary venous pressures. This condition mainly affects diastolic function.

The ventricle size often remains normal, but the thickening may block blood flow out of the ventricle. When this occurs, it is called obstructive hypertrophic cardiomyopathy. In other cases, the thickened heart muscle doesn't block blood flow out of the left ventricle and it is referred to as nonobstructive hypertrophic cardiomyopathy.

Sometimes the septum thickens and bulges into the left ventricle. This can also block blood flow out of the left ventricle, causing the ventricle to work harder to pump blood. Furthermore, hypertrophic cardiomyopathy can affect the heart's mitral valve, causing blood to leak backward through the valve during ventricular contraction.

Changes also occur to the cells in the damaged heart muscle, which can disrupt the heart's electrical signals and lead to dysrhythmias. It is rare, but some people with hypertrophic cardiomyopathy can have sudden cardiac arrest during very vigorous physical activity. The physical activity is believed to trigger dangerous dysrhythmias.

Hypertrophic cardiomyopathy is usually inherited but can also develop over time because of hypertension or aging.

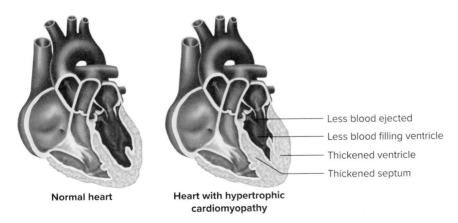

Normal heart   **Heart with hypertrophic cardiomyopathy**

- Less blood ejected
- Less blood filling ventricle
- Thickened ventricle
- Thickened septum

Figure 10-10
Comparison of normal heart to one with hypertrophic cardiomyopathy.

Figure 10-11
Heart with restrictive
cardiomyopathy is unable to
fill during diastole.

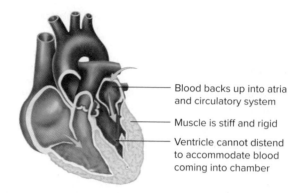

Blood backs up into atria
and circulatory system

Muscle is stiff and rigid

Ventricle cannot distend
to accommodate blood
coming into chamber

Restrictive Cardiomyopathy

This least common type of cardiomyopathy is restrictive cardiomyopathy. It is a severe, irreversible disease of the heart muscle fibers that leads to the heart muscle becoming stiff and rigid (Figure 10-11). It occurs because abnormal tissue, such as scar tissue, replaces the normal heart muscle.

In this disease, the contractile function of the heart and wall thicknesses are usually normal, but the filling phase of the heart is very abnormal. In restrictive cardiomyopathy, the heart muscle doesn't relax normally so one or both ventricles don't properly fill with blood. As such, it is a condition that mainly affects diastolic function. Because blood that would normally enter the ventricle backs up in the atria, it causes them to become enlarged. Blood continues to back up in the circulatory system, eventually leading to heart failure and dysrhythmias.

Restrictive cardiomyopathy tends to affect older adults.

Arrhythmogenic Right Ventricular Dysplasia

Arrhythmogenic right ventricular dysplasia (ARVD), also known as arrhythmogenic right ventricular cardiomyopathy (ARVC), is a rare but serious disorder that usually affects teens and young adults. It is often associated with sudden cardiac arrest death in young athletes. ARVD results in the right ventricular muscle being replaced by fibrous and fatty tissue (Figure 10-12) which can lead to heart failure and/or disrupt the heart's electrical signals leading to dysrhythmias. Most cases are inherited. For this reason, if a person is diagnosed with ARVD, then immediate family members (siblings, parents, and children) are tested for the condition. This condition is discussed in more depth in Chapter 20.

Figure 10-12
Arrhythmogenic right
ventricular dysplasia (ARVD)
with close-up view of tissue
changes in the right ventricle.

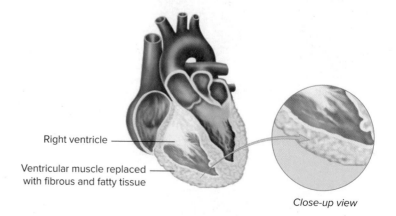

Right ventricle

Ventricular muscle replaced
with fibrous and fatty tissue

Close-up view

Heart Infection

Heart infections, such as **pericarditis**, **endocarditis**, and **myocarditis** (Figure 10-13), are caused when an irritant, such as a bacterium, virus, or chemical, reaches the heart muscle. The most common causes of heart infections include:

- Bacteria
- Viruses
- Parasites

Patients with these conditions can experience debilitating structural damage of the heart due to normal healing processes and the development of scar formation.

Pericarditis

Pericarditis is inflammation of the pericardium (Figure 10-14). Remember from our discussion in Chapter 1, the pericardium is a double-layered membrane that surrounds the heart. The potential space between the two layers holds 15 to 25 mL of serous, lubricating fluid.

The most common causes of pericarditis are viral and bacterial infections. Other causes include uremia, renal failure, rheumatic fever, connective tissue disease, and

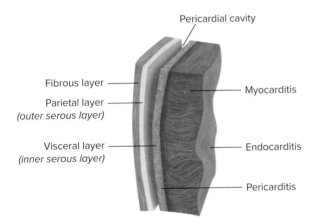

Figure 10-13
Types of heart infection.

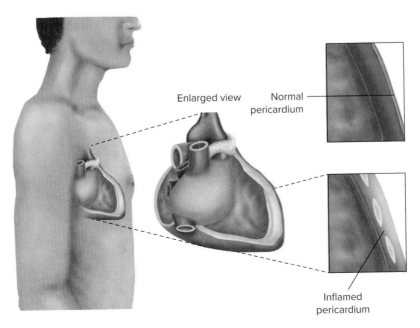

Figure 10-14
Effects of pericarditis on the heart.

cancer. Sometimes pericarditis can accompany myocardial infarction. When pericarditis occurs in conjunction with a myocardial infarction, it develops several days afterward.

Pericarditis can produce sharp, substernal chest pain. The pain has an abrupt onset that worsens on inspiration or with movement, particularly when the individual is lying down. The pain lessens when sitting up and leaning forward.

Pericardial Effusion

The inflammatory process associated with pericarditis can stimulate the body's immune response, resulting in white cells or serous, fibrous, purulent, and hemorrhagic exudates being sent to the injured area. This can lead to pericardial effusion (Figure 10-15). Pericardial effusion is a buildup of an abnormal amount of fluid and/or a change in the character of the fluid in the pericardial space. In extreme cases, pericardial effusion can lead to the development of cardiac tamponade.

Figure 10-15
Pericardial effusion is the buildup of fluid in the pericardial sac which can lead to cardiac tamponade.

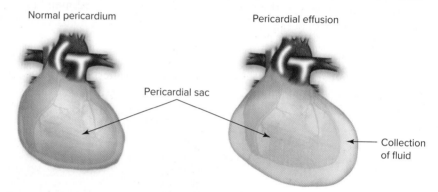

Normal pericardium

Pericardial effusion

Pericardial sac

Collection of fluid

Endocarditis

Endocarditis is an inflammation of the endocardium, the heart valves or the inner lining of the blood vessels attached to the heart (Figure 10-16). Endocarditis may be caused by many things including infection. When infection is the cause, it is referred to as infective endocarditis (IE). Infective endocarditis occurs if bacteria, fungi, or other germs invade the bloodstream and attach to abnormal areas of the heart. The infection can damage the heart and cause serious and sometimes fatal complications.

Infective endocarditis can develop quickly or slowly; it depends on what type of germ is causing it and whether there is an underlying heart problem. When it develops quickly, it's called acute infective endocarditis. When it develops slowly, it's called subacute infective endocarditis.

Figure 10-16
Endocarditis is an infection of the innermost layers of the heart. It may occur in people with congenital valve disease and those who have had rheumatic fever.

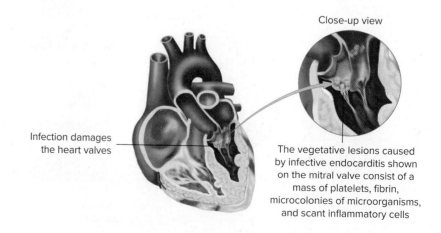

Close-up view

Infection damages the heart valves

The vegetative lesions caused by infective endocarditis shown on the mitral valve consist of a mass of platelets, fibrin, microcolonies of microorganisms, and scant inflammatory cells

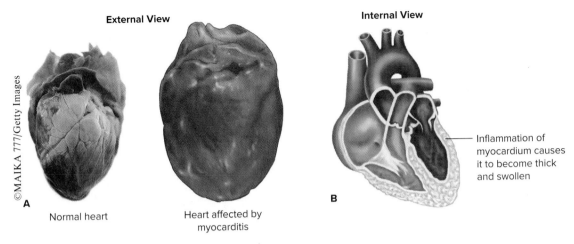

External View

Internal View

©MAIKA 777/Getty Images

A

Normal heart

Heart affected by
myocarditis

B

Inflammation of
myocardium causes
it to become thick
and swollen

Figure 10-17
(A) Comparison of external views of a normal heart and one with myocarditis. (B) Internal view of
heart with myocarditis.

Certain factors make it easier for bacteria to enter the bloodstream putting a person at greater risk for IE, including poor dental hygiene and unhealthy teeth and gums. Other risk factors include using intravenous (IV) drugs, having a catheter (tube) or another medical device in the body for long periods, and having a history of infective endocarditis.

Myocarditis

Myocarditis is focal or diffuse inflammation of the heart muscle (Figure 10-17). It can be acute or chronic and can occur at any age. Viruses such as Lyme disease, rheumatic fever, and others are often at the root of myocarditis. Other causes include bacterial infections, certain medications, toxins, and autoimmune disorders. Sometimes the cause is a systemic disorder such as celiac disease. The inflammation can lead to dysrhythmias, cardiomyopathy, or heart failure. Myocarditis may coincide with pericarditis.

Rheumatic Fever

Rheumatic fever is an infectious disease generally caused by a Streptococcus bacterium. The body develops antibodies to the streptococcal microorganism; then the antibodies attack the heart (particularly the valves), joints, central nervous system, skin, and subcutaneous tissues. Rheumatic fever can reoccur. If rheumatic fever isn't treated, scarring deformity of the cardiac structures results in rheumatic heart disease.

Valvular Heart Disease

In Chapter 1, we discussed how the heart has four valves: the tricuspid, pulmonary, mitral, and aortic valves. These heart valves consist of tissue flaps that open and close with each heartbeat. They help assure that blood flows one way only, forward, through the heart and into the pulmonary and systemic circulations.

Valvular heart disease is characterized by damage to or a defect in one or more of the valves, although it occurs more often in the mitral and aortic valves. It results mostly from aging with the diagnose being made when people are in their late fifties, but it can also result from birth defects, infections (such as rheumatic fever and infective endocarditis), connective tissue disorders, myocardial infarction, and others. The primary types of valvular heart disease are prolapse, insufficiency, and stenosis.

Prolapse

Prolapse occurs when the valve leaflets bulge or prolapse back into the chamber. In some cases, it may be a benign disorder. In other cases, the prolapsed valves form an imperfect closure and start to leak. Prolapse is the leading cause of insufficiency (Figure 10-18).

Figure 10-18
Prolapsed mitral valve.

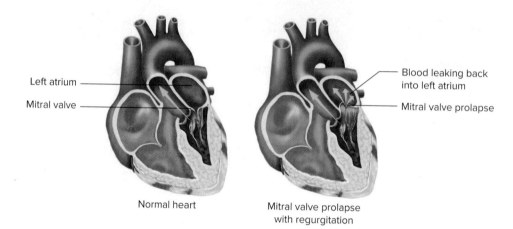

Left atrium

Mitral valve

Blood leaking back into left atrium

Mitral valve prolapse

Normal heart

Mitral valve prolapse with regurgitation

Insufficiency

Insufficiency occurs when one or more of the valves fails to close tightly and allows blood to leak back into the chamber. It is also called regurgitation, incompetence, or "leaky valve." Mitral valve insufficiency is the most common form of heart valve disease.

Mitral Insufficiency Mitral insufficiency results in blood from the left ventricle leaking back into the left atrium during ventricular contraction. The atrium enlarges to accommodate the backflow, leading to atrial dilation. It also causes the left ventricle to dilate to accommodate the increased volume from the atrium and to compensate for diminished cardiac output. This can lead to ventricular hypertrophy (Figure 10-19). The increased back pressure in the affected heart chambers can result in increased pulmonary artery pressure and left-sided and right-sided heart failure.

The enlarged atrium can also lead to the development of a cardiac dysrhythmia called atrial fibrillation. In atrial fibrillation, the atria quiver chaotically instead of beating effectively. This dysrhythmia further reduces the heart's ability to pump efficiently.

Figure 10-19
Valvular regurgitation is a condition in which blood leaks in the wrong direction because one or more heart valves closes improperly.
(A) Normal mitral valve.
(B) Mitral valve regurgitation.

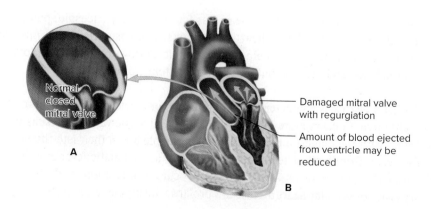

Normal closed mitral valve

Damaged mitral valve with regurgiation

Amount of blood ejected from ventricle may be reduced

A

B

Mild mitral regurgitation may be symptom free, while in more advanced regurgitation, signs of heart failure develop. In extreme cases, cardiovascular collapse with shock may result from papillary muscle rupture, rupture of a chorda tendinea, or infectious endocarditis of the mitral valve.

Aortic Insufficiency   Aortic insufficiency, also called aortic regurgitation or aortic incompetence, is a condition in which blood flows backward from a widened or weakened aortic valve (preventing its normal closure) into the left ventricle. Common causes of aortic regurgitation include endocarditis, rheumatic fever, collagen vascular disease, aortic dissection, and syphilis. In chronic aortic regurgitation, the stroke volume is increased, which leads to systolic hypertension, high pulse pressure, and increased afterload. The afterload seen with aortic regurgitation may be as high as that occurring in aortic stenosis.

Patients may be asymptomatic until they develop severe left ventricular dysfunction. The initial signs of aortic regurgitation are subtle and may include decreased functional capacity or fatigue. As the disease progresses, the typical presentation is that of left-sided heart failure: orthopnea, dyspnea, and fatigue.

Insufficiency can also occur in the tricuspid and pulmonary valves.

Stenosis

Stenosis occurs when a heart valve fails to fully open due to its flaps being thick, stiff, or fused together. This narrowed opening causes the heart to work harder to pump blood through the valve.

While stenosis most commonly affects the aortic and mitral valves, it can develop in all four valves. These conditions are called tricuspid stenosis, pulmonic stenosis, mitral stenosis, or aortic stenosis. Some valves present with both stenosis and regurgitation issues.

Mitral Stenosis   Mitral stenosis is a narrowing or blockage of the mitral valve. The narrowed valve causes blood to back up in the left atrium instead of flowing into the left ventricle. This leads to increased left atrium volume and pressure and dilation of the chamber. The increased resistance to blood flow causes pulmonary hypertension, right ventricular hypertrophy, and right-sided heart failure. Also, inadequate filling of the left ventricle produces low cardiac output. Most adults with mitral stenosis had rheumatic fever when they were younger. Mitral stenosis may also be associated with aging and a buildup of calcium on the ring around the valve where the leaflet and heart muscle meet.

Aortic Stenosis   Aortic stenosis is a narrowing or blockage of the aortic valve that results from calcification and degeneration of the aortic leaflets. The changed shape of the leaflets reduces blood flow through the valve. Left ventricular pressure rises to overcome the resistance of the narrowed valve causing the left ventricle to work harder to make up for the diminished cardiac output (Figure 10-20). Over time, the extra work can weaken the heart muscle and lead to left-sided heart failure. Furthermore, increased oxygen demand due to the added workload along with poor coronary perfusion due to diminished cardiac output leads to left ventricular ischemia. The survival of patients with aortic stenosis is nearly normal until the onset of symptoms, when survival rates decrease sharply.

Figure 10-20
Comparison of normal aortic
valve to aortic valve with
stenosis.

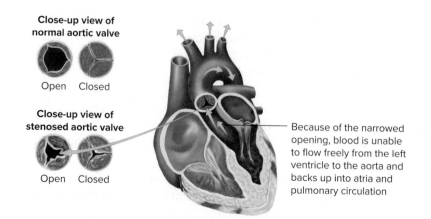

Because of the narrowed
opening, blood is unable
to flow freely from the left
ventricle to the aorta and
backs up into atria and
pulmonary circulation

Congenital Heart Defects

Congenital heart disease is a heart abnormality that is present at birth. The problem can affect the:

- heart walls
- heart valves (covered earlier)
- blood vessels

According to the Centers for Disease Control and Prevention, each year some 40,000 infants are born with congenital heart defects and there are currently one million adults and one million children in the United States living with congenital heart defects.

Congenital heart defects usually develop in utero, emerging as the heart develops, about a month after conception, changing the flow of blood in the heart. In some cases, the defect is not found until adulthood and can occur even if the defects were treated in childhood. This is because heart defects are seldom cured—they are often just repaired, so heart function is improved but it's often not completely normal. In other cases, problems which weren't serious enough to repair during childhood later worsen and require treatment. Some medical conditions, medications, and genes may play a role in causing heart defects.

While there are 35 different types of congenital heart defects, we only cover the more common ones (see Figure 10-21 and Table 10-1) in this chapter. They range from simple defects with no symptoms to complex defects with severe, life-threatening symptoms that require surgical repair.

Congenital heart disease is classified as either cyanotic congenital heart disease or acyanotic congenital heart disease. In both types, the heart isn't pumping blood as efficiently as it should. The main difference is that cyanotic congenital heart disease causes low levels of oxygen in the blood, and acyanotic congenital heart disease doesn't. Infants with reduced oxygen levels may experience breathlessness and a bluish tint to their skin. Infants who have enough oxygen in their blood don't display these symptoms, but they may still develop complications later in life, such as high blood pressure.

Over the past few decades, treatments and follow-up care for defects have improved, resulting in nearly all children with heart defects surviving into adulthood. Some need continuous care for their heart defect throughout their lives. However, many go on to have active and productive lives despite their condition.

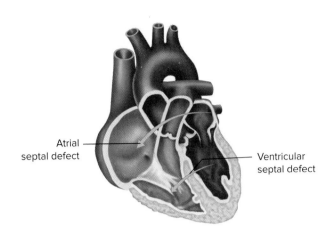

Figure 10-21
Two forms of congenital heart disease: Atrial septal defect (ASD) and ventricular septal defect (VSD).

Atrial septal defect

Ventricular septal defect

Congenital heart defects put patients at higher risk to develop dysrhythmias, heart failure, heart valve infections, and other problems. The table below lists the more common congenital heart defects.

Type of Defect	Mechanism
Ventricular Septal Defect (VSD)	With this condition, there is a hole within the membranous or muscular portions of the intraventricular septum that produces a left-to-right shunt (Figure 10-22) leading to blood not following the correct path through the heart. It is more severe with larger defects. Ventricular septal defect is the most common type of congenital heart disease, occurring in approximately 20% of cases. It can lead to left and/or right ventricular hypertrophy and pulmonary hypertension.
Atrial Septal Defect (ASD)	With this condition, there is a hole from a septum secundum (small) or septum primum (large) defect in the interatrial septum (Figure 10-22). It produces a modest left-to-right shunt and blood does not follow the correct path through the heart. It occurs in approximately 10% of cases and can lead to left and/or right ventricular hypertrophy and pulmonary hypertension.
Persistent Ductus Arteriosus (PDA)	In this condition, the ductus arteriosus, which normally closes soon after birth, remains open, allowing oxygen-rich blood from the aorta to mix with oxygen-poor blood from the pulmonary artery. This can put strain on the heart and increase blood pressure in the lung arteries. It occurs in approximately 10% of cases and can lead to left ventricular hypertrophy and pulmonary hypertension.
Valvular Stenosis	This defect occurs if the flaps of a valve thicken, stiffen, or fuse together. Thus, the valve cannot fully open, causing the heart to work harder to pump blood through the valve. It can lead to right atrial and right ventricular hypertrophy.
Tricuspid Atresia	In this condition, there's no tricuspid valve so blood can't flow normally from the right atrium to the right ventricle. Thus, the right ventricle is small and not fully developed. Survival depends on there being an opening in atria septal wall (ASD) and usually an opening in ventricular septal wall (VSD). As such, the low-oxygen blood that returns from the systemic veins to the right atrium flows through the atrial septal defect and into the left atrium. There it mixes with oxygen-rich blood from the lungs. Most of this partially oxygenated blood goes from the left ventricle into the aorta and on to the body. A smaller-than-normal amount of blood flows through the ventricular septal defect into the small right ventricle, through the pulmonary artery, and back to the lungs. Because of this abnormal circulation, the patient with this condition looks cyanotic until surgery can be performed.
Coarctation of Aorta (AoA or CoAo)	This condition is a narrowing of the aorta between the upper body branches and the lower body branches. It's typically in an isolated location just after the "arch" of the aorta. The blockage can increase blood pressure in the arms and head, yet reduce pressure in the legs. It results in the heart working harder to pump blood through the narrowed part of the aorta and can lead to left atrial and left ventricular hypertrophy. Coarctation of the aorta is generally present at birth and can range from mild to severe. In some cases, depending on the degree of narrowing of the aorta, it might not be detected until adulthood.

Continued

Type of Defect	Mechanism
Tetralogy of Fallot	This condition is a rare, complex defect that involves four heart defects: • A large ventricular septal defect (VSD) • Pulmonary stenosis • Right ventricular hypertrophy • An overriding aorta (in this condition, the aorta is located between the left and right ventricles, directly over the ventricular septal defect, resulting in oxygen-poor blood from the right ventricle flowing directly into the aorta instead of into the pulmonary artery) These defects, which affect the structure of the heart, cause oxygen-poor blood to flow out of the heart and to the rest of the body. Tetralogy of Fallot is often diagnosed during infancy or soon after. However, depending on the severity of the defects and symptoms, this condition might not be detected until later in life.
Transposition of Great Vessels	In this condition, the two main arteries carrying blood away from the heart are reversed. The aorta arises from the right ventricle and the pulmonic trunk from the left ventricle. Immediate surgery is needed. The only way to temporarily survive this condition after birth is to have leakages that allow some oxygen-rich blood to cross into the oxygen-poor blood for delivery to the body. This could occur through a VSD or an ASD with PDA. In this condition, there is right-to-left shunting.
Truncus Arteriosus (TA)	Truncus arteriosus is a rare heart defect that is present at birth. In this condition, one large blood vessel leads out of the heart instead of the normal anatomy in which two separate vessels come out of the heart. In addition, there is usually a VSD. Thus, oxygen-poor blood that should go to the lungs and oxygen-rich blood that should go to the rest of the body are mixed together. This creates severe circulatory problems. If left untreated, truncus arteriosus can be fatal. Surgery to repair truncus arteriosus is usually successful, especially if the repair occurs before 2 months of age.
Hypoplastic Left Heart Syndrome (HLHS)	This is a birth defect that affects normal blood flow through the heart. As the fetus develops during pregnancy, the left side of the heart does not correctly form. This condition affects several structures that do not fully develop, for example: • The left ventricle is underdeveloped and too small. • The mitral valve is not formed or is very small. • The aortic valve is not formed or is very small. • The ascending portion of the aorta is underdeveloped or is too small. Often, babies with hypoplastic left heart syndrome also have an ASD.
Total Anomalous Pulmonary Venous Connection (TAPVC)	In this condition, the pulmonary veins do not directly connect to the left atrium, but instead drain through abnormal connections to the right atrium. In the right atrium, oxygen-rich blood from the pulmonary veins mixes with low-oxygen blood from the body. Part of this mixture passes through an atrial septal defect into the left atrium. From there it goes into the left ventricle, then into the aorta and out to the body. The rest of the blood flows through the right ventricle, into the pulmonary artery, and on to the lungs. The blood passing through the aorta to the body lacks a normal amount of oxygen, which causes the patient to appear cyanotic.
Atrioventricular Septal Defect (AVSD) (also called atrioventricular canal defect or endocardial cushion defect)	In this is a condition, there is a large hole in the center of the heart and abnormalities of the valves that control the flow of blood. It is located where the atrial septum joins the ventricular septum. In AVSD, blood flows where it normally should not go, traveling across the holes from the left heart chambers to the right heart chambers and out into the lung arteries. The extra blood being pumped into the lung arteries makes the heart and lungs work harder, and the lungs can become congested. Sometimes there's regurgitation of the abnormal single valve. This can add to the heart failure symptoms.

Table 10-1
Common Congenital Heart Defects

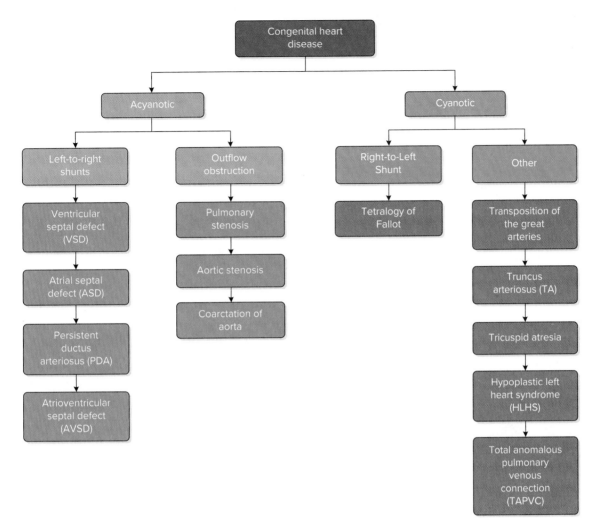

Figure 10-22
More Common congenital heart defects by category.

Key Points

LO 10.1	• Heart disease is defined as any medical condition of the heart or the blood vessels supplying it, or of muscles, valves, or internal electrical pathways, that impairs cardiac functioning.
LO 10.2	• Risk factors for heart disease include age, gender, family history, smoking, alcohol intake, poor diet, obesity, physical inactivity, high blood pressure, high cholesterol levels, diabetes, stress, and poor hygiene, among others.
LO 10.3	• There are many complications of heart disease, including dysrhythmias, angina, myocardial infarction, hypertrophy, heart failure, cardiogenic shock, stroke, aneurysm, peripheral artery disease, pulmonary embolism, and sudden cardiac arrest. • An acute blockage of one of the pulmonary arteries that leads to obstruction of blood flow to the lung segment supplied by the artery is called a *pulmonary embolism*.
LO 10.4	• Coronary artery disease is an umbrella term for the various diseases that reduce or stop blood flow through the coronary arteries. Atherosclerosis accounts for over 90% of the cases of coronary artery disease.

- Myocardial ischemia, injury, and death can occur with the interruption of coronary artery blood. Myocardial ischemia is a lack of oxygen and nutrients to the myocardium. Myocardial injury reflects a degree of cellular damage beyond ischemia. Myocardial infarction (MI) is the actual death of injured myocardial cells.

- Cardiomyopathy is any disease of the heart muscle that makes it harder for the heart to pump blood to the rest of the heart. The three types of cardiomyopathy are dilated cardiomyopathy, hypertrophic cardiomyopathy, and restrictive cardiomyopathy.

- Heart infections, such as pericarditis, endocarditis, and myocarditis, are caused when an irritant, such as a bacterium, a virus, or a chemical, reaches the heart muscle. Patients with these conditions can experience debilitating structural damage of the heart due to normal healing processes and the development of scar formation.

- Rheumatic fever is an infectious disease generally caused by a Streptococcus bacterium. If rheumatic fever isn't treated, scarring deformity of the cardiac structures results in rheumatic heart disease.

- Valvular heart disease is damage to or a defect in one or more of the valves, although it occurs more often in the mitral and aortic valves. Valvular heart disease interferes with normal flow of blood through the heart.

- Congenital heart defects are problems with the heart's structure that are present at birth.

 # Assess Your Understanding

The following questions give you a chance to assess your understanding of the material discussed in this chapter. The answers can be found in Appendix A.

1. Define the term "heart disease" and describe the benefit of learning about the pathophysiology of the various conditions. (LO 10.1)

2. People at greater risk for heart disease include (LO 10.2)
   a. young women
   b. people who use tobacco
   c. people who have diets containing fruits, vegetables, and fish
   d. persons who have excessive hand washing

3. Describe how hypertension contributes to heart disease. (LO 10.2)

4. Which of the following is true regarding cholesterol? (LO 10.2)
   a. Cholesterol travels through the bloodstream in small packages called lipoproteins.
   b. High levels of cholesterol in the blood lower the risk of formation of plaques and atherosclerosis.
   c. A high HDL level leads to a buildup of cholesterol in the arteries.
   d. The only source of cholesterol in the human body is what is acquired through food intake.

5. Describe how a person under continual stress at work is at greater risk for heart disease. (LO 10.2)

6. Angina has which of the following characteristics? (LO 10.3)
   a. It is defined as irregularities or abnormalities of the heart rhythm.
   b. It occurs when the heart muscle dies as a result of one or more coronary arteries being blocked.
   c. It is a disease of the coronary arteries.
   d. It may present as pressure or squeezing in the chest.

7. Describe how heart failure is different from cardiogenic shock. (LO 10.3)

8. Aneurysms are a condition in which of the following? (LO 10.3)
   a. The patient experiences sudden cardiac arrest.
   b. There is an acute blockage of one of the pulmonary arteries.
   c. There is a weakening and bulge in an artery or ventricle.
   d. The patient has leg pain when walking.

9. Explain how a pulmonary embolism diminishes blood flow to the lungs. (LO 10.3)

10. A pulmonary embolism (LO 10.3)
    a. is most commonly caused by a thrombus in one of the large neck veins.
    b. is an acute blockage of one of the coronary arteries.
    c. often produces chest pain unaffected by breathing efforts.
    d. can lead to acute cor pulmonale.

11. In sudden cardiac arrest, (LO 10.3)
    a. blood flow to the body is unaffected.
    b. treatment can be delayed until the patient is inside a hospital setting.
    c. immediate CPR and defibrillation may restore circulation.
    d. patients always experience ventricular fibrillation.

12. Describe how atherosclerosis develops. (LO 10.3)

13. Myocardial injury (LO 10.4)
    a. is caused by coronary artery vasodilation.
    b. is also known as *angina* or *angina pectoris*.
    c. is the death of injured myocardial cells.
    d. can be reversed if the supply of oxygen and nutrients can be increased to a level that meets the myocardial oxygen demand.

14. Myocardial _____ is the most severe result of an interruption of coronary artery blood flow. (LO 10.4)

    a. ischemia

    b. infarction

    c. injury

    d. interruption

15. Cardiomyopathy (LO 10.4)

    a. is an acute infection of the heart muscle.

    b. allows the heart to pump more efficiently than usual.

    c. can result in a hypertrophied heart.

    d. is usually caused by cancer treatments such as chemotherapy and radiation.

16. Inflammation of the pericardium is called (LO 10.4)

    a. preexcitation syndrome.

    b. myocardial ischemia.

    c. pericarditis.

    d. pulmonary embolism.

17. Describe how pericardial effusion develops. (LO 10.4)

18. Mitral valve insufficiency (LO 10.4)

    a. causes increased resistance to blood flow from the left ventricle.

    b. is characterized by blood leaking from the left ventricle back into the left atria.

    c. is characterized by the right atria becoming stiff and inflexible.

    d. is characterized by the aortic valve prolapsing back into the left atria.

19. Describe how aortic stenosis interferes with blood flow from the heart. (LO 10.4)

20. Congenital heart disease (LO 10.4)

    a. mostly occurs in adults.

    b. is limited to just a few conditions.

    c. often involves defects in the atria and ventricular septa.

    d. always requires surgical repair to correct the abnormality.

Origin and Clinical Aspects of Dysrhythmias

©rivetti/Getty Images

11 Overview of Dysrhythmias

©rivetti/Getty Images

Courtesy Physio-Control

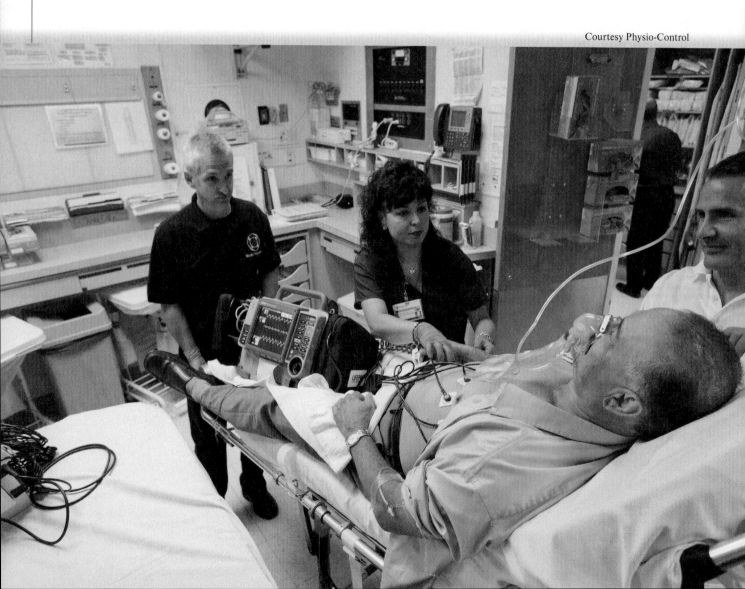

Learning Outcomes

LO 11.1   Describe the normal electrical activity of the heart.

LO 11.2   Define the term *dysrhythmia.*

LO 11.3   Describe the effects of dysrhythmias.

LO 11.4   Recall the different types of dysrhythmias.

LO 11.5   List and describe the mechanisms and causes of dysrhythmias.

LO 11.6   Categorize dysrhythmias by site of origin.

LO 11.7   Describe how to identify dysrhythmias when analyzing an ECG tracing.

LO 11.8   List the additional patient assessment tasks that should be completed as part of using an ECG to determine what medical condition a patient may be experiencing.

LO 11.9   Describe the treatments used to control or eliminate dysrhythmias.

Case History

A 58-year-old man is at the local grocery store, having just paid for his groceries, when he suddenly feels a pounding in his chest, becomes lightheaded, and then slumps to the floor. Two firefighter/paramedics who are in the store picking up food for dinner quickly come to his side. As the one positions his head into a head/tilt chin/lift position, he wakes up. Feeling embarrassed, he insists on sitting up. The ECG monitor is retrieved from the ambulance, and it is applied while vital signs are obtained, revealing a blood pressure of 90/60, a pulse rate of 220, and a respiratory rate of 22. The cardiac monitor shows a fast, regular rhythm with narrow QRS complexes. Although a T wave follows each QRS complex, P waves cannot be identified.

Within a few minutes the patient's heart rate slows to 82 beats per minute, and he appears more alert. He agrees to be transported to the hospital, where he undergoes evaluation and treatment for episodes of a rapid heart rate.

11.1 The Heart's Normal Electrical Activity

Here is a brief review of how the heartbeat originates and is conducted. Each heart-beat arises as an electrical impulse from the SA node. The impulse then spreads across the atria from right to left, depolarizing the tissue and causing both atria to contract. This is seen on the ECG as a P wave. The impulse then activates the AV node, which is normally the only electrical connection between the atria and the ventricles. There, the impulse is delayed slightly, allowing the atria to finish con-tracting and pushing any remaining blood from their chambers into the ventricles (referred to as the atrial kick). The impulse then spreads through both ventricles via the Bundle of His, right and left bundle branches, and the Purkinje fibers (seen as the QRS complex), causing a synchronized contraction of the primary pumping chambers of the heart, and thus, the pulse (Figure 11-1).

11.2 Dysrhythmias

In the previous chapters we talked about the anatomy and physiology of the heart and the machines and leads used to obtain an ECG tracing. We also discussed the nine-step process used to examine the ECG (for a brief review see page 88). We will continue to use this process to evaluate the ECGs throughout the remainder of this book.

One thing we look for when we examine the ECG is the presence of dysrhythmias. In this chapter, as well as the next five, we focus on dysrhythmias—the characteristics associated with them (which allow us to identify them), how they occur, their significance, and how to treat them (if necessary).

First, let's define what we mean by the term *dysrhythmia*. Simply stated, a dysrhyth-mia is an ECG rhythm that differs from normal sinus rhythm. More specifically, a dysrhythmia is a condition in which there is abnormal electrical activity in the heart.

Figure 11-1

Normal electrical activity in heart. (1) Electrical impulse originates in SA node and spreads across atria which contract; (2) impulse conduction slows as it passes through the AV node; (3) the impulse then travels down the bundle of His; (4) the impulse then spreads throughout ventricles with ventricular contraction and pumping of blood.

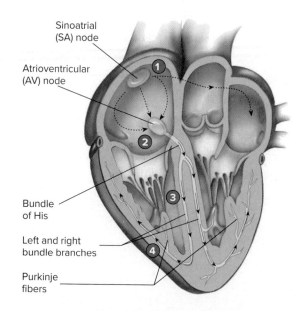

Sinoatrial
(SA) node

Atrioventricular
(AV) node

Bundle
of His

Left and right
bundle branches

Purkinje
fibers

The heartbeat may be slower or faster than normal, it may be irregular, or conduction through the heart may be delayed or blocked. The term *dysrhythmia* covers a large number of electrical abnormalities. Note that although the natural slowing and speeding up of the heart rate differs from the criteria for normal sinus rhythm, many do not consider it a dysrhythmia but, instead, the body's normal response to rest or exertion.

While the terms *dysrhythmia* and *arrhythmia* can be used interchangeably, *a*rrhythmia suggests there is an absence of a rhythm whereas *dys*rhythmia suggests there is a problem with the rhythm.

11.3 The Effects of Dysrhythmias

Some dysrhythmias are of little consequence and are simply annoying. During the course of their life, most people will occasionally feel their heart skip a beat or give an occasional extra strong beat; neither of these is usually a cause for concern. Other dysrhythmias are life-threatening medical emergencies that can lead to cardiac arrest and sudden death.

Palpitations, an abnormal sensation felt with the heartbeat as described in the paragraph above, are the most common symptom of dysrhythmias. Palpitations may be infrequent, frequent, or continuous. Another symptom syncope, is a transient, usually sudden, loss of consciousness due to inadequate perfusion of the brain. These symptoms may occur together or individually.

Palpitations and syncope result from cardiac dysrhythmias, other types of heart disease, nervous system disorders, anxiety, and thyroid disease.

Of greater significance are lightheadedness, dizziness, chest pain, shortness of breath, sweatiness, and/or pallor that may be seen in dysrhythmias which cause a lowering of the blood pressure. We refer to this as being *symptomatic*. When you see the description symptomatic bradycardia, symptomatic tachycardia, or any other that includes the word *symptomatic,* it is referring to a dysrhythmia that produces decreased cardiac output.

When considering the signs and symptoms seen with certain dysrhythmias, let's think back on our discussion in Chapter 1. Remember, heart rate × stroke volume = cardiac output. If either the heart rate or stroke volume is reduced, cardiac output is reduced (Figure 11-2). This can decrease the blood pressure and perfusion of the body's cells, leaving the patient in a perilous condition, and may even become life threatening.

Some dysrhythmias slow the heart rate, and, in that way, decrease cardiac output. Others decrease the stroke volume by making the heart beat too fast (which in turn decreases cardiac output by not allowing the heart to fill properly). Cardiac output can also be decreased when the atria don't contract properly or contract at all (eliminating the atrial kick which normally pushes blood into the ventricles; Figure 11-3). Incomplete contraction of the atria can also lead to clots being formed and passed along to the brain, resulting in the patient experiencing a stroke.

Another adverse effect of dysrhythmias is that they can reduce the oxygen supply to the heart cells by decreasing coronary blood flow or by causing the heart to beat so

Figure 11-2
Reduced cardiac output is
caused by a decrease in heart
rate or stroke volume.

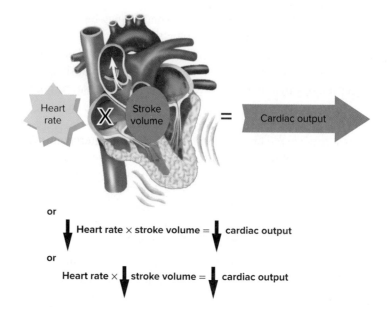

Figure 11-3
Decreased cardiac output due
to loss of atrial kick.

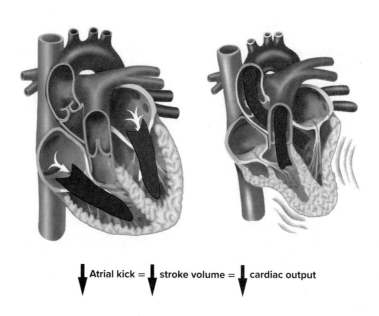

fast the heart expends its available oxygen due to the rapid heartbeat. The decrease in myocardial oxygen supply can cause chest pain. The sudden onset of a dysrhythmia in patients with underlying cardiac disease can also precipitate congestive heart failure.

Early beats that arise from the ventricles can also result in a decreased or absent heartbeat. If they occur frequently, cardiac output can be decreased. Dropped beats can also result in decreased cardiac output, particularly if it occurs frequently. Delays or blockage in conduction can lead to a worsening of the blockage or worse, a complete cessation of the heartbeat.

Sometimes the first clinical manifestation of a cardiac dysrhythmia is sudden death. This is particularly true in patients experiencing myocardial infarction. To prevent this, these patients should be hospitalized so that their heart rate and rhythm can be stabilized and monitored.

As we discuss the dysrhythmias throughout this textbook, we will review the effects each may have on the heart and subsequent cardiac output.

11.4 Types of Dysrhythmias

Dysrhythmias can be categorized many different ways. In this textbook, we describe them by type, causes, mechanisms, and site of origin. Not all dysrhythmias are described in this chapter, as it is meant to serve as an overview. However, all dysrhythmias are discussed in detail in subsequent chapters.

Bradycardia

A heart rate less than 60 beats per minute is called *bradycardia*. It can occur for many reasons and may or may not have an adverse affect on cardiac output. In the extreme, it can lead to severe reductions in cardiac output and eventually deteriorate into asystole.

Bradycardia can arise from the SA node. Sometimes this is normal, such as when the patient is at rest or is an athlete with a body that is more efficient and does not require as much oxygen. It can also be caused by increased parasympathetic (vagal) tone or a variety of medical conditions, including intrinsic sinus node disease, hypothermia, hypoxia, drug effects (e.g., digitalis, beta blockers, Ca++ channel blockers), myocardial infarction, and others.

Bradycardia can also be brought about by failure of the SA node, the heart's primary pacemaker. When this occurs, either the SA node will initiate an impulse, or an escape pacemaker from either the AV node or the ventricles should take over and initiate the heartbeat. Remember, the intrinsic rate of the AV node is 40 to 60 beats per minute whereas the ventricle's is 20 to 40 beats per minute (see Chapter 1). Whether the escape pacemaker arises from the AV node (called *junctional escape*) or ventricles (called *ventricular escape* or *idioventricular rhythm*), it is likely to produce bradycardia.

Blockage of the impulse traveling through the AV node can also be a cause of bradycardia. In one type, not all the sinus beats are conducted through to the ventricles; this results in a slower ventricular rate (as such, there are more P waves than QRS complexes). With a more severe form of AV heart block, there is complete blockage of the AV node, resulting in the sinus impulse not reaching the ventricles. An escape pacemaker then arises from somewhere below the AV node, resulting in a slower ventricular rate.

We may also see slower than normal ventricular rates in conditions where the atria repeatedly or chaotically depolarize and bombard the AV node so rapidly that not all the impulses are conducted through to the ventricles. If the number of atrial impulses reaching the ventricles falls to less than normal, it results in a slower than normal ventricular rate, or bradycardia.

Tachycardia

A heart rate greater than 100 beats per minute is called *tachycardia*. Tachycardia has many causes and leads to increased myocardial oxygen consumption, which can have an adverse effect on patients with coronary artery disease and other medical conditions. Extremely fast rates can have an adverse affect on cardiac output. Also, tachycardia that arises from the ventricles may lead to a chaotic quivering of the ventricles called *ventricular fibrillation*.

Sinus tachycardia is a fast rate, greater than 100 beats per minute, that arises from the SA node. Sinus tachycardia can occur with increased sympathetic nervous system stimulation due to exercise, exertion, stress, fear, anxiety, etc. Other common causes of sinus tachycardia include fever, hypoxia, shock, drug effects (such as from amphetamines or cocaine), ingestion of caffeine or alcohol, smoking, and myocardial infarction.

Tachycardia can also be brought about by a pacemaker that arises outside the SA node such as in the atria, AV junction, or ventricles. These tachycardias result from the rapid depolarization of a site that fires before the SA node can. Tachycardia that arises above the ventricles is called *supraventricular tachycardia,* whereas tachycardia that arises in the ventricles is called *ventricular tachycardia.* Atrial tachycardia and junctional tachycardia are supraventricular tachycardias (again, as they arise from above the ventricles). The term *supraventricular* is also used if the origin of the tachycardia cannot be determined as atrial or junctional (often due to the P waves being hidden in the T wave of the preceding beat in faster rates).

We may also see faster than normal ventricular rates in conditions in which the atria repeatedly or chaotically depolarize and bombard the AV node rapidly (see Figure 6-11). In some cases, the number of atrial impulses actually reaching the ventricles is greater.

Typically, tachycardias are referred to as one of five types (Figure 11-4)—narrow QRS complex regular rhythms, wide QRS complex regular rhythms, narrow QRS complex irregular rhythms, wide QRS complex irregular rhythms, and wide QRS complexes of unknown origin. Being able to differentiate between the five types of tachycardia is important because it will help determine what type of treatment should be employed.

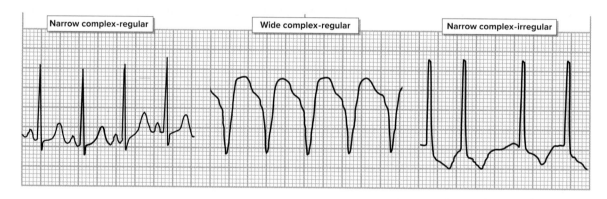

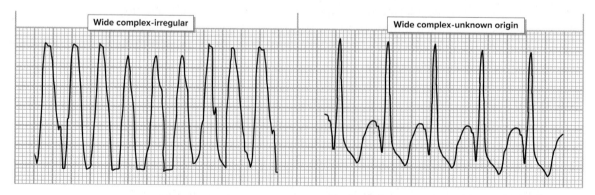

Figure 11-4
Five types of tachycardia.

Narrow QRS complexes indicate that the site of origin for the tachycardia is above the ventricles (supraventricular) whereas wide QRS complexes typically indicate a ventricular origin. However, some supraventricular tachycardias conduct abnormally through the ventricles, producing wide QRS complexes, and for this

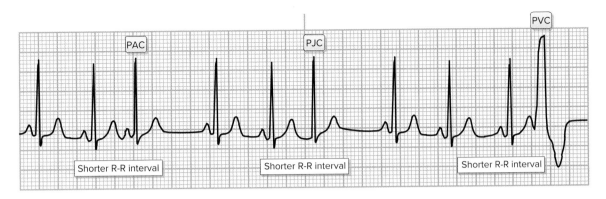

Figure 11-5
Early beats with a shortening of the R-R interval.

reason are referred to as *wide QRS complexes of unknown origin* (until otherwise determined). The tachycardias that are regular come from one site whereas irregular tachycardias originate from more than one site.

Early (Premature) Beats

Another cause of dysrhythmias is an impulse that appears early, called a *premature complex* or *ectopic focus*. As implied by its name, a premature complex fires early, before the SA node has a chance to initiate the impulse. These premature complexes can arise from any of the cells of the heart, including the atria, AV junction, or ventricles. The R-R interval between the normal complex and the premature complex is shorter than the interval between two normal complexes (Figure 11-5). The presence of a premature beat makes the rhythm irregular. The premature beats may occur only a few times a minute, or they can occur often. Frequent premature beats are more likely to progress to very fast atrial, junctional, or ventricular rates (tachycardia). Worse, the rhythm may deteriorate into a quivering of the heart muscle called *ventricular fibrillation*.

As described earlier, early beats that originate from above the ventricles (supraventricular) tend to have narrow QRS complexes (unless there is abnormal conduction through the ventricles) whereas those that originate from the ventricles have wide QRS complexes. Premature beats that arise from the atria are called *premature atrial complexes* (*PACs*), those that arise from the AV junction are called *premature junctional complexes* (*PJCs*), and those that arise from the ventricles are called *premature ventricular complexes* (*PVCs*).

Dropped Beats or QRS Complexes

Dropped beats occur when the SA node fails to initiate an impulse. This is seen as a pause in the ECG rhythm (Figure 11-6). Typically, the rhythm leading up to the pause looks normal. Then there is suddenly an absence of a P wave, QRS complex, and T wave. This creates a gap or pause. Usually, with this condition the SA node recovers and fires another impulse. If the SA node fails to fire, then an escape pacemaker from the atria, AV junction, or ventricles initiates an impulse.

Dropped QRS complexes can occur from a partial or intermittent block at the AV junction. In this condition, some impulses that originate from the SA node fail to conduct to the ventricles, resulting in one or more dropped ventricular beats. This causes more P waves than QRS complexes and R-R intervals that are longer wherever there is a dropped ventricular beat (as the QRS complex is absent). Dropped QRS complexes can also occur when a premature beat arises from the atria but fails to conduct to the ventricles. This produces an early beat with a P wave but no QRS complex following it.

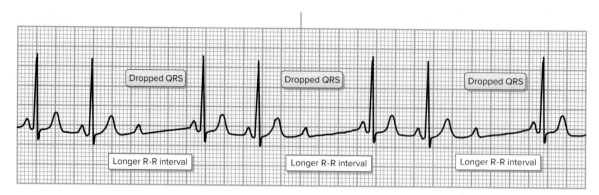

Figure 11-6
Dropped beats with longer R-R intervals as seen with an intermittent block at the AV junction.

Irregularity

An irregular rhythm is caused by some type of dysrhythmia. It includes several we have already discussed, such as early beats and dropped beats or QRS complexes. It also includes dysrhythmias that speed up and slow down in a cyclical manner, those that originate from more than one site (sometimes from many sites), and some types of AV heart block.

11.5 Causes and Mechanisms of Dysrhythmias

The primary causes and mechanisms believed to be responsible for dysrhythmias are increased parasympathetic tone, myocardial hypoxia, ischemia or infarction, increased automaticity, triggered activity (afterdepolarization), reentry, and conduction delay or block. Increased automaticity and triggered activity are disorders of impulse formation. Reentry is a problem with impulse conduction.

Increased Parasympathetic Tone

Stimulation of the parasympathetic nervous system causes a slowing of the heart rate and prolongs conduction of impulses through the AV node. This can lead to bradycardia, sinus arrest (a pause in the normal activity of the sinus node) and/or AV heart block (blocking of the electrical impulse on its way from the atria to the ventricles).

Myocardial Hypoxia, Injury, and Infarction

A myocardium deprived of oxygen is extremely susceptible to the development of dysrhythmias. Pulmonary disorders that interfere with the body's ability to bring in adequate supplies of oxygen (e.g., chronic obstructive lung disease, pulmonary embolism) are major causes of dysrhythmias as are myocardial ischemia and infarction. Also, if portions of the heart's conduction system are damaged, impulse formation and/or conduction may be blocked. Sometimes, myocarditis, an inflammation of the heart muscle, can precipitate dysrhythmias.

Increased Automaticity

Automaticity refers to the spontaneous depolarization of myocardial cells, in other words, their ability to fire on their own without an outside stimulus (Figure 11-7). All of the cells of the heart have the ability to initiate an impulse; however, only

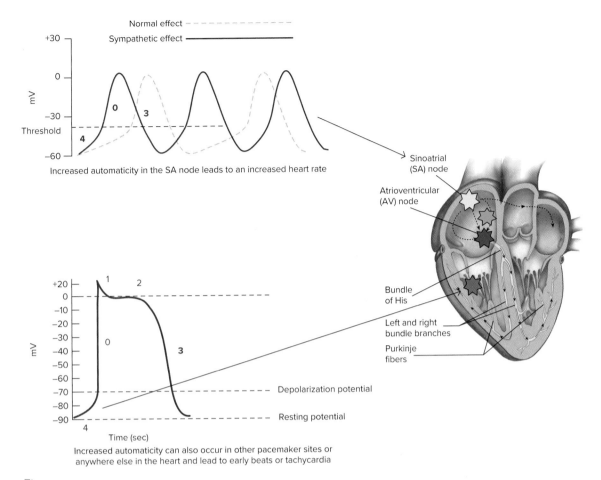

Figure 11-7
Increased automaticity as seen in the SA node as well as in other sites in the heart.

some are meant to regularly generate the heartbeat. These cells are found in the heart's conduction system and include the SA node, AV junction, Bundle of His, and Purkinje fibers. The SA node is the primary pacemaker, having a higher automaticity (quicker spontaneously depolarization) than the rest of cells, and therefore initiates each heartbeat and sets the heart rate.

Stimulation of the sympathetic nervous system increases automaticity. This causes the cells to spontaneously depolarize more quickly. The resulting heart rhythm depends on where the impulse originates; if it is the SA node, the rhythm remains normal but faster than 100 beats per minute; if it is from a location outside the SA node (called an *ectopic focus*), any number of different dysrhythmias may ensue. Ectopic means occurring in an unusual location, or off the normal conduction pathway. It could also mean arising from any of the pacing sites other than the SA node, like the AV junction or His Purkinje system.

An ectopic focus may cause a single, occasional early beat (also referred to as a *premature beat*), or, if the ectopic focus fires more often than the SA node, it can produce a sustained abnormal rhythm. Sustained rhythms produced by an ectopic focus in the atria or atrioventricular junction are less dangerous than those that arise from the ventricles; however they can still produce a decrease in the pumping efficiency of the heart. This is because depolarization of the atria and ventricles does not occur in the usual manner and myocardial contraction may be poorly coordinated.

The most common causes of increased automaticity include emotional stress or physical exercise, caffeine, amphetamines, ischemia, hypoxia, atrial stretching or dilation (usually due to congestive heart failure, mitral valve disease, increased pulmonary artery pressures, or a combination of these conditions), hyperthyroidism (an overactive thyroid gland), or a myriad of other medical conditions, such as hypovolemia, congestive heart failure, etc.

Reentry

Reentry occurs when an electrical impulse reenters a conduction pathway (often in a retrograde or reverse manner), rather than moving from one end of the heart to the other and then terminating. To understand this, let's review how impulses normally travel through the heart. Every cardiac cell is able to transmit impulses in every direction but only do so once and just briefly. Normally, the impulse spreads through the heart quickly enough that each cell responds only once. However, if conduction is abnormally slow and/or blocked somewhere along the conductive pathway, part of the impulse arrives late (during myocardial repolarization) and can be responded to as a new impulse.

Some hearts have an accessory pathway (bundle of connecting tissue) located between either the right atrium and the right ventricle or the left atrium and the left ventricle. These accessory pathways allow electrical impulses to bypass the AV node and depolarize the ventricles. Another abnormality in some people is a dual conduction pathway through the AV node (Figure 11-8). In both accessory pathways and dual conduction pathways through the AV node, under the right circumstances, reentry can occur.

Depending on the timing, this mechanism can generate a sustained abnormal circuit rhythm. Reentry circuits are responsible for a number of dysrhythmias, including atrial flutter, most paroxysmal supraventricular tachycardia, and ventricular tachycardia (to be discussed later).

A unique form of reentry is referred to as *fibrillation*. Fibrillation results when there are multiple micro-reentry circuits in the chambers of the heart and portions or all of it is quivering due to chaotic electrical impulses. Fibrillation that occurs in the atria is called *atrial fibrillation*. Fibrillation that occurs in the ventricles is called *ventricular fibrillation*.

Triggered Beats

Triggered beats occur when problems at the level of the ion channels in individual heart cells lead to partial repolarization. Partial repolarization causes repetitive ectopic firing called *triggered activity*. The depolarization produced by triggered activity is known as *afterdepolarization* and can bring about atrial or ventricular tachycardia. Triggered beats are relatively rare but can result from the action of antidysrhythmic drugs (i.e., digoxin toxicity), cell injury, and other conditions.

Proarrhythmia

The term *proarrhythmia* refers to the development of new or a more frequent occurrence of pre-existing dysrhythmias that are caused by antidysrhythmic therapy or drugs used to treat other conditions. In other words, it is a side effect associated with the use of these medications.

Some people have more than one conduction pathway through the AV node. Under the right conditions, these pathways can form an electrical circuit or loop.

Here an early beat from the atria triggers the impulse to indefinitely spin around in the circuit producing a very rapid and regular rhythm.

Enlarged view

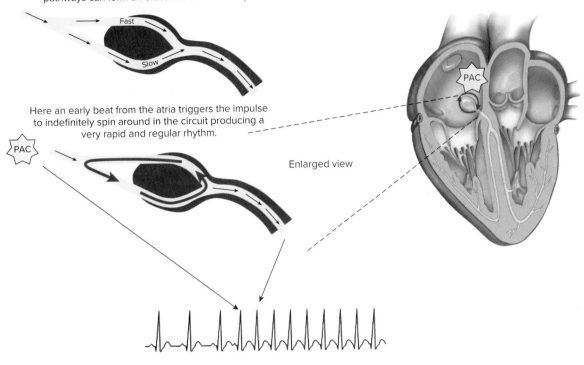

Figure 11-8
One example of how reentry can occur.

11.6 Site of Origin

It is also appropriate to classify dysrhythmias by site of origin (Figure 11-9). As you will see in the next five chapters, this how we present dysrhythmias in this textbook.

Sinus Dysrhythmias

Dysrhythmias that originate from the SA node are called *sinus dysrhythmias* and include

- Sinus bradycardia
- Sinus tachycardia
- Sinus dysrhythmia
- Sinus arrest
- Sinoatrial exit block
- Sick sinus syndrome

Figure 11-9
Sites where dysrhythmias originate.

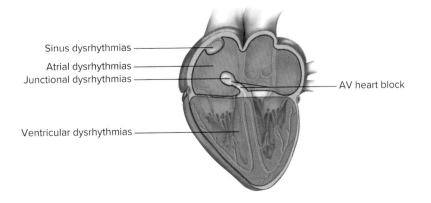

Atrial Dysrhythmias

Dysrhythmias that originate from the atria are called *atrial dysrhythmias* and include

- PACs
- Wandering atrial pacemaker
- Atrial tachycardia
- Supraventricular tachycardia (SVT)
- Multifocal atrial tachycardia
- Atrial flutter
- Atrial fibrillation

Junctional Dysrhythmias

Dysrhythmias that originate from the AV junction are called *junctional dysrhythmias* and include

- PJCs
- Junctional escape
- Accelerated junctional rhythm
- Junctional tachycardia
- AV nodal reentrant tachycardia
- AV reentrant tachycardia

Ventricular Dysrhythmias

Dysrhythmias that originate from the ventricles are called *ventricular dysrhythmias* and include

- PVCs
- Ventricular escape beats
- Idioventricular rhythm
- Accelerated idioventricular rhythm
- Ventricular tachycardia
- Ventricular fibrillation
- Asystole

AV Heart Block

Dysrhythmias that occur due to a delay or blockage in the AV node are called *AV heart blocks* and include

- 1st degree AV block
- 2nd degree AV block, Types I and II
- 3rd degree AV block
- Atrioventricular dissociation

We can use the term supraventricular dysrhythmias to describe dysrhythmias that arise from above the ventricles. This includes dysrhythmias whose origins are the SA node, atria, and AV junction. Ventricular dysrhythmias originate from the ventricles. Again, this is how we present dysrhythmias over the next five chapters of this book.

11.7 Identifying Dysrhythmias

Typically, dysrhythmias can be identified through the use of one lead, most commonly lead II. This lead has an excellent view of the normal conduction of the impulse through the heart. With lead II the positive electrode is on the left leg (or on the left midclavicular line, below the last palpable rib), resulting in the waveforms appearing upright or positive because depolarization of the heart flows toward the positive electrode. It is particularly helpful in identifying sinus node and atrial dysrhythmias (Figure 11-10).

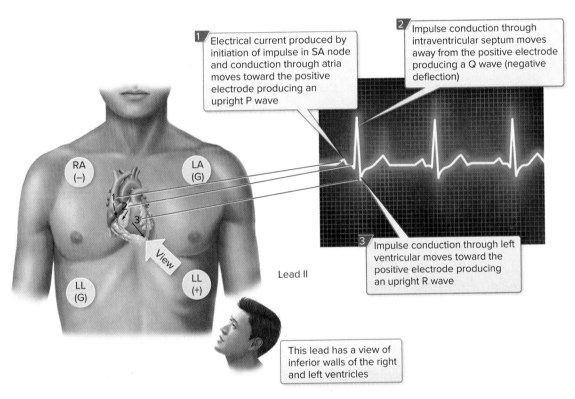

Figure 11-10
Leads II is a commonly used lead for detecting dysrhythmias.

To identify dysrhythmias, you need to look across the ECG tracing and analyze it using a systematic and organized approach. Although six-second tracings are effective for teaching how to analyze and interpret dysrhythmias, they don't give you much information from which to see emerging dysrhythmias. Don't hesitate to print off more of the tracing in order to identify the presence of dysrhythmias. This is particularly important when there are patterns that can be seen only over several seconds.

An effective way of looking at the tracing is to scan it from left to right, picking out the things that are different than normal and then further analyzing them (Figure 11-11). We look for the presence or absence of key characteristics and then match those up against the characteristics we know to be common with the various dysrhythmias.

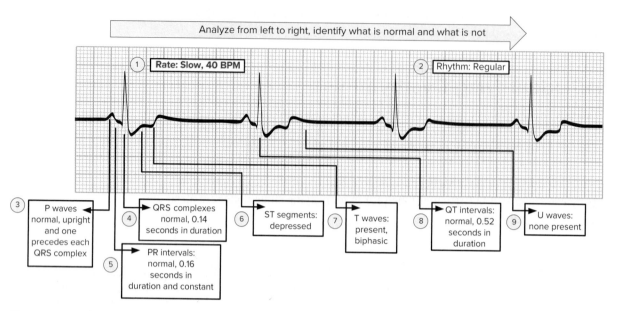

Figure 11-11

Using a systematic, step-by-step process to analyze an ECG tracing.

Unless the findings are extremely subtle, it is usually easy to identify the presence of a dysrhythmia. Determining what the dysrhythmia is, however, can be challenging and hence the reason for using a systematic, organized approach for ECG analysis.

11.8 Patient Assessment

While we use the ECG machine to identify dysrhythmias as well as various cardiac conditions, we are seeing only one part of the patient's condition. To get the whole picture of what is occurring with our patient, we need to do a patient assessment. Sometimes it is the patient assessment that gives us the information needed to identify what condition the patient is suffering. As an example, a faster than normal heart rate can occur with shock. If all we are relying on is the ECG, we know it is fast but we don't know why. Patient assessment often helps us determine the why.

This section provides a *brief* overview of patient assessment. However, it does not include all the elements that can be assessed in the clinical setting. For this reason, other references should be consulted to gain a more in-depth understanding. Again, a patient assessment should be done as a part of obtaining any ECG tracing.

Primary Assessment (ABCDEs)

To assess the patient, we first need to determine if they are experiencing a life-threatening or serious condition. We refer to as this as the primary assessment. While we normally use the acronym ABCDE, if the patient is unresponsive and may be in cardiac arrest, we instead use the acronym CAB (circulation, airway, breathing).

A—Assessing the Airway

The first step of the primary assessment is to assess if the patient has an open, patent airway. The presence of sounds such as snoring (likely caused by the tongue relaxing backward against the posterior oropharynx) or gurgling (likely caused by fluid accumulating in the posterior oropharynx) indicate the airway is compromised. Problems found at this step need to be promptly corrected. Applying a head tilt-chin lift procedure and inserting an oropharyngeal or nasopharyngeal airway should move the tongue anteriorly away from the posterior oropharynx while suctioning the upper airway will clear fluids and secretions.

B—Assessing Breathing

The next thing to do is determine if the patient is breathing or not. If breathing is present, then determine if it is adequate. Ask this question: does the respiratory rate and volume deliver the needed amount of air over the course of a minute to sustain life? If breathing is inadequate, steps must be promptly taken to support the patient's breathing such as assisting or taking over breathing with a bag-valve-mask device.

C—Assessing Circulation

Again, if the patient is unresponsive and may be experiencing cardiac arrest, immediately after assessing for responsiveness, check for a pulse. If no pulse is found, immediately begin CPR. If the patient is in ventricular fibrillation, defibrillate the patient as soon as possible. If the pulse is present, is it strong or weak? Is severe hemorrhage present? Identify and control sources of external bleeding by applying direct pressure and applying a tourniquet if necessary.

D—Disability

What is the patient's mental status? Are they awake and able to answer questions appropriately? If so, what can they do? Do they know the day, time, and place? Are they confused? Can they verbalize appropriately? If their eyes are closed, do they arouse with verbal stimuli? If not, do they respond to painful stimuli? If the patient fails to respond to verbal or painful stimuli, they are categorized as unresponsive. Any decrease in mentation can indicate an underlying medical condition that is causing cerebral hypoxia or other serious medical conditions that interfere with or compromise brain function.

E—Exposure

Exposure means removing sufficient clothing to adequately assess the patient. Sometimes that involves opening their shirt to look for signs of respiratory or circulatory compromise or cutting away their pants to look for fractures or bleeding. The degree of exposure depends on whether the patient is experiencing a medical condition or has been traumatically injured. The bottom line is you must be able to see enough of the patient's body to identify significant problems.

Secondary Assessment

Once we have assessed the patient for life-threatening and serious problems and corrected any that are found, we can move onto the secondary assessment. In short, the secondary assessment is used to gather pertinent information that can help us identify what medical or traumatic emergency(ies) the patient is experiencing. This includes such elements as vital signs, Focused History (SAMPLE, OPQRST), Focused Physical Exam, and ongoing reassessment.

Vital Signs

Assessing vital signs includes determining the rate and quality of the pulse and breathing, evaluating the skin condition (and capillary refill, as age appropriate), determining pulse oximetry (SpO2), waveform capnography (end tidal CO2), blood glucose levels and temperature, and assessing pupillary response. These vital signs provide us a glimpse into how the patient's body is responding to the presenting medical condition or traumatic injury.

Focused History—SAMPLE

The SAMPLE history is a combination of the Present Medical History and Past Medical History and stands for the following:

- S—signs and symptoms the patient presents with
- A—any allergies the patient may have
- M—medications the patient is taking
- P—past medical history
- L—last oral intake
- E—events leading up to problem

We use the acronym SAMPLE to help us remember those important components of the assessment and approach each patient in a systematic way.

Focused History—OPQRST

OPQRST is most effective for assessing pain or other symptoms and stands for the following:

- O—onset: when did it start and what was the patient doing when it started? It is helpful if the onset can be correlated to their activity or lack of activity.
- P—provokes: what things either provoke it or lessen it? Sometimes, sitting up or laying down, or being in a certain position eases some symptoms.
- Q—quality: how can the pain (or other symptoms) be described, i.e., "a dull aching pain," or "can only speak in short sentences" (relating to breathing difficulty).
- R—radiation: does the pain or other symptom radiate to other parts of the body?
- S—severity: how bad is it? In dealing with pain, we often ask the patient to rate it on a scale of 1 to 10; 1 being the least and 10 being the worst. Because each person's tolerance to pain is different, it is not always the most reliable indicator. Many are now using simple cartoon faces that show various expressions ranging from a smiling face to one that is in tears to assess severity.
- T—time: how long has the symptom been present?

Again, using an acronym such as OPQRST helps us remember to include all the needed information for each patient assessment we perform.

Focused Physical Exam

The physical exam can yield a tremendous amount of valuable information in all types of medical conditions and/or traumatic injury. However, the degree and focus of the physical exam really depends on the condition or problem which the patient is experiencing. As an example, a trauma patient will typically require a more detailed physical exam than does a patient experiencing an isolated medical condition. The following are considered part of the physical exam:

- Head, eyes, ears, nose, and throat/neck: look for absence or presence of edema or swelling around the eyes (periorbital edema), pupillary response, discoloration or swelling of mucous membranes (angioedema), dysphagia, jugular vein distension, crepitation, tracheal tugging or shifting, etc.
- Chest: absence or presence and character of lung sounds, heart sounds, intercostal retraction, accessory muscle usage, symmetry of chest wall expansion, apical pulse, etc.

- Abdomen: look for absence or presence of symmetry, distension, ascites, rigidity, organ enlargement, palpable pulsations, rebound tenderness, presence and quality of bowel sounds, etc.
- Extremities: absence or presence of peripheral pulses, sensory and motor function, appropriate range of motion, edema, color changes, etc.

Ongoing Assessment

Patient assessment is an ongoing process that must be continued throughout care of the patient. In the stable patient, reassessment can be repeated every 15 minutes. In an unstable patient, reassessment is repeated every 5 minute (can be done more frequently as needed). Ongoing assessment includes the following parameters:

- Reassess the patient's mental status
- Monitor the airway
- Monitor the breathing rate and quality
- Reassess the pulse rate and quality
- Monitor the skin for color, temperature, and condition
- Realign patient priorities as needed
- Reassess vital signs
- Repeat the focused examination regarding the complaint or injuries
- Check the efficiency of the treatments or interventions provided

11.9 Treatment of Dysrhythmias

Many dysrhythmias require no treatment as they are either benign or resolve on their own. An example is tachycardia that arises from the SA node in order to resolve a body's increased need for oxygen and nutrients and to remove metabolic waste. Here, there is no need to treat the tachycardia; it will naturally resolve on its own. In other cases, we treat the cause, and that resolves the dysrhythmia. Still at other times, we need to administer a treatment to suppress or eliminate the dysrhythmia. Whenever the dysrhythmia produces a life-threatening condition, it must be corrected promptly if the patient is to survive.

The method used to manage dysrhythmias depends on whether or not the patient is stable or unstable. Patients who are symptomatic are considered unstable as symptoms indicate decreased cardiac output.

Several options are available to treat dysrhythmias and are employed on the basis of the cause or mechanism of the dysrhythmia. Commonly used treatments include physical maneuvers, electricity therapy, and the administration of certain medications. More advanced treatment of dysrhythmias includes electro- or cryocautery. This procedure is performed in specialized catheter laboratories where cardiologists use very small probes inserted through the blood vessels to map electrical activity from within the heart. This allows abnormal areas of conduction to be located and subsequently eliminated with heat or cold or electrical or laser probes. This can completely resolve some dysrhythmias but not others.

Physical Maneuvers

One way to treat dysrhythmias, specifically tachycardias that arise from above the ventricles, is to employ one of several physical maneuvers. These maneuvers are collectively called *vagal maneuvers* and act to stimulate the parasympathetic nervous system. This slows the rate at which the SA node discharges and how quickly impulses are conducted through the AV node. It is the slowing of AV node conduction that slows the heart rate in supraventricular tachycardias.

Electrical Therapy

Electrical therapy is sometimes used to manage certain dysrhythmias. One type of therapy is the delivery of an electrical shock across the heart—either externally, on the surface of the chest, or internally to the heart via implanted electrodes. It is delivered as either synchronized cardioversion or as defibrillation. Synchronized cardioversion, used to treat patients experiencing symptomatic tachydysrhythmias, is the delivery of the electrical shock timed with the R wave of the QRS complex on the ECG. This helps prevent delivering the shock during the relative refractory period of the cardiac cycle, which could induce ventricular fibrillation. When used electively, the patient is usually sedated or lightly anesthetized for the procedure. Defibrillation differs in that the shock is not synchronized. It is used to treat ventricular fibrillation (a completely chaotic rhythm) and ventricular tachycardia when the patient is pulseless. Often, a higher level of electricity is required for defibrillation than for synchronized cardioversion. With defibrillation, most patients have lost consciousness. Because of this, sedation is not required. Defibrillation or cardioversion may also be accomplished by an implantable cardioverter-defibrillator (ICD).

Electrical treatment of dysrhythmias also includes cardiac pacing. Temporary pacing may be necessary for reversible causes of bradycardia. A permanent pacemaker may be inserted in situations in which the patient is not expected to recover from the bradycardia.

Medications

Many different medications can be used to treat dysrhythmias. These drugs are referred to as antidysrhythmics.

Antidysrhythmic drugs are used to restore normal rhythm and conduction. When returning the rhythm to normal sinus rhythm is not possible, antidysrhythmic drugs are often used to prevent the onset of more serious and possibly lethal dysrhythmias.

Remember our discussion in Chapter 1, where we covered the action potential and depolarization of both the pacemaker cells, such as the SA node, and AV node, and the nonpacemaker myocytes of the atria, ventricles, and Purkinje tissue. Depolarization of the nonpacemaker myocytes is dependent on the rapid entry of sodium ions through the fast sodium channels into the cell. This occurs during phase 0. These channels determine how fast the membrane depolarizes during an action potential. Depolarization of the pacemaker cells (nodal tissue) is instead carried by calcium currents. Repolarization of the cells is driven largely by the efflux of potassium from the cell during phase 3.

Three primary uses of antidysrhythmic drugs include:

- decreasing or increasing conduction velocity
- altering the excitability of cardiac cells by changing the duration of the effective refractory period (ERP)
- suppressing abnormal automaticity

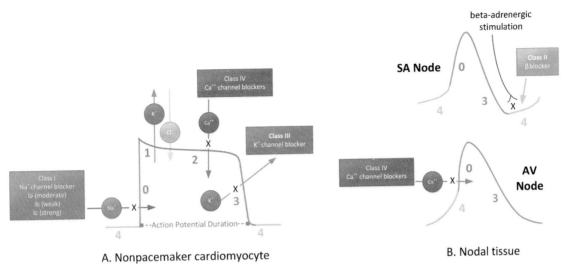

Figure 11-12
Effect of various antidysrhythmic agents on (A) Nonpacemaker cardiomyocyte and (B) nodal tissue.

Antidysrhythmic drugs work by directly or indirectly altering the movement of ions through the cell membrane which then changes the physical characteristics of the cardiac action potentials (Figure 11-12).

Some antidysrhythmic drugs block the fast sodium channels. These channels determine how fast the membrane depolarizes during an action potential. Since conduction velocity is related to how fast the membrane depolarizes, a drug that blocks sodium channels reduces conduction velocity and can effectively eradicate tachyarrhythmias caused by reentry circuits.

Other antidysrhythmic drugs affect the duration of action potentials, particularly the effective refractory period. They do this by controlling the influx or efflux of potassium across the potassium channels. These drugs typically work by delaying repolarization of action potentials in phase 3. By prolonging the effective refractory period, reentry tachyarrhythmias can often be eliminated.

Some drugs are used to inhibit sympathetic beta-adrenergic receptor stimulatory effects on the heart. Because beta-adrenergic receptors are coupled to ion channels, beta-blockers indirectly alter membrane ion influx or efflux, particularly the movement of calcium and potassium ions.

Other drugs are used to block slow inward calcium channels thereby reducing pacemaker firing rate by slowing how quickly depolarizing pacemaker potentials of the SA node and AV node (phases 4 and 0 depolarization) rise. Slowing conduction through the AV node helps to block reentry mechanisms, which can cause supraventricular tachycardia.

While there are many different antidysrhythmic drugs, each having its own unique mechanisms of action, we can place them in five generally recognized categories, Class I through Class V. Below are descriptions of each category. It is important to note that many of the drugs placed into these five classes have considerable overlap in their pharmacologic properties.

Class I

Class I antidysrhythmic drugs interfere with the sodium channel by inhibiting sodium ion influx during phase 0 of the action potential. This minimizes the chance of sodium reaching its threshold potential and causing depolarization. Because of this action, these medications are often referred to as sodium channel blockers or fast channel blockers. This type of action potential is found in nonpacemaker cardiomyocytes such as those found in the atria and ventricles and Purkinje tissue. For this reason, sodium channel blockers have no direct effect on nodal tissue, at least through the blockage of the fast sodium channels

Class I medications are divided into three groups Ia, Ib, and Ic, based on their effect on the action potential duration (APD).

- Class Ia medications have the following effects (Figure 11-13A): (1) bind to and produce a moderate block of the sodium channels which depresses the rate of phase 0 depolarization prolonging conduction; (2) prolong repolarization and duration of the action potential; and (3) increase the effective refractory period (ERP). A negative effect of slowing repolarization is the QT interval lengthens. This can lead to torsades de pointes, a form of ventricular tachycardia. Examples of drugs in this class include quinidine, procainamide, and disopyramide.

- Class Ib medications have the following effects (Figure 11-13B): (1) produce a mild block of the sodium channels which slows phase 0 depolarization; (2) accelerate phase 3 repolarization by increasing the efflux of potassium; and (3) shorten the action potential duration and effective refractory period. Examples of drugs in this class include lidocaine, phenytoin, mexiletine, and tocainide.

- Class 1c medications markedly block sodium channels which prolongs phase 0 depolarization and slows conduction (Figure 11-13C). These drugs, however, have little effect on repolarization and do not change action potential duration or effective refractory period. Examples of drugs in this class include flecainide, propafenone, and aprindine.

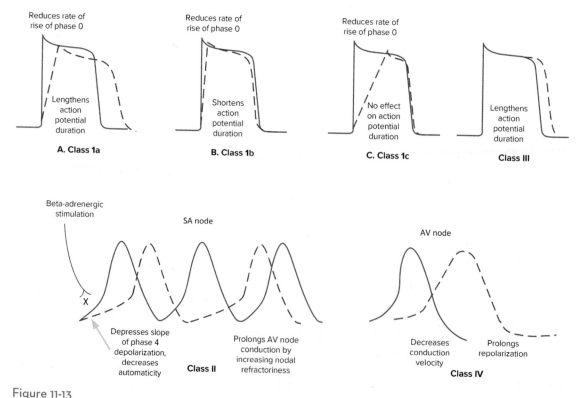

Figure 11-13
Specific antidysrhythmic action on nonpacemaker cardiomyocyte and nodal tissue, A. Class 1a, B. Class 1b, Class, 1c, C. Class II, D. Class III, Class E

Class II

Class II drugs block the sympathetic nervous system. Therefore, they are sympatholytic drugs. Most agents in this class are beta blockers. They depress SA node automaticity and produce increased AV nodal refractoriness (making it resistant to stimulation) (Figure 11-13D). Class II medications include atenolol, esmolol, propranolol, and metropolol.

Class III

Class III drugs block the efflux of potassium during phase 3 of the action potential and prolong repolarization and the effective refractory period (Figure 11-13E). Class III drugs include amiodarone, sotalol, bretylium, ibutilide, dofetilide, and dronedarone.

Class IV

Class IV drugs slow the movement of calcium ions through the L-type calcium channels located on the vascular smooth muscle, cardiomyocytes, and SA and AV nodes. These channels are responsible for regulating the influx of calcium into muscle cells during phase 2 of the action potential, which in turn stimulates smooth muscle contraction and cardiomyocyte contraction and plays an important role in pacemaker currents and in phases 4 and 0 of the action potentials. By blocking calcium entry into the cell, these drugs cause vascular smooth muscle relaxation (vasodilation), decreased myocardial force generation (negative inotropy), decreased heart rate (negative chronotropy), and decreased conduction velocity and prolonged repolarization within the heart (negative dromotropy), particularly at the AV node (Figure 11-13F). Class IV medications include verapamil and diltiazem.

Class V

Class V drugs work by other or unknown mechanisms. They include adenosine (Adenocard), atropine, digoxin (Lanoxin), epinephrine, and magnesium sulfate.

For slow heart rates, a drug that blocks the effects of the parasympathetic nervous system, such as atropine, may be initially used. These drugs are called **anticholinergic** or **parasympatholytic** drugs. The drug's action allows the heart to resume its inherent rate or sympathetic activity to increase the heart rate. The bradycardic patient may also be treated by directly stimulating the sympathetic nervous system using drugs that have sympathomimetic properties, such as epinephrine or dopamine.

Although drug therapy is used to prevent dysrhythmias, nearly every antidysrhythmic drug has the potential to act as a proarrhythmic (in other words it promotes dysrhythmias), so each must be selected and used carefully.

A number of other drugs can have a role in managing cardiac dysrhythmias. Several groups of drugs slow conduction through the heart, without actually preventing the dysrhythmia. These drugs are used to control the rate of a tachycardia and make it tolerable for the patient.

Some dysrhythmias promote blood clotting within the heart and increased risk of embolus and stroke. Anticoagulant medications and antiplatelet drugs can reduce the risk of clotting.

The treatments used for specific dysrhythmias are listed in the next five chapters.

11.1	• Each heartbeat arises as an electrical impulse from the SA node and spreads across the atria, depolarizing the tissue, causing both atria to contract. The impulse then activates the AV node, where it is delayed slightly. The impulse then spreads through both ventricles via the Bundle of His, right and left bundle branches, and the Purkinje fibers, causing a synchronized contraction of the primary pumping chambers of the heart.
11.2	• One of the things we look for during examination of the ECG is the presence of dysrhythmias.
	• A dysrhythmia is an ECG rhythm that differs from normal sinus rhythm. The heartbeat may be slower or faster than normal, it may irregular, or conduction through the heart may be delayed or blocked.
11.3	• Some dysrhythmias are of little consequence and simply annoying while others are life-threatening medical emergencies that can lead to cardiac arrest and sudden death.
	• The most common symptom of dysrhythmias is palpitations, an abnormal sensation felt with the heartbeat.
	• Symptoms such as lightheadedness, dizziness, fainting, chest pain, shortness of breath, sweatiness, and/or pallor may be seen in dysrhythmias that cause decreased cardiac output. We refer to this as being symptomatic.
	• Sometimes the first clinical manifestation of a cardiac dysrhythmia is sudden death.
11.4	• Types of dysrhythmias include bradycardia, tachycardia, early (premature) beats, dropped beats, or QRS complexes and irregular rhythms.
11.5	• Causes and mechanisms of dysrhythmias include increased parasympathetic tone, myocardial hypoxia, injury and infarction, increased automaticity, reentry, triggered beats, and proarrhythmia.
11.6	• Dysrhythmias can originate from the SA node, atria, AV junction, or ventricles and can occur due to AV heart block.
11.7	• Typically, dysrhythmias can be identified using one lead, most commonly lead II. This lead has an excellent view of the normal conduction of the impulse through the heart.
	• Examination of the ECG rhythm on the ECG monitor must be done in a systematic, organized way.
11.8	• To get the whole picture of what is occurring with our patient, we need to do a patient assessment. Sometimes, it is the patient assessment that gives us the information needed to identify what condition the patient is suffering.
11.9	• Many dysrhythmias require no treatment as they are either benign or resolve on their own.
	• The method used to manage dysrhythmias depends on whether the patient is stable or unstable. Patients who are symptomatic are considered unstable because symptoms indicate decreased cardiac output.
	• Several treatment options are available to treat dysrhythmias and are employed based on the mechanism or etiology of the dysrhythmia.
	• Commonly used treatments include physical maneuvers, electrical therapy, and the administration of certain medications.

Assess Your Understanding

The following questions give you a chance to assess your understanding of the material discussed in this chapter. The answers can be found in Appendix A.

1. Which of the following describes the generation and conduction of the normal heartbeat? (LO 11.1)
   a. The impulse that generates the heartbeat originates from the AV junction.
   b. Impulse conduction is delayed as it passes through the AV node.
   c. Impulses spread through both ventricles by way of accessory pathways.
   d. The impulse that generates the heartbeat progresses through the heart from the ventricles to the atria.

2. Describe what occurs when there is decreased cardiac output. (LO 11.3)

3. The primary pacemaker during normal heart activity is the (LO 11.1)
   a. bundle of His
   b. SA node
   c. AV node
   d. AV junction

4. Define the term dysrhythmia. (LO 11.2)

5. The most common symptom of cardiac dysrhythmias is/are (LO 11.3)
   a. low blood pressure.
   b. lightheadedness.
   c. shortness of breath.
   d. palpitations.

6. If the patient has a slow heart rate that is producing a low blood pressure, we would describe it as (LO 11.3)
   a. sinus bradycardia.
   b. symptomatic bradycardia.
   c. bradycardia with low blood pressure.
   d. vagal response.

7. Match the following types of dysrhythmias with the corresponding characteristics. (LO 11.4)

Types		Characteristics
a.	Bradycardia	_____ A heart rate greater than 100 beats per minute.
b.	Tachycardia	_____ Impulses that appear early.
c.	Premature complexes	_____ Occur when the SA node fails to initiate an impulse or from a partial or intermittent block at the AV node.
d.	Dropped beats or QRS complexes	_____ A heart rate less than 60 beats per minute.

8. Describe how increased parasympathetic tone effects the heart. (LO 11.5)

9. Match the following types of causes or mechanisms with the corresponding characteristics. (LO 11.5)

	Types		Characteristics
a.	Increased automaticity	_____	A new or more frequent occurrence of preexisting dysrhythmias that are brought about by antidysrhythmic therapy.
b.	Reentry	_____	Occurs with problems at the level of the ion channels in individual heart cells and leads to partial repolarization which can cause repetitive ectopic firing.
c.	Triggered beats	_____	Spontaneous depolarization of the myocytes that occurs more quickly than normal.
d.	Proarrhythmia	_____	Occurs when an electrical impulse goes back into a conduction pathway rather than moving from one end of the heart to the other and then terminating.

10. Dysrhythmias that arise from the atria include (LO 11.6)
    a. junctional escape.
    b. atrial fibrillation.
    c. sinus arrest.
    d. idioventricular rhythm.

11. AV heart blocks include (LO 11.6)
    a. 3rd degree AV block.
    b. asystole.
    c. wandering atrial pacemaker.
    d. sinoatrial exit block.

12. The preferred lead for identifying cardiac dysrhythmias is lead (LO 11.7)
    a. I.
    b. II.
    c. MCL6.
    d. V_3.

13. Physical maneuvers (LO 11.9)
    a. act to stimulate the sympathetic nervous system.
    b. increase the speed impulses travel through the AV node.
    c. are used to treat AV heart block.
    d. slow the rate at which the SA node fires.

14. Synchronized cardioversion is used to treat (LO 11.9)
    a. bradycardia.
    b. symptomatic tachydysrhythmias.
    c. ventricular fibrillation.
    d. AV heart block.

15. Drugs that block the effects of the parasympathetic nervous system are used to treat (LO 11.9)

    a. bradycardia.

    b. rapid heart rates.

    c. ventricular fibrillation.

    d. premature beats.

16. Match the following antidysrhythmic medications with their actions. (LO 11.9)

Medications		Characteristics
a. Class I	_____	Block the efflux of K+ during Phase 3 of the action potential and prolong repolarization and the refractory period.
b. Class II	_____	Inhibit the movement of Ca++ ions during Phase 2 of the action potential, prolonging conductivity and increasing the refractory period at the AV node.
c. Class III	_____	Block the sympathetic nervous system. Most agents in this class are beta blockers. They depress SA node automaticity and produce increased AV nodal refractoriness (making it resistant to stimulation).
d. Class IV	_____	Reduce the influx of Na+ ions into the cells during phase 0 of the action potential, which minimizes the chance of Na+ reaching its threshold potential and causing depolarization.

Referring to the scenario at the beginning of this chapter, answer the following questions.

17. What type of dysrhythmia did the patient first experience? (LO 11.5)

    a. Bradycardia

    b. Frequent premature beats

    c. Tachycardia

    d. Fibrillation

18. During the dysrhythmia the patient was (LO 11.3)

    a. symptomatic.

    b. asymptomatic.

    c. in cardiac arrest.

    d. a and c.

19. Which of the following was the likely cause of this dysrhythmia? (LO 11.5)

    a. Decreased automaticity

    b. Reentry

    c. Fibrillation

    d. Increased parasympathetic tone

20. Describe how to identify cardiac dysrhythmias while analyzing the ECG tracing. (LO 11.8)

12 Sinus Dysrhythmias

©rivetti/Getty Images

Courtesy Physio-Control

Learning Outcomes

LO 12.1 Describe the origin of sinus dysrhythmias.

LO 12.2 Recall the key features of dysrhythmias originating from the sinus node.

LO 12.3 Describe the appearance of normal sinus rhythm.

LO 12.4 List the causes, effects, appearance, and treatment of sinus bradycardia.

LO 12.5 List the causes, effects, appearance, and treatment of sinus tachycardia.

LO 12.6 Recall the causes, effects, appearance, and treatment of sinus dysrhythmia.

LO 12.7 Identify the causes, effects, appearance, and treatment of sinus arrest.

LO 12.8 Identify the causes, effects, appearance, and treatment of sinoatrial exit block.

LO 12.9 Identify the causes, effects, appearance, and treatment of sick sinus syndrome.

LO 12.10 Describe sinus rhythm as the underlying rhythm.

Case History

A 26-year-old woman is visiting her dying grandmother in the nursing home when she begins to cry and hyperventilate. She faints and collapses to the floor. The nursing staff brings the code cart to the woman's side while 911 is summoned. The nursing supervisor feels for a pulse and comments that the woman's heart is beating slowly. The cardiac monitor is applied while vital signs are obtained, revealing a blood pressure of 60/30, pulse of 44, and respiratory rate of 16 and unlabored.

The cardiac monitor shows a slow narrow QRS complex rhythm with a normal P wave before every QRS complex and a QRS complex after every P wave and a normal PR interval. The nurses elevate the woman's legs and place a cool cloth on her forehead.

By the time EMS arrives, the patient is awake and alert. Her pulse is now 68 and her blood pressure normal. She refuses transport to the hospital after disclosing that she has not eaten in 24 hours, which she is certain contributed to her fainting episode.

12.1 Rhythms Originating from the Sinus Node

In the previous chapters we talked about the anatomy and physiology of the heart and the machines and leads used to obtain an ECG tracing. We also discussed the nine-step process (see Figure 3-2) used to examine the ECG. We will continue to use this process to evaluate the ECGs through the rest of this book. One of the things we look for when we examine the ECG is the presence of dysrhythmias. This chapter focuses on sinus dysrhythmias—the characteristics associated with them (which allow us to identify them), how they occur, their significance, and treatments.

Rhythms originating from the SA node are called sinus rhythms (Figure 12-1). These include

- Normal sinus rhythm
- Sinus bradycardia
- Sinus tachycardia
- Sinus dysrhythmia
- Sinus arrest
- Sinoatrial exit block

A condition, called *sick sinus syndrome*, is a group of abnormal rhythms that occur with malfunction of the sinus node. It will also be discussed in this chapter.

12.2 ECG Appearance of Sinus Rhythms

The key characteristics for sinus rhythms include a normal P wave preceding each normal QRS complex and PR intervals within normal duration of 0.12 to 0.20 seconds and constant. The ST segments and T waves should also be normal. The QT intervals may vary depending on the dysrhythmia. U waves may or may not be present.

In any dysrhythmia where P waves or QRS complexes should be normal, such as with sinus bradycardia, they can instead appear different but yet not change what the dysrhythmia is called. For example, the P waves might be notched or wider than normal due to atrial enlargement, or the QRS complexes might be wide or unusual looking if there is intraventricular conduction defect. This does not change what we categorize the dysrhythmia as; instead, we might refer to it as *sinus bradycardia with wide QRS complexes* (although it would also be correct to call it *sinus bradycardia*).

During normal heart activity, the SA (sinoatrial) node acts as the primary pacemaker. It assumes this role because the rate at which it fires automatically is faster than that of the other pacemakers in the heart. The right coronary artery and the circumflex branch of the left coronary artery supply the SA node with oxygen and nutrients. The vagus nerve and several sympathetic nerves (of the autonomic nervous system) richly innervate the sinus node. Stimulation of the vagus nerve decreases the firing rate of the SA node while stimulation of the sympathetic system increases it.

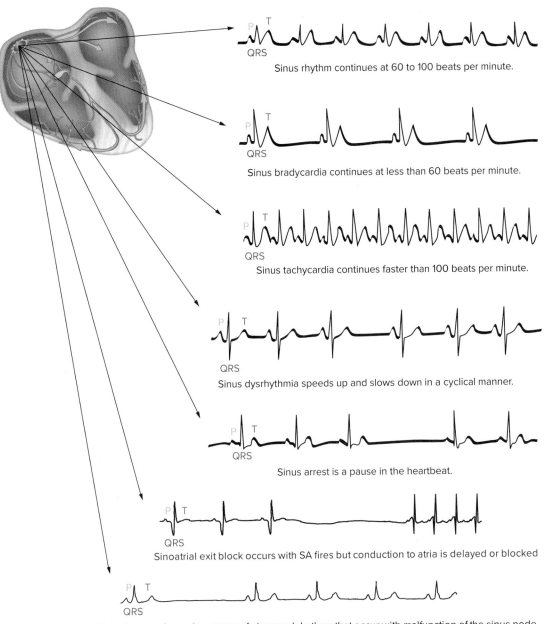

Sinus rhythm continues at 60 to 100 beats per minute.

Sinus bradycardia continues at less than 60 beats per minute.

Sinus tachycardia continues faster than 100 beats per minute.

Sinus dysrhythmia speeds up and slows down in a cyclical manner.

Sinus arrest is a pause in the heartbeat.

Sinoatrial exit block occurs with SA fires but conduction to atria is delayed or blocked

Sick sinus syndrome is a group of abnormal rhythms that occur with malfunction of the sinus node

Figure 12-1
Sinus dysrhythmias originate from the SA node.

12.3 Normal Sinus Rhythm

Normal sinus rhythm (**NSR**) is considered the normal electrical activity of the heart. It is the rhythm with which we compare all other rhythms. Normal sinus rhythm has a heart rate of 60 to 100 beats per minute (in the average adult). It is a regular rhythm with a normal P wave preceding each normal QRS complex. The PR intervals are within the range of 0.12 to 0.20 seconds in duration and are constant (each PR interval is the same). Each normal QRS complex (0.10 seconds or less in duration) is followed by a normal ST segment and T wave (Figure 12-2). The QT interval is within normal limits. U waves may or may not be present.

Normal sinus rhythm arises from the SA node. Each impulse travels down through the conduction system in a normal manner.

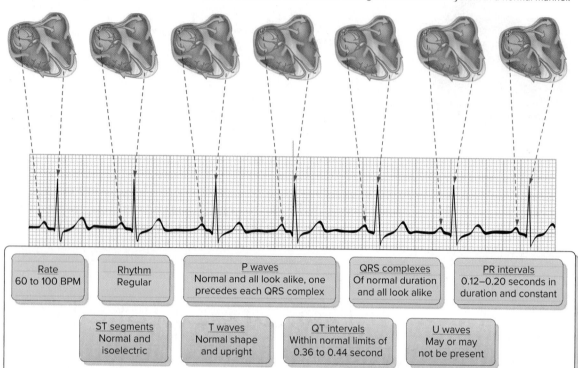

Figure 12-2
Normal sinus rhythm.

In some normal sinus rhythms there may be some minor variances, such as a slightly irregular rhythm or an elevated ST segment; you may see these described as regular sinus rhythm or sinus rhythm because they don't fit the textbook description of NSR.

12.4 Sinus Bradycardia

Description

The word part "brady" means slow while "cardia" refers to the heart. Put together, *bradycardia* means slow heart rate. As its name implies, sinus bradycardia is a slower than normal rhythm that arises from the SA node (Figure 12-3). Sinus bradycardia has all the characteristics of normal sinus rhythm, but the heart rate is less than 60 beats per minute (BPM). It looks like a slow NSR. It results from a slowing of the rate at which the heart's pacemaker, the SA node, fires.

Causes

Sinus bradycardia often occurs naturally as the body's way to conserve energy during times of reduced demand for blood flow. In this case, parasympathetic stimulation increases and sympathetic stimulation decreases, resulting in decreased SA node stimulation and the heart rate slowing. This occurs normally during sleep as a result of circadian variations in the heart rate.

Sinus bradycardia arises from the SA node. Each impulse travels down through the conduction system in a normal manner.

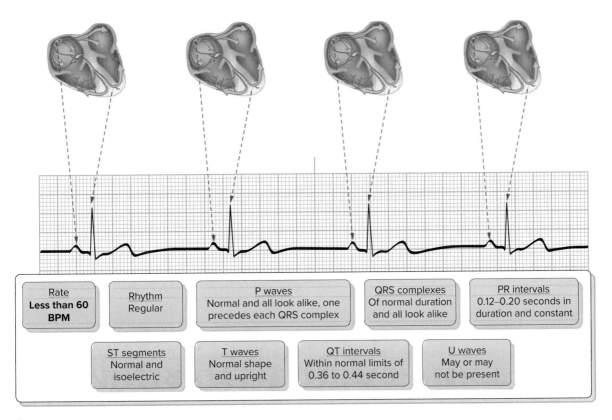

Rate **Less than 60 BPM**	Rhythm Regular	P waves Normal and all look alike, one precedes each QRS complex	QRS complexes Of normal duration and all look alike	PR intervals 0.12–0.20 seconds in duration and constant

ST segments Normal and isoelectric	T waves Normal shape and upright	QT intervals Within normal limits of 0.36 to 0.44 second	U waves May or may not be present

Figure 12-3
Sinus bradycardia.

Sinus bradycardia is considered normal in aerobically conditioned persons or in athletes, whose cardiovascular systems are well-conditioned and whose hearts can maintain a normal stroke volume with less than normal effort. They often have slower heart rates of 40 to 50 beats per minute.

Sinus bradycardia can also occur as a result of various conditions. It is commonly seen in patients who have suffered an inferior-wall MI (this involves the right coronary artery, which supplies blood to the SA node). It is also thought to be one of the body's protective mechanisms in acute myocardial infarction—the heart rate slows to reduce the myocardial oxygen consumption, thereby decreasing the severity of myocardial ischemia and injury (provided that the patient is otherwise well perfused). Sinus bradycardia may also result from a host of other conditions, including noncardiac disorders, conditions involving excessive parasympathetic stimulation or decreased sympathetic stimulation, cardiac diseases, and the use of certain drugs. You may remember from our discussion in Chapter 1 that the autonomic nervous system has a branch, the parasympathetic division, responsible for the vegetative functions, and a branch, the sympathetic division, responsible for helping the body address its needs during times of excitement, work, and stress. The parasympathetic division, supplied by the vagus nerve, slows the heart rate and decreases the AV conduction. Both branches exert their effects via neurotransmitters that bind specific receptors.

Causes of Sinus Bradycardia	
Cause	**Examples**
Cardiac diseases	SA node disease, cardiomyopathy, myocarditis, myocardial ischemia immediately following inferior-wall MI, and heart block
Use of certain drugs	Beta-adrenergic blockers (metoprolol, propranolol), calcium channel blockers (diltiazem, verapamil), antidysrhythmics (amiodarone, propafenone, quinidine, sotalol), lithium, digoxin
Excessive parasympathetic tone or decreased sympathetic stimulation	Carotid sinus massage, Valsalva's maneuver, deep relaxation, sleep, and vomiting
Noncardiac disorders	Hypoxia, hypothermia, hyperkalemia, stroke, increased intracranial pressure, hypothyroidism, and glaucoma

Effects

Often sinus bradycardia is insignificant and the patient is asymptomatic. Typically, most adults are able to tolerate a heart rate of 50 to 59 beats per minute. They are less tolerant of rates less than 50 beats per minute, particularly if they have underlying heart disease that prevents them from compensating for the slow heart rate with an increase in stroke volume or change in vascular resistance.

In some patients the slower rate associated with sinus bradycardia can compromise cardiac output, resulting in hypotension; angina pectoris; cool, clammy skin; crackles; dyspnea; S_3 heart sound (indicating heart failure); and central nervous system symptoms such as lightheadedness, blurred vision, vertigo, and syncope (Stokes–Adams attack). Recall the bradycardic patient in the case history at the beginning of this chapter. She experienced a syncopal episode (she fainted) due to her bradycardia. If a patient is symptomatic (e.g., short of breath, chest pain, dizziness), you should consider the presence of sinus bradycardia to be abnormal and potentially significant. Bradycardia can predispose some patients to more serious dysrhythmias such as ventricular tachycardia and ventricular fibrillation. Also, bradycardia is considered a diagnostic indicator of the seriously ill child because it denotes significant hypoxia and inadequate perfusion.

ECG Appearance

The first thing you will note with sinus bradycardia is the slower than normal heart rate. This causes longer R-R intervals and P-P intervals (making it appear as if there are large spaces between the groups of complexes). The reason for the longer R-R intervals and P-P intervals is because there are fewer R waves and P waves, making the distance between each much greater. Other than that, all the features are essentially the same as what we would expect to see with normal sinus rhythm.

Treatment

Treatment for sinus bradycardia depends on whether or not the patient is symptomatic. Patients who are asymptomatic, also referred to as stable, require no treatment. However, when medical conditions cause the problem, it is essential to observe the heart rhythm for the progression of the bradycardia. Patients who are symptomatic, also called unstable, may improve with the administration of atropine, 0.5 mg

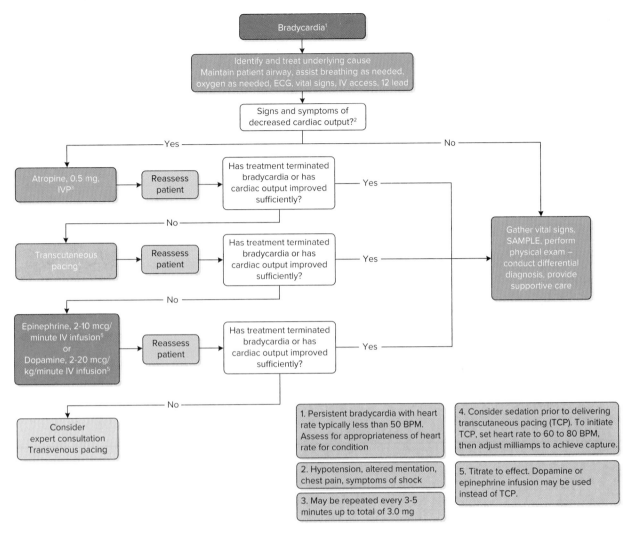

Figure 12-4
Treatment algorithm for bradycardia.

(acts to block the effects of increased vagal tone), IV push every 3 to 5 minutes to a total of 3.0 mg, transcutaneous pacing (acts to directly stimulate the heart to beat faster), or the administration of an IV infusion of dopamine, 2 to 20 mcg/kg/minute or epinephrine, 2 to 10 mcg/minute (both act to stimulate the sympathetic nervous system) (Figure 12-4).

12.5 Sinus Tachycardia

Description

The opposite of bradycardia is tachycardia. "Tachy" means fast. Tachycardia means a fast heart rate. As its name implies, sinus tachycardia is a faster than normal rhythm that results from an increase in the rate of sinus node discharge (Figure 12-5). It has the same characteristics as normal sinus rhythm but has a rate greater than 100 beats per minute.

It is widely held that the SA node can fire only so fast; generally, it is believed that the upper limit is between 160 and 180 beats per minute. However, each person has a built-in maximum rate so, if in doubt, get a 12-lead ECG to obtain more information.

Sinus tachycardia arises from the SA node. Each impulse travels down through the conduction system in a normal manner.

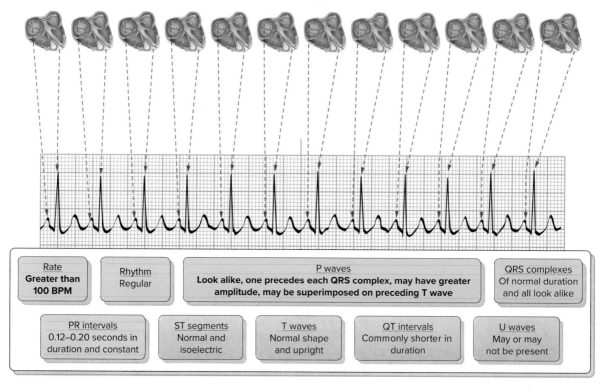

Rate Greater than 100 BPM	Rhythm Regular	P waves Look alike, one precedes each QRS complex, may have greater amplitude, may be superimposed on preceding T wave			QRS complexes Of normal duration and all look alike
PR intervals 0.12–0.20 seconds in duration and constant	ST segments Normal and isoelectric	T waves Normal shape and upright	QT intervals Commonly shorter in duration	U waves May or may not be present	

Figure 12-5
Sinus tachycardia.

Causes

Sinus tachycardia often occurs naturally as a means of increasing delivery of oxygen and nutrients and removing waste products during times of exertion, exercise, or stress. We often see sinus tachycardia with physical labor, exercise, pain, fear, excitement, anxiety, and where increased sympathetic stimulation occurs in response to the need for more oxygen and nutrients at the cellular level.

We also see it with the ingestion of caffeine or alcohol, smoking, or fever, or where there is some type of underlying illness such as hypovolemia, respiratory distress, hyperthyroidism, or anemia. When the stimulus for the tachycardia is removed, the dysrhythmia spontaneously resolves.

Causes of Sinus Tachycardia	
Cause	**Examples**
Cardiac diseases	Congestive heart failure, cardiogenic shock, and pericarditis
Use of certain drugs	Administration of pharmacologic agents that dramatically increase sinus rate, such as sympathomimetic drugs (epinephrine, isoproterenol, dopamine, dobutamine), parasympatholytic drugs (atropine), or other stimulants and amphetamines
Increased sympathetic stimulation	Strenuous exercise, pain, stress, fever, fear, anxiety, hypermetabolism, or as a compensatory mechanism in shock
Other causes	Respiratory distress, hypoxia, pulmonary embolism, anemia, sepsis, hyperthyroidism, alcohol, caffeine, and nicotine

Effects

Although sinus tachycardia may be of no clinical significance, it can reflect an underlying cause that needs to be identified, evaluated, and treated. A significant problem associated with sinus tachycardia is that it increases myocardial oxygen consumption, which can aggravate ischemia (bringing on chest pain) and lead to infarction. It can also predispose the patient to more serious rhythm disturbances, particularly in persons with coronary artery disease (CAD) or obstructive types of heart conditions such as aortic stenosis and hypertrophic cardiomyopathy.

ECG Appearance

The most obvious characteristic you will see with sinus tachycardia is a faster than normal rate. This causes shorter R-R intervals and P-P intervals (making it appear as if there are smaller spaces between the groups of complexes). With the faster rates, the P wave may be **superimposed** on the T wave of the preceding beat, making it harder to identify, and the QT interval is commonly shorter than with normal sinus rhythm. Also, with sinus tachycardia, the P waves may increase in amplitude.

Treatment

No treatment is indicated if the patient is asymptomatic. However, the patient should be encouraged to abstain from triggers such as alcohol, caffeine, and nicotine. Treatment for symptomatic sinus tachycardia is directed at treating the cause. In the presence of an underlying medical or traumatic condition, continued monitoring is indicated.

For patients experiencing myocardial ischemia, consideration may be given to the use of beta-adrenergic receptor blockers, i.e., metropolol to slow the heart rate, or nitrates to vasodilate the coronary vessels.

12.6 Sinus Dysrhythmia

Description

With the exception of a patterned irregularity, sinus dysrhythmia is characteristically the same as sinus rhythm (normal P waves, QRS complexes, and T waves) (Figure 12-6). It can be described as a repeating cycle of "slowing, then speeding up, then slowing of the heart rate again." The beat-to-beat variation produced by the irregular firing of the pacemaker cells of the SA node corresponds with the respiratory cycle and changes in intrathoracic pressure. The heart rate increases during inspiration (decreased parasympathetic stimulation, increased venous return) and decreases during expiration (increased parasympathetic stimulation, decreased venous return).

There is also evidence to show that sinus dysrhythmia may play an active physiologic role. At the peak of inspiration, the greatest amount of oxygen is present in the alveolar/capillary membrane, so the heart rate speeds up to maximize the perfusion to this area. During exhalation, the heart rate drops, thus saving energy expenditure by suppressing unnecessary heartbeats.

Sinus dysrhythmia arises from the SA node. Each impulse travels down through the conduction system in a normal manner.

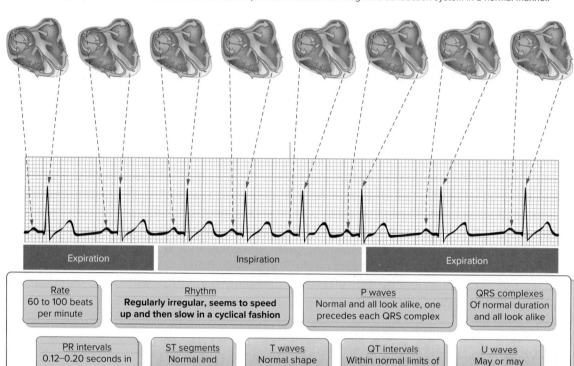

Rate	Rhythm	P waves	QRS complexes
60 to 100 beats per minute	**Regularly irregular, seems to speed up and then slow in a cyclical fashion**	Normal and all look alike, one precedes each QRS complex	Of normal duration and all look alike

PR intervals	ST segments	T waves	QT intervals	U waves
0.12–0.20 seconds in duration and constant	Normal and isoelectric	Normal shape and upright	Within normal limits of 0.36 to 0.44 second	May or may not be present

Figure 12-6
Sinus dysrhythmia.

Causes

Sinus dysrhythmia can occur naturally in athletes, children, and older adults but is uncommon in infants. It can also occur in patients with heart disease or inferior wall MI, advanced age, or multiple sclerosis; in those receiving certain drugs such as digitalis and morphine; and in conditions where there is increased intracranial pressure. It is usually of no clinical significance and produces no symptoms. However, in some patients and conditions, it may be associated with palpitations, dizziness, and syncope. A marked variation in P-P intervals in the elderly may indicate sick sinus syndrome (a condition we discuss later).

Causes of Sinus Dysrhythmia	
Cause	**Examples**
Cardiac diseases	Inferior-wall MI
Use of certain drugs	Digoxin, morphine, sedatives
Autonomic nervous system stimulation	Inhibition of reflex parasympathetic activity (tone)
Other causes	Increased intracranial pressure, multiple sclerosis

ECG Appearance

As you assess this dysrhythmia, the most distinguishing characteristic is the patterned irregular rhythm. When the rate speeds up, the complexes are closer together; this shortens the P-P and R-R intervals. When the rate slows down, the complexes are farther away, lengthening the P-P and R-R intervals. The other features are essentially the same as normal sinus rhythm, particularly the round normal P wave preceding each QRS complex. The QRS complexes should be normal. The QT interval may vary slightly.

Treatment

Usually, there is no treatment for sinus dysrhythmia provided that the patient is asymptomatic. If unrelated to respirations, consideration may be given to treating the underlying cause.

The next two dysrhythmias we discuss result from either the SA node transiently failing to generate an impulse or an impulse generated by the SA node being delayed or blocked during its conduction through the atria. The anatomical basis for this is the SA node consists of two main groups of cells: a central core of pacemaking cells ("P cells") that produce the sinus impulses and an outer layer of transitional cells ("T cells") that transmit the sinus impulses into the right atrium. This produces a pause in the cycle of electrical activity. Remember, sinus node activity is not recorded on the surface ECG. For this reason it is difficult to determine if sinus node automaticity or sinoatrial conduction abnormalities are responsible for the sudden disappearance of sinus P waves for variable intervals. We discuss both types of conditions below.

12.7 Sinus Arrest

Description

Sinus arrest is an abnormality in automaticity. It occurs when the P cells of the SA node transiently stops firing (Figure 12-7), causing a short period of cardiac stand-still until a lower-level pacemaker discharges, creating an escape beat, or the sinus node resumes its normal function. The usual effect of sinus arrest is a brief pause in all electrical activity and cardiac output. The next impulse occurs randomly and not necessarily when the next cardiac cycle would be expected.

This dysrhythmia is referred to as **sinus pause** when one or two beats are not formed and sinus arrest when three or more beats are not formed.

Causes

Sinus arrest is an uncommon rhythm disturbance. It results from a marked depression in SA node automaticity. It is occasionally seen in elderly patients. Causes include an increase in parasympathetic tone on the SA node, hypoxia or ischemia, digitalis or beta-adrenergic blocker toxicity, hyperkalemia, degenerative fibrotic disease, or damage to the SA node such as with MI (particularly with acute inferior or true posterior MI).

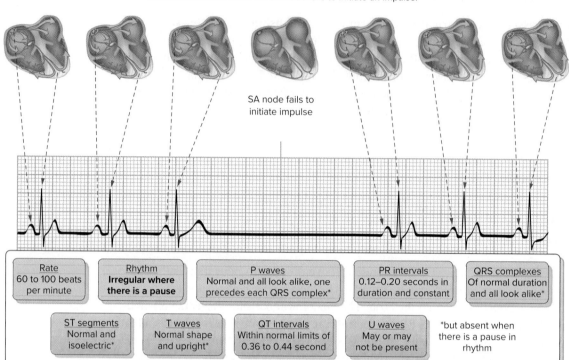

Figure 12-7
Sinus arrest.

Causes of Sinus Arrest	
Cause	**Examples**
Cardiac diseases	Inferior-wall MI, acute myocarditis, coronary artery disease (CAD), cardiomyopathy, fibrosis, hypertensive heart disease, sinus node disease, sick sinus syndrome
Use of certain drugs	Amiodarone, beta-adrenergic blockers (bisoprolol, metapronol, propranolol), calcium channel blockers (verapamil, diltiazem), digoxin, procainamide, quinidine, salicylates (particularly in toxic doses)
Autonomic nervous system stimulation	Increased parasympathetic tone, carotid sinus sensitivity
Other causes	Acute infection, hyperkalemia, hypoxia

Effects

Sinus arrest becomes clinically significant when there is an extended pause or when there are frequent occurrences of the pause or arrest. This can lead to decreased heart rate and elimination of the atrial contribution to ventricular filling, resulting in a drop in cardiac output and decreased blood pressure and tissue perfusion. There is also a danger that SA node activity will completely cease and an escape pacemaker may not take over pacing (resulting in asystole).

ECG Appearance

The ECG tracing looks like normal sinus rhythm except that there is a pause in the rhythm resulting in an absence of the P, QRS, and T waveforms. This produces an occasional irregularity in the rhythm that is isolated to the event of automaticity failure in the SA node. The rhythm typically resumes its normal appearance after this transient pause unless an escape pacemaker resumes the rhythm.

Depending on how frequently it occurs and how long the sinus arrest lasts, the rate may be normal to slow (less than 60 beats per minute). Again, because the rhythm arises from the SA node, the P waves are normal and each is followed by a QRS complex.

Treatment

If the patient is asymptomatic, no treatment is indicated. If the patient is symptomatic, treatment includes administration of atropine or temporary or permanent ventricular pacing. Reasons to pace sinus arrest include the development of an AV junctional or ectopic ventricular pacemaker that is slow enough to result in such problems as syncope, CHF, angina, or frequent ventricular ectopic beats. As needed, drugs affecting SA node discharge or conduction, such as beta-adrenergic blockers, calcium channel blockers, and digoxin, should be discontinued.

12.8 Sinoatrial Exit Block

Description

Sinoatrial exit block occurs when the SA node fires but the T cells fail to or slowly conduct the impulse to the atria (Figure 12-8). The mechanism of such blocking is not fully understood. While there are four types of sinoatrial block, only two can be identified by the surface ECG.

Causes

Sinoatrial exit block results from a block in the conduction of the impulse from SA node to the atria. It is very uncommon. Causes include an increase in parasympathetic tone on the SA node, hypoxia or ischemia, excessive administration of

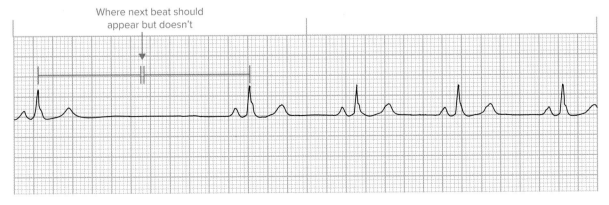

Figure 12-8
Example of sinoatrial exit block.

digitalis or propranolol (Inderal), hyperkalemia, and damage to the SA node (such as myocardial infarction [MI] or degenerative fibrotic disease).

Causes of Sinoatrial Exit Block	
Cause	**Examples**
Cardiac diseases	Acute inferior-wall MI, acute myocarditis, CAD, cardiomyopathy, fibrosis, hypertensive heart disease, sinus node disease, sick sinus syndrome
Use of certain drugs	Amiodarone, beta-adrenergic blockers (bisoprolol, metapronol propranolol), calcium channel blockers (verapamil, diltiazem), digoxin, procainamide, quinidine, salicylates (particularly in toxic doses)
Autonomic nervous system stimulation	Increased parasympathetic tone, carotid sinus sensitivity
Other causes	Acute infection, hyperkalemia, hypoxia

Effects

Sinoatrial exit block is usually insignificant. It can become clinically significant when there is an extended pause or when there are frequent occurrences of the dropped P waves (and subsequent QRS complexes). This can lead to decreased heart rate, a drop in cardiac output, and decreased blood pressure and tissue perfusion.

ECG Appearance

In one form of sinoatrial exit block, there is a pause resulting from the absence of a P wave (and associated QRS complex), but then the P wave (and associated QRS complex) reoccurs at the next expected interval. In other words, the cycle of the dropped P wave is exactly equal to two of the basic sinus cycles. If two or more consecutive sinus impulses are blocked, it creates a considerable pause. However, the P-P intervals have a pattern and are a multiple of the basic sinus cycle.

With one form of sinoatrial exit block, the P-P cycle progressively shortens until there is a pause. The pause is due to the dropped P wave and measures less than twice the P-P cycle. The cycle is repeated. In this form of sinoatrial exit block, the PR interval remains constant until the dropped QRS complex.

As a point of information, blocked atrial premature beats (discussed in the next chapter) sometimes mimic sinoatrial exit block.

Depending on how frequently it occurs and how long the sinoatrial exit block lasts, the rate may be normal to slow (less than 60 beats per minute). Again, because the rhythm arises from the SA node, the P waves are normal, and each is followed by a QRS complex.

Treatment

No treatment is indicated provided that the patient is asymptomatic. If the patient is symptomatic, treatment includes administration of atropine or temporary or permanent ventricular pacing. As needed, drugs affecting SA node discharge or conduction, such as beta-adrenergic blockers, calcium channel blockers, and digoxin, should be discontinued.

12.9 Sick Sinus Syndrome

Description

Sick sinus syndrome, also referred to as *sinus node dysfunction,* is a group of abnormal rhythms that occur with malfunction of the sinus node. Bradycardia-tachycardia syndrome is a variant of sick sinus syndrome in which slow dysrhythmias and fast dysrhythmias alternate.

Causes

Sick sinus syndrome is a moderately uncommon disorder. It can result in many abnormal rhythms, including sinus arrest, sinus node exit block, sinus bradycardia, and other types of bradycardia. It can also be associated with tachycardias such as paroxysmal supraventricular tachycardia (PSVT) and atrial fibrillation (which are discussed in the next chapter). Tachycardias that occur with sick sinus syndrome are characterized by a long pause after the tachycardia.

These abnormal rhythms are often caused or worsened by medications such as digitalis, calcium channel blockers, beta-adrenergic blockers, sympatholytic medications, and antidysrhythmic agents. Sick sinus syndrome can also be caused by disorders that produce conduction system scarring, degeneration, or damage. It is more common in elderly adults, in whom the cause is often a nonspecific, scar-like degeneration of the heart's conduction system. In children, cardiac surgery (particularly to the atria) is a common cause of sick sinus syndrome. Coronary artery disease, high blood pressure, and aortic and mitral valve diseases may be associated with sick sinus syndrome, although this association may only be incidental.

Causes of Sick Sinus Syndrome	
Cause	**Examples**
Cardiac diseases	CAD, high blood pressure, and aortic and mitral valve diseases
Scarring of conduction system	Cardiomyopathy, sarcoidosis, and amyloidosis
Use of certain drugs	Digitalis, beta-adrenergic blockers, calcium channel blockers, sympatholytic medications, and antidysrhythmic agents
Others	Cardiac surgery in children

Effects

Although many types of sick sinus syndrome are symptom free, patients may present with Stokes–Adams attacks, fainting, dizziness or lightheadedness, palpitations, chest pain, shortness of breath, fatigue, headache, and nausea.

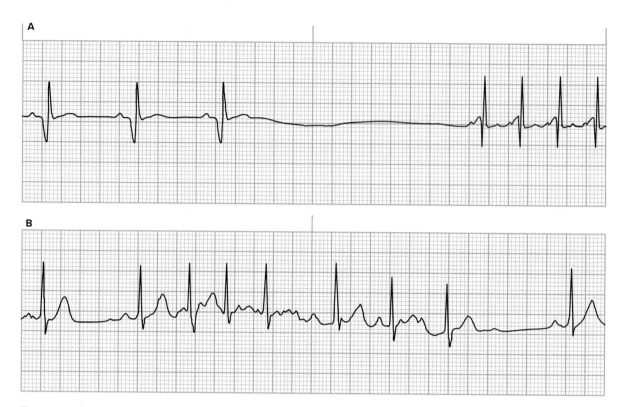

Figure 12-9
Examples of sick sinus syndrome.

ECG Appearance

The ECG may show a variety of dysrhythmias, including inappropriate sinus brady-cardia, sinus arrest, and sinoatrial exit block, as well as several that are discussed in upcoming chapters, including atrial fibrillation with slow ventricular response, a prolonged period of asystole after a period of tachycardias, atrial flutter, and atrial tachycardia (Figure 12-9).

Treatment

Bradydysrhythmias are well controlled with pacemakers whereas tachydysrhyth-mias respond well to medications. However, because both bradydysrhythmias and tachydysrhythmias may be present, drugs to control the rapid heart rates may worsen bradydysrhythmia. For this reason, a pacemaker is implanted before drug therapy is begun for the tachydysrhythmia.

12.10 Sinus Rhythm as the Underlying Rhythm

Sinus rhythm may be what is referred to as an *underlying rhythm*. What that means is the sinus rhythm is seen, but then there is another dysrhythmia or cardiac condi-tion seen as well. For example, if there is a delay in conduction through the AV node (referred to as fir*st degree AV block,* in chapter 15) then we call it *sinus rhythm with first degree AV block* (Figure 12-10A). Similarly, if there are early beats, we call it *sinus rhythm with early beats* (Figure 12-10B). We will cover this in more depth in subsequent chapters.

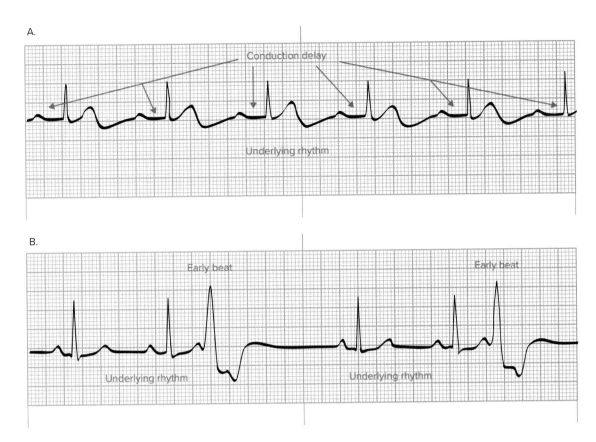

Figure 12-10
Examples of sinus rhythm as an underlying rhythm.

Practice Makes Perfect

For each of the 57 ECG tracings on the following pages, practice the Nine-Step Process for analyzing ECGs. To achieve the greatest learning, you should practice assessing and interpreting the ECGs immediately after reading this chapter. Below are questions you should consider as you access each tracing. Your answers can be written into the area below each ECG marked "ECG Findings." Your findings can then be compared to the answers provided in Appendix A. All tracings are six seconds in length.

1. Determine the heart rate. Is it slow? Normal? Fast? What is the ventricular rate? What is the atrial rate?

2. Determine if the rhythm is regular or irregular. If it is irregular, what type of irregularity is it? Occasional or frequent? Slight? Sudden acceleration or slowing in heart rate? Total? Patterned? Does it have a variable conduction ratio?

3. Determine if P waves are present. If so, how do they appear? Do they have normal height and duration? Are they tall? Notched? Wide? Biphasic? Of differing morphology? Inverted? One for each QRS complex? More than one preceding some or all the QRS complexes? Do they have a sawtooth appearance? Is there an indiscernible chaotic baseline?

4. Determine if QRS complexes are present. If so, how do they appear? Narrow with proper amplitude? Tall? Low amplitude? Delta wave? Notched? Wide? Bizarre-looking? With chaotic waveforms?

5. Determine the presence of PR intervals. If present, how do they appear? Constant? Of normal duration? Shortened? Lengthened? Progressively longer? Varying?

6. Evaluate the ST segments. Do they have normal duration and position? Are they elevated? (If so, are they flat, concave, convex, arched?) Depressed? (If so, are they normal, flat, downsloping or upsloping?)

7. Determine if T waves are present. If so, how do they appear? Of normal height and duration? Tall? Wide? Notched? Inverted?

8. Determine the presence of QT intervals. If present, what is their duration? Normal? Shortened? Prolonged?

9. Determine if U waves are present. If present, how do they appear? Of normal height and duration? Inverted?

10. Identify the rhythm or dysrhythmia.

1.

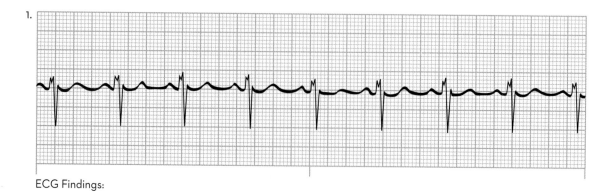

ECG Findings:

2.

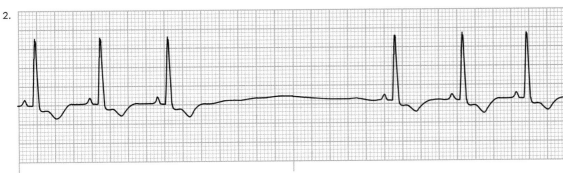

ECG Findings:

3.

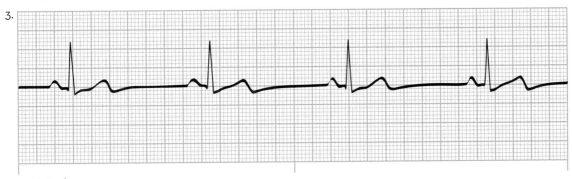

ECG Findings:

4.

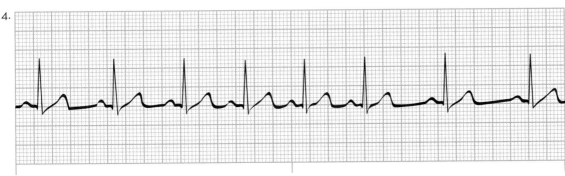

ECG Findings:

5.

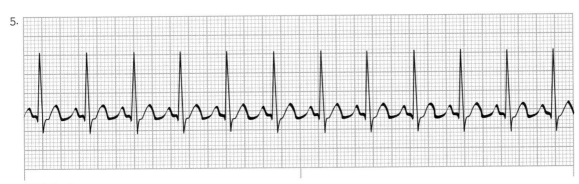

ECG Findings:

6.

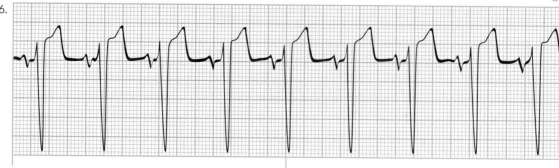

ECG Findings:

7.

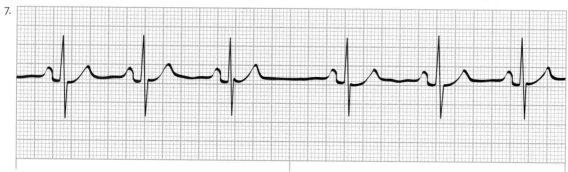

ECG Findings:

8.

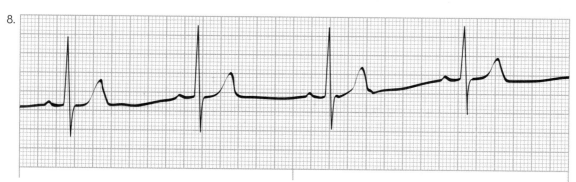

ECG Findings:

9.

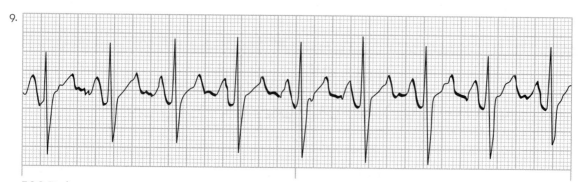

ECG Findings:

10.

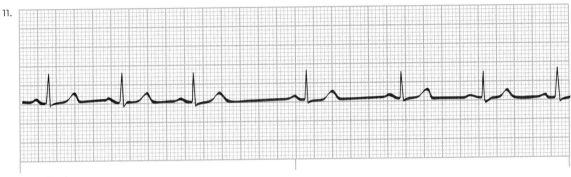

ECG Findings:

11.

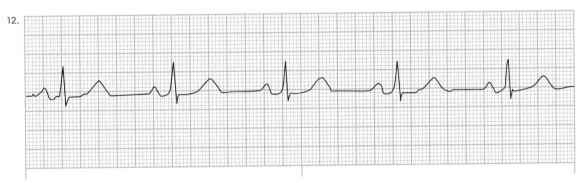

ECG Findings:

12.

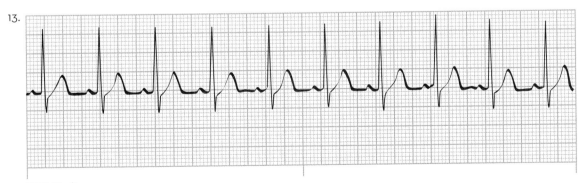

ECG Findings:

13.

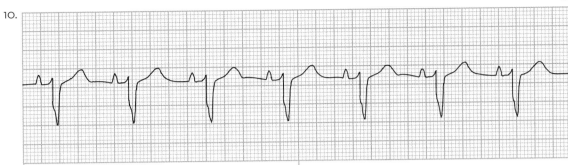

ECG Findings:

14.

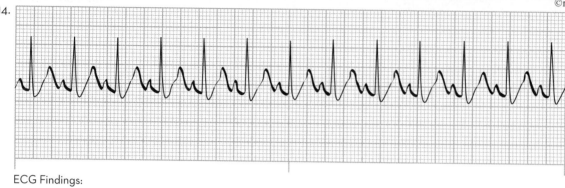

ECG Findings:

15.

ECG Findings:

16.

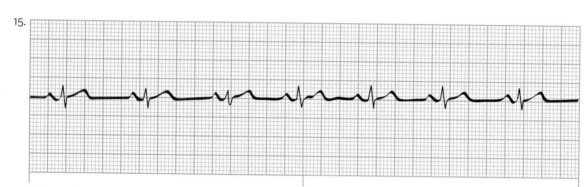

ECG Findings:

17.

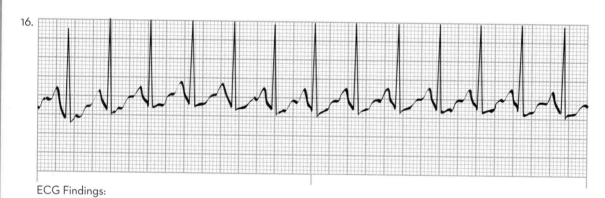

ECG Findings:

18.

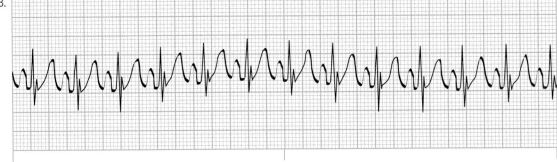

ECG Findings:

19.

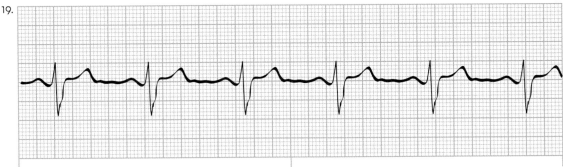

ECG Findings:

20.

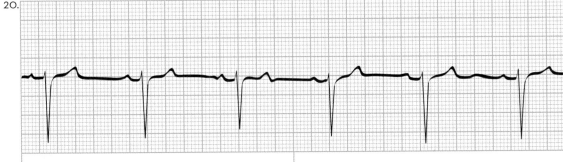

ECG Findings:

21.

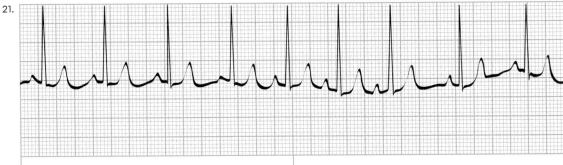

ECG Findings:

22.

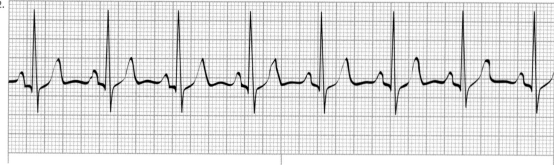

ECG Findings:

23.

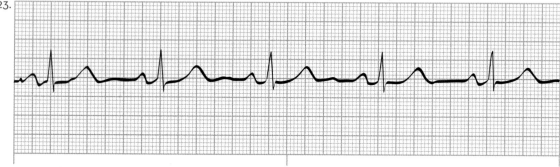

ECG Findings:

24.

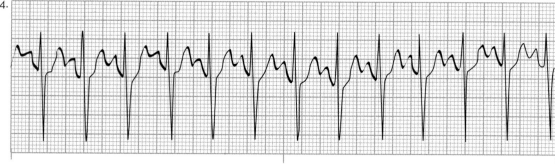

ECG Findings:

25.

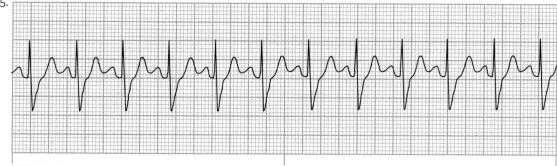

ECG Findings:

26.

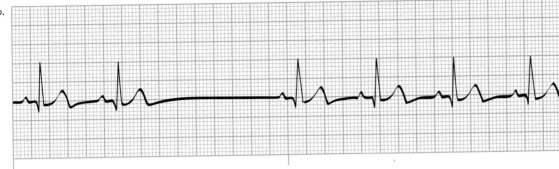

ECG Findings:

27.

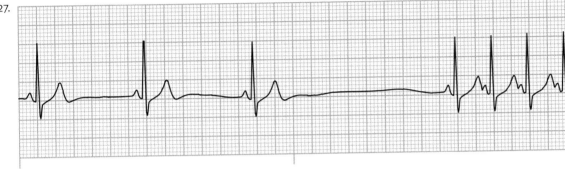

ECG Findings:

28.

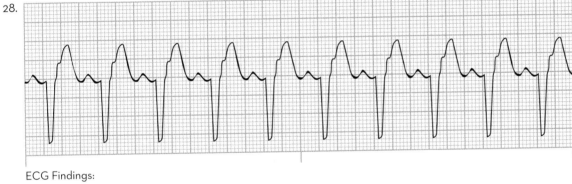

ECG Findings:

29.

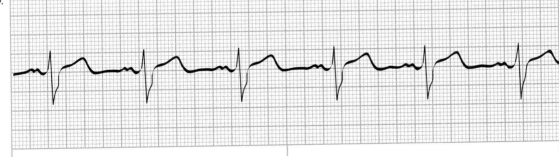

ECG Findings:

30.

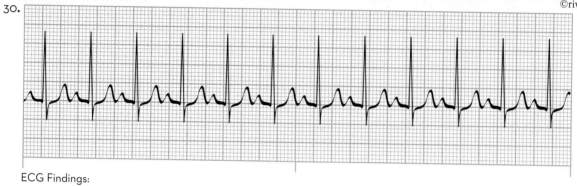

ECG Findings:

31.

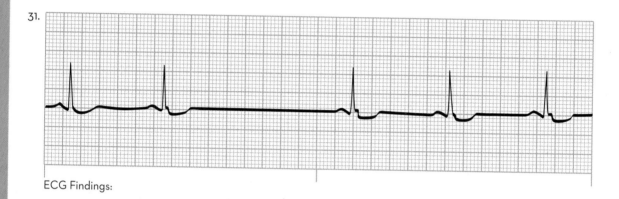

ECG Findings:

32.

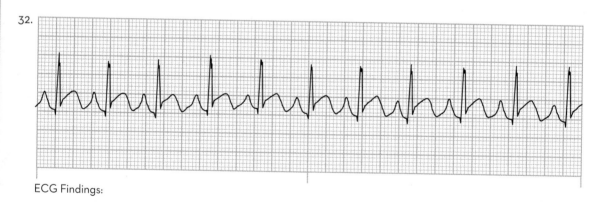

ECG Findings:

33.

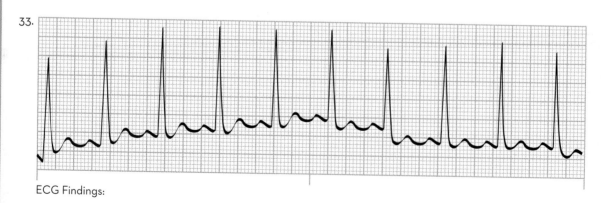

ECG Findings:

34.

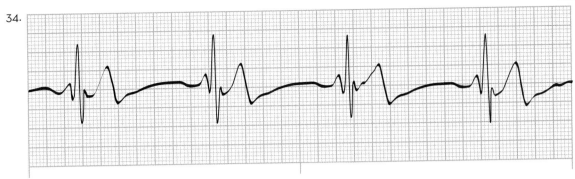

ECG Findings:

35.

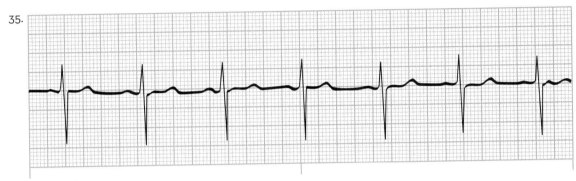

ECG Findings:

36.

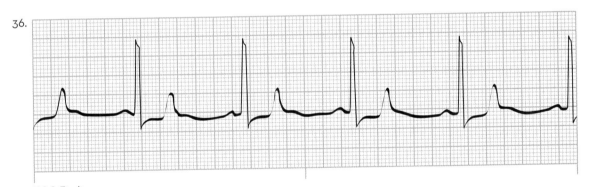

ECG Findings:

37.

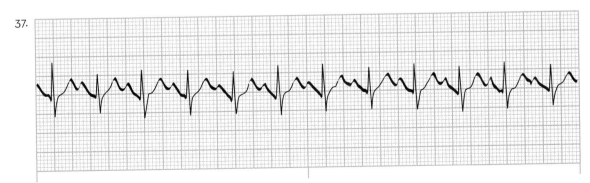

ECG Findings:

38.

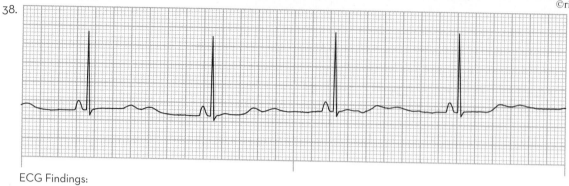

ECG Findings:

39.

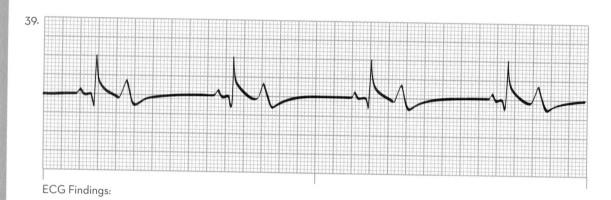

ECG Findings:

40.

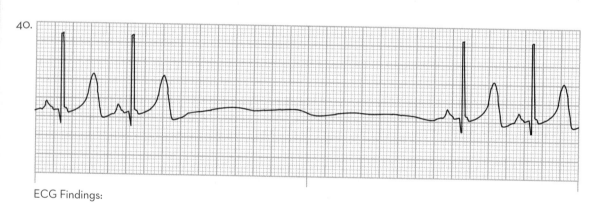

ECG Findings:

41.

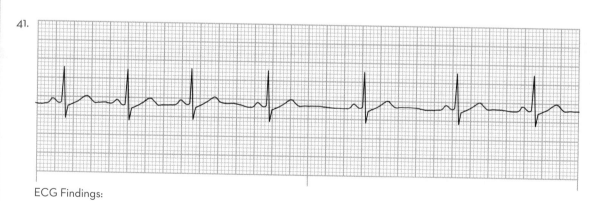

ECG Findings:

42.

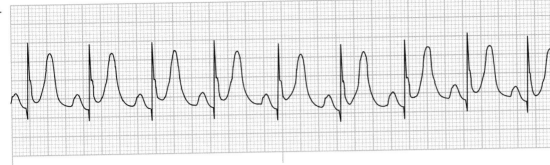

ECG Findings:

43.

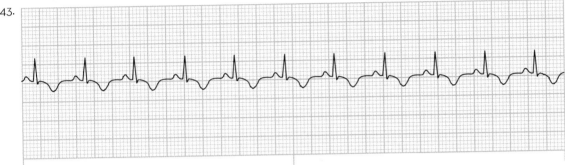

ECG Findings:

44.

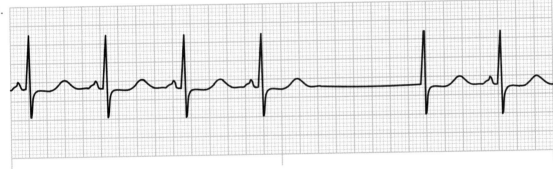

ECG Findings:

45.

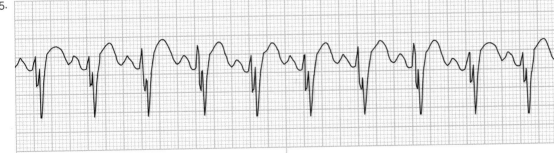

ECG Findings:

46.

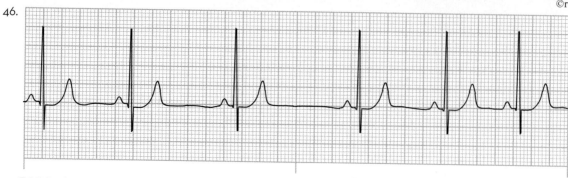

ECG Findings:

47.

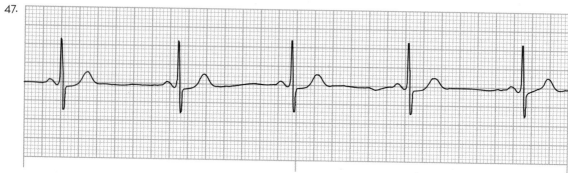

ECG Findings:

48.

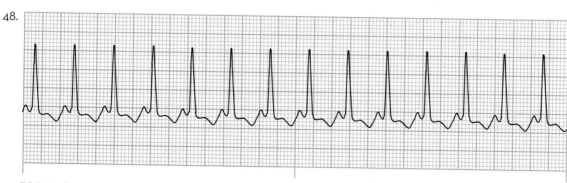

ECG Findings:

49.

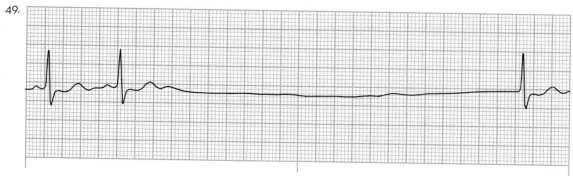

ECG Findings:

50.

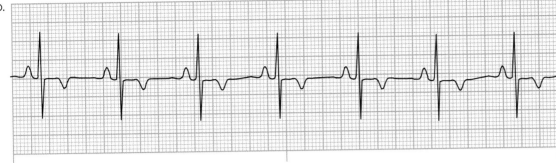

ECG Findings:

51.

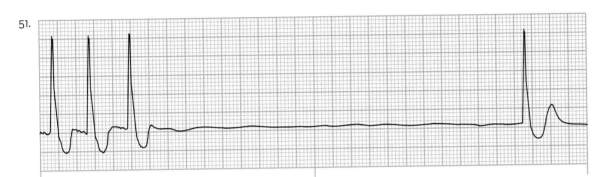

ECG Findings:

52.

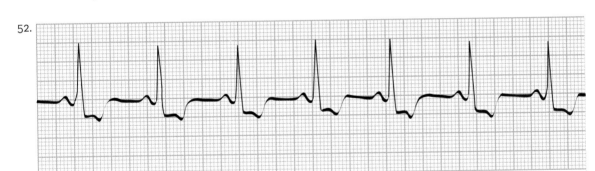

ECG Findings:

53.

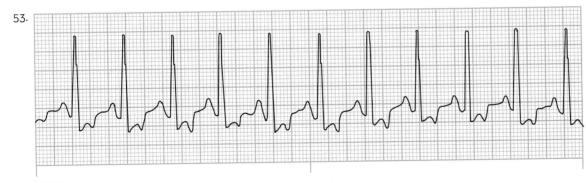

ECG Findings:

54.

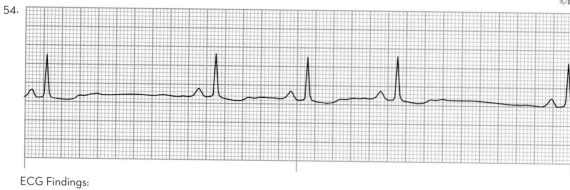

ECG Findings:

55.

ECG Findings:

56.

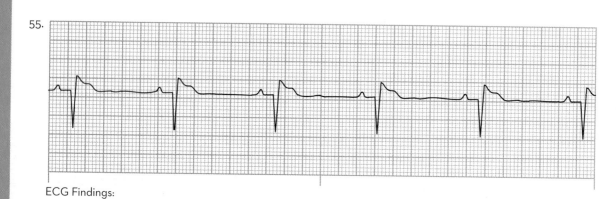

ECG Findings:

57.

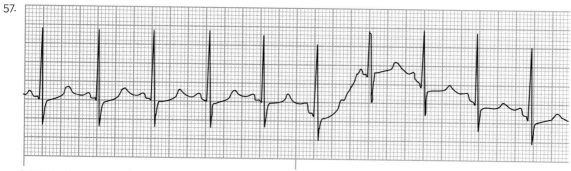

ECG Findings:

Key Points

LO 12.1	• One of the things we look for during examination of the ECG is the presence of dysrhythmias. • Rhythms originating from the SA node are called sinus rhythms. • Sinus rhythms include: normal sinus rhythm, sinus bradycardia, sinus tachycardia, sinus dysrhythmia, sinus arresrt, sinoatrial exit block.
LO 12.2	• Key characteristics for sinus rhythms include a normal P wave preceding each normal QRS complex and PR intervals within normal duration of 0.12 to 0.20 seconds and constant. • During normal heart activity, the SA (sinoatrial) node acts as the primary pacemaker.
LO 12.3	• Normal sinus rhythm has a heart rate of 60 to 100 beats per minute. Normal sinus rhythm is the rhythm against which we measure dysrhythmias.
LO 12.4	• Sinus bradycardia has all the characteristics of normal sinus rhythm but the heart rate is less than 60 beats per minute (BPM). It often occurs naturally as a means of conserving energy during times of rest or sleep.
LO 12.5	• Sinus tachycardia has the same characteristics as normal sinus rhythm but has a rate of greater than 100 beats per minute. It often occurs naturally as a means of increasing delivery of oxygen and nutrients and removing waste products during times of exertion, exercise, or stress.
LO 12.6	• Sinus dysrhythmia is the same as sinus rhythm except that there is the presence of a patterned irregularity. It can be described as "slowing, then speeding up, then slowing again." This cycle continually repeats.
LO 12.7	• With sinus arrest, the ECG rhythm looks like normal sinus rhythm except that there is a pause in the rhythm with an absence of the P, QRS, and T waveforms until a pacemaker site reinitiates the rhythm.
LO 12.8	• With sinoatrial exit block, the ECG rhythm looks like normal sinus rhythm except that there is a pause in the rhythm with an absence of the P, QRS, and T waveforms. Then the P wave (and associated QRS complex) reoccurs at the next expected interval.
LO 12.9	• Sick sinus syndrome, also referred to as *sinus node dysfunction*, is a group of abnormal rhythms that occur with malfunction of the sinus node.
LO 12.10	• Sinus rhythms may be what are referred to as *an underlying rhythm*. What that means is that the sinus rhythm is seen but then another dysrhythmia or cardiac condition is also seen as well.

Assess Your Understanding

The following questions give you a chance to assess your understanding of the material discussed in this chapter. The answers can be found in Appendix A.

1. A dysrhythmia that arises from the SA node is   (LO 12.1)
   a. Wandering atrial pacemaker.
   b. Sinus arrest.
   c. Junctional escape.
   d. Ventricular tachycardia.

2. Key characteristics of sinus rhythms are   (LO 12.2)
   a. upright and wide QRS complexes.
   b. normal P waves.
   c. PR intervals within normal duration of 0.12 to 0.20 seconds and varying with each beat.
   d. Both b and c.

3. The primary pacemaker during normal heart activity is   (LO 12.2)
   a. bundle of His.
   b. SA node.
   c. AV node.
   d. AV junction.

4. Characteristics of normal sinus rhythm include   (LO 12.3)
   a. an irregular rhythm.
   b. a rate of between 60 and 100 beats per minute.
   c. an inverted P wave that follows each QRS complex.
   d. progressively longer PR intervals.

5. The heart rate characteristic of sinus bradycardia is _____ than _____ beats per minute.   (LO 12.4)
   a. less, 80
   b. greater, 100
   c. less, 60
   d. less, 40

6. Sinus bradycardia   (LO 12.4)
   a. often occurs naturally.
   b. is considered abnormal in athletes.
   c. is often seen in Wolff-Parkinson-White syndrome.
   d. is caused by increased sympathetic stimulation.

7. Characteristics of sinus bradycardia include   (LO 12.4)
   a. PR intervals that change.
   b. inverted P waves in lead II.

c. a regular rhythm.

d. wide, bizarre QRS complexes.

8. Sinus tachycardia (LO 12.5)

   a. has mostly the same characteristics as normal sinus rhythm except that it has a rate of greater than 120 beats per minute.

   b. is produced by stimulation of the parasympathetic branch of the autonomic nervous system.

   c. may be caused by ingestion of caffeine or alcohol, smoking, or fever.

   d. continues even after the stimulus causing it is removed.

9. In sinus tachycardia, the P waves, PR intervals, and QRS complexes are

(LO 12.5)

   a. widened.

   b. normal.

   c. shortened.

   d. delayed.

10. The type of irregularity seen with sinus dysrhythmia is called _____ irregularity. (LO 12.6)

   a. patterned

   b. slight

   c. total

   d. frequent

11. Sinus dysrhythmia (LO 12.6)

   a. is common in infants.

   b. occurs only in patients who have significant heart disease.

   c. usually produces symptoms.

   d. has round and upright P waves in lead II.

12. Which of the following is true of sinus arrest? (LO 12.7)

   a. It occurs when there is complete blockage of the SA node.

   b. The rhythm is regular.

   c. It has QRS complexes greater than 0.12 seconds.

   d. It usually results in a brief pause in all electrical activity.

13. With sinus arrest, the escape mechanism that follows the pause in electrical activity may arise from the (LO 12.7)

   a. SA node.

   b. AV junction.

   c. ventricles.

   d. All of the above.

14. Match the following sinus dysrhythmias with the correct description.

(LO 12.4, 12.5, 12.6, 12.7)

	Types		Characteristics
a.	Sinus bradycardia	_____	Has a rate of greater than 100 beats per minute
b.	Sinus dysrhythmia	_____	Occurs when the SA node transiently stops firing
c.	Sinus arrest	_____	Is a dysrhythmia with patterned irregularity.
d.	Sinus tachycardia	_____	Has a rate of less than 60 beats per minute

15. Sinoatrial exit block occurs when the   (LO 12.8)

    a. impulse originating in the SA node is blocked from reaching the atria.

    b. heart rate slows to less than 60 beats per minute.

    c. heart rate speeds up and slows down in a cyclical manner.

    d. heartbeat ceases completely.

16. Treatment for symptomatic bradycardia includes   (LO 12.4)

    a. vagal maneuvers.

    b. atropine with a fluid bolus of 250 cc.

    c. temporary or permanent pacing.

    d. defibrillation.

Referring to the scenario at the beginning of this chapter, answer the following questions.

17. What rhythm is the patient in?   (LO 12.4)

    a. Sinus bradycardia

    b. Junctional

    c. 2nd-degree AV heart block

    d. 3rd-degree AV heart block

18. The stimulation of what nerve causes this rhythm?   (LO 12.4)

    a. Phrenic

    b. Vagus

    c. Carotid

    d. Aortic

19. Which physiologic sensor noted the patient's lowered blood pressure?

(LO 1.8)

    a. Baroreceptor

    b. Cerebral cortex

    c. Vagus nerve

    d. Chemoreceptor

13 Atrial Dysrhythmias

©rivetti/Getty Images

Courtesy Physio-Control

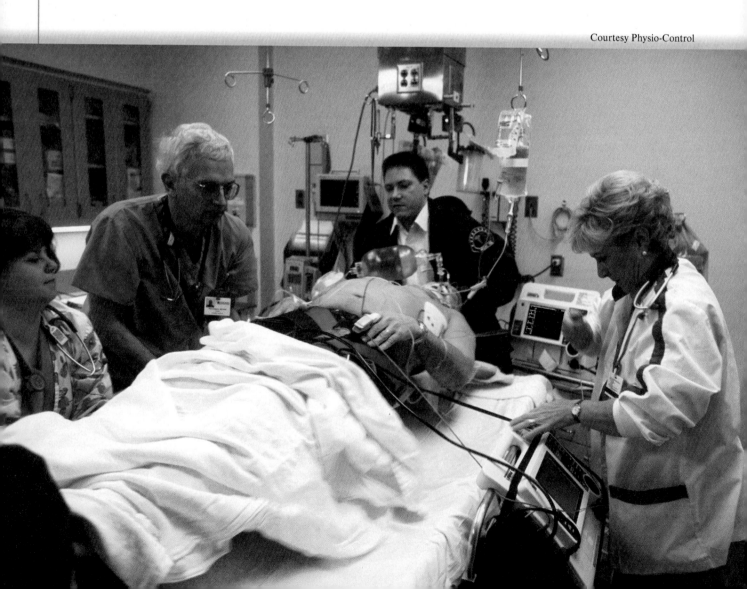

Learning Outcomes

LO 13.1 Describe the origin of atrial dysrhythmias.

LO 13.2 Identify the key features of dysrhythmias originating from the atria.

LO 13.3 Identify the causes, effects, appearance, and treatment of premature atrial complexes.

LO 13.4 Identify the causes, effects, appearance, and treatment of wandering atrial pacemaker.

LO 13.5 Identify the causes, effects, appearance, and treatment of atrial tachycardia.

LO 13.6 Identify the causes, effects, appearance, and treatment of multifocal atrial tachycardia.

LO 13.7 Define the term *supraventricular tachycardia*.

LO 13.8 Identify the causes, effects, appearance, and treatment of atrial flutter.

LO 13.9 Identify the causes, effects, appearance, and treatment of atrial fibrillation.

Case History

EMS is called to evaluate an elderly man who passed out in church. The man has a history of chronic obstructive pulmonary disease (COPD) and hypertension. Upon their arrival, the EMS team is met by a physician attending services with the patient. He tells them that the gentleman was sitting in the pew when he arose for prayer and then slumped back down and fell over on his side. The doctor then rushed to the man's side and checked for a pulse, which he says was fast and regular but barely palpable. He says that the patient never stopped breathing or became cyanotic and that once he lay flat he returned to his normal self.

The paramedics introduce themselves and begin their assessment. Oxygen is applied and vital signs are taken, showing a blood pressure of 100/90, pulse 150, respirations of 16 per minute, and oxygen saturation of 92% on room air. They apply the cardiac monitor to the patient and see a fast, regular heart rate with narrow QRS complexes. One of the paramedics comments that the rhythm looks like sinus tachycardia. But the physician says, "No, look closely. See the sawtooth pattern?"

The paramedics nod in agreement after confirming that, indeed, there are no normal P waves and that the QRS complexes are preceded by F waves.

13.1 Rhythms Originating in the Atria

Atrial dysrhythmias originate outside the SA node in the atrial tissue or in the internodal pathways. They are among the most common types of dysrhythmias, particularly in persons older than 60 years of age. Most atrial dysrhythmias are benign, but tachycardias that originate from the atria can be dangerous. The three mechanisms thought to cause atrial dysrhythmias are enhanced automaticity, circus reentry, and triggered beats (afterdepolarization).

Atrial dysrhythmias (Figure 13-1) include the following:

- Premature atrial complex (PAC)
- Wandering atrial pacemaker
- Atrial tachycardia
- Multifocal atrial tachycardia (MAT)
- Supraventricular tachycardia (SVT)
- Atrial flutter
- Atrial fibrillation

13.2 Key Features of Atrial Dysrhythmias

P Wave Appearance

The key characteristics of atrial dysrhythmias include P′ waves (if present) that differ in appearance from normal sinus P waves and normal, abnormal, shortened, or prolonged P′R intervals.

The appearance of the P′ waves will depend on the site where the atrial impulses originate. Typically, the closer the site of origin is to the SA node, the more it looks like a normal P wave. If the ectopic pacemaker arises from the upper- or middle-right atrium, depolarization occurs in a normal direction—from right to left and then downward. If it is initiated from the upper-right atrium, the P′ wave should be upright in lead II, resembling a normal sinus P wave. If it is initiated from the middle of the right atrium, the P′ wave is less positive than one that originates from the upper-right atrium.

If the impulse arises from the lower-right atrium near the AV node or in the left atrium, depolarization occurs in a *retrograde* direction—from left to right and then upward, resulting in the P′ wave being inverted in lead II.

QRS Complex Appearance

As with all dysrhythmias that arise from above the ventricles (sinus, atrial, and junctional dysrhythmias), the QRS complexes should appear narrow (0.06 to 0.10 seconds in duration) and normal (unless there is an intraventricular conduction disturbance, aberrancy, or ventricular preexcitation).

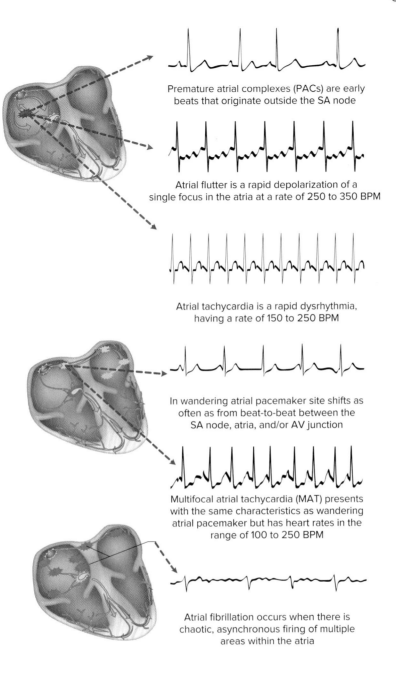

Figure 13-1
Dysrhythmias originating from
the atria.

Premature atrial complexes (PACs) are early
beats that originate outside the SA node

Atrial flutter is a rapid depolarization of a
single focus in the atria at a rate of 250 to 350 BPM

Atrial tachycardia is a rapid dysrhythmia,
having a rate of 150 to 250 BPM

In wandering atrial pacemaker site shifts as
often as from beat-to-beat between the
SA node, atria, and/or AV junction

Multifocal atrial tachycardia (MAT) presents
with the same characteristics as wandering
atrial pacemaker but has heart rates in the
range of 100 to 250 BPM

Atrial fibrillation occurs when there is
chaotic, asynchronous firing of multiple
areas within the atria

Effects of Atrial Dysrhythmias

While many atrial dysrhythmias have little effect on cardiac output, others can reduce ventricular filling time and diminish the strength of the atrial "kick," the atrial contraction that normally supplies the ventricles with about 30% of their blood. These effects can lead to decreased cardiac output, decreased blood pressure, and, ultimately, decreased tissue perfusion.

Decreased tissue perfusion can have dire consequences. First, there is not enough oxygen and nutrients for the body to satisfy its energy needs. Second, if the waste products of metabolism are not removed, they accumulate in the tissues and create a toxic environment. If not corrected, both factors can lead to shock and, ultimately, death of the patient.

Premature atrial complexes arise from somewhere in the atrium.

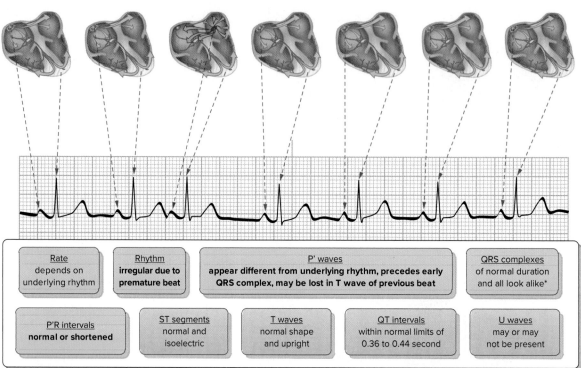

Figure 13-2
Premature atrial complex.

13.3 Premature Atrial Complexes

Description

PACs are early beats that originate outside the SA node (Figure 13-2) somewhere in the atria. The SA node fires an impulse, but then an irritable focus jumps in, firing its own impulse before the SA node can fire again. PACs can originate from a single ectopic site or from multiple sites in the atria. Early beats like PACs aren't really a dysrhythmia. Rather, they are an abnormality that occurs within an underlying rhythm. When referring to early beats, we say, as an example, "we have sinus rhythm with occasional PACs." With premature beats (atrial, junctional, or ventricular), the term complex or contraction may be used. For example, we can call an early ectopic atrial beat a premature atrial complex or a premature atrial contraction. The term contraction is the appropriate term if the premature beat produces a pulse. The term complex may be more fitting because not all premature beats produce a pulse.

Causes

The most common cause of PACs is enhanced automaticity. Patients with healthy hearts can experience PACs due to excessive use of caffeine, nicotine, or alcohol, or in the presence of anxiety, fatigue, or fever. Other common causes of PACs are listed in the "Causes of Premature Atrial Complexes" box. In some cases there is no apparent cause.

Causes of Premature Atrial Complexes	
Cause	**Examples**
Cardiac diseases	Coronary heart disease, heart failure, valvular heart disease
Use of certain drugs	Digitalis toxicity, drugs that prolong the absolute refractory period of the SA node (procainamide, quinidine)
Increased sympathetic tone	Increased catecholamines
Noncardiac disorders	Acute respiratory failure, alcohol, anxiety, caffeine, certain electrolyte imbalances, nicotine, fatigue, fever, hyperthyroidism, hypoxia, infectious disease, COPD

Effects

Isolated PACs seen in patients with healthy hearts are considered insignificant. Commonly they cause no symptoms and can go unrecognized for years. Asymptomatic patients usually require only observation. The patient may perceive PACs as "palpitations" or skipped beats.

However, in patients with heart disease, frequent PACs may predispose the patient to serious atrial dysrhythmias such as atrial tachycardia, atrial flutter, or atrial fibrillation. In a patient experiencing an acute MI, PACs can serve as an early indicator of an electrolyte imbalance or congestive heart failure. PACs with aberrant ventricular conduction may cause wide QRS complexes and be confused with premature ventricular complexes.

ECG Appearance

When PACs are present, the first thing you see as you look across the ECG tracing is an irregular rhythm. As you review it more closely, you can see that, wherever there is irregularity, the P′-P and R′-R intervals are shorter than the P-P and R-R intervals of the underlying rhythm. For example, if you look across the ECG tracing at P-P intervals and R-R intervals of three consecutive normal beats, the distance should be relatively the same. Where there is an ectopic beat, the P′-P and R′-R interval is shorter. This tells you that the beat(s) producing the irregularity occurred early.

Now that we know there are early beats, it is a matter of identifying from where they are arising. This can be done by looking at the P waves and QRS complexes associated with the early beats. There should be a P′ wave (in most leads) preceding the QRS complex, but it will have a different morphology (appearance) than the P waves in the underlying rhythm. Although an upright P′ wave is common in PACs, it is dependent upon where in the atria the impulse originates. For example, if the impulse forms near the SA node, the PAC will have a near normal shape and an upright P wave. But if the impulse originates in the atria or near the AV node, the P wave morphology will be very different from the normal P waves and likely be either biphasic or negative in direction. PACs that originate in the atria near the AV node may also have a shortened PR interval in addition to their altered P wave morphology.

Sometimes the P wave is buried in the T wave of the preceding beat, thus changing its appearance somewhat. The QRS complexes are typically normal (although with aberrant conduction they will look different). Another characteristic that allows us to conclude that early beats are PACs is that they are followed by a noncompensatory pause.

The pause that follows a premature beat is called a **noncompensatory pause** if the beat following the premature complex is normal and occurs before it was expected. In other words, the space between the complex before and after the premature beat is "less than" the sum of two R-R intervals. To determine the presence of a noncompensatory pause, we measure the R-R interval preceding the early beat. We then measure to the R wave of the first normal beat that follows it. If the combined distance is less than the sum of two R-R intervals, the pause is considered a noncompensatory pause (Figure 13-3). The reason a noncompensatory pause occurs is that the electrical activity of the premature atrial beat can enter the sinus node and reset its timing, thus allowing it to fire before its next scheduled beat. This creates an irregularity in the R-R pattern.

In contrast, compensatory pause occurs when there are two full R-R intervals between the R wave of the normal beat that precedes the early beat and the R wave of the first normal beat that follows it. A compensatory pause is characteristically seen with premature beats that arise from the ventricles. These are referred to as *premature ventricular complexes*. The reason a compensatory pause occurs is that the electrical activity of a premature ventricular complex is usually not conducted through the AV node towards the atria; thus the sinus node will not be reset. So the SA node fires as planned and on schedule.

Another way to describe PACs is how they are intermingled among the normal beats. When every other beat is a PAC, it is called *bigeminal PACs* or *atrial bigeminy*. If every third beat is a PAC, it is called *trigeminal PACs* or *atrial trigeminy*. Likewise, if PACs occur every fourth beat, they are called *quadrigeminal PACs* or

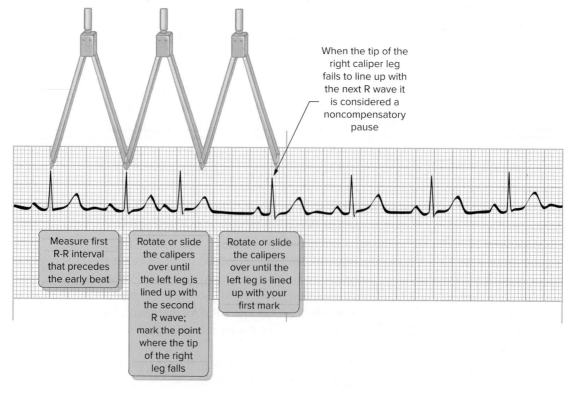

When the tip of the right caliper leg fails to line up with the next R wave it is considered a noncompensatory pause

Measure first R-R interval that precedes the early beat

Rotate or slide the calipers over until the left leg is lined up with the second R wave; mark the point where the tip of the right leg falls

Rotate or slide the calipers over until the left leg is lined up with your first mark

(Continued)

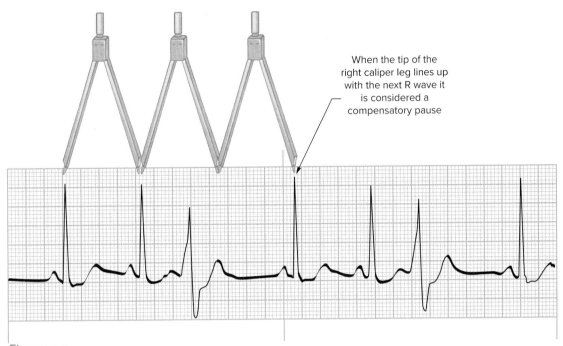

When the tip of the right caliper leg lines up with the next R wave it is considered a compensatory pause

Figure 13-3
Noncompensatory pause and compensatory pause.

atrial quadrigeminy. Regular PACs at greater intervals than every fourth beat have no special name (Figure 13-4). Remember, bi = 2, tri = 3, and quad = 4.

Remember, in PACs with aberrant ventricular conduction, the QRS complexes are preceded by a P′ wave and a constant P′R interval because the pacemaker site is above the ventricles. This is one of the characteristics we can use to differentiate between PACs with aberrancy and premature ventricular complexes.

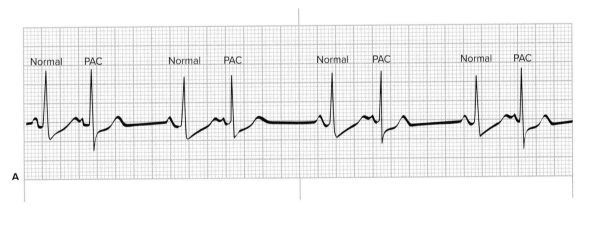

A

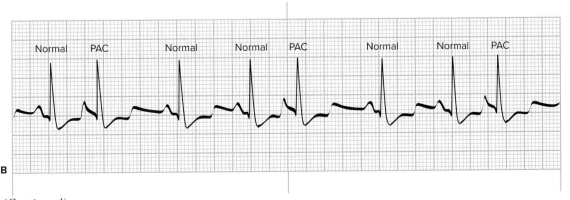

B

(Continued)

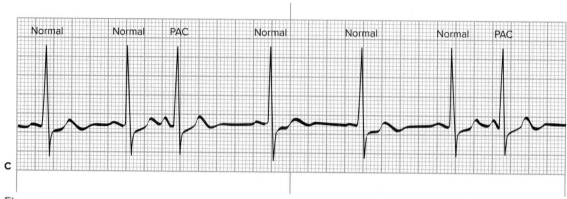

Figure 13-4
Premature atrial complexes occurring in a (A) bigeminal, (B) trigeminal, and (C) quadrigeminal pattern.

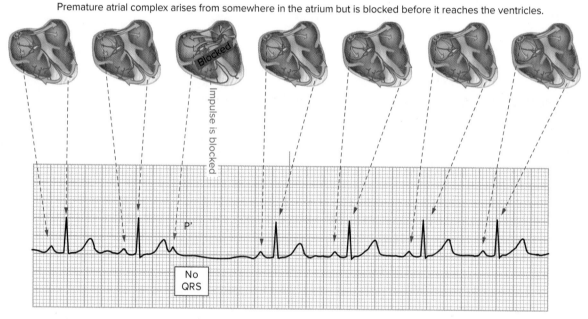

Figure 13-5
Blocked PACs are seen as P′ waves not followed by a QRS complex.

Nonconducted PACs or **blocked PACs** occur when an atrial impulse arrives too early, before the AV node has a chance to repolarize. As a result, the P′ wave fails to conduct to the ventricles. You will know this has occurred when you see a premature P′ wave not followed by a QRS complex (Figure 13-5). Sometimes it is necessary to look closely because they can be buried within the T wave of the preceding beat.

Differentiating Blocked PACs from Sinus Arrest

Nonconducted or blocked PACs can sometimes be misinterpreted as sinus arrest. With sinus arrest there should be an absence of P waves following the T wave that precedes the pause. In contrast, with a blocked PAC there should be a nonconducted P wave, which occurs before, during, or just after the T wave preceding the pause. If the P wave is not obviously apparent, compare the T waves that precede the pause with the other T waves in the tracing. A P wave that is buried in the T wave should distort its height or shape. Again, with sinus arrest you shouldn't see the P wave around or buried in the T wave preceding the pause whereas with a blocked PAC you should see a P wave before, during, or just after the T wave that precedes the pause but is not followed by a QRS complex.

Treatment

PACs generally do not require treatment. PACs caused by the use of caffeine, tobacco, or alcohol or by anxiety, fatigue, or fever can be controlled by eliminating the underlying cause. Frequent PACs may be treated with drugs that increase the atrial refractory time. This includes beta-adrenergic blockers and calcium channel blockers.

13.4 Wandering Atrial Pacemaker

Description

Wandering atrial pacemaker is a rhythm in which the pacemaker site shifts location as often as from beat to beat between the SA node, atria, and/or AV junction (Figure 13-6). It is also called *multifocal atrial rhythm.*

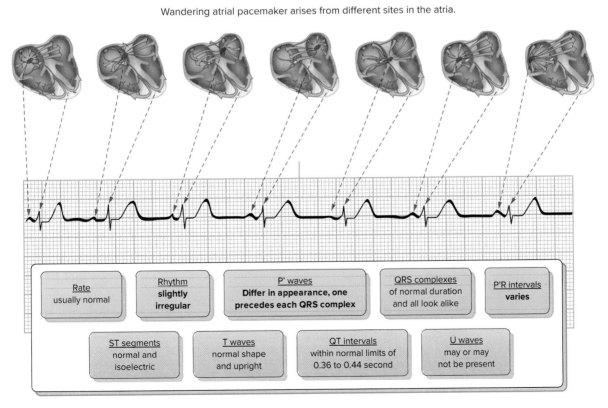

Wandering atrial pacemaker arises from different sites in the atria.

Rate	Rhythm	P' waves	QRS complexes	P'R intervals
usually normal	**slightly irregular**	**Differ in appearance, one precedes each QRS complex**	of normal duration and all look alike	**varies**

ST segments	T waves	QT intervals	U waves
normal and isoelectric	normal shape and upright	within normal limits of 0.36 to 0.44 second	may or may not be present

Figure 13-6
Wandering atrial pacemaker.

Causes

Wandering atrial pacemaker is generally caused by the inhibitory vagal effect of respiration on the SA node and AV junction. With increased vagal stimulation, the SA node slows, allowing faster (briefly) pacemakers in the atria or AV junction to initiate the heartbeat. After the vagal tone decreases, the SA node becomes the dominant pacemaker again. Other common causes are listed in the "Causes of Wandering Atrial Pacemaker" box.

Causes of Wandering Atrial Pacemaker	
Cause	**Examples**
Cardiac diseases	Organic heart disease (e.g., rheumatic carditis), valvular heart disease
Use of certain drugs	Digoxin toxicity
Vagal stimulation	Inhibitory effect of vagal stimulation of respiration on the SA and AV nodes
Noncardiac disorders	Pulmonary disease with hypoxemia

Effects

Wandering atrial pacemaker is rarely serious, having no effect on cardiac output. It is a normal finding in children, older adults, and well-conditioned athletes and is not usually of any clinical significance.

ECG Appearance

The most characteristic feature of wandering atrial pacemaker is P′ waves that change in appearance, sometimes as often as from beat to beat. Changes in appearance often involve changes in shape (morphology), which may include either positive elements or negative elements of the waveform or both (biphasic). At least three different P wave configurations (seen in the same lead) are needed to diagnose a dysrhythmia as wandering atrial pacemaker. Also, due to the changing pacemaker site, the P′R interval and the regularity of the rhythm often varies. Because the impulse arises above the ventricles, the QRS complexes are normal. The heart rate with this dysrhythmia is within normal rates—60 to 100 beats per minute, but it may also be slow. Wandering atrial pacemaker may be difficult to identify because the dysrhythmia is commonly transient.

Treatment

No treatment is necessary for patients experiencing wandering atrial pacemaker. However, chronic dysrhythmias are a sign of heart disease and should be monitored.

13.5 Atrial Tachycardia

Description

Atrial tachycardia is a rapid dysrhythmia, having a rate of 150 to 250 beats per minute (rarely exceeding 250 beats per minute) that arises from the atria (Figure 13-7). It is so fast that it overrides the SA node. It looks like an extremely fast rate with narrow (unless seen in the presence of preexcitation, ventricular conduction disturbance or aberrant conduction causing the QRS complexes to be wide) QRS complexes. Atrial tachycardia can occur at any age.

Atrial tachycardia may occur in short bursts or it may be sustained. Short bursts of the tachycardia are well tolerated in otherwise normally healthy people. With sustained rapid ventricular rates, ventricular filling may be decreased during diastole.

When the onset of the tachycardia is sudden (and typically witnessed) with a finite duration and abrupt termination, it is called **paroxysmal**. Paroxysmal tachycardia may originate in the atria (paroxysmal atrial tachycardia) or AV junction (paroxysmal junctional tachycardia).

Causes

Digitalis toxicity is the most common cause of atrial tachycardia. Also, sudden onset atrial tachycardia is common in patients who have Wolff–Parkinson–White syndrome.

Atrial tachycardia arises from a single focus in the atria.

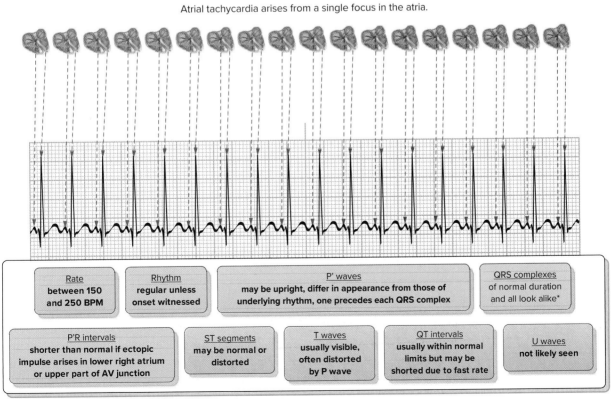

Rate between 150 and 250 BPM	Rhythm regular unless onset witnessed	P' waves may be upright, differ in appearance from those of underlying rhythm, one precedes each QRS complex	QRS complexes of normal duration and all look alike*

P'R intervals shorter than normal if ectopic impulse arises in lower right atrium or upper part of AV junction	ST segments may be normal or distorted	T waves usually visible, often distorted by P wave	QT intervals usually within normal limits but may be shorted due to fast rate	U waves not likely seen

*unless aberrantly conducted in which case they may be wide and bizarre-looking

Figure 13-7
Atrial tachycardia.

In patients with healthy hearts, excessive use of caffeine or other stimulants, marijuana use, electrolyte imbalances, hypoxia, and physical or psychological stress can cause PACs, which can precipitate atrial tachycardia. Common causes of atrial tachycardia are listed in the "Causes of Atrial Tachycardia" box.

Effects

Symptoms can develop abruptly and may go away without treatment. They may last a few minutes or as long as one to two days, sometimes continuing until treatment is delivered.

With the rapid heartbeat seen with atrial tachycardia, there is less time for the ventricles to fill. This can reduce stroke volume and lead to decreased cardiac output and a drop in blood pressure. Although atrial tachycardia can often be tolerated in healthy persons, it can significantly compromise cardiac output in patients with underlying heart disease. Decreased cardiac output can lead to vertigo (dizziness), lightheadedness, syncope, hypotension, congestive heart failure, and loss of consciousness (in serious cases).

Causes of Atrial Tachycardia	
Cause	**Examples**
Cardiac diseases	MI, cardiomyopathy, congenital anomalies, systemic hypertensive, valvular heart disease, Wolff–Parkinson–White syndrome
Use of certain drugs	Albuterol, cocaine, digitalis toxicity (the most common cause), theophylline
Excessive sympathetic stimulation	Physical or psychological stress
Noncardiac disorders	Alcohol, caffeine, cor pulmonale, hyperthyroidism, electrolyte imbalance, nicotine

Further, fast heart rates increase oxygen requirements. This can increase myocardial ischemia and may lead to MI.

Other symptoms seen with atrial tachycardia include anxiety, nervousness, palpitations, pounding chest, shortness of breath, rapid breathing, and numbness of various body parts.

ECG Appearance

The most striking feature of atrial tachycardia is its rapid rate of 150 to 250 beats per minute. For this reason you should be able to see it almost immediately. It arises from one site, so the rhythm should be regular unless you see the onset or termination, as with paroxysmal atrial tachycardia. Although there is one P′ wave (unless there is a block) preceding each QRS complex, it deviates in appearance from the normal P wave and is typically buried in the T wave of the preceding beat. If present, the P′ waves may be flattened or notched. The P′R intervals are typically indeterminable because the P′ waves tend to be buried. If visible, the P′R interval is often

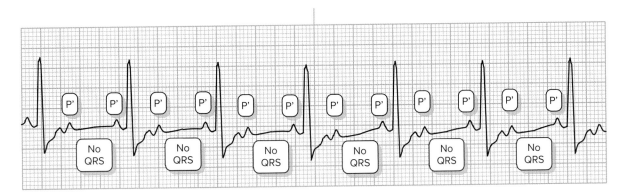

Figure 13-8
Atrial tachycardia with block.

shortened, but it may also be normal. The dysrhythmia arises above the ventricles, so the QRS complexes are normal unless there is preexcitation, ventricular conduction disturbance or aberrant conduction.

With the rapid atrial rates seen with atrial tachycardia, the AV junction is sometimes unable to carry all the impulses, and an atrial tachycardia with block occurs. Most commonly, only one of every two beats (a 2 to 1 block) is conducted to the ventricles. This then results in more than one P′ wave preceding each QRS complex (Figure 13-8). If the block is constant, it will have the same number of P′ waves preceding each QRS complex, but, if the block is variable, the number of P′ waves will change.

Treatment

Treatment is dependent on the type of tachycardia and symptom severity (Figure 13-9). It is directed at eliminating the cause and decreasing ventricular rate. Patients who are symptomatic (e.g., chest pain, hypotension) should receive oxygen, an IV infusion of normal saline administered at a keep-open rate, and prompt delivery of synchronized cardioversion, use of vagal maneuvers or medication administration.

Synchronized cardioversion starting at 50–100 joules is indicated if the patient is symptomatic. In the conscious patient, consider sedation before cardioversion. However, do not delay cardioversion. If this fails to convert the rhythm, the energy level may be increased.

If the patient is stable, vagal maneuvers and adenosine at 6 mg rapid (over 1 to 3 seconds) IV push may be used. A second bolus of adenosine, at 12 mg rapid IV push, can be administered within 1 to 2 minutes if the first dose of adenosine fails to convert the tachydysrhythmia. Each bolus of adenosine is immediately followed by a 20 mL flush. If these treatments fail to resolve the tachycardia, calcium channel blockers (verapamil, diltiazem) and beta-adrenergic blockers (metroplol, propanolol) (if no contraindications exist) may be considered.

When atrial tachycardia is due to some underlying abnormality (e.g., electrolyte imbalance, drug toxicity), the conversion to a normal rhythm with cardioversion may be temporary because the causative factor has not been eliminated. In such cases, a recurrence of atrial tachycardia soon after successful electrical cardioversion

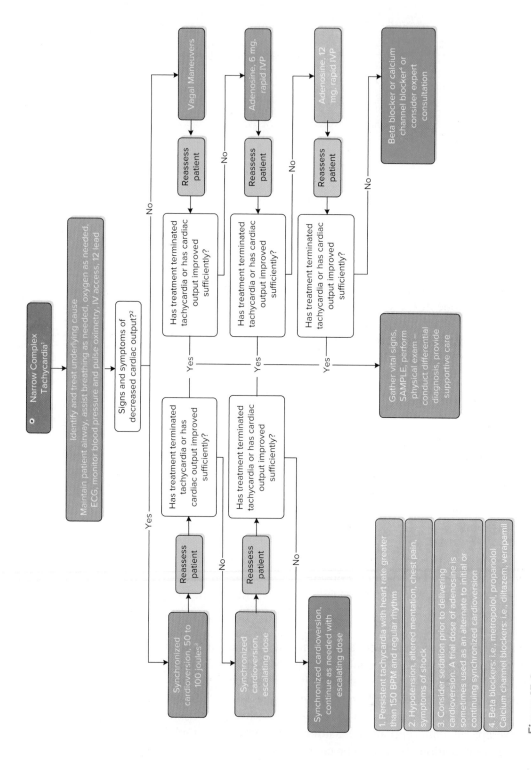

Figure 13-9
Treatment algorithm for tachycardia.

does not warrant additional electrical cardioversion attempts until the underlying problem has been resolved. In the patient with COPD, correction of hypoxia and electrolyte imbalances is indicated.

Atrial overdrive pacing may also be employed to stop this dysrhythmia. If the dysrhythmia is related to WPW syndrome, catheter ablation may be indicated. Procainamide, amiodarone, or sotalol may be considered in wide complex tachycardias.

13.6 Multifocal Atrial Tachycardia

Description

MAT is a pathological condition that presents with the same characteristics as wandering atrial pacemaker (Figure 13-10) but with heart rates in the range of 100 to 250 beats per minute (usually less than 160 beats per minute).

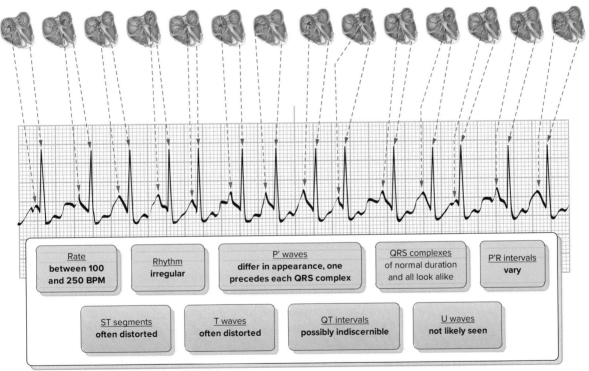

In multifocal atrial tachycardia the pacemaker site shifts between the SA node, atria, and/or the AV junction.

| Rate between 100 and 250 BPM | Rhythm irregular | P' waves differ in appearance, one precedes each QRS complex | QRS complexes of normal duration and all look alike | P'R intervals vary |

| ST segments often distorted | T waves often distorted | QT intervals possibly indiscernible | U waves not likely seen |

Figure 13-10
Multifocal atrial tachycardia.

Causes

MAT is very rare in healthy patients. It is more common in the elderly. It is usually precipitated by acute exacerbation (with resultant hypoxia) of COPD, elevated atrial pressures, or heart failure.

Causes of Multifocal Atrial Tachycardia	
Cause	Examples
Cardiac diseases	Acute coronary syndromes, elevated atrial pressures, heart failure, rheumatic heart disease
Use of certain drugs	Digitalis, theophylline toxicity
Noncardiac disorders	COPD (most common cause), electrolyte imbalance (hypokalemia, hypomagnesemia), hypoxia

MAT can also occur after MI. It is sometimes associated with digitalis toxicity in patients with heart disease. Common causes are listed in the "Causes of Multifocal Atrial Tachycardia".

Effects

The patient may complain of palpitations. Signs and symptoms of decreased cardiac output, such as hypotension, syncope, and blurred vision, may be seen.

ECG Appearance

MAT is often misdiagnosed as atrial fibrillation with rapid ventricular response but can be distinguished by looking closely for the clearly visible but changing P′ waves. The P′ waves change in morphology as often as from beat to beat. It results in three or more different-looking P waves. The P waves may be upright, rounded, notched, inverted, biphasic, or buried in the QRS complex. It also results in varying PR intervals and narrow QRS complexes.

Sometimes with SVT we see QRS complexes that appear wide. This is due to an intraventricular conduction disturbance or other condition such as aberrant conduction. This makes the assessment of SVT difficult as it appears to be ventricular tachycardia. This is called *wide complex tachycardia of unknown origin.*

Treatment

MAT usually represents either severe underlying heart or lung disease; the appropriate therapy is treatment of the underlying condition. In symptomatic patients treatment may include administration of medications such as calcium channel blockers (verapamil, diltiazem). Beta-adrenergic blockers are typically contraindicated because of the presence of severe underlying pulmonary disease.

13.7 Supraventricular Tachycardia

In very fast rates with narrow QRS complexes, it may be difficult to determine whether the tachycardia arises from the atria or AV junction (Figure 13-11) because the P′ waves are often buried in the T wave of the preceding beat. We describe these

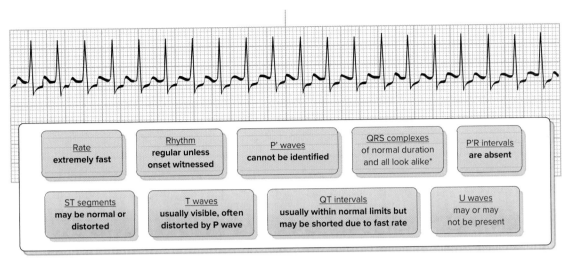

Rate	Rhythm	P' waves	QRS complexes	P'R intervals
extremely fast	regular unless onset witnessed	cannot be identified	of normal duration and all look alike*	are absent

ST segments	T waves	QT intervals	U waves
may be normal or distorted	usually visible, often distorted by P wave	usually within normal limits but may be shorted due to fast rate	may or may not be present

* unless aberrantly conducted in which case they may be wide and bizzare-looking

Figure 13-11

The term supraventricular tachycardia is used when the rate is so fast you cannot determine whether or not P' waves are present.

as **supraventricular tachycardia**, meaning that it is arising from above the ventricles. This group of tachycardias includes paroxysmal SVT (PSVT), nonparoxysmal atrial tachycardia, MAT, AV nodal reentrant tachycardia (AVNRT), atrioventricular reentrant tachycardia, and junctional tachycardia. AVNRT, atrioventricular reentrant tachycardia, and junctional tachycardia are discussed in Chapter 14. Distinguishing among these tachycardias is often difficult as the ventricular rate is so fast that it is hard to tell if there are P' waves or P'R intervals. The term narrow complex tachycardia can also be used when referring to atrial tachycardia and/or SVT.

13.8 Atrial Flutter

Description

Atrial flutter is a rapid depolarization of a single focus in the atria at a rate of 250 to 350 beats per minute (usually 300; Figure 13-12). It results from circus reentry, a condition during which the impulse from the SA node circles back through the atria, returning to the SA node region and repeatedly restimulating the AV node over and over. Another cause is increased automaticity. On the ECG, the P waves lose their distinction due to the rapid atrial rate. The waves blend together in a sawtooth or picket fence pattern. They are called flutter waves, or F waves.

Causes

Although atrial flutter occasionally occurs in patients with healthy hearts, it is usually caused by conditions that elevate atrial pressures and enlarge the atria. Also, it is commonly seen in patients with severe mitral valve disease, hyperthyroid disease, pericardial disease, and primary myocardial disease. The "Causes of Atrial Flutter" box lists causes of atrial flutter.

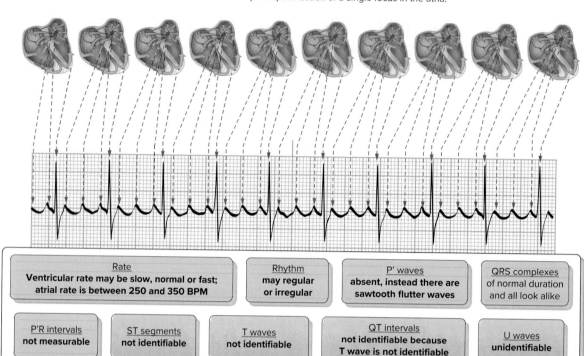

Atrial flutter arises from rapid depolarization of a single focus in the atria.

| Rate Ventricular rate may be slow, normal or fast; atrial rate is between 250 and 350 BPM | Rhythm may regular or irregular | P' waves absent, instead there are sawtooth flutter waves | QRS complexes of normal duration and all look alike |

| P'R intervals not measurable | ST segments not identifiable | T waves not identifiable | QT intervals not identifiable because T wave is not identifiable | U waves unidentifiable |

Figure 13-12
Atrial flutter.

Effects

Atrial flutter is often well tolerated. However, the number of impulses conducted through the AV node, expressed as the conduction ratio, determines the ventricular rate. Slower ventricular rates (fewer than 40 beats per minute) or faster ventricular rates (greater than 150 beats per minute) can seriously compromise cardiac output. For example, if the atrial flutter rate is 300 and the ventricular response is at a rate of 150, the conduction ratio is 300:150 or 2:1 (atrial to ventricular beats). This may result in loss of atrial kick, decreased ventricular filling time, and decreased coronary artery perfusion leading to angina, heart failure, pulmonary edema, hypotension, and syncope.

Causes of Atrial Flutter	
Cause	**Examples**
Cardiac diseases	MI, cardiomyopathy, conditions that enlarge atrial tissue and elevate atrial pressures, following cardiac surgery, mitral or tricuspid valve disease, pericarditis/myocarditis
Use of certain drugs	Digitalis or quinidine toxicity
Noncardiac disorders	COPD, hypoxia, hyperthyroidism, pneumonia, pulmonary embolism, systemic arterial hypoxia, pulmonary hypertension

ECG Appearance

The key feature of atrial flutter is the presence of sawtooth flutter waves. These flutter waves correspond to the rapid atrial rate of 250 to 350 beats per minute. The atrial rhythm is regular, and depending on conduction ratio, the ventricular rhythm may be regular or irregular. The ventricular rate depends on ventricular response; it may be normal, slow, or fast. A 1:1 atrial-ventricular conduction is rare; it is usually 2:1, 3:1, 4:1, or variable. The PR interval is not measurable. QRS complexes are usually normal.

Treatment

Vagal maneuvers may make flutter waves more visible by transiently increasing the degree of the block. In patients experiencing atrial flutter and an associated rapid ventricular rate who are symptomatic but stable, treatment is directed at controlling the rate or converting the rhythm to sinus rhythm. Patients who have serious signs and symptoms (e.g., hypotension, signs of shock, or heart failure) should receive oxygen, an IV infusion of normal saline administered at a keep-open (TKO) rate, and prompt treatment. Synchronized cardioversion should be considered in unstable patients. If necessary, the energy may be increased with subsequent shocks.

13.9 Atrial Fibrillation

Description

Atrial fibrillation occurs when there is chaotic, asynchronous firing of multiple areas within the atria (Figure 13-13). This completely suppresses normal SA node output and causes the atria to quiver instead of contract. This results in an ineffective contraction of the atria and no contribution to filling the ventricles during systole—resulting in a reduction in cardiac output. Atrial fibrillation stems from

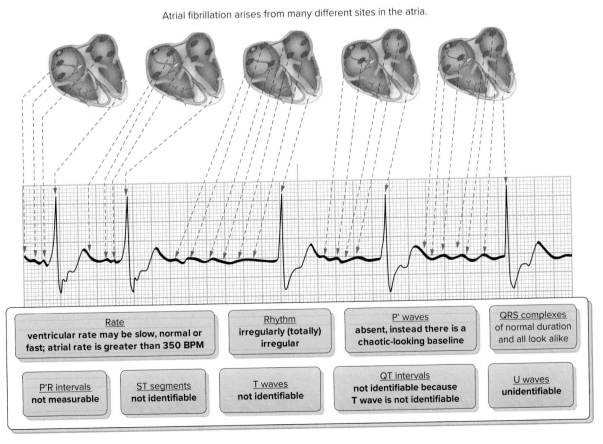

Figure 13-13
Atrial fibrillation.

multiple foci of impulse formation (greater than 350 beats per minute). These impulses bombard the AV node but are only conducted to the ventricles in an irregular and sporadic fashion.

Causes

Atrial fibrillation is more common than atrial tachycardia or atrial flutter. It can occur in healthy persons after excessive caffeine, alcohol, or tobacco ingestion or because of fatigue and acute stress. Atrial fibrillation may also be seen with other conditions, including rheumatic heart disease, CHF (atrial dilation), ischemic heart disease, and pulmonary diseases (COPD). It is rarely caused by digitalis toxicity, but atrial fibrillation with a very slow, regular ventricular response should raise suspicion of digitalis toxicity. Causes of atrial fibrillation are listed in the "Causes of Atrial Fibrillation" box.

Effects

As with atrial flutter, atrial fibrillation leads to a loss of the atrial kick. This decreases ventricular filling by up to 30%. Patients experiencing atrial fibrillation may develop intraatrial thrombi (clots), as the atria are not contracting and blood stagnates in the atrial chambers. If the thrombi are dislodged from the atria, this predisposes the patient to systemic emboli—particularly stroke. A ventricular response within the normal rate (60 to 100 beats per minute) is often well tolerated whereas a fast ventricular response can result in decreased cardiac output that leads to heart failure, angina, or syncope. Patients who have preexisting heart disease, such as hypertrophic obstructive cardiomyopathy, mitral stenosis, and rheumatic heart disease, or who have mitral prosthetic valves are less tolerant of atrial fibrillation and may experience shock and severe heart failure.

Causes of Atrial Fibrillation

Cause	Examples
Cardiac diseases	MI, atrial septal defects, CAD, cardiomyopathy, congestive heart failure, congenital heart disease, ischemic heart disease, mitral regurgitation, mitral stenosis, pericarditis, rheumatic heart disease. May also be caused by cardiac surgery
Use of certain drugs	Digitalis toxicity, aminophylline
Excessive sympathetic stimulation	Endogenous catecholamines released during exercise
Noncardiac disorders	Advanced age, COPD, diabetes, electrocution, hypoxia, hypokalemia, hypoglycemia, hypertension, hyperthyroidism, pulmonary embolism, WPW syndrome, idiopathic (no clear cause); may also occur in healthy people who use coffee, alcohol, or nicotine to excess or who are fatigued and under stress

ECG Appearance

The standout feature in atrial fibrillation is a totally (grossly) irregular rhythm (also referred to as *irregularly irregular*). An irregularly irregular supraventricular rhythm is atrial fibrillation until proven otherwise. Another key characteristic that is there are no discernible P waves. Instead, a chaotic baseline of fibrillatory waves, or f waves, represents the atrial activity. The absence of discernable P waves occurs because the atria have impulse formations from multiple foci with such high frequency (greater than 350 beats per minute) that no atrial depolarization can occur. The ventricular rate depends on how many impulses bombarding the AV node are conducted through to the ventricles. It may be normal, slow, or fast. There are no PR intervals present because of the absence of P waves. The QRS complexes are usually normal.

Treatment

If the rate of ventricular response is normal (often seen in patients taking digitalis), the dysrhythmia is usually well tolerated and requires no immediate intervention. In patients experiencing atrial fibrillation and an associated rapid ventricular rate who are symptomatic but stable, treatment is directed at controlling the rate or converting the rhythm to sinus rhythm. Patients who have serious signs and symptoms (e.g., hypotension, signs of shock, or heart failure) should receive oxygen, an IV infusion of normal saline administered at a TKO rate, and prompt synchronized cardioversion at 120 to 200 joules. If necessary, the energy level may be increased with subsequent shocks. Other treatments may include employing certain antidysrhythmic agents and seeking expert consultation.

Practice Makes Perfect

For each of the 57 ECG tracings on the following pages, practice the Nine-Step Process for analyzing ECGs. To achieve the greatest learning, you should practice assessing and interpreting the ECGs immediately after reading this chapter. Below are questions you should consider as you access each tracing. Your answers can be written into the area below each ECG marked "ECG Findings." Your findings can then be compared to the answers provided in Appendix A. All tracings are six seconds in duration.

1. Determine the heart rate. Is it slow? Normal? Fast? What is the ventricular rate? What is the atrial rate?

2. Determine if the rhythm is regular or irregular. If it is irregular, what type of irregularity is it? Occasional or frequent? Slight? Sudden acceleration or slowing in heart rate? Total? Patterned? Does it have a variable conduction ratio?

3. Determine if P waves are present. If so, how do they appear? Do they have normal height and duration? Are they tall? Notched? Wide? Biphasic? Of differing morphology? Inverted? One for each QRS complex? More than one preceding some or all the QRS complexes? Do they have a sawtooth appearance? Is there an indiscernible chaotic baseline?

4. Determine if QRS complexes are present. If so, how do they appear? Narrow with proper amplitude? Tall? Low amplitude? Delta wave? Notched? Wide? Bizarre-looking? With chaotic waveforms?

5. Determine the presence of PR intervals. If present, how do they appear? Constant? Of normal duration? Shortened? Lengthened? Progressively longer? Varying?

6. Evaluate the ST segments. Do they have normal duration and position? Are they elevated? (If so, are they flat, concave, convex, arched?) Depressed? (If so, are they normal, flat, downsloping or upsloping?)

7. Determine if T waves are present. If so, how do they appear? Of normal height and duration? Tall? Wide? Notched? Inverted?

8. Determine the presence of QT intervals. If present, what is their duration? Normal? Shortened? Prolonged?

9. Determine if U waves are present. If present, how do they appear? Of normal height and duration? Inverted?

10. Identify the rhythm or dysrhythmia.

1.

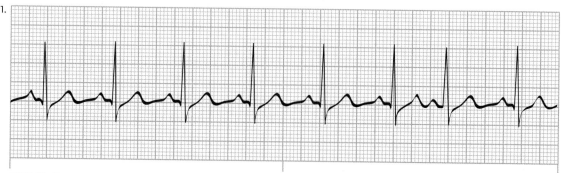

ECG Findings:

2.

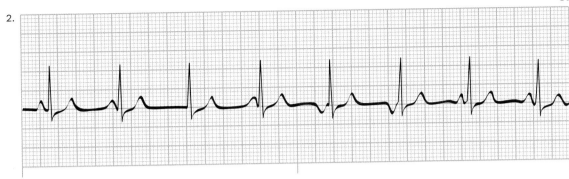

ECG Findings:

3.

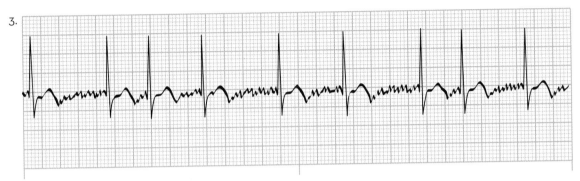

ECG Findings:

4.

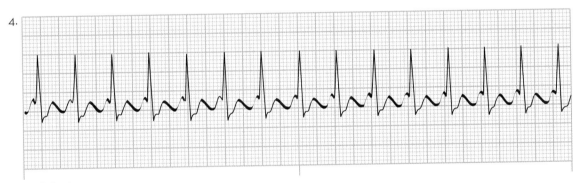

ECG Findings:

5.

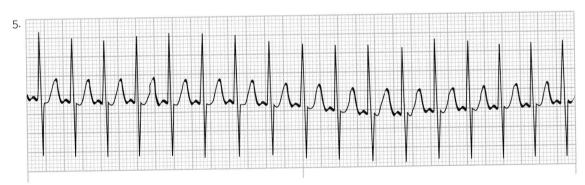

ECG Findings:

6.

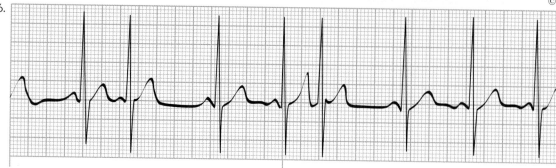

ECG Findings:

7.

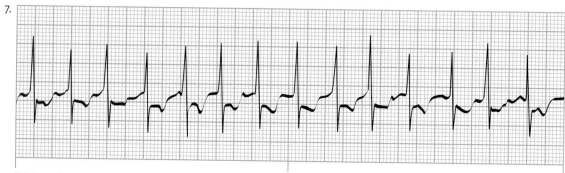

ECG Findings:

8.

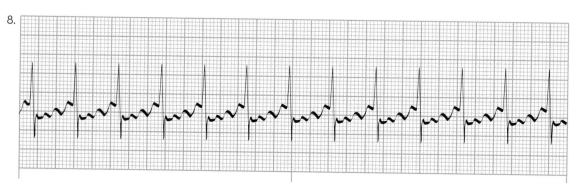

ECG Findings:

9.

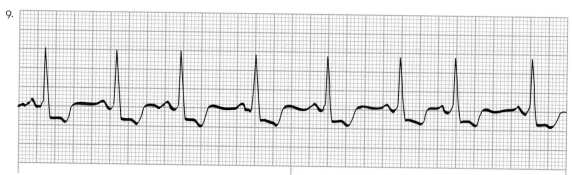

ECG Findings:

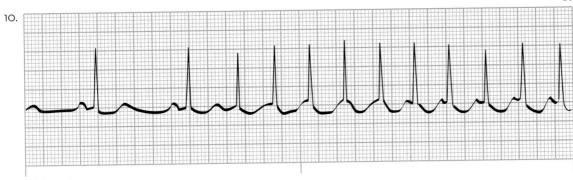

10.

ECG Findings:

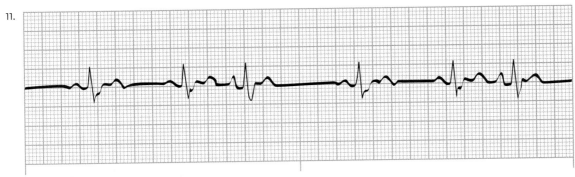

11.

ECG Findings:

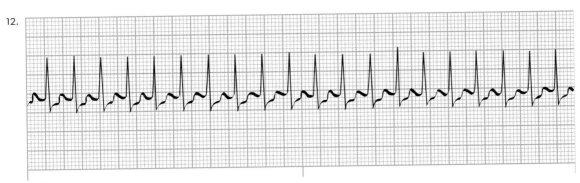

12.

ECG Findings:

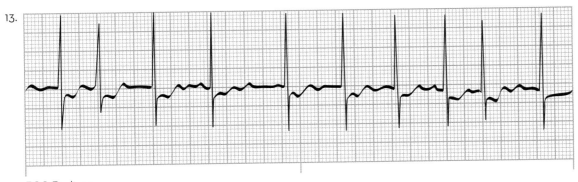

13.

ECG Findings:

14.

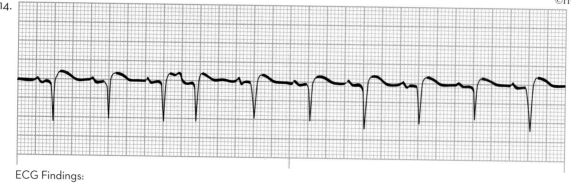

ECG Findings:

15.

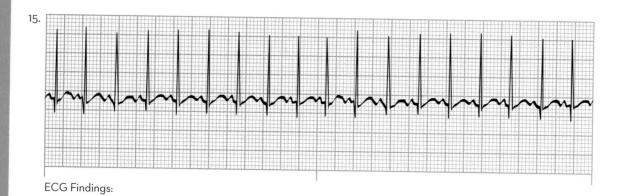

ECG Findings:

16.

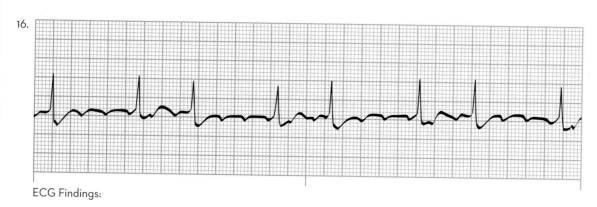

ECG Findings:

17.

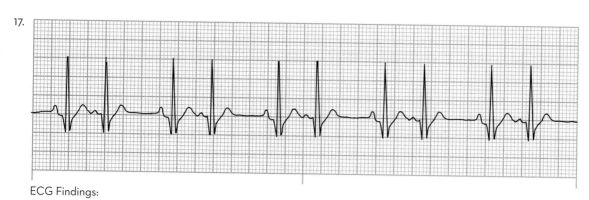

ECG Findings:

18.

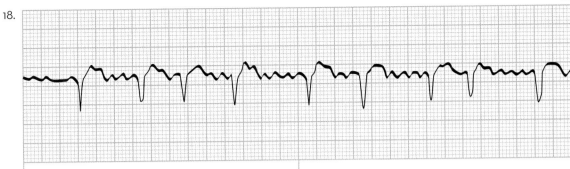

ECG Findings:

19.

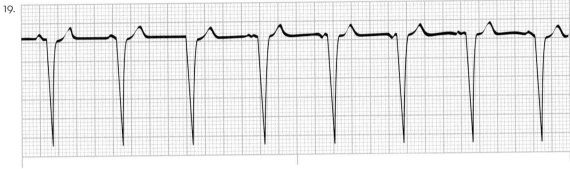

ECG Findings:

20.

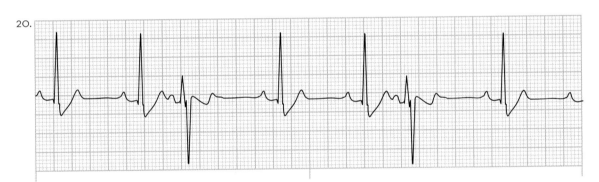

ECG Findings:

21.

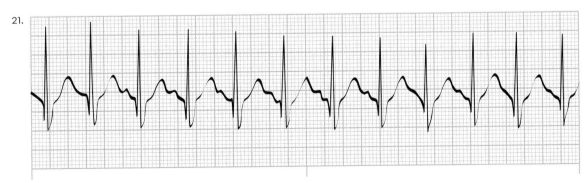

ECG Findings:

22.

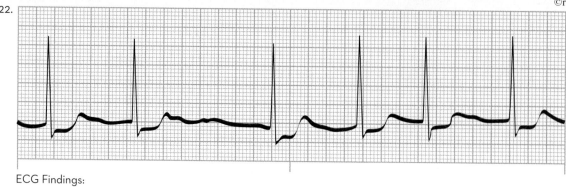

ECG Findings:

23.

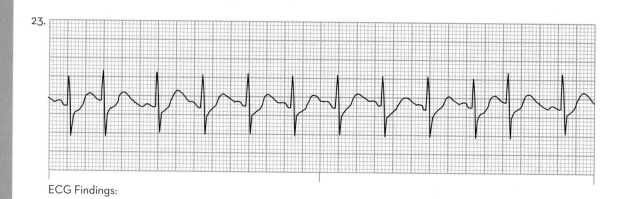

ECG Findings:

24.

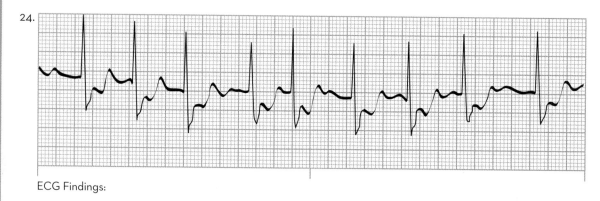

ECG Findings:

25.

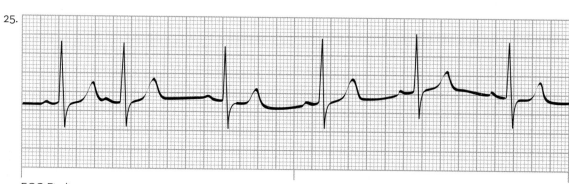

ECG Findings:

26.

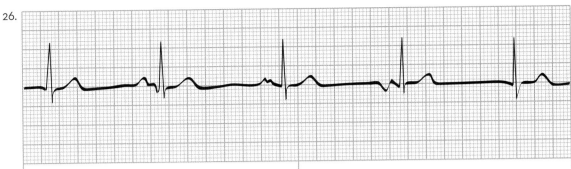

ECG Findings:

27.

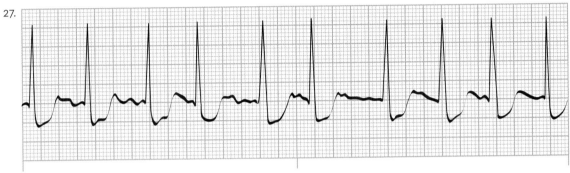

ECG Findings:

28.

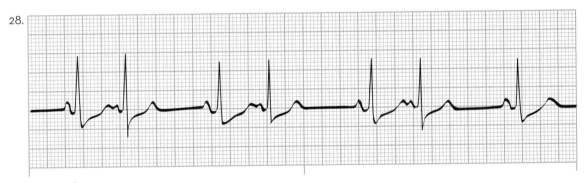

ECG Findings:

29.

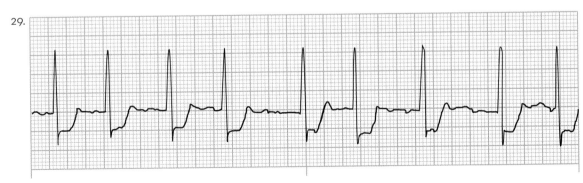

ECG Findings:

30.

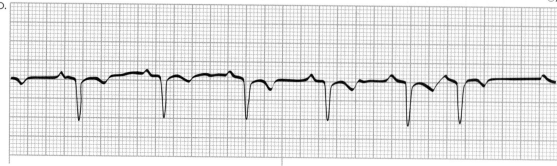

ECG Findings:

31.

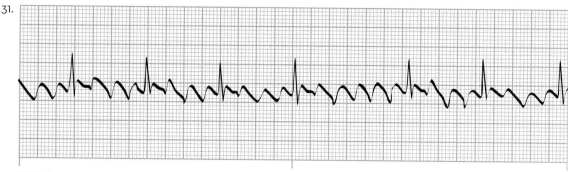

ECG Findings:

32.

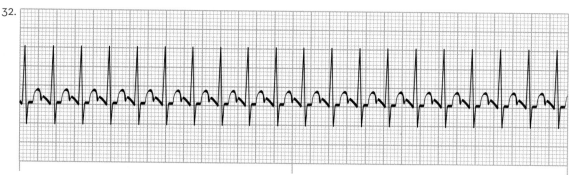

ECG Findings:

33.

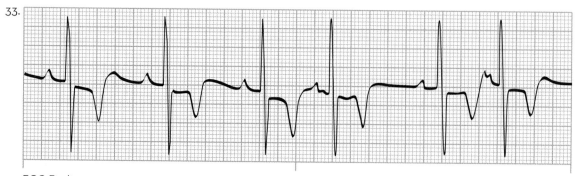

ECG Findings:

34.

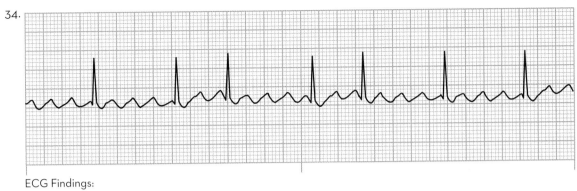

ECG Findings:

35.

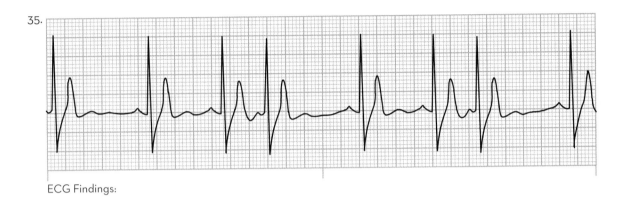

ECG Findings:

36.

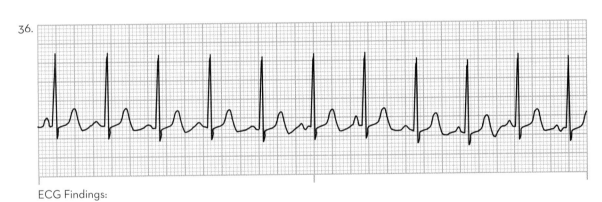

ECG Findings:

37.

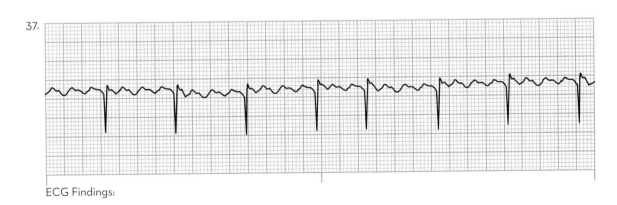

ECG Findings:

38.

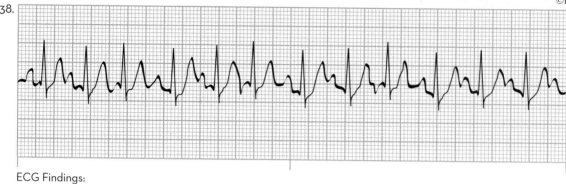

ECG Findings:

39.

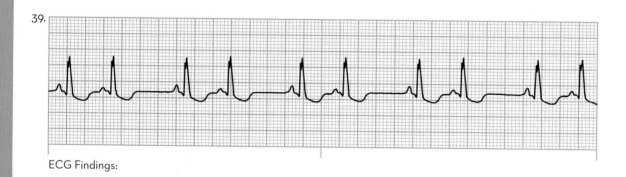

ECG Findings:

40.

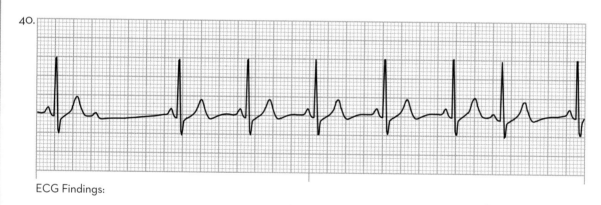

ECG Findings:

41.

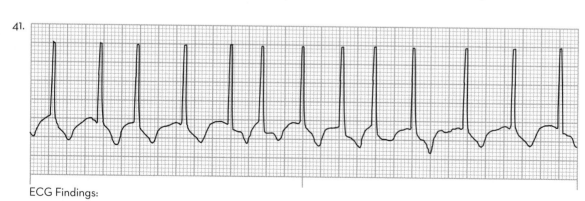

ECG Findings:

42.

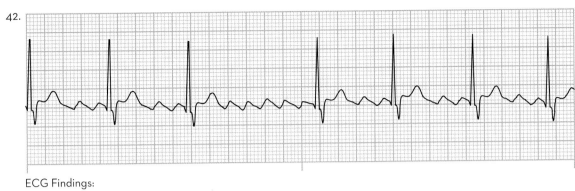

ECG Findings:

43.

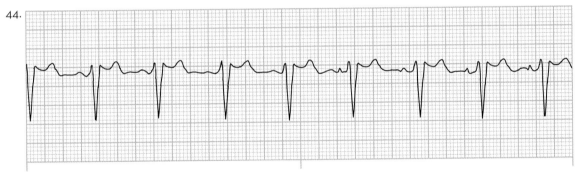

ECG Findings:

44.

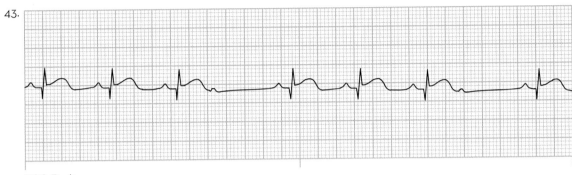

ECG Findings:

45.

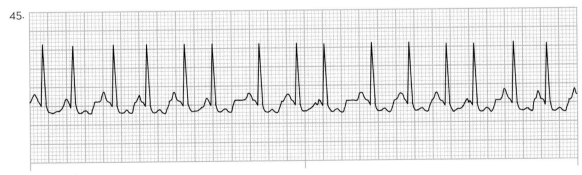

ECG Findings:

46.

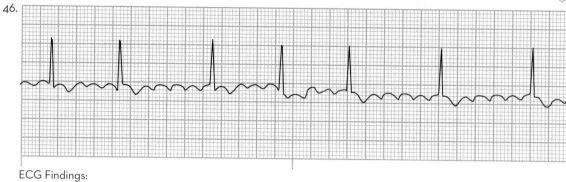

ECG Findings:

47.

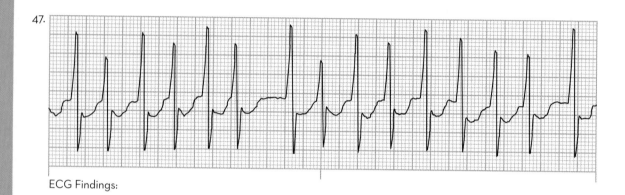

ECG Findings:

48.

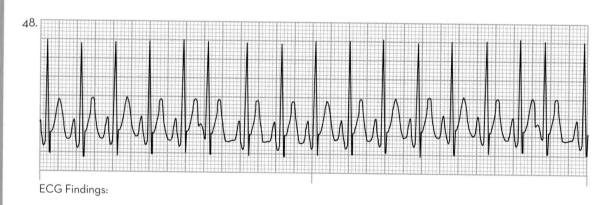

ECG Findings:

49.

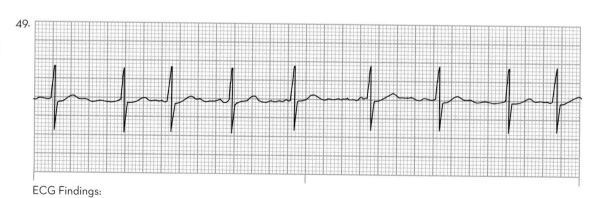

ECG Findings:

50.

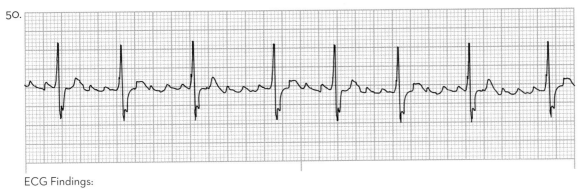

ECG Findings:

51.

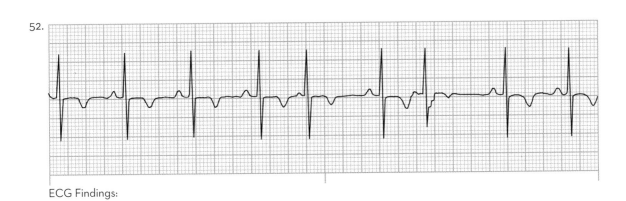

ECG Findings:

52.

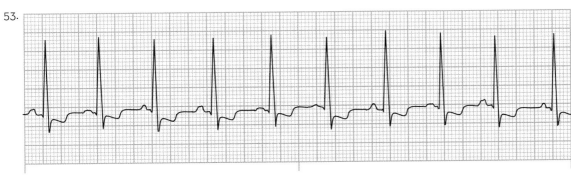

ECG Findings:

53.

ECG Findings:

54.

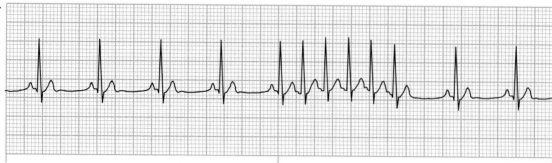

ECG Findings:

55.

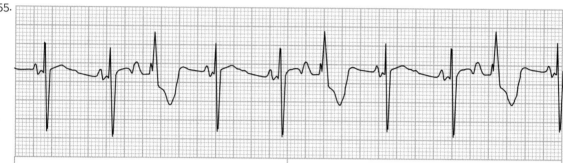

ECG Findings:

56.

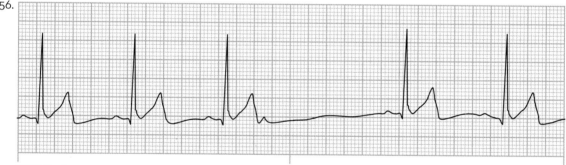

ECG Findings:

57.

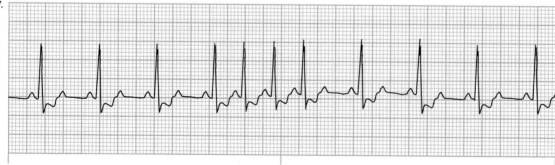

ECG Findings:

LO 13.1	• Dysrhythmias originating outside the SA node in the atrial tissue or in the internodal pathways are called atrial dysrhythmias.
	• Three mechanisms responsible for atrial dysrhythmias are increased automaticity, triggered activity, and reentry.
LO 13.2	• Key characteristics for atrial dysrhythmias are P′ waves (if present) that differ in appearance from normal sinus P waves; abnormal, shortened, or prolonged P′R intervals; and QRS complexes that appear narrow and normal (unless there is an intraventricular conduction defect, aberrancy, or preexcitation).
LO 13.3	• PACs are early ectopic beats that originate outside the SA node but above the AV node.
	• PACs produce an irregularity in the rhythm where the P′-P and R′-R intervals are shorter than the P-P and R-R intervals of the underlying rhythm. The P′ waves should be an upright (in lead II) preceding the QRS complex but will have a different morphology (appearance) than the P waves in the underlying rhythm.
LO 13.4	• Wandering atrial pacemaker is a rhythm in which the pacemaker site shifts between the SA node, atria, and/or AV junction producing its most characteristic features, P9 waves that change in appearance and a rhythm that varies slightly.
LO 13.5	• Atrial tachycardia is a rapid dysrhythmia (having a rate of 150 to 250 beats per minute) that arises from the atria. It is so fast that it overrides the SA node.
	• When the onset of the atrial tachycardia is sudden (and typically witnessed) with a finite duration, it is called paroxysmal.
	• Atrial tachycardia is regular unless you see the onset or termination, as with paroxysmal atrial tachycardia, and there is one P′ wave preceding each QRS complex, although it is typically buried in the T wave of the preceding beat.
LO 13.6	• MAT is a pathological condition that presents with the same characteristics as wandering atrial pacemaker but has heart rates in the 100 to 250 beats per minute range.
LO 13.7	• SVT arises from above the ventricles but cannot be definitively identified as atrial or junctional because the P′ waves cannot be seen with any real degree of certainty.
	• SVTs include PSVT, nonparoxysmal atrial tachycardia, and MAT.
LO 13.8	• Atrial flutter is a rapid depolarization of a single focus in the atria at a rate of 250 to 350 beats per minute.
	• Atrial flutter produces atrial waveforms that have a characteristic sawtooth or picket fence appearance. They are called flutter waves, or F waves.
LO 13.9	• Atrial fibrillation occurs when there is chaotic, asynchronous firing of multiple areas within the atria. It stems from the rapid firing of ectopic impulses (greater than 350 beats per minute) in circus reentry pathways.
	• Atrial fibrillation is a totally (grossly) irregular rhythm in which there are no discernible P waves. Instead, a chaotic baseline of fibrillatory waves or f waves represents the atrial activity. Atrial fibrillation occurs when there is chaotic, asynchronous firing of multiple areas within the atria. It stems from the rapid firing of ectopic impulses (greater than 350 beats per minute) in circus reentry pathways.

Assess Your Understanding

The following questions give you a chance to assess your understanding of the material discussed in this chapter. The answers can be found in Appendix A.

1. With atrial dysrhythmias, the   (LO 13.1, 13.2)
   a. atrial waveforms differ in appearance from normal sinus P waves.
   b. P'R intervals are almost always prolonged.
   c. QRS complexes are wider than normal.
   d. site of origin is in the bundle of His.

2. Why is it important to check for a pulse when there are premature beats in the rhythm? (LO 13.3)

3. Premature atrial complexes (LO 13.3)
   a. are typically preceded by an inverted P' wave.
   b. are always followed by a compensatory pause.
   c. have a wide and bizarre QRS complex.
   d. may have normal P'R intervals.

4. Describe the effect that a PAC has on the regularity of the underlying rhythm. (LO 13.3)

5. In wandering atrial pacemaker, the pacemaker site shifts between the SA node, atria, and/or (LO 13.4)
   a. Purkinje fibers.
   b. ventricles.
   c. AV junction.
   d. bundle of His.

6. Your patient is a 67-year-old female with a history of cardiac problems. After attaching her to the monitor, you see a slightly irregular rhythm with normal QRS complexes, but each P' wave is different. This rhythm is (LO 13.4)
   a. frequent PACs.
   b. sinus arrest.
   c. sinus dysrhythmia.
   d. wandering atrial pacemaker.

7. The heart rate characteristic of atrial tachycardia is _____ beats per minute. (LO 13.5)
   a. 60 to 100
   b. 100 to 150
   c. 150 to 250
   d. 300 to 350

8. The QRS complexes seen with atrial tachycardia are normally _____ seconds in duration. (LO 13.5)
   a. 0.06 to 0.10
   b. 0.10 to 0.20
   c. 0.12 to 0.20
   d. 0.20 to 0.24

9. List the treatments for stable SVT. (LO 13.5)

10. List the treatments for unstable SVT. (LO 13.5)

11. With MAT, the (LO 13.6)
    a. P′ waves have the same morphology from beat to beat.
    b. heart rate is 100 to 250 beats per minute.
    c. QRS complexes are usually greater than 0.12 seconds.
    d. Both b and c.

12. Which rhythm has a characteristic sawtooth pattern? (LO 13.8)
    a. Sinus dysrhythmia
    b. Ventricular flutter
    c. Atrial flutter
    d. Atrial tachycardia

13. Atrial flutter has an atrial rate of _____ beats per minute. (LO 13.9)
    a. between 40 and 60
    b. between 100 and 150
    c. between 250 and 350
    d. greater than 350

14. Atrial fibrillation typically has (LO 13.9)
    a. a regular rhythm.
    b. an atrial rate of between 250 and 350 beats per minute.
    c. normal QRS complexes.
    d. a PR intervals of greater than 0.20 seconds.

15. The atrial waveforms associated with atrial fibrillation are _____ and the PR intervals are _____. (LO 13.9)
    a. referred to as saw toothed; variable
    b. indiscernible; nonexistent
    c. inverted; less than 0.12 seconds
    d. dissociated; between 0.12 and 0.20 seconds

16. Which dysrhythmia is totally irregular? (LO 13.9)
    a. Sinus bradycardia
    b. 3rd-degree AV heart block
    c. Ventricular tachycardia
    d. Atrial fibrillation

17. Atrial fibrillation has an atrial rate of _____ beats per minute. (LO 13.9)
    a. between 40 and 60
    b. between 100 and 150
    c. between 250 and 350
    d. greater than 350

Referring to the scenario at the beginning of this chapter, answer the following questions.

18. What dysrhythmia is the patient experiencing? (LO 13.8)
    a. Accelerated junctional
    b. Atrial fibrillation
    c. Atrial flutter
    d. Ventricular tachycardia

19. The patient passed out because of a decrease in his (LO 13.2)
    a. diastolic blood pressure.
    b. cardiac output.
    c. epinephrine levels.
    d. heart rate.

20. The formula for cardiac output is (LO 13.2)
    a. heart rate × systolic blood pressure.
    b. systolic blood pressure/diastolic blood pressure.
    c. stroke volume × heart rate.
    d. heart rate × respiratory rate.

14 Junctional Dysrhythmias

©rivetti/Getty Images

Courtesy Physio-Control

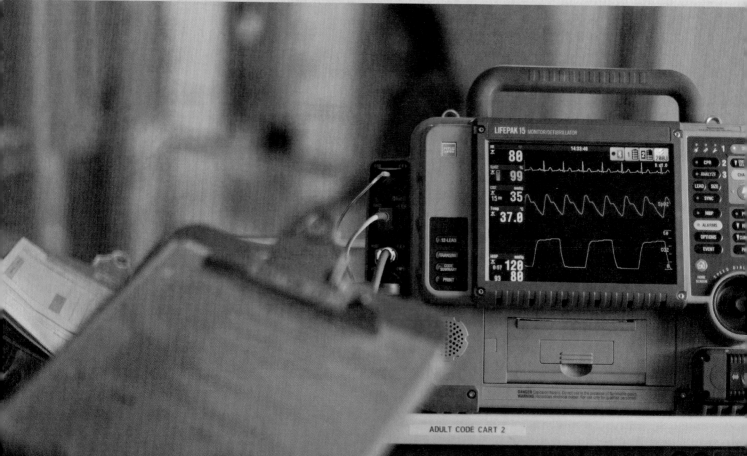

Learning Outcomes

Case History

A 48-year-old woman presents to the urgent care center complaining of a rapid heart rate. She tells the triage nurse that she has had this before and the condition is called *paroxysmal supraventricular tachycardia* (PSVT). The patient is quickly taken back to the exam room, and the physician is summoned. Her vital signs are normal except for a pulse rate of 180 beats per minute. When the physician enters the exam room, the woman states that she has had this before but usually it goes away if she performs vagal maneuvers such as bearing down and holding her breath. But that it is not working today. The physician orders a 12-lead ECG, which he then examines while the nurse establishes an IV. The physician notes that the patient's heart rate is fast and the rhythm is regular with narrow QRS complexes. He is unable to distinguish discernable P waves because of the fast rate. The nurse administers a medication to slow the heart rate. Within seconds the patient's rhythm converts to normal sinus rhythm.

After further questioning, the physician determines that the woman has not been compliant with her medications and that she continues to smoke and drink a lot of caffeinated beverages, all of which have contributed to the return of her dysrhythmia. He schedules her an appointment with her family physician and refills her medication of beta-blockers. He encourages her to remove nicotine and caffeine from her life.

14.1 Dysrhythmias Originating in the Atrioventricular Junction

Rhythms that originate in the AV junction, the area around the AV node, and the bundle of His, are called *junctional rhythms* (sometimes referred to as *nodal rhythms*). They can result from suppression or blockage of the SA node, which results in its failure to generate electrical impulses. Without an impulse from the SA node, the next lower pacemaker (the AV junction) should act as a back-up pacemaker and take over. Junctional rhythms may also result from increased automaticity of the AV junction or a reentry mechanism.

Junctional dysrhythmias include (Figure 14-1):

- Premature junctional complex (PJC)
- Junctional escape complexes or rhythms
- Accelerated junctional rhythm
- Junctional tachycardia
- AV nodal reentrant tachycardia (AVNRT)
- AV reentrant tachycardia

14.2 Key Features of Junctional Dysrhythmias

P Wave Appearance

The AV junction is located in the middle of the heart; therefore the impulse originating here travels upward and causes backward or retrograde depolarization of the atria (see Figure 6-12), resulting in an inverted P′ wave (when it would otherwise be upright) with a short P′R interval (less than 0.12 seconds in duration). At the same time, the impulse travels down to the ventricles and depolarizes them.

Alternatively, the P′ wave may be absent. This occurs when the electrical impulse, originating in the middle of the heart near the AV node, reaches the atria and ventricles at the same time, resulting in simultaneous depolarization and causing the P′ wave to be buried in the QRS complex (which can change the morphology of the QRS complex). Another reason for an absent P′ wave is that retrograde conduction of the impulse to the atria can be blocked or not conducted at all. As a result, the atrial tissue is not depolarized and a P′ wave is not formed. The impulse can be blocked due to diseased or damaged AV nodal issue or MI, or the AV node is just incapable of retrograde depolarization. Lastly, the P wave may follow the QRS complex. This occurs when retrograde atrial depolarization follows ventricular depolarization.

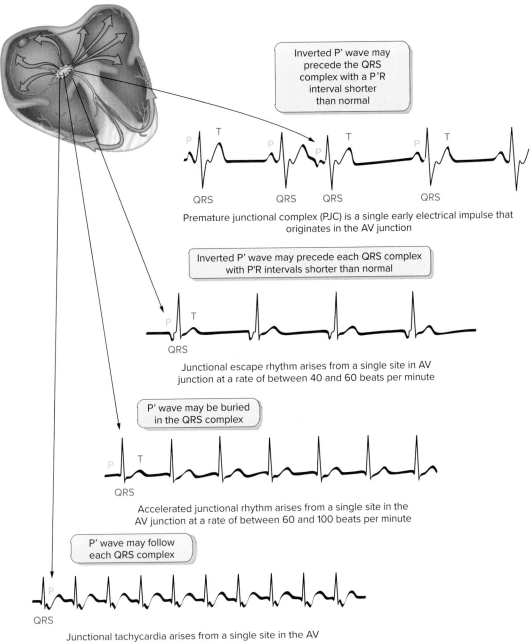

Inverted P' wave may precede the QRS complex with a P'R interval shorter than normal

Premature junctional complex (PJC) is a single early electrical impulse that originates in the AV junction

Inverted P' wave may precede each QRS complex with P'R intervals shorter than normal

Junctional escape rhythm arises from a single site in AV junction at a rate of between 40 and 60 beats per minute

P' wave may be buried in the QRS complex

Accelerated junctional rhythm arises from a single site in the AV junction at a rate of between 60 and 100 beats per minute

P' wave may follow each QRS complex

Junctional tachycardia arises from a single site in the AV junction at a rate of between 100 and 180 beats per minute

Figure 14-1
Junctional dysrhythmias originate from the AV junction.

PR Interval

PR intervals seen with premature junctional complexes or junctional rhythms are shorter than normal (less than 0.12 seconds in duration) or absent if the P wave is buried in the QRS complex. If the P' wave follows the QRS complex, it is referred to as the RP' interval and is usually less than 0.20 seconds.

QRS Complex Appearance

In junctional rhythms, whereas the atria are depolarized in an abnormal way, the ventricles depolarize normally. Electrical impulses travel in a normal pathway from the AV junction through the bundle of His and bundle branches to the Purkinje fibers, ending in the ventricular muscle. This results in the QRS complexes usually being within normal limits of 0.06 to 0.10 seconds. However, in any dysrhythmia where the QRS complexes should be normal, they can instead be wide or unusual looking if there is intraventricular conduction disturbance, aberrancy, or ventricular preexcitation.

Effects

If the atria are depolarized at the same time or after the ventricles, the atria are forced to pump against the contracting ventricles, which contract with much greater force. This can result in a loss in atrial kick, decreased stroke volume, and, ultimately, decreased cardiac output. Decreased cardiac output can also occur with slow or fast junctional dysrhythmias.

As a reminder, dysrhythmias arising from the SA node, atria, or AV junction can be collectively called *supraventricular dysrhythmias*.

14.3 Premature Junctional Complex

Description

A PJC is a single early electrical impulse that originates in the AV junction (Figure 14-2). It occurs before the next expected sinus impulse, interrupting the regularity of the underlying rhythm. Because the impulse arises from the middle of the heart, the atria are depolarized in a retrograde fashion. This causes the P' wave

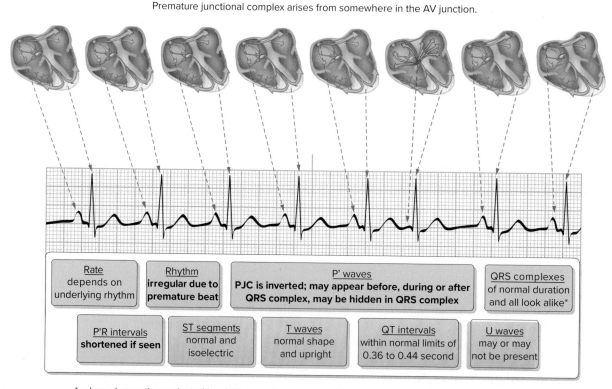

Premature junctional complex arises from somewhere in the AV junction.

| Rate depends on underlying rhythm | Rhythm **irregular due to premature beat** | P' waves **PJC is inverted; may appear before, during or after QRS complex, may be hidden in QRS complex** | QRS complexes of normal duration and all look alike* |

| P'R intervals **shortened if seen** | ST segments normal and isoelectric | T waves normal shape and upright | QT intervals within normal limits of 0.36 to 0.44 second | U waves may or may not be present |

*unless aberrantly conducted in which case they may be wider than normal and bizarre-looking

Figure 14-2
Premature junctional complex.

(if visible) to be inverted. The ventricles are depolarized in the normal manner (measuring 0.06 to 0.10 seconds in duration).

Causes

PJCs are thought to result from enhanced automaticity. Other causes are shown in the box following box.

Causes of Premature Junctional Complexes	
Cause	**Examples**
Cardiac diseases	Cardiomyopathy, CAD, congestive heart failure, congenital abnormalities, inferior-wall MI, myocardial ischemia, pericarditis, rheumatic heart disease, swelling of the AV node after surgery, valvular heart disease
Use of certain drugs	Digitalis toxicity, cocaine, excessive amphetamine intake
Noncardiac disorders	Acute respiratory failure, alcohol, anxiety, caffeine, certain electrolyte imbalances, COPD, electrolyte imbalance (particularly magnesium and potassium), fatigue, fear or stress, hyperthyroidism, hypoxia, nicotine, vigorous exercise

Effects

In the healthy heart, isolated PJCs are of little clinical significance. However, frequent PJCs (more than 4 to 6 per minute) warn of more serious conditions.

The patient may be asymptomatic or may experience palpitations. Your assessment will likely reveal the presence of an irregular pulse. Frequent PJCs may cause the patient to experience hypotension due to a transient decrease in cardiac output. Be sure to check for the presence of a pulse in rhythms that have premature beats to determine if they are perfusing beats. Typically, PJCs produce a pulse, but premature beats that arise from the ventricles may not. You can detect this by watching the ECG in the dynamic mode while feeling for a pulse.

When looking at the ECG rhythm, PJCs appear as "early beats." This causes the R-R interval to be shorter between the preceding beat and the early beat (as compared with the underlying rhythm).

ECG Appearance

As with any premature beats, you first see an irregular rhythm as the underlying rhythm is disrupted by the presence of early beats. When you look at the area(s) of irregularity you see a shorter R-R interval between the beat of the underlying rhythm and the premature beat. P′ waves may or may not be seen with the PJCs. If P′ waves are present with the premature beats, they will differ from the P waves of the underlying rhythm and are usually inverted. P′R intervals, if present with the premature junctional beats, will be less than 0.12 seconds in duration. The impulse arises above the ventricles; therefore the QRS complexes of PJCs are normal and look the same as QRS complexes of the underlying rhythm. The T waves of the premature beats deflect in the same direction as the R waves. PJCs are normally followed by a noncompensatory pause.

PJCs intermingled between normal beats are named depending on how frequent they occur. When every other beat is a PJC, it is called *bigeminal PJCs,* or *junctional bigeminy.* If every third beat is a PJC, it is called *trigeminal PJCs,* or *junctional trigeminy.* Likewise, if a PJC occurs every fourth beat, it is called *quadrigeminal PJCs,* or *junctional quadrigeminy.* Regular PJCs at greater intervals than every fourth beat have no special name. Remember, bi = 2, tri = 3, and quad = 4.

Treatment

PJCs generally do not require treatment. PJCs caused by the use of caffeine, tobacco, or alcohol, or with anxiety, fatigue, or fever can be controlled by eliminating the underlying cause.

14.4 Junctional Escape Rhythm

Description

Junctional escape rhythm is slow, steady rhythm (40 to 60 beats per minute) with narrow QRS complexes, inverted or absent P′ waves, and shorter than normal P′R intervals (when the P′ wave precedes the QRS complex). It typically occurs when the rate of the primary pacemaker (SA node) falls below that of the AV junctional tissue (Figure 14-3). Remember, if the SA node fails to fire or slows down, the AV junction (or ventricles) in their role as backup pacemakers, should initiate the heart-

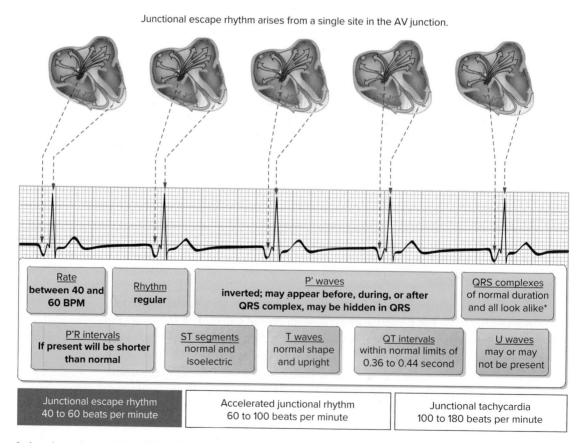

Junctional escape rhythm arises from a single site in the AV junction.

Rate between 40 and 60 BPM	Rhythm regular	P′ waves inverted; may appear before, during, or after QRS complex, may be hidden in QRS		QRS complexes of normal duration and all look alike*
P′R intervals If present will be shorter than normal	ST segments normal and isoelectric	T waves normal shape and upright	QT intervals within normal limits of 0.36 to 0.44 second	U waves may or may not be present

Junctional escape rhythm 40 to 60 beats per minute	Accelerated junctional rhythm 60 to 100 beats per minute	Junctional tachycardia 100 to 180 beats per minute

*unless aberrantly conducted in which case they may be wider than normal and bizarre-looking

Figure 14-3
Junctional escape rhythm.

beat. This serves as a safety mechanism to prevent cardiac standstill. Junctional escape may also be seen as an isolated junctional complex (escape complex) that occurs when the SA node fails to initiate an impulse. This isolated beat is then followed by the SA node resuming the heart beat and sinus rhythm is seen again.

Causes

Junctional escape rhythm is brought about by AV heart block or conditions that interfere with SA node function. Causes of this dysrhythmia are listed in the following box.

Causes of Junctional Escape Rhythm	
Cause	Examples
Cardiac diseases	Cardiomyopathy, disease of the SA node (sick sinus syndrome), heart failure, inferior wall MI, myocarditis, postcardiac surgery, SA node ischemia, rheumatic heart disease, valvular heart disease
Use of certain drugs	Digoxin, quinidine, beta-blockers, calcium channel blockers
Increased autonomic nervous system stimulation	Increased vagal tone on the SA node
Noncardiac disorders	Hypoxia, electrolyte imbalance

Effects

The normal firing rate of the AV junction is 40 to 60 beats per minute. Rates of greater than 50 beats per minute are usually well tolerated. Slower rates can cause decreased cardiac output and may lead to symptoms (chest pain or pressure, syncope, altered level of consciousness, hypotension).

Remember from our earlier discussion, decreased cardiac output leads to decreased tissue perfusion. Decreased tissue perfusion can have dire consequences. First, there is insufficient oxygen and nutrients for the body to satisfy its energy needs. Second, if the waste products of metabolism are not removed, they accumulate in the tissues and create a toxic environment. If not corrected, both factors can lead to shock and, ultimately, death of the patient.

ECG Appearance

The key features of this dysrhythmia are a slow heart rate of between 40 and 60 beats per minute; inverted P′ waves that precede, are lost in (buried), or follow the QRS complexes; and, if present, a P′R interval less than 0.12 seconds in duration. The rhythm is regular if it is an escape rhythm but irregular if it is an isolated escape beat. The QRS complexes are normal.

If the junctional escape rhythm has a heart rate of less than 40 beats per minute, it is described as *slow* or *bradycardic junctional escape rhythm or junctional escape rhythm with bradycardia.*

Treatment

When medical conditions cause the problem, it is essential to observe the heart rhythm for a worsening of the bradycardia. Treatment for junctional escape is directed at identifying and correcting the underlying cause and reestablishing the SA node activity at a fast enough rate to prevent the escape rhythm. For example, you should withhold digoxin or correct electrolyte imbalances if serum levels point to inappropriate levels of either. Also, deliver oxygen, place the patient on an ECG monitor, monitor the blood pressure and pulse oximetry, and establish an IV.

Patients who are symptomatic and unstable may require the administration of atropine (which acts to block the effects of parasympathetic nervous system stimulation), IV push, transcutaneous pacing (which acts to directly stimulate the heart to beat faster), or the administration of an IV infusion of dopamine or epinephrine (both of which act to stimulate the sympathetic nervous system; see Figure 12-4). A permanent pacemaker may be needed if the patient has little or no chance of regaining a normal rhythm at a normal rate.

14.5 Accelerated Junctional Rhythm

Description

Accelerated junctional rhythm is a dysrhythmia that arises from the AV junction (Figure 14-4). At a rate of 60 to 100 beats per minute, it speeds up to take over as

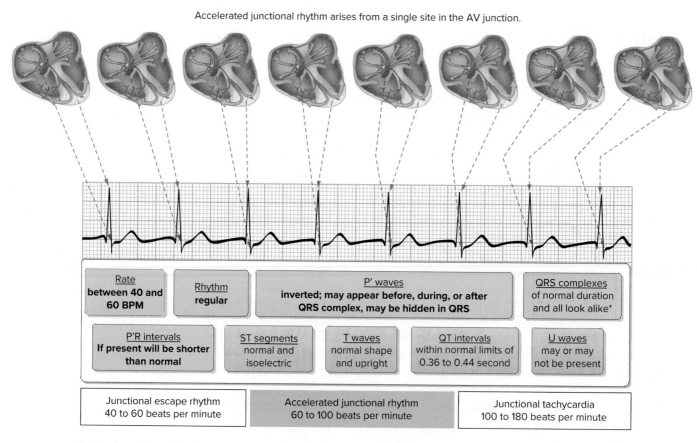

Accelerated junctional rhythm arises from a single site in the AV junction.

Rate	Rhythm	P' waves	QRS complexes
between 40 and 60 BPM	regular	inverted; may appear before, during, or after QRS complex, may be hidden in QRS	of normal duration and all look alike*

P'R intervals	ST segments	T waves	QT intervals	U waves
If present will be shorter than normal	normal and isoelectric	normal shape and upright	within normal limits of 0.36 to 0.44 second	may or may not be present

Junctional escape rhythm 40 to 60 beats per minute	Accelerated junctional rhythm 60 to 100 beats per minute	Junctional tachycardia 100 to 180 beats per minute

*unless aberrantly conducted in which case they may be wider than normal and bizarre-looking

Figure 14-4
Accelerated junctional rhythm.

the heart's pacemaker. Remember that the term *tachycardia* is applied to heart rates above 100 beats per minute. The inherent rate of the AV node/junction is 40 to 60 beats per minute. A heart rate above 60 beats per minute cannot correctly be classified as tachycardia until it reaches a rate of 100 beats per minute, so we use the term *accelerated* to describe this dysrhythmia. The word *accelerated* emphasizes that it is faster than the inherent rate.

Causes

Accelerated junctional rhythm is due to increased automaticity or irritability of the AV junction. The following box lists the causes of accelerated junctional rhythm.

Causes of Accelerated Junctional Rhythm	
Cause	**Examples**
Cardiac diseases	Heart failure, inferior or posterior wall MI, myocardial ischemia, pericarditis, rheumatic heart disease, post open-heart surgery, valvular heart disease
Use of certain drugs	Digitalis (common cause)
Noncardiac disorders	COPD, electrolyte imbalance (hypokalemia)

Effects

Accelerated junctional rhythm has the same rate as normal sinus rhythm, so it is usually well tolerated. However, it may predispose patients with myocardial ischemia to more serious dysrhythmias. Also, because the atria are depolarized by way of retrograde conduction and may actually follow ventricular depolarization, blood ejection from the atria into the ventricles (atrial kick) may be prevented. This may cause decreased cardiac output. If cardiac output is low, the patient may exhibit signs of decreased tissue perfusion.

ECG Appearance

The key characteristic of accelerated junctional rhythm is a regular rhythm (as it originates from one site) with P′ waves that precede, are lost in (buried), or follow the QRS complex. When present, the P′ waves are inverted. The P′R intervals, if present, are less than 0.10 seconds. The rate is 60 to 100 beats per minute. Accelerated junctional rhythm arises from above the ventricles, so the QRS complexes are normal.

Treatment

Given the heart rate seen with accelerated junctional rhythm, the patient is typically asymptomatic. Treatment is directed at identifying and correcting the underlying cause. For example, digoxin should be withheld or electrolyte imbalances corrected if serum levels point to inappropriate levels of either. The patient should be continually observed for signs of decreased cardiac output. Temporary pacing may be indicated if the patient is symptomatic.

14.6 Junctional Tachycardia

Description

Junctional tachycardia is a fast ectopic rhythm that originates from an irritable focus (increased automaticity) in the bundle of His and overrides the SA node (Figure 14-5). It is said to be present when there are three or more premature junctional complexes in a row or there is a junctional rhythm occurring at a rate of between 100 and 180 (and rarely as fast as 200) beats per minute. As with other junctional rhythms, the atria are depolarized by retrograde conduction while conduction through the ventricles occurs in the normal manner.

When the rate in junctional tachycardia exceeds 150 beats per minute, the rhythm becomes hard to distinguish from atrial tachycardia and is often referred to as *supraventricular tachycardia*. The term *paroxysmal* is added to describe junctional tachycardia that starts and ends suddenly. It is frequently precipitated by a premature junctional complex.

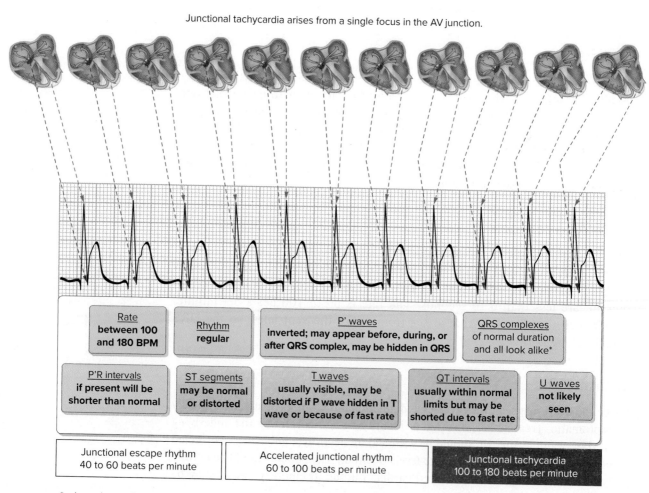

Junctional tachycardia arises from a single focus in the AV junction.

| Rate between 100 and 180 BPM | Rhythm regular | P' waves inverted; may appear before, during, or after QRS complex, may be hidden in QRS | QRS complexes of normal duration and all look alike* |

| P'R intervals if present will be shorter than normal | ST segments may be normal or distorted | T waves usually visible, may be distorted if P wave hidden in T wave or because of fast rate | QT intervals usually within normal limits but may be shorted due to fast rate | U waves not likely seen |

| Junctional escape rhythm 40 to 60 beats per minute | Accelerated junctional rhythm 60 to 100 beats per minute | Junctional tachycardia 100 to 180 beats per minute |

*unless aberrantly conducted in which case they may be wide and bizarre-looking

Figure 14-5
Junctional tachycardia.

Causes

Junctional tachycardia is believed to be caused by enhanced automaticity and is commonly the result of digitalis toxicity. It may also be brought about by myocardial ischemia or infarction and CHF. It can occur at any age without a patient history of underlying heart disease. Other causes of junctional tachycardia are listed in the following box.

Causes of Junctional Tachycardia	
Cause	**Examples**
Cardiac diseases	Swelling of the AV junction after heart surgery, damage to AV junction from inferior or posterior wall MI or rheumatic fever
Use of certain drugs	Digoxin (particularly in the presence of hypokalemia), theophylline
Noncardiac disorders	Excessive catecholamine administration, anxiety, hypoxia, electrolyte imbalance (particularly hypokalemia)

Effects

Palpitations, nervousness, anxiety, vertigo, lightheadedness, and syncope frequently accompany this dysrhythmia. Short bursts of junctional tachycardia are well tolerated in otherwise healthy people. With sustained rapid ventricular rates and retrograde depolarization of the atria, ventricular filling is not as complete during diastole, leading to compromised cardiac output in patients with underlying heart disease. The loss of the atrial kick may cause up to a 30% reduction in cardiac output. Increases in cardiac oxygen requirements may increase myocardial ischemia and the frequency and severity of the patient's chest pain. It can extend the size of MI; cause congestive heart failure, hypotension, and cardiogenic shock; and possibly predispose the patient to ventricular dysrhythmias.

ECG Appearance

The most evident feature of junctional tachycardia is a rapid heart rate with narrow QRS complexes. The rate is 100 to 180 beats per minute. P' waves may precede, be lost in (buried), or follow the QRS complexes. When present, the P' waves will be inverted. The P'R intervals, if present, will be less than 0.12 seconds. Junctional tachycardia is coming from one site, so the rhythm is regular (except if there is an onset or termination as seen with PSVT). The QRS complexes are normal.

14.7 Atrioventricular Nodal Reentrant Tachycardia

Description

In Chapter 11, we pointed out that some people have an abnormal extra anatomical pathway (congenital in nature) within or just next to the AV node. Formed from tissue that acts similarly to the AV node, this circuit involves two pathways: a fast pathway and a slow pathway. These pathways conduct impulses at different rates and recover at different speeds. The fast pathway conducts impulses rapidly but recovers more slowly. The slow pathway conducts impulses more slowly but recovers more quickly.

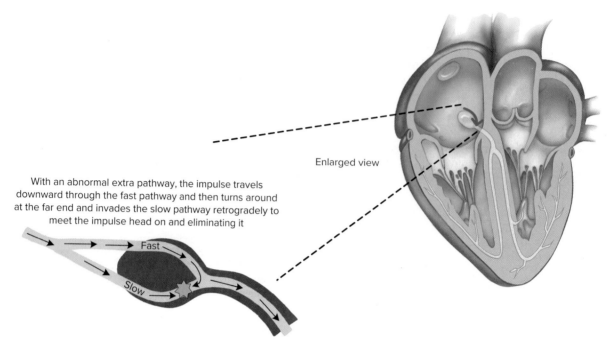

With an abnormal extra pathway, the impulse travels downward through the fast pathway and then turns around at the far end and invades the slow pathway retrogradely to meet the impulse head on and eliminating it

Enlarged view

Figure 14-6
Conduction of impulse through an extra AV pathway in sinus rhythm.

In sinus rhythm, the slow pathway does not transmit impulses as the depolarization wave passes through the fast pathway **anterogradely** (moving downward) and then turns around at the far end and invades the slow pathway **retrogradely** (moving backward), meeting the slow depolarization impulse head on and eliminating it (Figure 14-6).

However, because of its quick recovery period, under the right circumstances, such as the introduction of a PAC, the slow pathway can be stimulated first. Then, at its far end, in addition to stimulating the ventricles, the depolarizing impulse can turn into the fast pathway passing retrogradely. When the impulse reaches the proximal end, in addition to depolarizing the atria retrogradely, it can reenter the slow circuit which has already recovered. The impulse can become trapped within this circuit (of slow and fast pathways), resulting in a continuous circular reentrant pattern that rapidly depolarizes the ventricles and then the atria (Figure 14-7).

This type of supraventricular tachycardia is called *AV nodal reentrant tachycardia* (AVNRT), or *atrioventricular nodal reentrant tachycardia*. AVNRT, the most common regular supraventricular tachycardia, occurs more often in women than men (approximately 75% of cases occurring in females).

ECG Appearance

AVNRT is typically characterized by an abrupt onset and termination. Episodes may last from seconds to minutes to days. Most commonly, because the ventricles are stimulated by the normal route via the AV node, the QRS complexes are narrow. However, the QRS complexes can be broad due to preexisting bundle branch block (discussed in Chapters 8 and 20). The QRS rate is usually 140 to 240 BPM. Because the atria are depolarized retrogradely from the AV node, the P′ wave is often buried in the QRS complex. Sometimes it is seen following the QRS complex.

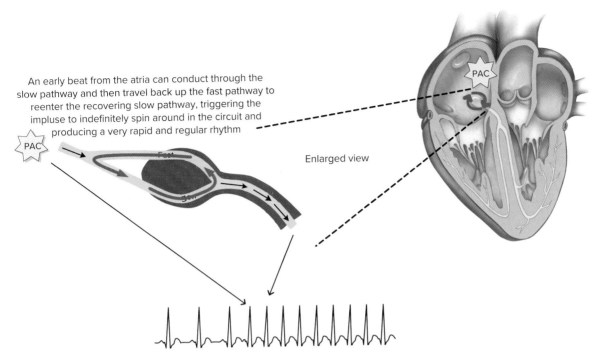

An early beat from the atria can conduct through the slow pathway and then travel back up the fast pathway to reenter the recovering slow pathway, triggering the impluse to indefinitely spin around in the circuit and producing a very rapid and regular rhythm

PAC

PAC

Fast

Slow

Enlarged view

Figure 14-7
AV nodal reentrant tachycardia.

The ST segment depression is common at high heart rates even in the absence of coronary artery disease.

Effects

In the absence of structural heart disease, AVNRT is usually well tolerated. Common symptoms include palpitations, nervousness, anxiety, lightheadedness, pounding sensation in the neck, chest discomfort, and/or dyspnea. Polyuria can occur after termination of the episode (due to the release of atrial natriuretic factor).

AVNRT may cause or worsen heart failure in patients with poor left ventricular function. It may cause angina or MI in patients with coronary artery disease. Syncope may occur in patients with a rapid ventricular rate or prolonged tachycardia due to poor ventricular filling, decreased cardiac output, hypotension, and reduced cerebral perfusion. Syncope may also occur because of transient asystole when the tachycardia terminates, due to tachycardia-induced depression of the sinus node.

14.8 Preexcitation

In Chapters 8 and 11 we talked about accessory pathways and how they can lead to supraventricular tachycardia. Normally, conduction of the impulse from the atria to the ventricles is delayed in the AV node for about 0.1 second. This is long enough for the atria to contract and empty their contents into the ventricles.

Some people have accessory conduction pathways that provide a direct connection between the atria and ventricles, thereby bypassing the AV node and bundle of His.

They allow atrial impulses to depolarize the ventricles earlier than usual, a condition called *preexcitation.* The term *bypass tract* is used when one end of an accessory pathway is attached to normal conductive tissue.

Because these accessory pathways do not possess the rate-slowing properties of the AV node, they can conduct electrical activity at a much higher rate than the AV node. For example, if the atrial rate is 300 beats per minute, the accessory bundle may conduct all the electrical impulses from the atria to the ventricles, causing the ventricles to beat up to 300 times per minute. Extremely fast heart rates can be extremely dangerous and cause hemodynamic instability. In some cases, the combination of an accessory pathway and dysrhythmias can trigger ventricular fibrillation (discussed in Chapter 15).

Accessory pathways and bypass tracts are seen in a small percentage of the general population. They sometimes occur in normal healthy hearts as an isolated finding, or may be in conjunction with mitral valve prolapse or hypertrophic cardiomyopathy, as well as various congenital disorders.

The most common preexcitation syndromes are Wolff-Parkinson-White (WPW) syndrome and Lown-Ganong-Levine (LGL) syndrome. Both of these disorders can be diagnosed by the ECG and can present as supraventricular tachycardia.

Another preexcitation syndrome occurs as a result of the Mahaim fibers. Instead of bypassing the AV node, these fibers originate below the AV node and attach into the ventricular wall, circumventing part or all of the bundle of His, bundle branches, and Purkinje fibers.

Wolff-Parkinson-White Syndrome

In WPW syndrome, the bundle of Kent, an accessory pathway, connects the atria to the ventricles, bypassing the AV node (refer to Figure 8-7). It can be left-sided (connecting the left atrium and left ventricle) or right-sided (connecting the right atrium and right ventricle). ECG features include the following:

- Rate is normal
- Rhythm is regular
- P waves are normal
- QRS complexes are widened with slurring of the initial portion (delta wave)
- PR interval is usually shortened (less than 0.12 seconds)

In contrast to bundle branch block, where the QRS complex is widened due to delayed ventricular activation, in WPW it is widened because of premature activation of the ventricles. The QRS complex associated with WPW is a fusion beat. Although most of the ventricular myocardium is stimulated through the normal conduction pathways, a small area is depolarized early via the bundle of Kent. This produces a characteristic slurred initial upstroke, called the *delta wave,* in the QRS complex. Be sure to scan the entire ECG as a true delta wave may be seen in only a few leads.

Lown-Ganong-Levine Syndrome

In LGL syndrome, the accessory pathway, referred to as the *James fibers,* is within the AV node (Figure 14-8). This accessory pathway bypasses the normal delay

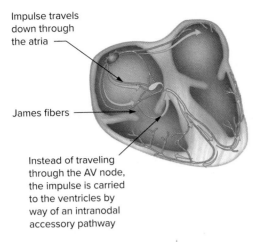

Impulse travels down through the atria —

James fibers —

Instead of traveling through the AV node, the impulse is carried to the ventricles by way of an intranodal accessory pathway

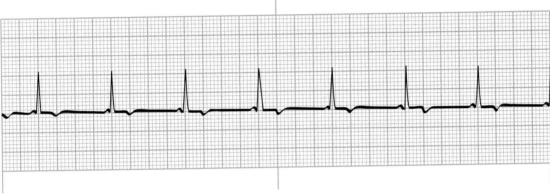

Figure 14-8
In LGL, the impulse travels through an intranodal accessory pathway, referred to as the *James fibers*, bypassing the normal delay within the AV node. This produces a shortening of the PR interval but no widening of the QRS complex.

within the AV node, but unlike WPW there is no small region of ventricular myocardium depolarized independently of the rest of the ventricles. Instead, all ventricular conduction occurs through the usual ventricular conduction pathways. For this reason, there is no delta wave and the QRS complex is not widened. The only indication of LGL on the ECG is shortening of the PR interval as a result of the accessory pathway bypassing the delay within the AV node.

Instead of traveling through the AV node, the impulse is carried to the ventricles by way of an intranodal accessory pathway; the James fibers. The impulse travels down through the atria.

ECG criteria for LGL include the following:

- The PR interval is less than 0.12 seconds
- The QRS complex is not widened
- There is no delta wave

Unless tachycardia is present, WPW and LGL are usually of no clinical significance. However, preexcitation (specifically WPW) can predispose the patient to various tachydysrhythmias. The most common tachydysrhythmia is atrioventricular reentrant tachycardia (AVRT), followed by atrial fibrillation and atrial flutter.

Atrial fibrillation and atrial flutter seen with WPW are extremely dangerous. As mentioned earlier, very rapid ventricular rates can result from conduction of the atrial impulses directly into the ventricles.

14.9 Atrioventricular Reentrant Tachycardia

Description

AVRT, also known as *circus movement tachycardia* (CMT), results from a reentry circuit that includes the AV node and an accessory pathway from the atria to the ventricle such as the bundle of Kent. This reentry circuit is physically much larger than the one associated with AVNRT.

Appearance

Reentry through an accessory pathway can take one of two directions, orthodromic conduction or antedromic conduction.

Orthodromic AVRT

Most often the AV node serves as the forward portion of the reentry circuit, and the accessory pathway serves as the retrograde portion. What this means is that atrial impulses are conducted down through the AV node and retrogradely reenter the atrium via the accessory pathway (Figure 14.9A). Because of the normal conduction

a. Impulse travels antegradely through AV node but then retrogradely reenters the atria through accessory pathway. It continues to spin around in the circuit producing a very rapid and regular rhythm

b. Impulse travels antegradely through accessory pathway but then retrogradely reenters the atria through AV node. It continues to spin around in the circuit producing a very rapid and regular rhythm

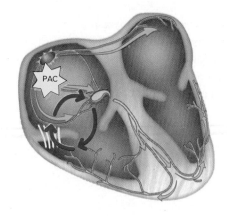

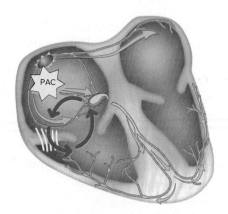

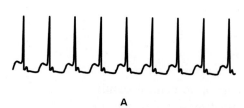

A

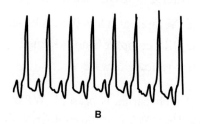

B

Figure 14-9

Atrioventricular reentrant tachycardia: (A) Orthodromic (B) Antedromic.

to the ventricles, regular, narrow QRS complexes are seen. Further, a P′ wave may follow each QRS complex due to retrograde conduction.

Antedromic AVRT

When a reentry circuit circulates in the opposite direction, atrial impulses are conducted down through the accessory pathway and reenter the atrium retrogradely via the AV node (Figure 14.9B). Because the accessory pathway initiates conduction in the ventricles outside of the bundle of His, the QRS complex in antidromic AVRT is often wider than usual, with a delta wave.

Effects

Common signs and symptoms associated with a rapid ventricular rate include palpitations, lightheadedness, shortness of breath, anxiety, weakness, dizziness, chest discomfort, and signs of shock.

14.10 Treatment of Supraventricular Tachycardia

As they are the same, the treatment for the three forms of supraventricular tachycardia (SVT) discussed in this chapter (junctional tachycardia, AVNRT, and AVRT) are presented here.

When assessing patients with tachycardia, attempt to determine whether the tachycardia is the primary cause of the presenting symptoms or secondary to an underlying condition that is causing both the presenting symptoms and the faster heart rate.

As with all patients, maintain patent airway and assist breathing as necessary. If oxygenation is inadequate or the patient shows signs of increased breathing effort, provide supplementary oxygen. Attach an ECG monitor to the patient, evaluate blood pressure and oximetry, and establish IV access. If available, obtain a 12-lead ECG to better define the rhythm, but do not delay immediate cardioversion if the patient is unstable.

If the patient demonstrates rate-related decreased cardiac output with signs and symptoms such as acute altered mental status, ischemic chest discomfort, acute heart failure, hypotension, or other signs of shock suspected to be due to the tachy-dysrhythmia, proceed to immediate synchronized cardioversion. If the patient is conscious, consider establishing IV access before cardioversion and administering sedation. Avoid any delay in cardioversion if the patient is extremely unstable.

If he or she is not hypotensive, the patient with a regular narrow-complex SVT may be treated with adenosine while preparations are made for synchronized cardioversion. Stable patients may await expert consultation because treatment has the potential for harm.

Vagal maneuvers and adenosine are the preferred initial therapeutic choices for the termination of stable supraventricular tachycardia. If stable SVT does not respond to vagal maneuvers, adenosine, 6 mg, rapid IV push (followed by a 20 mL saline flush) should be administered. A second, higher dose of 12 mg can be administered if the rhythm does not convert within 1 to 2 minutes. Because of the possibility of initiating atrial fibrillation with rapid ventricular rates in a patient with WPW, a defibrillator should be available when adenosine is administered to any patient in whom WPW is suspected. As with vagal maneuvers, the effect of adenosine on other SVTs (such as atrial fibrillation or flutter) is to transiently slow the ventricular rate (which may be useful diagnostically) but not result in their termination or lasting rate control.

If adenosine or vagal maneuvers fail to convert SVT, if it recurs after such treatment, or if these treatments reveal a different form of SVT (such as atrial fibrillation or flutter), the use of longer-acting AV nodal blocking agents, such as the nondihydropyridine calcium channel blockers (verapamil and diltiazem) or beta-blockers, may be considered (see Figure 14-9).

Frequent attacks may require radiofrequency ablation, in which the abnormally conducting tissue in the heart is destroyed. In the clinical setting, patients who have had a recent MI or heart surgery may need a temporary pacemaker to reset the heart's rhythm. A child with a permanent junctional tachycardia may be treated with ablation therapy, followed by permanent pacemaker insertion.

In patients with stable wide-QRS complex tachycardia where it is unclear as to whether it is SVT or ventricular tachycardia (discussed in Chapter 15), a reasonable approach is to try to identify which it is and treat on the basis of the algorithm for that rhythm. If the etiology of the rhythm cannot be determined, the rate is regular, and the QRS is monomorphic, recent evidence suggests that IV adenosine is relatively safe for both treatment and diagnosis.

Practice Makes Perfect

©rivetti/Getty Images

For each of the 53 ECG tracings on the following pages, practice the Nine-Step Process for analyzing ECGs. To achieve the greatest learning, you should practice assessing and interpreting the ECGs immediately after reading this chapter. Below are questions you should consider as you assess each tracing. Your answers can be written into the area below each ECG marked "ECG Findings." Your findings can then be compared to the answers provided in Appendix A. All tracings are six seconds in length.

1. Determine the heart rate. Is it slow? Normal? Fast? What is the ventricular rate? What is the atrial rate?

2. Determine if the rhythm is regular or irregular. If it is irregular, what type of irregularity is it? Occasional or Irregular? Slight? Sudden acceleration or slowing in heart rate? Total? Patterned? Does it have a variable conduction ratio?

3. Determine if P waves are present. If so, how do they appear? Do they have normal height and duration? Are they tall? Notched? Wide? Biphasic? Of differing morphology? Inverted? One for each QRS complex? More than one preceding some or all the QRS complexes? Do they have a sawtooth appearance? An indiscernible chaotic baseline?

4. Determine if QRS complexes are present. If so, how do they appear? Narrow with proper amplitude? Tall? Low amplitude? Delta wave? Notched? Wide? Bizarre? With chaotic waveforms?

5. Determine the presence of PR intervals. If present, how do they appear? Constant? Of normal duration? Shortened? Lengthened? Progressively longer? Varying?

6. Evaluate the ST segments. Do they have normal duration and position? Are they elevated? (If so, are they flat, concave, convex, arched?) Depressed? (If so, are they normal, flat, downsloping or upsloping?)

7. Determine if T waves are present. If so, how do they appear? Of normal height and duration? Tall? Wide? Notched? Inverted?

8. Determine the presence of QT intervals. If present, what is their duration? Normal? Shortened? Prolonged?

9. Determine if U waves are present. If present, how do they appear? Of normal height and duration? Inverted?

10. Identify the rhythm or dysrhythmia.

1.

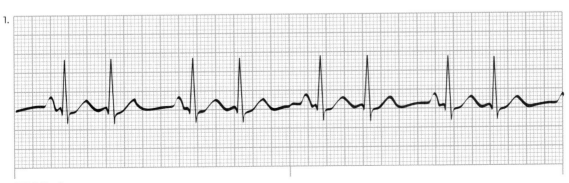

ECG Findings:

Chapter 14 ▲ Junctional Dysrhythmias **375**

2.

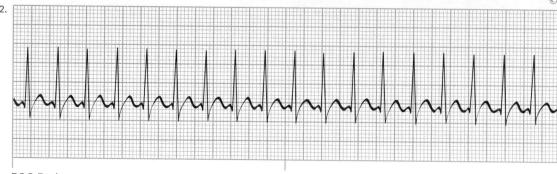

ECG Findings:

3.

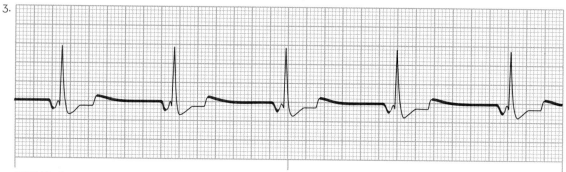

ECG Findings:

4.

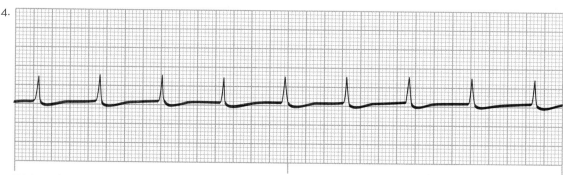

ECG Findings:

5.

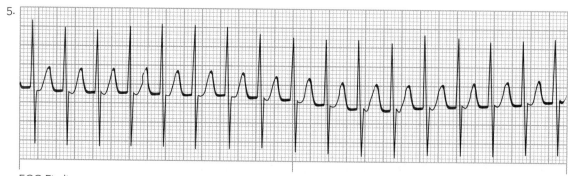

ECG Findings:

6.

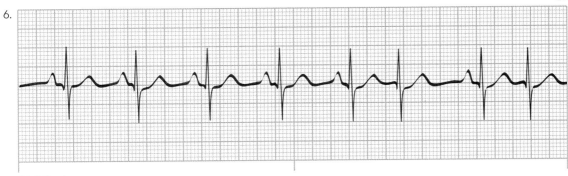

ECG Findings:

7.

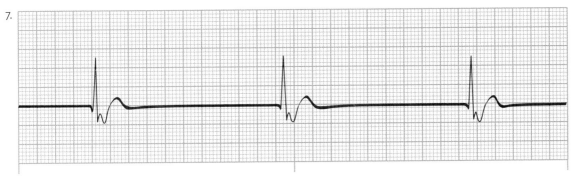

ECG Findings:

8.

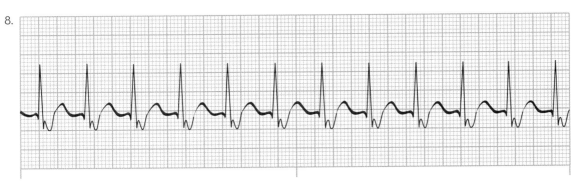

ECG Findings:

9.

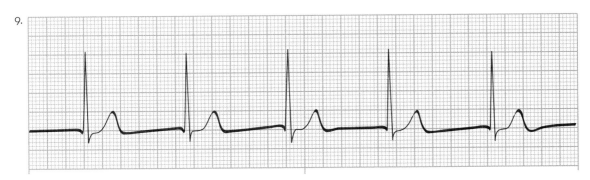

ECG Findings:

10.

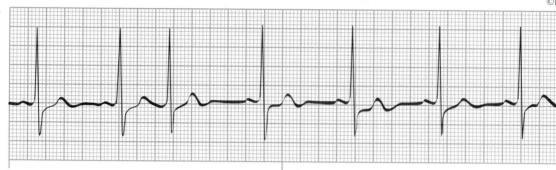

ECG Findings:

11.

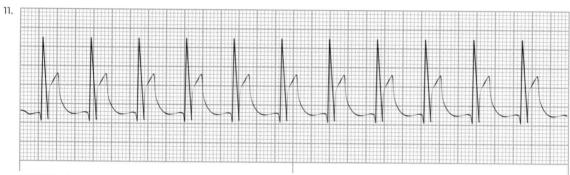

ECG Findings:

12.

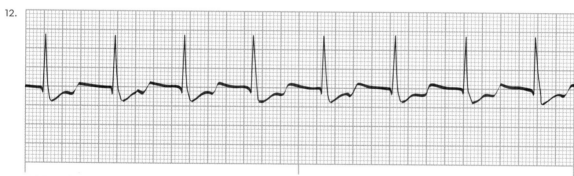

ECG Findings:

13.

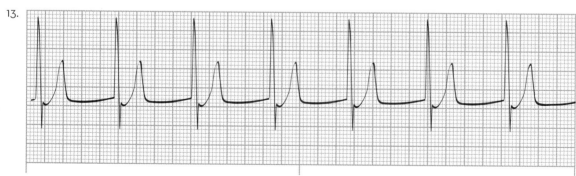

ECG Findings:

14.

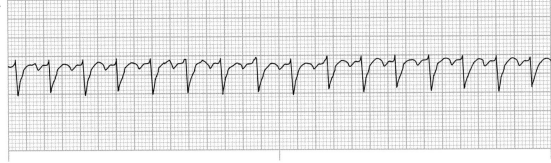

ECG Findings:

15.

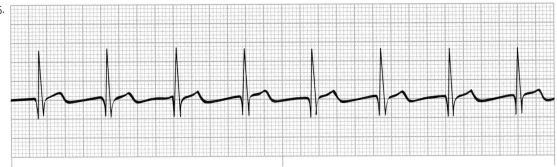

ECG Findings:

16.

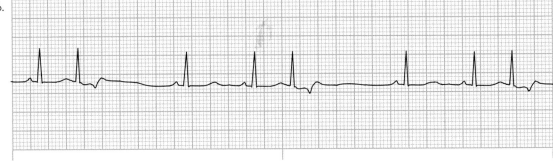

ECG Findings:

17.

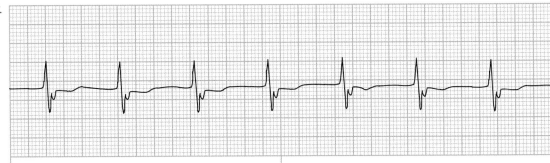

ECG Findings:

18.

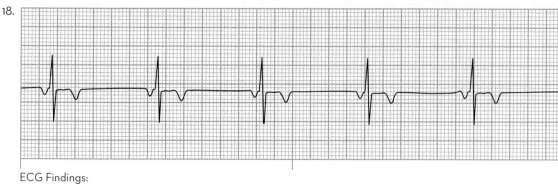

ECG Findings:

19.

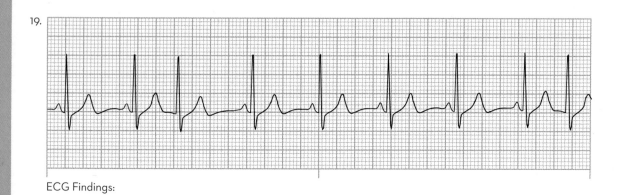

ECG Findings:

20.

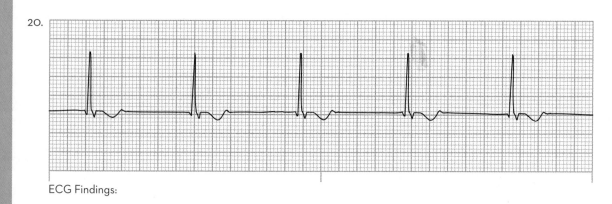

ECG Findings:

21.

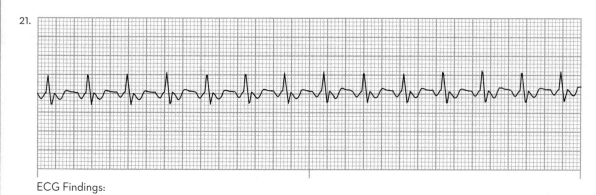

ECG Findings:

22.

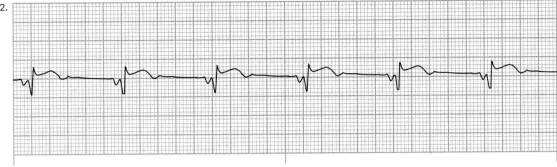

ECG Findings:

23.

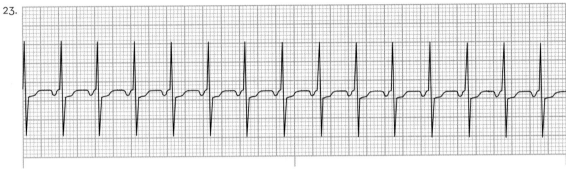

ECG Findings:

24.

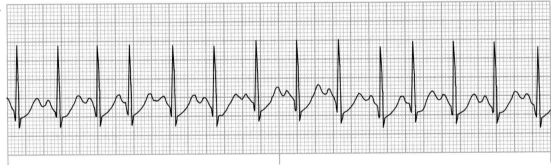

ECG Findings:

25.

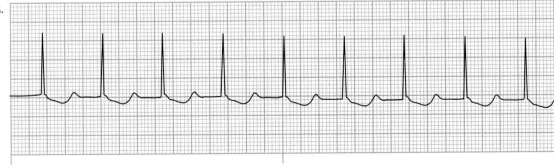

ECG Findings:

26.

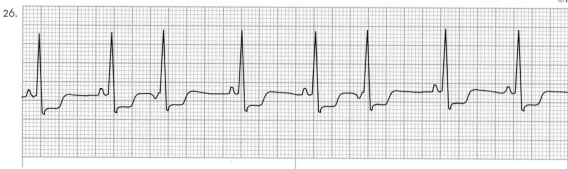

ECG Findings:

27.

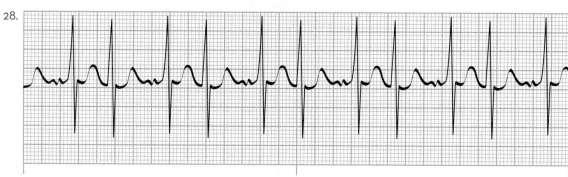

ECG Findings:

28.

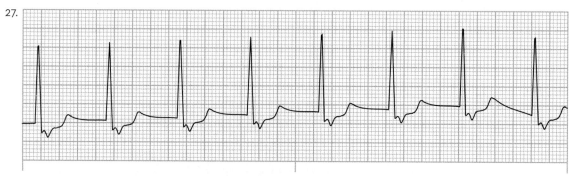

ECG Findings:

29.

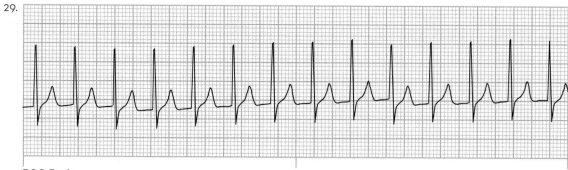

ECG Findings:

30.

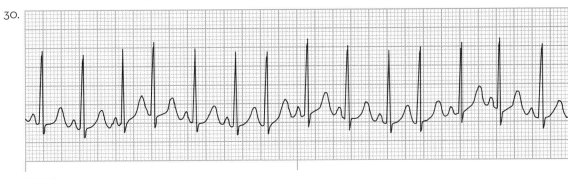

ECG Findings:

31.

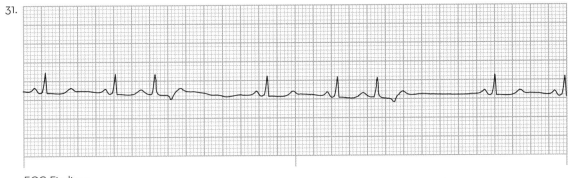

ECG Findings:

32.

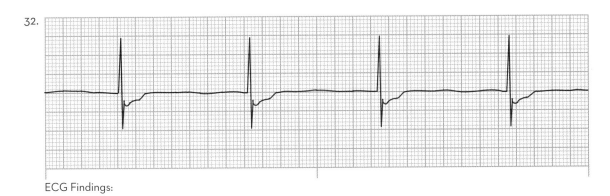

ECG Findings:

33.

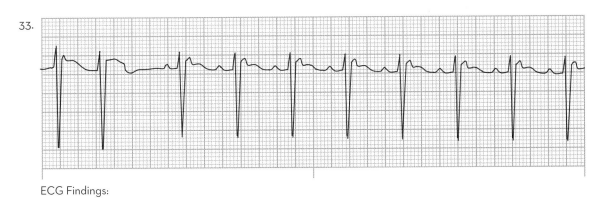

ECG Findings:

34.

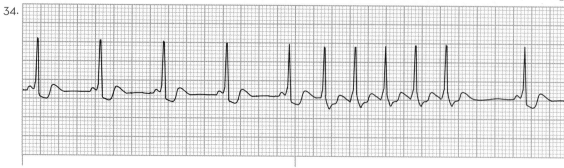

ECG Findings:

35.

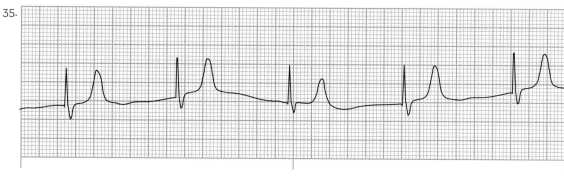

ECG Findings:

36.

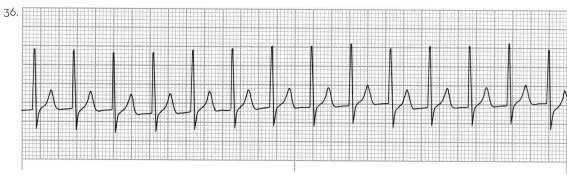

ECG Findings:

37.

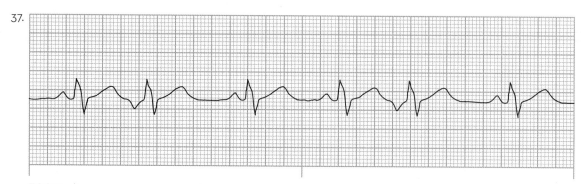

ECG Findings:

38.

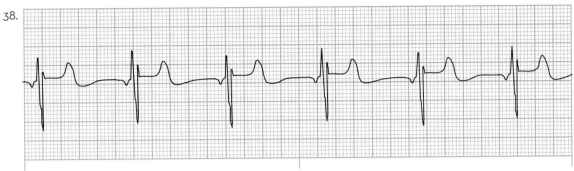

ECG Findings:

39.

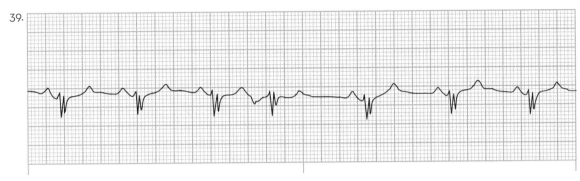

ECG Findings:

40.

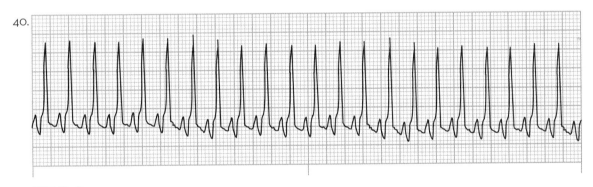

ECG Findings:

41.

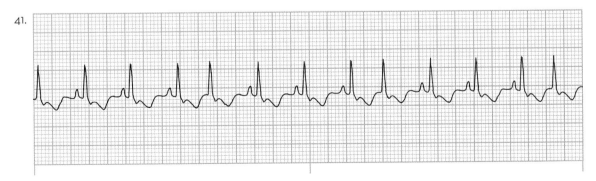

ECG Findings:

42.

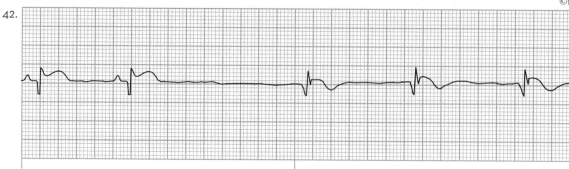

ECG Findings:

43.

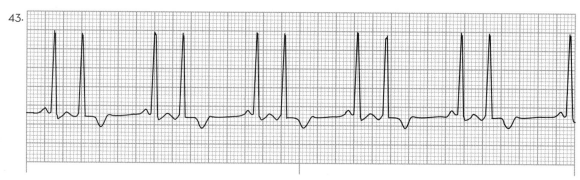

ECG Findings:

44.

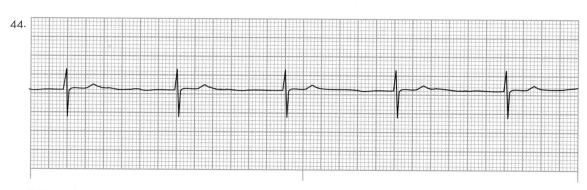

ECG Findings:

45.

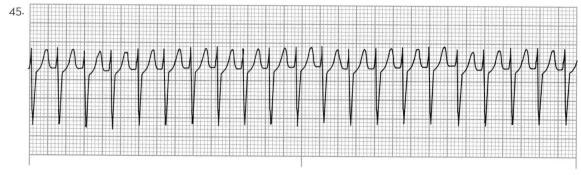

ECG Findings:

46.

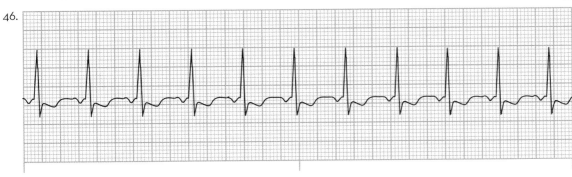

ECG Findings:

47.

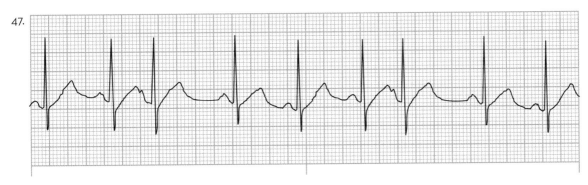

ECG Findings:

48.

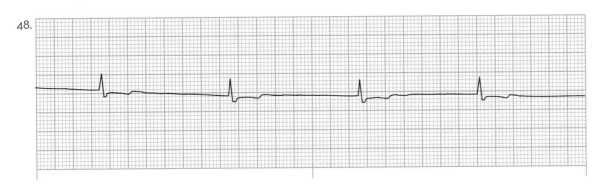

ECG Findings:

49.

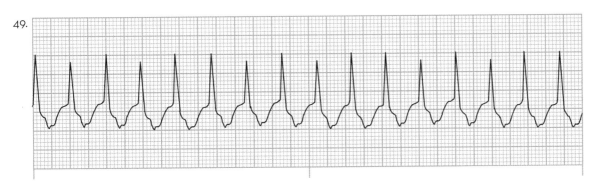

ECG Findings:

50.

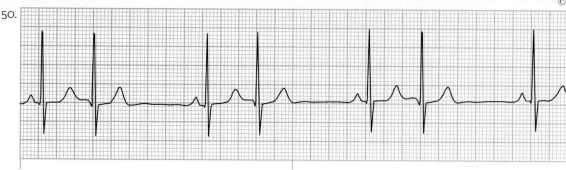

ECG Findings:

51.

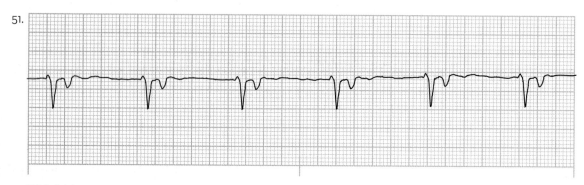

ECG Findings:

52.

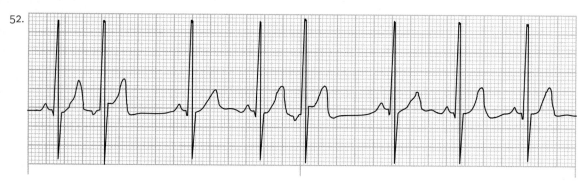

ECG Findings:

53.

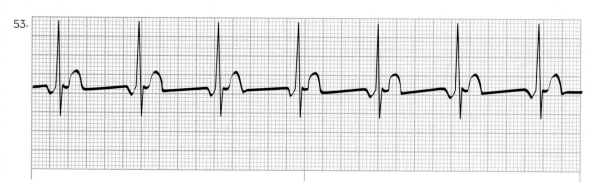

ECG Findings:

Key Points

LO 14.1	• Rhythms that originate in the AV junction, the area around the AV node and the bundle of His, are referred to as *junctional rhythms*.
LO 14.2	• Impulses originating in the AV junction travel upward and cause backward or retrograde depolarization of the atria, resulting in inverted P′ waves in lead II with a short P′R interval, absent P′ waves (as they are buried by the QRS complex), or retrograde P′ waves.
	• With junctional dysrhythmias, the QRS complexes are usually normal (unless there is an intraventricular conduction defect, aberrancy, or preexcitation)
LO 14.3	• A PJC is a single early electrical impulse that arises from the AV junction.
LO 14.4	• Junctional escape rhythm arises from the AV junction at a rate of 40 to 60 beats per minute.
LO 14.5	• Accelerated junctional rhythm arises from the AV junction at a rate of 60 to 100 beats per minute.
LO 14.6	• Junctional tachycardia is a fast, ectopic rhythm that arises from the bundle of His at a rate of between 100 and 180 beats per minute.
LO 14.7	• In AVNRT, fast and slow pathways are located within the right atrium near or within the AV node and exhibit electrophysiologic properties like AV nodal tissue. These pathways can allow for development of SVT.
LO 14.8	• Preexcitation syndromes occur when accessory conduction pathways exist between the atria and ventricles that bypass the AV node and bundle of His and allow the atria to depolarize the ventricles earlier than usual.
	• Preexcitation is diagnosed by looking for a short PR interval.
	• Criteria for WPW include a PR interval less than 0.12 seconds, wide QRS complexes due to a delta wave (seen in some leads). Patients with WPW are vulnerable to PSVT.
	• In LGL, there is an intranodal accessory pathway that bypasses the normal delay within the AV node. Criteria for LGL include a PR interval less than 0.12 seconds and a normal QRS complex.
LO 14.9	• AVRT, also known as *circus movement tachycardia* (CMT), results from a reentry circuit that includes the AV node and an accessory pathway from the atria to the ventricle such as the bundle of Kent. This reentry circuit is physically much larger than the one associated with AVNRT.
LO 14.10	• Vagal maneuvers and adenosine are the preferred initial treatments of stable supraventricular tachycardia. If stable SVT fails to respond to vagal maneuvers, 6 mg of IV adenosine should be delivered. A second, 12 mg dose can be administered if the rhythm does not convert within 1 to 2 minutes.
	• If the patient has decreased cardiac output because of a tachydysrhythmia with signs and symptoms such as acute altered mental status, ischemic chest discomfort, acute heart failure, hypotension, or other signs of shock suspected to be due to the tachydysrhythmia, immediately deliver synchronized cardioversion. If the patient is conscious, consider establishing an IV line before cardioversion and administering sedation. Avoid delays in delivering synchronized cardioversion if the patient is extremely unstable.

Assess Your Understanding

The following questions give you a chance to assess your understanding of the material discussed in this chapter. The answers can be found in Appendix A.

1. The normal PR interval is _____ seconds.　　　(LO 14.2)
   a. 0.06 to 0.10
   b. 0.10 to 0.20
   c. 0.12 to 0.20
   d. 0.20 to 0.24

2. PR intervals of less than 0.12 seconds in duration indica a(n)　　(LO 14.2)
   a. intermittent blocking of the impulse as it passes through the AV node.
   b. delay in conduction of the impulse between the SA node and the ventricles.
   c. increased rate of impulse discharge from the SA node.
   d. pacemaker site, either in or close to the AV junction.

3. Your patient shows normal sinus rhythm with occasional ectopic beats on the ECG. The ectopic beats appear with absent P waves along with a normal QRS complex. These are premature _____ complexes.　　(LO 14.3)
   a. atrial
   b. bundle
   c. junctional
   d. ventricular

4. Which of the following originates from the AV junction?　　(LO 14.1)
   a. Atrial tachycardia
   b. Junctional escape rhythm
   c. Premature atrial complex
   d. Sinus bradycardia

5. Match the following junctional rhythms with the appropriate heart rate.
   (LO 14.4, 14.5, 14.6)

Dysrhythmias	Beats per minute
a. Junctional tachycardia	_____ 40 to 60
b. Junctional escape rhythm	_____ 100 to 180
c. Accelerated junctional rhythm	_____ 60 to 100

6. P'R intervals seen with junctional rhythms will appear　　(LO 14.2)
   a. longer than 0.20 seconds in duration.
   b. shorter than 0.12 seconds in duration.
   c. within normal ranges.
   d. varied.

7. Which rhythm may be preceded by an inverted P′ wave?　　　(LO 14.4)
    a.　Atrial flutter
    b.　Junctional escape rhythm
    c.　Atrial tachycardia
    d.　Idioventricular rhythm

8. Match the following types of PJCs with the correct description.　(LO 14.3)

Types		Descriptions
a.	Bigeminal	_____ PJCs that occur every third beat
b.	Quadrigeminal	_____ PJCs that occur every other beat
c.	Trigeminal	_____ PJCs that occur every fourth beat

9. In junctional escape rhythm, the pacemaker site is in the　　(LO 14.4)
    a.　SA node.
    b.　atrial and internodal conduction pathways.
    c.　AV junction.
    d.　bundle of His.

10. Accelerated junctional rhythm　　　　　　　　　　　　　(LO 14.5)
    a.　is regularly irregular.
    b.　has upright P′ waves in all leads.
    c.　has normal P′R intervals.
    d.　has normal QRS complexes.

11. With preexcitation syndromes,　　　　　　　　　　　　(LO 14.8)
    a.　conduction of the impulse from the atria to the ventricles is delayed in the AV node.
    b.　the heart is stimulated to beat faster.
    c.　wide, bizarre QRS complexes are generated.
    d.　impulses are conducted through accessory conduction pathways between the atria and ventricles

12. In WPW,　　　　　　　　　　　　　　　　　　　　　(LO 14.8)
    a.　the T wave is inverted.
    b.　impulses bypass the AV node by traveling from the atria to the ventricles via the bundle of Kent.
    c.　a small area of myocardium that is depolarized early produces a characteristic slurred initial downstroke in the QRS complex.
    d.　the PR interval is prolonged.

Referring to the scenario at the beginning of this chapter, answer the following questions.

13. PSVT stands for　　　　　　　　　　　　　　　　　　(LO 13.7)
    a.　premature ventricular tachycardia.
    b.　premature atrial tachycardia.

    c. paroxysmal supraventricular tachycardia.

    d. paradoxical supraventricular tachycardia.

14. The P waves in PSVT are not discernable because   (LO 13.7)

    a. they are not there.

    b. they are hidden in the T wave.

    c. f waves are present instead.

    d. F waves are present instead.

15. PSVT is caused by an electrical impulse circulating in the   (LO 13.7)

    a. Atria

    b. AV node

    c. Bundle of His

    d. Purkinje fibers

15 Ventricular Dysrhythmias

Chapter Outline

Learning Outcomes

Case History

It's two P.M. when pagers go off, alerting your EMS crew of a male who collapsed outside his home.

Upon arrival you find the patient lying next to the sidewalk. You immediately determine an absence of a pulse and initiate chest compressions. Your partner verifies he can feel a carotid pulse with compressions. She then begins ventilating the patient with a bag-valve-mask (BVM) and 100% oxygen. Shortly afterward a fire department first responder unit arrives on scene to help.

After the defibrillator has been turned on and the pads applied to the patient's chest, you briefly stop CPR and examine the monitor. You see a chaotic and random waveforms with no discernable P waves or QRS complexes. "It's ventricular fibrillation," you say to your team. "Let's charge up and shock." You set the power to 360 joules and push the charge button. The machine makes a characteristic whine as the capacitors prepare the energy for delivery. "I'm clear. You're clear. Everyone is clear," you say while visually confirming that no one is touching the patient. You then push the shock button and the energy is delivered. You direct CPR be continued and have a team member intubate the patient and another place an IV. After delivering two minutes of CPR, you reassess the rhythm to see that the ventricular fibrillation is still present. You then defibrillate the patient again, after which the rhythm appears to be a flat line. CPR is continued as you administer epinephrine per protocol. Despite your efforts, the patient is pronounced dead after arrival at the hospital.

15.1 Dysrhythmias Originating in the Ventricles

Ventricular dysrhythmias originate in the ventricles below the bundle of His. This results in the heart being depolarized and contracting much differently than normal. Although some ventricular dysrhythmias may be benign, many ventricular dysrhythmias are potentially life-threatening because the ventricles are ultimately responsible for cardiac output and, if they are not pumping effectively, cardiac output is diminished or even ceases.

Ventricular dysrhythmias typically result when the atria, AV junction, or both are unable to initiate an electrical impulse or when there is enhanced automaticity of the ventricular myocardium. Myocardial ischemia is a common cause of enhanced automaticity.

Quick recognition and treatment of ventricular dysrhythmias improves the chance of successful resuscitation. Many respond well to medications, although some medications used to treat ventricular dysrhythmias can actually cause them in certain circumstances. Electrical therapy is the treatment choice for some ventricular dysrhythmias, but in others the condition is lethal despite aggressive treatment.

Ventricular dysrhythmias include (Figure 15-1):

- Premature ventricular complex (PVC)
- Idioventricular rhythm
- Accelerated idioventricular rhythm
- VT
- VF
- Asystole

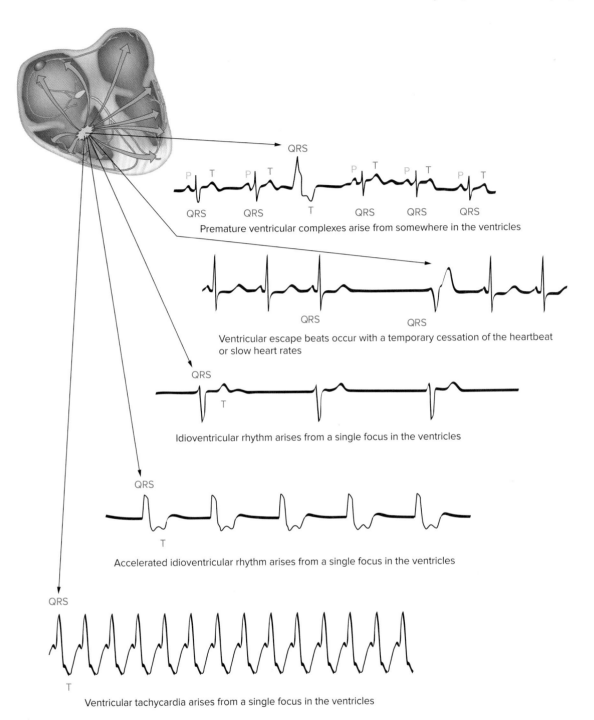

Premature ventricular complexes arise from somewhere in the ventricles

Ventricular escape beats occur with a temporary cessation of the heartbeat or slow heart rates

Idioventricular rhythm arises from a single focus in the ventricles

Accelerated idioventricular rhythm arises from a single focus in the ventricles

Ventricular tachycardia arises from a single focus in the ventricles

Figure 15-1
Ventricular dysrhythmias originate from the ventricles.

15.2 Key Features of Ventricular Dysrhythmias

P Wave Appearance

With the impulse originating below the bundle of His, the atria typically do not depolarize. As a result, the P waves are absent. If the impulse does travel backward to depolarize the atria, the P wave is buried in the abnormal QRS complex and is still not visible. If P waves are seen, one of two things is occurring; either the P waves are disassociated from the QRS complexes or the impulse originated from above the ventricles and there is aberrant or delayed conduction through the ventricles.

QRS Complex Appearance

Because ventricular dysrhythmias originate in the ventricular tissue and, instead of rapidly moving through the conduction pathways, they travel slowly from cell to cell, producing wide (greater than 0.12 seconds in duration), bizarre-looking QRS complexes. Further, the T waves take an opposite direction to the R wave (see Figure 7-15 in Chapter 7). The ST segment will either be elevated (if the R wave takes a negative deflection) or depressed (if the R wave takes a positive deflection).

Effects

With the loss in atrial kick (as the atria do not depolarize) and decreased stroke volume, ventricular dysrhythmias may produce little or no cardiac output. Some ventricular dysrhythmias fail to provide adequate perfusion because the rate is so slow it does not meet the metabolic needs of the body. In others, such as with ventricular tachycardia (VT), there is insufficient time to allow for ventricular filling. Although some ventricular dysrhythmias may be well tolerated by the patient, many cause symptoms of decreased cardiac output or, even worse, cardiac standstill.

15.3 Premature Ventricular Complexes

Description

Premature ventricular complexes (PVCs) are early ectopic beats that interrupt the normal rhythm and originate from an irritable focus in the ventricular conduction system or muscle tissue (Figure 15-2). These beats occur earlier than the next expected sinus beat.

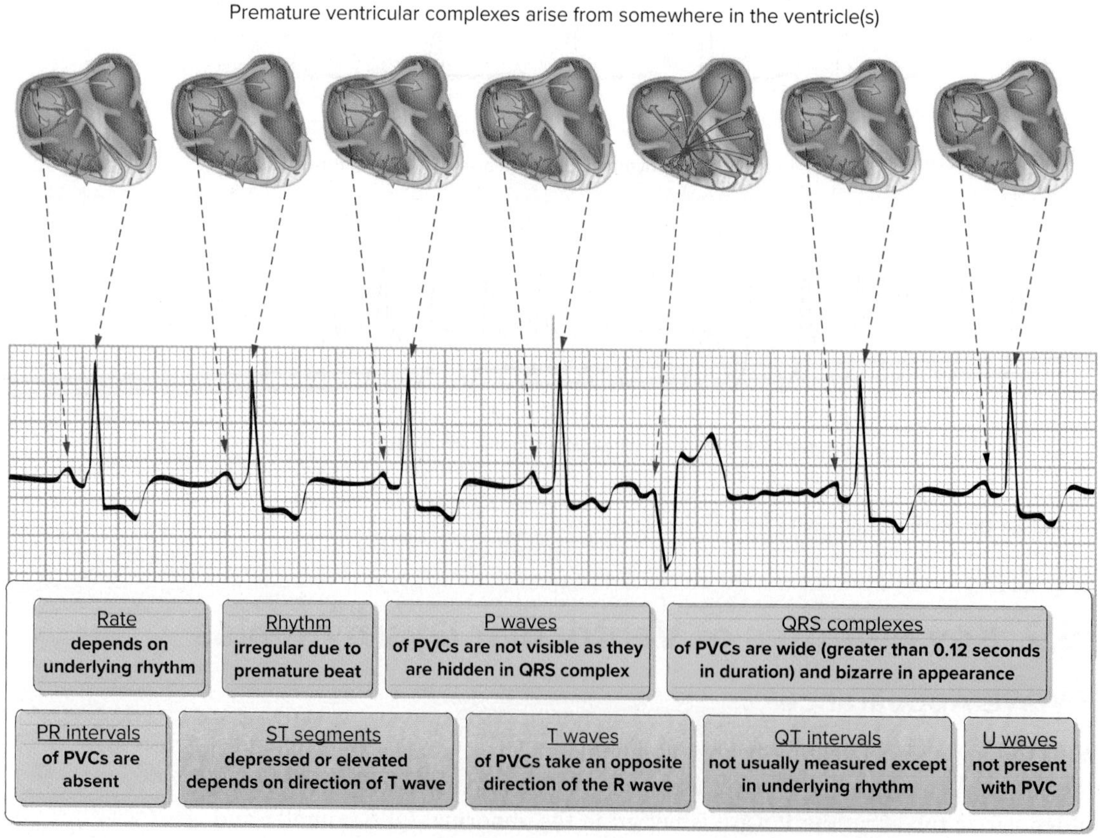

Premature ventricular complexes arise from somewhere in the ventricle(s)

Rate	Rhythm	P waves	QRS complexes
depends on underlying rhythm	irregular due to premature beat	of PVCs are not visible as they are hidden in QRS complex	of PVCs are wide (greater than 0.12 seconds in duration) and bizarre in appearance

PR intervals	ST segments	T waves	QT intervals	U waves
of PVCs are absent	depressed or elevated depends on direction of T wave	of PVCs take an opposite direction of the R wave	not usually measured except in underlying rhythm	not present with PVC

Figure 15-2
Characteristics of PVCs.

Causes

PVCs are caused by enhanced automaticity or reentry, brought about by disruption of the normal electrolyte shifts during cell depolarization and repolarization and other causes listed in the following box.

Causes of Premature Ventricular Complex	
Cause	**Examples**
Cardiac diseases	Myocardial ischemia and MI, congestive heart failure, irritation of ventricles by pacemaker electrodes or PA catheter, enlargement of the ventricular chambers, mitral valve prolapse, myocarditis
Use of certain drugs	Particularly amphetamine intoxication; cocaine, digoxin, quinidine, phenothiazines, tricyclic antidepressants; sympathomimetic drugs such as aminophylline, dopamine, epinephrine, isoproterenol; thyroid supplements
Autonomic nervous system stimulation	Increased sympathetic stimulation, exercise, emotional stress
Noncardiac disorders	Use of stimulants such as alcohol, caffeine, tobacco; electrolyte imbalance such as hypokalemia, hyperkalemia, hypomagnesemia, and hypocalcemia; hypoxia; metabolic acidosis

PVCs can occur for no apparent reason in individuals who have healthy hearts. They are often of no significance. The incidence of PVCs increases with age and can occur during exercise or at rest.

Effects

PVCs can be significant for two reasons. First, they can precipitate more serious dysrhythmias such as VT or VF. The danger is greater in patients experiencing myocardial ischemia and/or MI. Second, PVCs can result in decreased cardiac output due to reduced diastolic filling time and a loss of atrial kick. This leads to a diminished or nonpalpable pulse. When no pulse is felt during a PVC, it is called a **nonperfusing PVC.** Patients frequently experience the sensation of "skipped beats." Whether PVCs result in decreased cardiac output hinges on how long the abnormal rhythm lasts. Given this information, be sure to count and document how many PVCs are seen each minute.

With PVCs, it is important to check for a pulse to determine if they are perfusing PVCs. The absence of a pulse may require treatment if the underlying heart rate is slow or there are many PVCs. For example, with a rhythm in which there are PVCs every other beat (bigeminal) and the heart rate is 60 beats per minute, cardiac output will likely be adequate if the PVCs are producing a pulse. Conversely, cardiac output will likely be inadequate if the PVCs are not producing a pulse and the actual heart rate is 30 beats per minute.

Be sure to pay special attention to these rhythm disturbances in persons with acute myocardial ischemia. PVCs that occur with myocardial ischemia may indicate the presence of enhanced automaticity, a reentry mechanism, or both and may trigger lethal ventricular dysrhythmias. If the PVCs are frequent and occur early enough in the cardiac cycle, cardiac output is compromised. PVCs that serve as warning signs of the potential development of serious ventricular dysrhythmias in patients with myocardial ischemia include the following:

- Frequent PVCs
- Multifocal PVCs
- Early PVCs (R-on-T)
- Patterns of grouped PVCs

ECG Appearance

Like other early ectopic beats, the first thing you see with PVCs is an irregular rhythm. When you look at the area(s) of irregularity, the PVCs appear earlier in the cycle than the normal set of complexes would be expected to occur. Another characteristic that stands out is a wide (greater than 0.12 seconds), bizarre-looking QRS complex with a T wave that takes an opposite direction to the R wave of the PVC.

The QRS complexes of the PVCs differ from QRS complexes of the underlying rhythm. P waves and PR intervals are typically present with the underlying rhythm but not the premature beats. A pause, called a **compensatory pause,** typically follows the PVC.

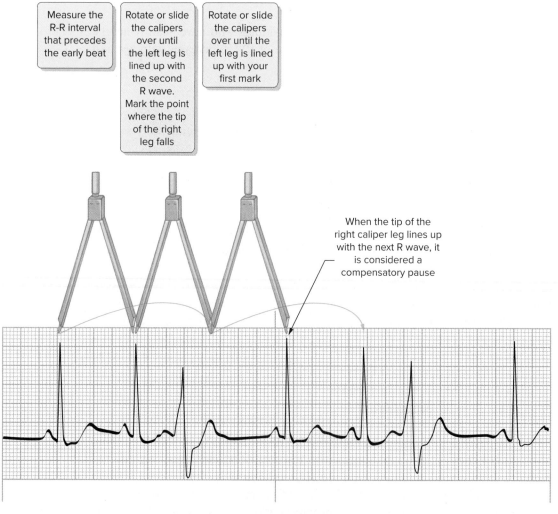

Measure the R-R interval that precedes the early beat

Rotate or slide the calipers over until the left leg is lined up with the second R wave. Mark the point where the tip of the right leg falls

Rotate or slide the calipers over until the left leg is lined up with your first mark

When the tip of the right caliper leg lines up with the next R wave, it is considered a compensatory pause

Figure 15-3
A compensatory pause is seen with premature ventricular complexes.

As discussed in previous chapters, the pause that follows a premature beat is called a **compensatory pause** if the normal beat following the premature complex occurs when it was expected (Figure 15-3). In other words, the period between the complex before and after the premature beat is the same as two normal R-R intervals. The pause occurs because the ventricle is refractory and cannot respond to the next regularly scheduled P wave from the sinus node. To determine if there is a compensatory pause, we measure from the R wave of the beat that precedes the PVC to the R wave of the first beat that follows it. If there are two full R-R intervals between those R waves (with the PVC between), it is considered a compensatory pause.

PVCs may occur as a single beat, in clusters of two or more, or in repeating patterns. Sometimes, PVCs originate from only one location in the ventricle. These beats look the same and are called *uniform* (also referred to as *unifocal*) PVCs. Other times, PVCs arise from different sites in the ventricles. These beats look different from each other (Figure 15-4) and are called *multiformed (multifocal)* PVCs. Multiform PVCs indicate an irritable myocardium and should prompt you to take corrective action.

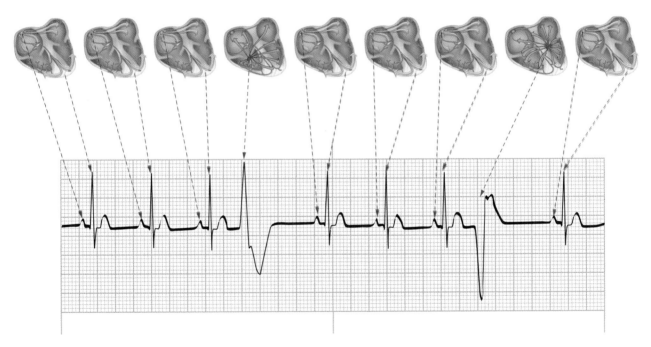

Figure 15-4
Multiform PVCs arise from different sites within the ventricles and look different from each other.

PVCs may occur as a single, early beat or more frequently (Figure 15-5). PVCs intermingled between normal beats are named depending on their frequency. Bigeminal PVCs are said to be present when every other beat is a PVC, regardless if unifocal or multifocal. If every third beat is a PVC, the condition is called *trigeminal PVCs*, or *ventricular trigeminy*. Similarly, a PVC every fourth beat is called *ventricular quadrigeminy*.

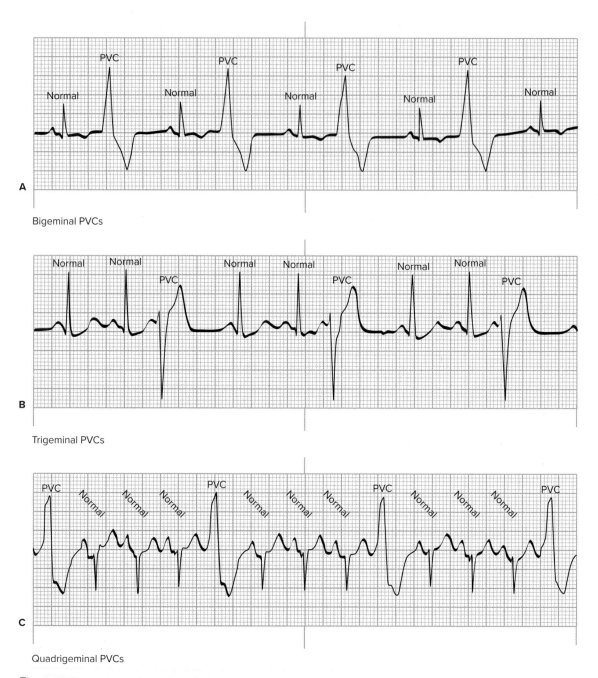

Figure 15-5
Premature ventricular complexes occurring in a (A) bigeminal, (B) trigeminal, and (C) quadrigeminal pattern.

Regular PVCs at greater intervals than every fourth beat have no special name and are referred to as *frequent PVCs*.

PVCs may also occur one after the other. Two PVCs in a row are called a *couplet,* or *pair.* A couplet indicates the ventricles are extremely irritable. Three or more PVCs in a row at a ventricular rate of at least 100 beats per minute constitute an abnormal rhythm known as *ventricular tachycardia.* It may be called a **salvo, run,** or **burst** of VT (Figure 15-6).

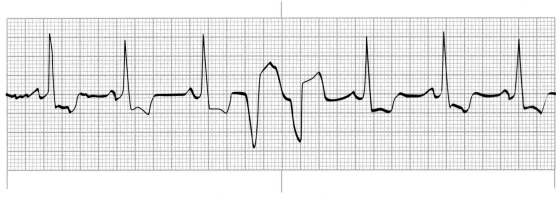

Two PVCs in succession are called a couplet

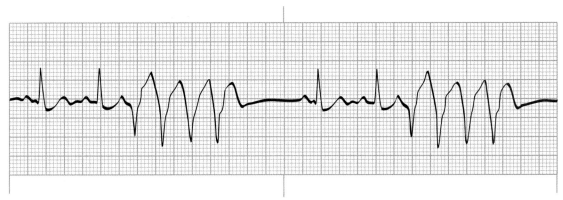

Three or more PVCs in a row are called a run of ventricular tachycardia or a run of PVCs

Figure 15-6
Couplet of PVCs and run of VT.

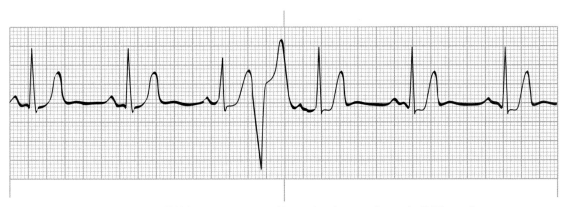

An interpolated PVC between two normal beats that does not disrupt the R-R interval

Figure 15-7
An interpolated PVC.

Another type of PVC is called an *interpolated PVC* (Figure 15-7). It occurs when a PVC does not disrupt the normal cardiac cycle. It appears as a PVC squeezed between two regular complexes. Another feature is the PR interval of the cardiac cycle that follows the PVC may be longer than normal.

Frequent PVCs, especially if bigeminal, trigeminal, or couplets or runs of VT, may foretell the deterioration of the rhythm into VF.

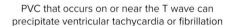

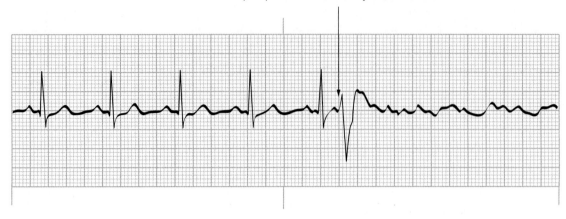

Figure 15-8
R-on-T PVC.

PVCs occurring on or near the previous T wave also are likely to progress to more life-threatening ventricular dysrhythmias (such as VF) because a portion of the T wave is vulnerable to electrical stimulation during the relative refractory period (Figure 15-8). A PVC occurring during this period of the cardiac cycle may precipitate VT or VF.

A PVC is not an entire rhythm—it is a single beat. To describe it, refer to the underlying rhythm along with the ectopic beat(s), such as "sinus rhythm at a rate of 70 beats per minute with frequent PVCs."

Treatment

Asymptomatic patients seldom require treatment. In patients with myocardial ischemia, treatment of frequent PVCs includes administration of oxygen and placement of an IV line. Care for the patient is focused on identifying and correcting the underlying factor causing the PVCs. In some settings, you may administer lidocaine or other antidysrhythmics. Follow your established protocols or consult with the appropriate medical authority.

Other treatments for the patient may include administering potassium chloride intravenously to correct hypokalemia or magnesium sulfate intravenously to correct hypomagnesemia; adjusting drug therapy; or correcting acidosis, hypothermia, and/or hypoxia.

15.4 Ventricular Escape Beats

Description

Ventricular escape beats occur when there is temporary cessation of the heartbeat such as with sinus arrest or when the rate of the underlying rhythm falls to less than that the inherent rate of the ventricles (Figure 15-9). It is a self-generated electrical impulse initiated by and causing contraction of the ventricles. The ventricular escape beat follows a long pause, typically 2 to 3 seconds after an electrical impulse fails to reach the ventricles, and acts to prevent cardiac arrest. As discussed in

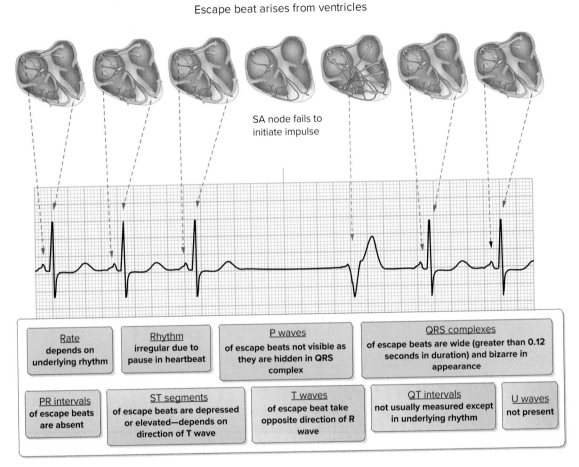

Figure 15-9
Sinus pause with a ventricular escape beat that acts to initiate a heartbeat.

Chapter 11, the ventricles spontaneously depolarize more slowly than the SA node and AV junction, so one would not expect to see ventricular escape beats unless there is a failure of the heart's electrical conduction system to stimulate the ventricles. A persistent absence of a higher pacemaker causes a continuum of beats, called *idioventricular rhythm,* to arise.

Causes

Ventricular escape beats are compensatory in nature; in other words, they occur when a higher pacemaker fails to initiate a heartbeat. They are meant to preserve cardiac output. Ventricular escape beats can occur in sinoatrial pause/arrest, by a failure of the conductivity from the SA node to the AV node, or by atrioventricular block (especially 3rd-degree AV block).

Effects

Because of the low heart rate associated with the event that leads up to the appearance of ventricular escape beats, there can be a drop in blood pressure and syncope. Ventricular escape beats should be temporary as the heart should resume its normal electrical activity.

Shorter than normal R-R interval with PVCs

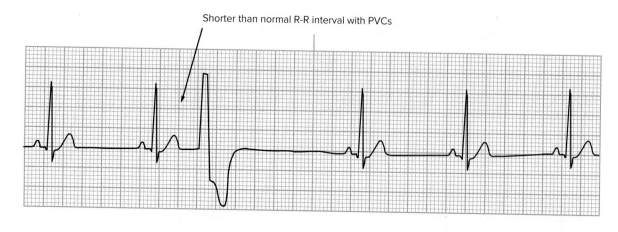

Longer than normal R-R interval with escape beats

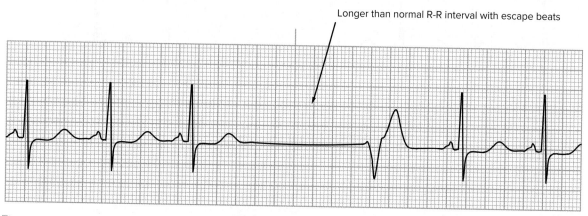

Figure 15-10
PVCs result in a shorten R-R interval, while ventricular escape beats result in a longer than normal R-R interval.

ECG Appearance

Ventricular escape beats appear as only one or two ectopic beats that follow a period of electrical inactivity (cardiac standstill). The ventricular escape beats are wide (greater than 0.12 seconds in duration) and bizarre-looking; there is an absence of a P wave, and the T wave takes a direction opposite of the R wave.

In contrast to PVCs, which occur earlier than normal, resulting in a shortened R-R interval, ventricular escapes come later than normal, resulting in a longer than normal R-R interval (Figure 15-10).

Treatment

Treating patients experiencing ventricular escape beats centers on treating the underlying condition. Support the circulation, airway, and breathing; deliver oxygen; monitor the ECG, blood pressure, and pulse oximetry; and establish an IV infusion. Continually reassess the clinical status and correct any reversible conditions. A primary goal of treatment is to maintain a ventricular rate sufficient to produce adequate cardiac output. Atropine (a drug used to speed the heart rate), transcutaneous pacing, or a dopamine or epinephrine infusion (see Figure 12-4) can be used to treat the symptomatic patient. Drugs used to suppress ventricular activity such as lidocaine and amiodarone are contraindicated and may be lethal (by suppressing any remaining cardiac activity).

15.5 Idioventricular Rhythm

Description

Idioventricular rhythm is a dysrhythmia that occurs when stimuli from higher pacemakers, such as the SA node or AV junction, fail to reach the ventricles (Figure 15-11) or there is a block in the heart's conduction system.

In these instances, the cells of the His-Purkinje system can take over and act as the heart's pacemaker to generate electrical impulses. Idioventricular rhythm normally fires at the inherent rate of the ventricles, 20 to 40 beats per minute. When the event occurs as an escape complex or rhythm, it is considered a safety mechanism to prevent cardiac standstill. Idioventricular rhythm is common during cardiac arrest because it represents the final escape rhythm to be generated in an attempt to perfuse the body. Also, idioventricular rhythm is often the first organized rhythm following defibrillation.

Sometimes idioventricular rhythm is also called *agonal rhythm*. The term *agonal rhythm* usually indicates that the heart rate is less than 10 beats per minute. Patients in the final dying stage from a terminal disease often exhibit this rhythm prior to dying. Agonal rhythm has no cardiac output.

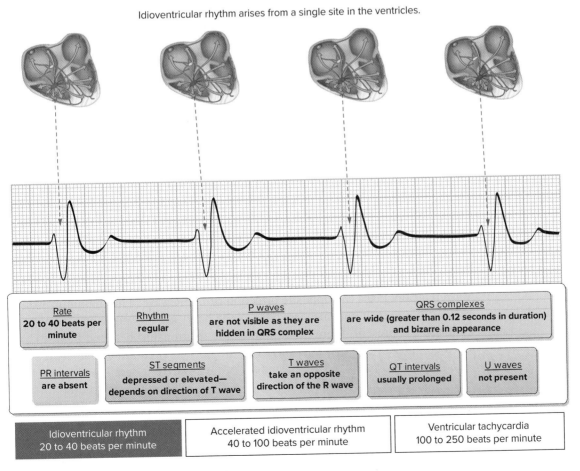

Idioventricular rhythm arises from a single site in the ventricles.

Rate	Rhythm	P waves	QRS complexes
20 to 40 beats per minute	regular	are not visible as they are hidden in QRS complex	are wide (greater than 0.12 seconds in duration) and bizarre in appearance

PR intervals	ST segments	T waves	QT intervals	U waves
are absent	depressed or elevated— depends on direction of T wave	take an opposite direction of the R wave	usually prolonged	not present

Idioventricular rhythm 20 to 40 beats per minute	Accelerated idioventricular rhythm 40 to 100 beats per minute	Ventricular tachycardia 100 to 250 beats per minute

Figure 15-11
Idioventricular rhythm.

Causes

Idioventricular rhythm can be caused by massive myocardial ischemia or MI, digoxin toxicity, pacemaker failure, and metabolic imbalances. Other causes are listed in the following box.

Causes of Idioventricular Rhythm	
Cause	**Examples**
Cardiac diseases	Myocardial ischemia and MI, failure of all of the heart's higher pacemakers, failure of supraventricular impulses to reach the ventricles due to a block in the conduction system, pacemaker failure, 3rd-degree AV block, sick sinus syndrome
Use of certain drugs	Digoxin beta-adrenergic blockers, calcium channel blockers, tricyclic antidepressants
Autonomic nervous system stimulation	Increased vagal tone
Noncardiac disorders	Metabolic acidosis, hypoxia

Effects

Patient assessment is essential because escape beats and/or idioventricular rhythm may be perfusing or nonperfusing. Ventricular escape beats are usually transient, with a higher pacemaker taking over the heartbeat, in contrast to idioventricular rhythm, which is more likely to continue.

The slow ventricular rate of idioventricular rhythm and the loss of atrial kick can significantly decrease cardiac output. The resultant decreased cardiac output will likely cause the patient to be symptomatic. You can expect to see signs of decreased cardiac output such as disorientation, unconsciousness, hypotension, and/or syncope, and the patient may complain of dizziness, chest pain, and/or shortness of breath. In extreme cases, the patient is pulseless.

ECG Appearance

Wide QRS complexes in a slow regular rhythm are key features of idioventricular rhythm. The heart rate is usually 20 to 40 beats per minute although it may be slower, and the QRS complexes are wide (greater than 0.12 seconds). Also, the T wave typically takes the opposite direction of the R wave. The rhythm is usually regular; however, it becomes irregular as the heart dies. P waves may be present or absent. If present, there is no predictable relationship between P waves and QRS complexes (AV dissociation). This would be called *3rd-degree heart block with an idioventricular escape*. Heart blocks are covered in detail in Chapter 16.

Treatment

To treat patients experiencing symptomatic idioventricular rhythm (e.g., chest pain, hypotension), first focus on supporting the airway and breathing. Deliver oxygen, place the patient on an ECG monitor, monitor the blood pressure and pulse oximetry, and place an IV infusion. Continually reassess the clinical status, and correct any reversible conditions. A primary goal of treatment is to maintain a ventricular rate sufficient to produce adequate cardiac output. Atropine, transcutaneous pacing, or a dopamine or epinephrine infusion (see Figure 12-4) can be used to treat the symptomatic patient. In unresolved idioventricular rhythm, the atropine dose may be repeated. Lidocaine and amiodarone are contraindicated and may be lethal (by suppressing any remaining cardiac activity). If there is no pulse, treat the dysrhythmia as if it is pulseless electrical activity (PEA).

15.6 Accelerated Idioventricular Rhythm

Description

When the rate of the idioventricular rhythm exceeds the inherent rate of the ventricles and is between 40 and 100 beats per minute, it is called *accelerated idioventricular rhythm* (Figure 15-12).

Causes

This dysrhythmia can be caused by the same conditions that lead to idioventricular rhythm. Also, it is very common following acute MI. Further, it is often seen after the administration of thrombolytic medications. For that reason, it is considered a *reperfusion dysrhythmia*.

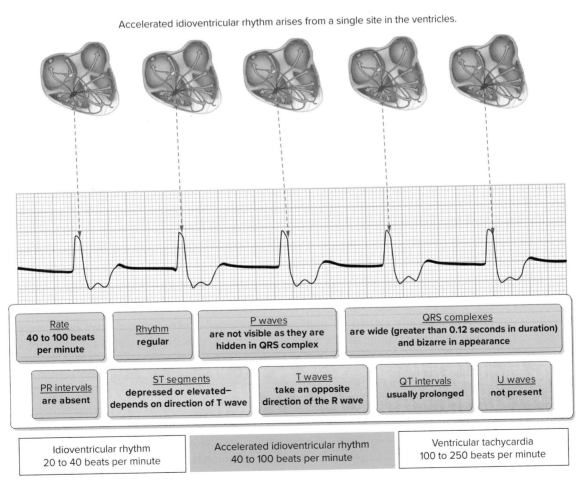

Accelerated idioventricular rhythm arises from a single site in the ventricles.

| Rate 40 to 100 beats per minute | Rhythm regular | P waves are not visible as they are hidden in QRS complex | QRS complexes are wide (greater than 0.12 seconds in duration) and bizarre in appearance |

| PR intervals are absent | ST segments depressed or elevated—depends on direction of T wave | T waves take an opposite direction of the R wave | QT intervals usually prolonged | U waves not present |

| Idioventricular rhythm 20 to 40 beats per minute | Accelerated idioventricular rhythm 40 to 100 beats per minute | Ventricular tachycardia 100 to 250 beats per minute |

Figure 15-12
Accelerated idioventricular rhythm.

Effects

While accelerated idioventricular rhythm is usually short lived and because the heart rate is close to normal, no ill effect may be seen. However, with accelerated idioventricular rhythm there may be decreased cardiac output, particularly when the rate is slower. Remember what has been emphasized throughout this text: heart rate × stroke volume = cardiac output. Slow dysrhythmias, because of the slower heart rate, can cause decreased cardiac output.

ECG Appearance

Because it is a rhythm that originates from the ventricles, it has the same characteristics as idioventricular rhythm, except that it is faster. The key difference that separates this dysrhythmia from idioventricular rhythm and VT is the heart rate.

Treatment

Treatment of the patient depends on whether the rhythm is sustained and whether or not cardiac output is diminished due to a slow heart rate (less than 50 beats per minute). Atropine, transcutaneous pacing, or an infusion of epinephrine or dopamine (see Figure 12.5) can be used to treat the symptomatic patient. A permanent pacemaker may be required in sustained symptomatic accelerated idioventricular rhythm. Lidocaine and amiodarone are contraindicated and may be lethal (by suppressing any remaining cardiac activity). If there is no pulse, treat the dysrhythmia as if it were pulseless electrical activity (PEA).

15.7 Ventricular Tachycardia

Description

VT is a fast dysrhythmia that arises from the ventricles (Figure 15-13). It is said to be present when there are three or more PVCs in a row. It may come in bursts of 6 to 10 complexes or may persist (sustained VT). It can occur with or without pulses, and the patient may be stable or unstable with this rhythm. Usually this dysrhythmia occurs in the presence of myocardial ischemia or significant cardiac disease. VT often precedes VF and sudden death. It is a common rhythm encountered in out-of-hospital cardiac arrest, and it is commonly called *VT, V-Tach* or wide complex tachycardia.

Causes

VT is usually caused by increased myocardial irritability that may be triggered by enhanced automaticity, PVCs that occur during the downstroke of the preceding T wave, and reentry in the Purkinje system. Other causes are listed in the following page.

Effects

Clinically, VT is always significant. It may be perfusing or nonperfusing. Even if the rhythm produces a pulse, it should be considered as potentially unstable because patients are likely to develop more life-threatening rhythms, which progress into cardiac arrest. Usually VT indicates significant underlying cardiovascular disease. The rapid rate and concurrent loss of atrial kick associated with VT results in

Ventricular tachycardia arises from a single site in the ventricles.

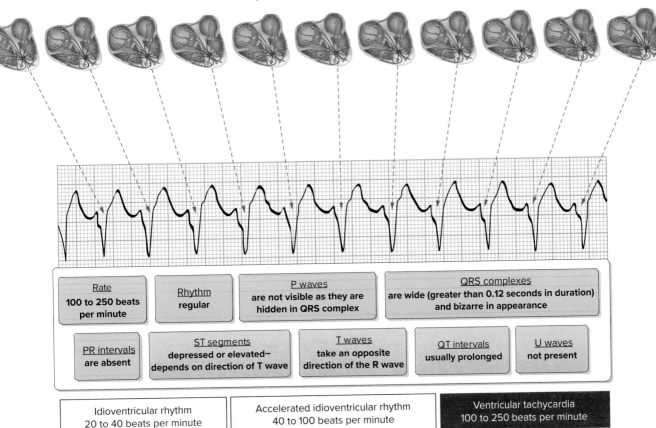

| Rate 100 to 250 beats per minute | Rhythm regular | P waves are not visible as they are hidden in QRS complex | QRS complexes are wide (greater than 0.12 seconds in duration) and bizarre in appearance |

| PR intervals are absent | ST segments depressed or elevated– depends on direction of T wave | T waves take an opposite direction of the R wave | QT intervals usually prolonged | U waves not present |

| Idioventricular rhythm 20 to 40 beats per minute | Accelerated idioventricular rhythm 40 to 100 beats per minute | Ventricular tachycardia 100 to 250 beats per minute |

Figure 15-13
Characteristics of Ventricular tachycardia.

Causes of Ventricular Tachycardia

Cause	Examples
Cardiac diseases	Myocardial ischemia and MI, coronary artery disease, congestive heart failure, cardiomyopathy, valvular heart disease, pulmonary embolism, rheumatic heart disease
Use of certain drugs	Amphetamines, cocaine, digoxin, quinidine, sympathomimetic drugs such as epinephrine, thyroid medications, theophylline drugs
Autonomic nervous system	Increased sympathetic stimulation, exercise, emotional stress
Noncardiac disorders	Electrolyte imbalance, such as hypokalemia, acid-base imbalance, hypoxia, trauma, ingestion of stimulants (alcohol, caffeine, tobacco)

compromised cardiac output and decreased coronary artery and cerebral perfusion. Remember what has been emphasized throughout this text: heart rate × stroke volume = cardiac output. Fast dysrhythmias, because of the decreased stroke volume associated with the fast heart rate, cause decreased cardiac output. The severity of symptoms varies with the rate of the VT and the presence and degree of underlying myocardial dysfunction. Again, the gravest concern is that it may initiate or deteriorate into VF.

ECG Appearance

The most visible characteristics of VT are a rapid rate of between 100 and 250 beats per minute and wide (greater than 0.12 seconds), bizarre-looking QRS complexes. The T waves may or may not be present and typically are of the opposite direction of the R waves. If the rate is between 100 and 150 beats per minute, it is referred to as *slow VT*. Rarely, the rate may be greater than 250 and is then referred to as *ventricular flutter*. Typically, P waves are not discernable (if seen, they are dissociated). The rhythm is usually regular.

Treatment

Treatment of a patient experiencing VT (see Figure 15-14) includes maintaining a patent airway, administering oxygen, and placing an IV line.

The stable patient can be treated with antidysrhythmics (such as procainamide, amiodarone, or sotalol).

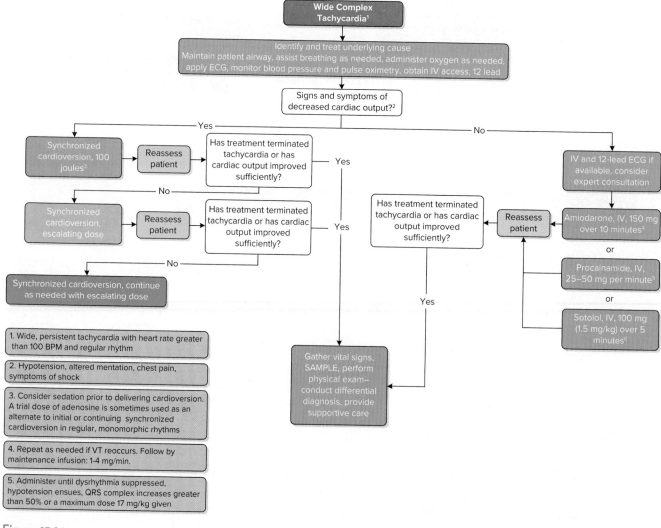

Figure 15-14
Treatment for ventricular tachycardia

The patient experiencing decreased level of consciousness, ongoing chest pain, shortness of breath, low blood pressure, or other signs of shock should be considered clinically unstable. Initially these patients are managed with immediate synchronized cardioversion (100 J). Contributing causes should be considered and treated. The initial energy level may be increased if the tachycardia does not convert with initial treatments (100, 200, 300, 360 J, or the biphasic equivalent). Patients with pulseless VT should be treated as though they are in VF.

15.8 Polymorphic Ventricular Tachycardia

Description

VT may be monomorphic, where the appearance of each QRS complex is similar, or polymorphic, where the appearance varies considerably from complex to complex. Either is potentially life-threatening. A specific, rare variety of **polymorphic ventricular tachycardia** is called **torsades de pointes,** which means twisting about the points (Figure 15-15). This form of VT exhibits distinct characteristics on the ECG.

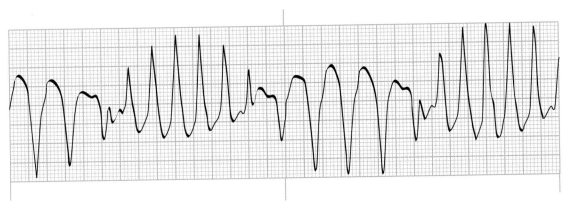

Figure 15-15
Polymorphic VT. Torsades de pointes is characterized by QRS complexes that alternate between upright deflections and downward deflections.

Causes

Torsades de pointes is associated with long QT syndrome, a condition whereby prolonged QT intervals are visible on the ECG. Long QT intervals predispose the patient to an R-on-T phenomenon, where the R wave representing ventricular depolarization occurs simultaneously to the relative refractory period at the end of repolarization (represented by the latter half of the T wave). Sometimes pathologic T-U waves may be seen in the ECG before the initiation of torsades. Long QT syndrome can either be inherited as congenital mutations of ion channels carrying the cardiac impulse (action potential) or acquired as a result of drugs that block cardiac ion currents. Causes of torsades de pointes are shown in the box on the following page.

Causes of Torsades de Pointes	
Cause	**Examples**
Cardiac diseases	AV heart block, hereditary QT prolongation syndrome, myocardial ischemia, Prinzmetal's angina, SA node disease resulting in severe bradycardia, hypoxia, acidosis, heart failure, left ventricular hypertrophy
Use of certain drugs	Amiodarone, quinidine, procainamide, sotalol toxicity, psychotropic drugs (phenothiazines, tricyclic antidepressants, lithium), chloroquine, clarithromycin (Biaxin) or haloperidol (Haldol) taken concomitantly with a specific cytochrome P450 inhibitor such as fluoxetine (Prozac) or cimetidine (Tagamet), some dietary supplements such as St. John's wort
Noncardiac disorders	Electrolyte imbalance such as hypocalcemia, hypokalemia, hypomagnesemia, diarrhea (commonly seen in malnourished individuals and chronic alcoholics), hypothermia, subarachnoid hemorrhage

Effects

The effects of torsades de pointes depend on the rate and duration of tachycardia and the degree of cerebral hypoperfusion. Findings include rapid pulse, low or normal blood pressure, or syncope or prolonged loss of consciousness. This could be preceded by bradycardia or premature ventricular contractions (PVCs), also called *leading palpitations.* Pallor and diaphoresis may be noted, especially with a sustained episode. It can degenerate into VF, which will lead to sudden death in the absence of medical intervention.

ECG Appearance

Torsades de pointes is characterized by QRS complexes that alternate (usually gradually) between upright deflections and downward deflections. The ventricular rate can range from 150 to 300 beats per minute (usually 200 to 250 beats per minute).

Treatment

Polymorphic VT cannot be reliably synchronized and should be managed like ventricular fibrillation (VF), with an initial unsynchronized shock. Standard antidysrhythmic drugs (such as procainamide) can worsen the condition, leading to cardiac arrest. Amiodarone may be effective in stable patients with normal QT interval, whereas magnesium sulfate should be administered in stable patients with prolonged QT interval. Also, offending drug(s) should be discontinued and electrolyte imbalance corrected. Patients with pulseless VT should be treated as though they are in VF, with the treatment of choice being prompt defibrillation.

15.9 Ventricular Fibrillation

Description

VF results from chaotic firing of multiple sites in the ventricles (Figure 15-16). It is the most common cause of prehospital cardiac arrest in adults.

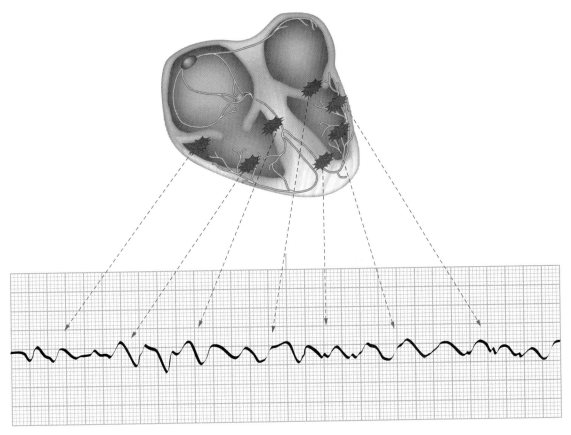

Figure 15-16
VF results from chaotic firing of multiple sites in the ventricles.

Causes

VF is most commonly associated with significant cardiovascular system disease. Causes are listed below.

Causes of Ventricular Fibrillation	
Cause	**Examples**
Cardiac diseases	Myocardial ischemia/infarction, coronary artery disease, cardiomyopathy, untreated VT, R on T phenomenon
Use of certain drugs	Digoxin [rarely], epinephrine, procainamide, quinidine; drug overdose (cocaine, tricyclics)
Noncardiac disorders	Acid-base imbalance, drowning, electric shock, severe hypothermia, electrolyte imbalances such as hypokalemia, hyperkalemia, hypercalcemia

Effects

VF causes the heart muscle to quiver, much like a handful of worms, rather than contracting efficiently. It produces ineffective muscular contraction and an absence of cardiac output. Within 10 seconds, the amount of blood pumped by the heart diminishes to zero, causing all life-supporting physiological functions to cease because of lack of circulating blood flow. Death occurs if the patient is not promptly treated (with CPR and defibrillation).

ECG Appearance

VF is generally easy to recognize on the cardiac monitor; it appears like a wavy line, totally chaotic, without any logic. There are no coordinated ventricular complexes present. The unsynchronized ventricular impulses occur at rates from 300 to 500 beats per minute. There are no discernible P waves, QRS complexes, or PR intervals.

Treatment

Treatment (Figure 15-17) includes prompt initiation of CPR and defibrillation as soon as possible. It is critical to provide high-quality CPR (with chest compressions of adequate rate and depth, allowing complete chest recoil after each compression,

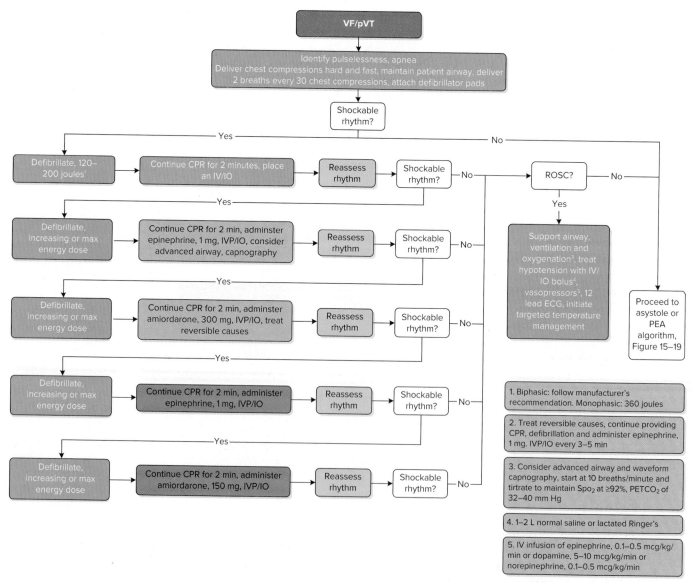

Figure 15-17
Treatment algorithm for ventricular fibrillation/pulseless ventricular tachycardia (VF/pVT)

minimizing interruptions in chest compressions, and avoiding excessive ventila-
tion). Rotate the person delivering chest compressions every 2 minutes. ACLS
actions should be organized around uninterrupted periods of CPR. Securing the
airway and placing an IV line should be initiated in the course of treatment. After
one shock and a 2-minute period of CPR, deliver a second shock then administer
epinephrine. Amiodarone may be considered when VF/VT is unresponsive to CPR,
defibrillation, and epinephrine administration. If amiodarone is unavailable, lidocaine may be considered if allowed by local protocol. Continue CPR, stopping only
to defibrillate and reassess rhythm. Minimize interruptions in chest compressions
before and after shock; resume CPR beginning with compressions immediately
after each shock. If the rhythm is successfully converted to an effective electromechanical rhythm (with a pulse and good perfusion), assess vital signs, support
airway and breathing, provide IV fluids and, if necessary, vasopressors to support
blood pressure, heart rate, and rhythm and to prevent reoccurrence.

15.10 Asystole

Description

Asystole is the absence of any cardiac activity (Figure 15-18). It may be the primary
event in cardiac arrest, or it may occur in complete heart block when there is an
absence of a functional escape pacemaker. It is usually associated with global myocardial ischemia or necrosis and often follows VT, VF, PEA, or an agonal escape
rhythm in the dying heart. An ominous dysrhythmia, asystole often represents a
predictor of death, as the likelihood of resuscitation from asystole is dismal.

Causes

Some conditions that cause asystole are listed in the following box.

Causes of Asystole	
Cause	**Examples**
Cardiac diseases	MI, cardiac tamponade, massive pulmonary embolism, underlying heart disease, ruptured ventricular aneurysm
Noncardiac disorders	Acute respiratory failure, massive pulmonary embolism, prolonged hypoxemia, severe uncorrected acid-base imbalance (particularly metabolic acidosis), severe electrolyte imbalances (especially hyperkalemia and hypokalemia), tension pneumothorax, hypothermia, drug overdose

Effects

With asystole, there is no electrical activity. It produces complete cessation of cardiac output. It is a terminal rhythm, and once a person has become asystolic, the
chances of recovery are extremely low. Some use the term **ventricular standstill** to
define a condition where the atria continue to beat but the ventricles have stopped.
A rhythm strip of ventricular standstill will have only P waves present without QRS
complexes.

ECG Appearance

Asystole appears as a flat (or nearly flat) line on the monitor screen.

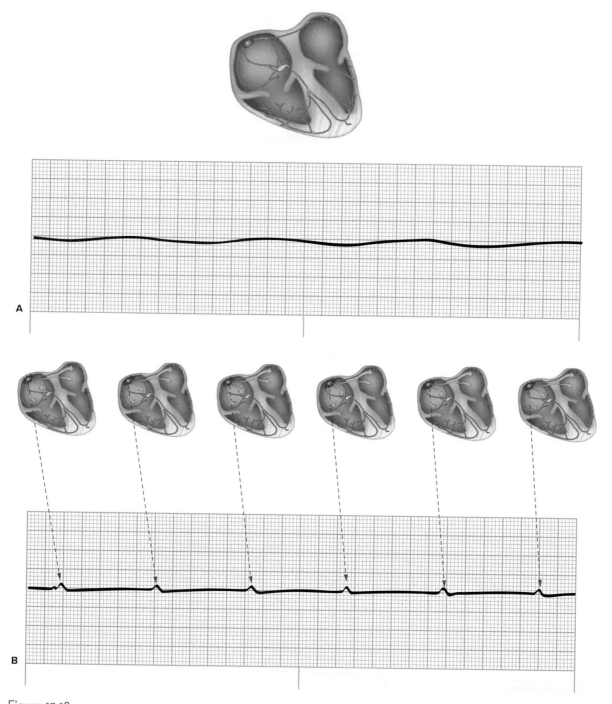

Figure 15-18

Asystole and ventricular standstill. (A) Asystole is a complete absence of cardiac activity. (B) Ventricular standstill occurs when there is electrical activity from the atria but none from the ventricles.

Treatment

Treatment (see Figure 15-19) of asystole includes prompt initiation of CPR, high-concentration oxygen, placing an IV line, and securing the airway. Administer epinephrine, repeating its administration every 3 to 5 minutes. Continue CPR throughout the resuscitation effort, stopping periodically to reassess for a change in the rhythm and to check for the presence of a pulse. Follow local protocols for terminating resuscitation efforts.

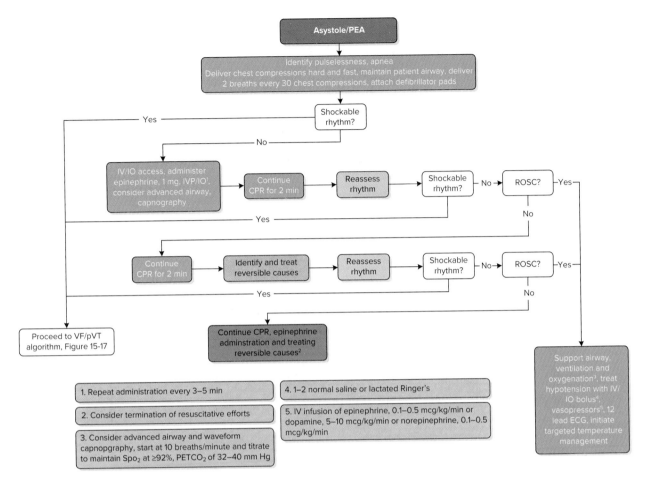

Figure 15-19
Treatment algorithm for asystole/pulseless electrical activity (PEA)

Always verify the presence of asystole in two leads prior to initiating treatment. Misplacement of an ECG lead or a loose wire can mimic asystole (or VF) on the monitor. If a patient is in cardiac arrest and appears to be in asystole, be sure to check the following:

- The leads are attached in the proper places to both the machine and the patient.
- The correct lead (e.g., I, II, III) is selected on the monitor.
- The rhythm appears as asystole in more than one lead.
- The monitor batteries are functioning appropriately.

15.11 Pulseless Electrical Activity

Description

Pulseless electrical activity (PEA) is a condition rather than a dysrhythmia. We cover it here for the sake of completeness in our discussion of the various causes of cardiac arrest. With PEA, we see an organized electrical rhythm on the ECG monitor that should result in contraction of the heart muscle, but there is no corresponding pulse. This results in the patient being pulseless and apneic. In other words, there is organized electrical activity in the heart that fails to generate effective cardiac contraction (Figure 15-20).

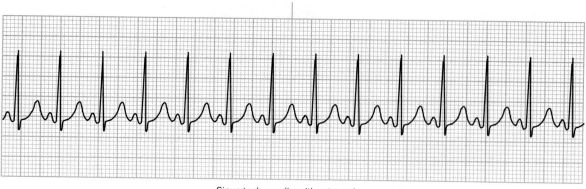

Sinus tachycardia without a pulse

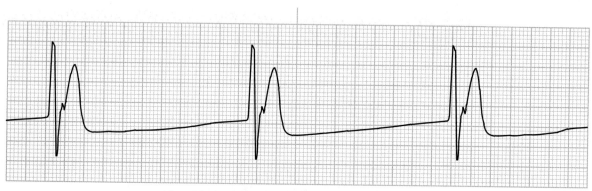

Idioventricular rhythm without a pulse (also called slow PEA)

Figure 15-20
Examples of PEA.

Causes

This condition is often associated with severe underlying heart disease, but you should always consider reversible causes of PEA, which include acidosis, drug overdose (such as opioids or tricyclic antidepressants), hypovolemia, hypoxia, hypothermia, hyperkalemia, hypokalemia, pericardial tamponade, MI, hypoglycemia, tension pneumothorax, and thrombosis.

Effects

With PEA there is no cardiac output. The patient will be unconscious, apneic, and pulseless; in other words, they are in cardiac arrest.

ECG Appearance

The electrical activity seen with PEA may be sinus rhythm, sinus tachycardia, idioventricular rhythm, or any rhythm that is generally expected to generate a pulse.

Treatment

Treatment of (see Figure 15-19) PEA includes prompt initiation of CPR, high-concentration oxygen, placing an IV line, securing the airway and confirming placement. Administer epinephrine, repeating its administration every 3 to 5 minutes. Continue CPR throughout the resuscitation effort, stopping periodically to reassess for a change in the rhythm and to check for the presence of a pulse. Be sure to look for reversible causes and provide needed treatments.

Practice Makes Perfect

For each of the 57 ECG tracings on the following pages, practice the Nine-Step Process for analyzing ECGs. To achieve the greatest learning, you should practice assessing and interpreting the ECGs immediately after reading this chapter. Below are questions you should consider as you assess each tracing. Your answers can be written into the area above each ECG marked "ECG Findings." Your findings can then be compared to the answers provided in Appendix A. All tracings are 6 seconds in length.

1. Determine the heart rate. Is it slow? Normal? Fast? What is the ventricular rate? What is the atrial rate?

2. Determine if the rhythm is regular or irregular. If it is irregular, what type of irregularity is it? Occasional or Irregular? Slight? Sudden acceleration or slowing in heart rate? Total? Patterned? Does it have a variable conduction ratio?

3. Determine if P waves are present. If so, how do they appear? Do they have normal height and duration? Are they tall? Notched? Wide? Biphasic? Of differing morphology? Inverted? One for each QRS complex? More than one preceding some or all the QRS complexes? Do they have a sawtooth appearance? An indiscernible chaotic baseline?

4. Determine if QRS complexes are present. If so, how do they appear? Narrow with proper amplitude? Tall? Low amplitude? Delta wave? Notched? Wide? Bizarre? With chaotic waveforms?

5. Determine the presence of PR intervals. If present, how do they appear? Constant? Of normal duration? Shortened? Lengthened? Progressively longer? Varying?

6. Evaluate the ST segments. Do they have normal duration and position? Are they elevated? (If so, are they flat, concave, convex, arched?) Depressed? (If so, are they normal, flat, downsloping or upsloping?)

7. Determine if T waves are present. If so, how do they appear? Of normal height and duration? Tall? Wide? Notched? Inverted?

8. Determine the presence of QT intervals. If present, what is their duration? Normal? Shortened? Prolonged?

9. Determine if U waves are present. If present, how do they appear? Of normal height and duration? Inverted?

10. Identify the rhythm or dysrhythmia.

1.

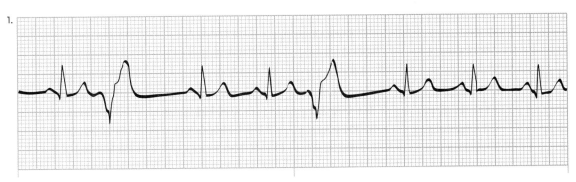

ECG Findings:

2.

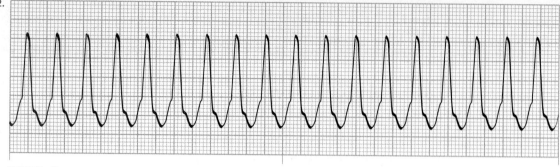

ECG Findings:

3.

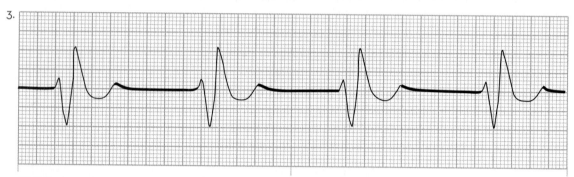

ECG Findings:

4.

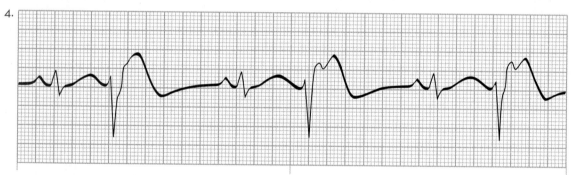

ECG Findings:

5.

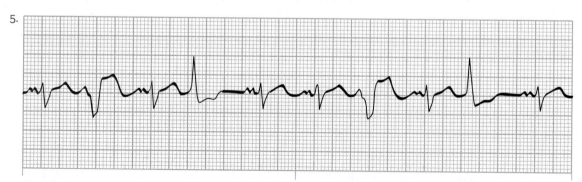

ECG Findings:

6.

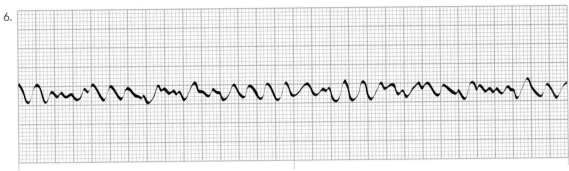

ECG Findings:

7.

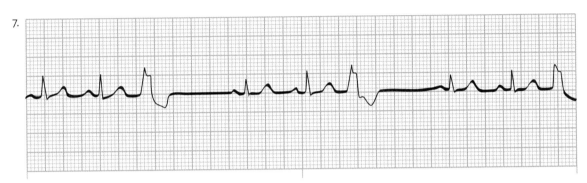

ECG Findings:

8.

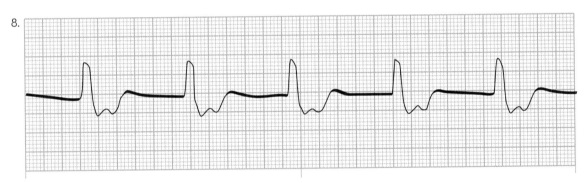

ECG Findings:

9.

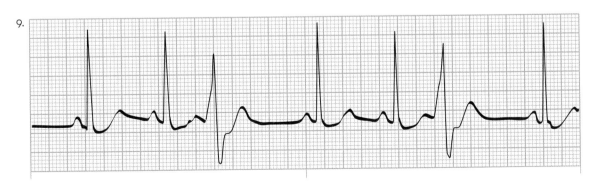

ECG Findings:

©rivetti/Getty Images

©rivetti/Getty Images

10.

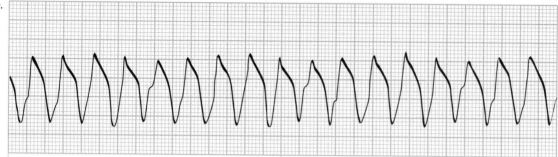

ECG Findings:

11.

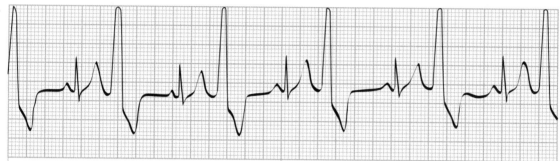

ECG Findings:

12.

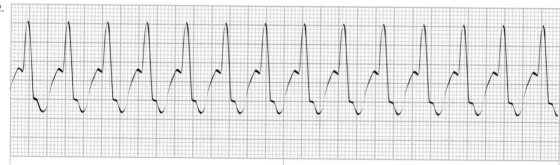

ECG Findings:

13.

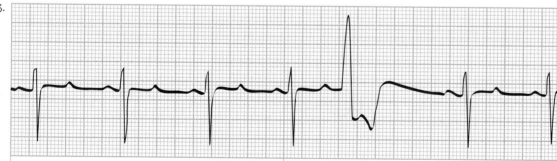

ECG Findings:

14.

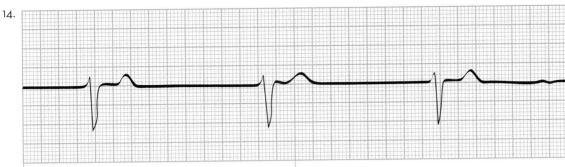

ECG Findings:

15.

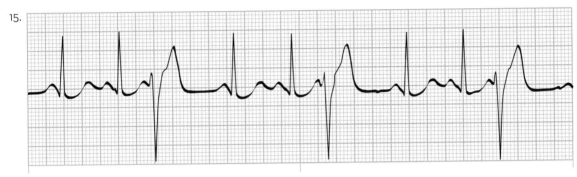

ECG Findings:

16.

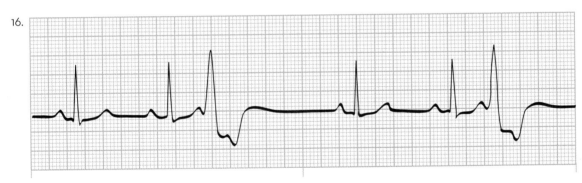

ECG Findings:

17.

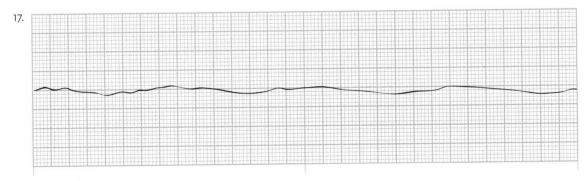

ECG Findings:

18.

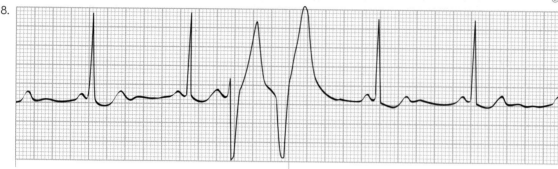

ECG Findings:

19.

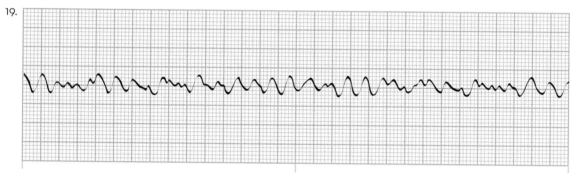

ECG Findings:

20.

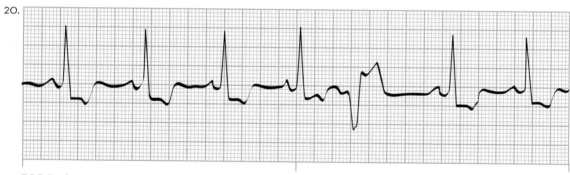

ECG Findings:

21.

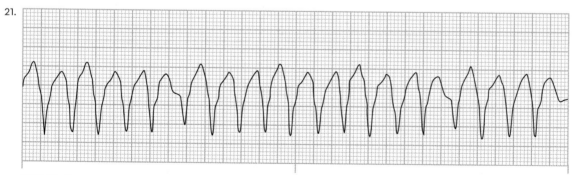

ECG Findings:

22.

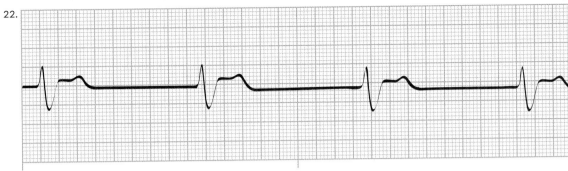

ECG Findings:

23.

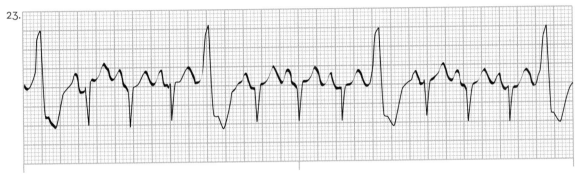

ECG Findings:

24.

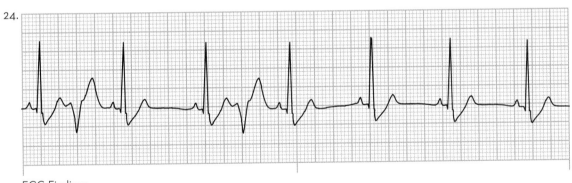

ECG Findings:

25.

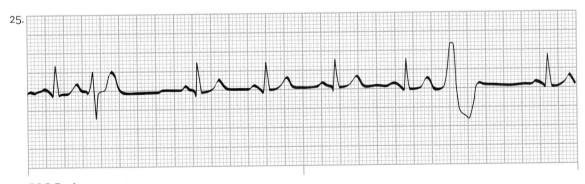

ECG Findings:

26.

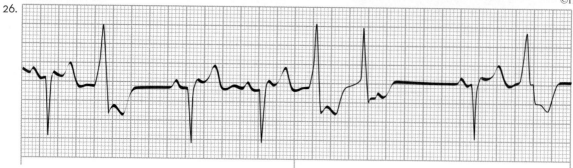

ECG Findings:

27.

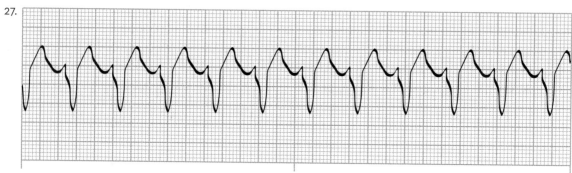

ECG Findings:

28.

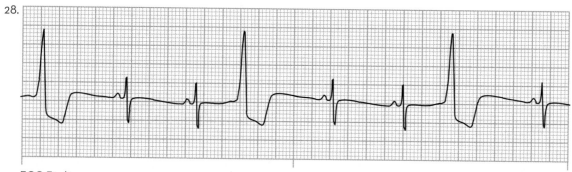

ECG Findings:

29.

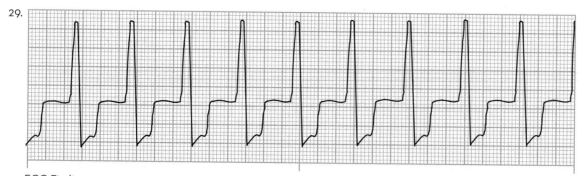

ECG Findings:

30.

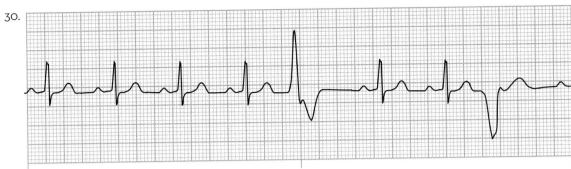

ECG Findings:

31.

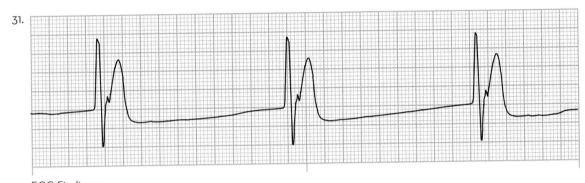

ECG Findings:

32.

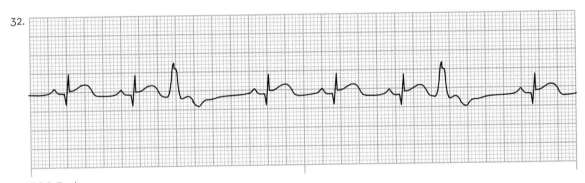

ECG Findings:

33.

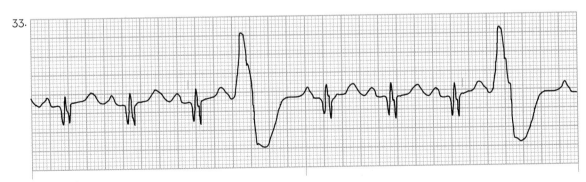

ECG Findings:

34.

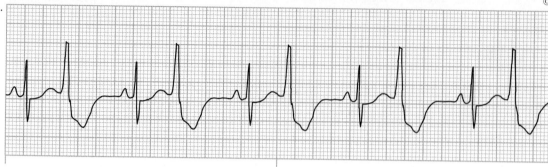

ECG Findings:

35.

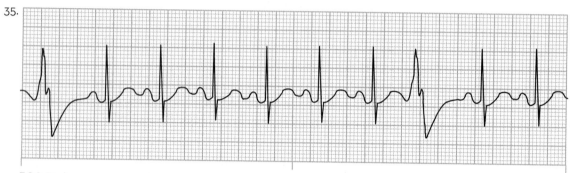

ECG Findings:

36.

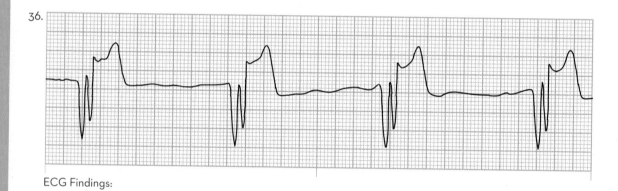

ECG Findings:

37.

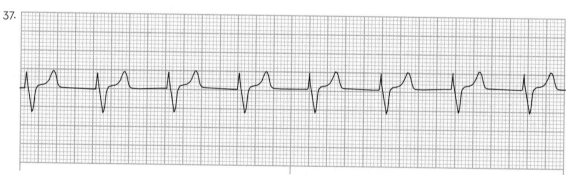

ECG Findings:

38.

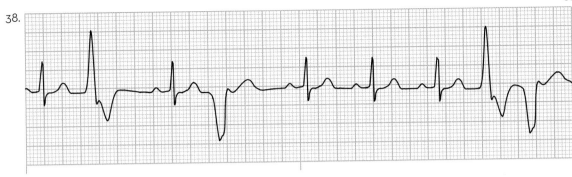

ECG Findings:

39.

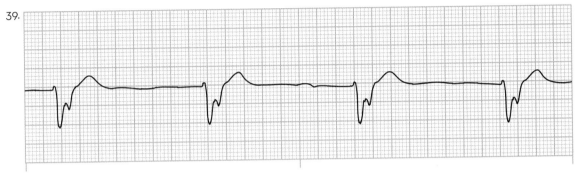

ECG Findings:

40.

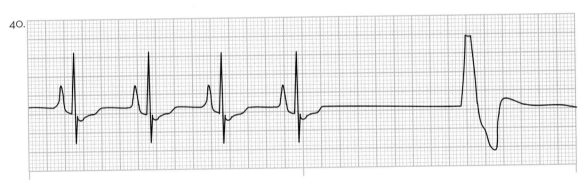

ECG Findings:

41.

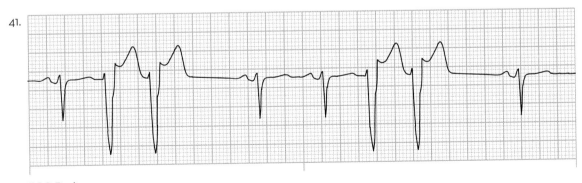

ECG Findings:

42.

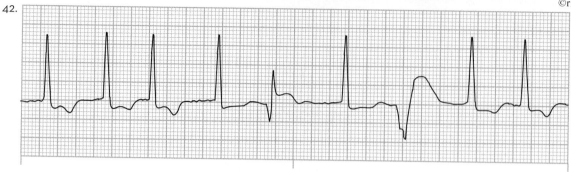

ECG Findings:

43.

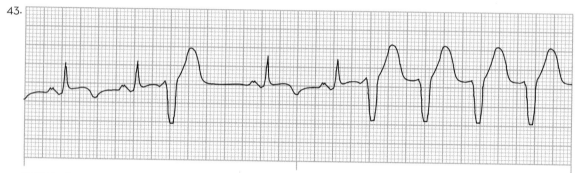

ECG Findings:

44.

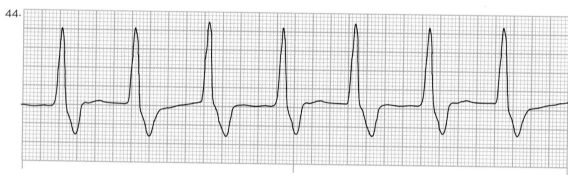

ECG Findings:

45.

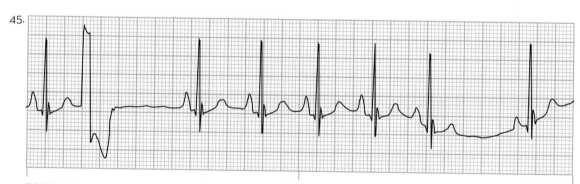

ECG Findings:

46.

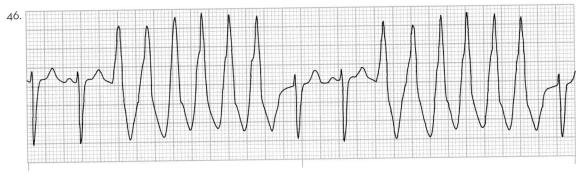

ECG Findings:

47.

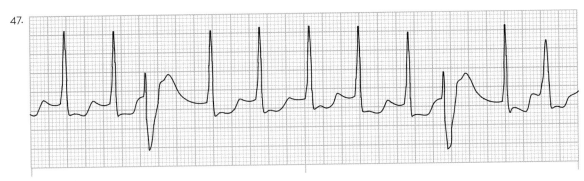

ECG Findings:

48.

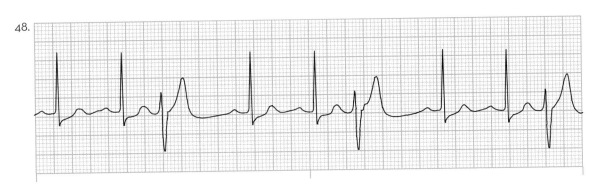

ECG Findings:

49.

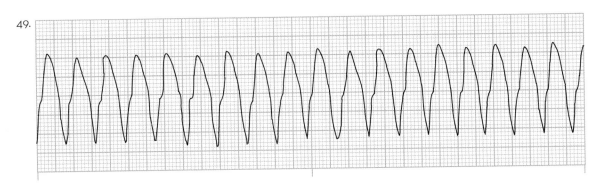

ECG Findings:

50.

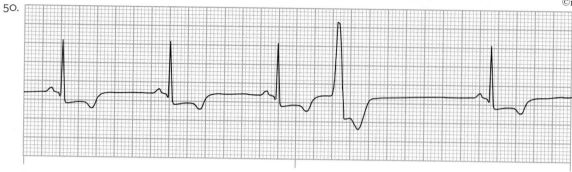

ECG Findings:

51.

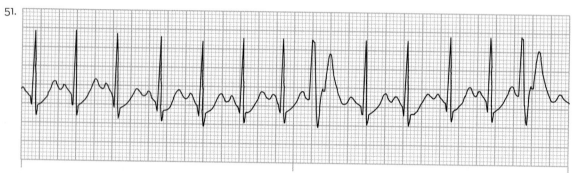

ECG Findings:

52.

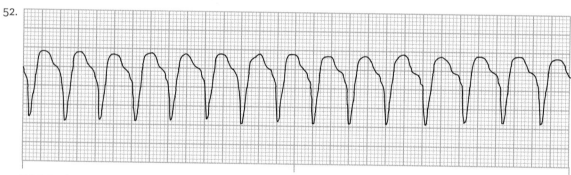

ECG Findings:

53.

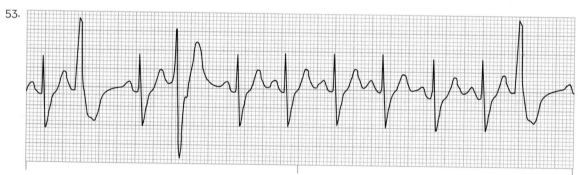

ECG Findings:

54.

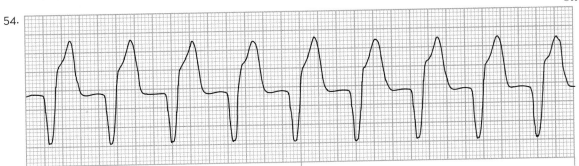

ECG Findings:

55.

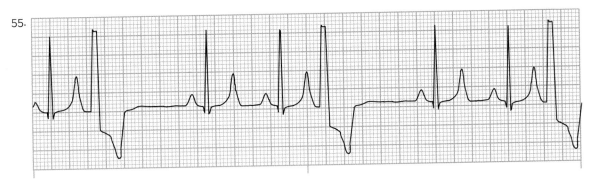

ECG Findings:

56.

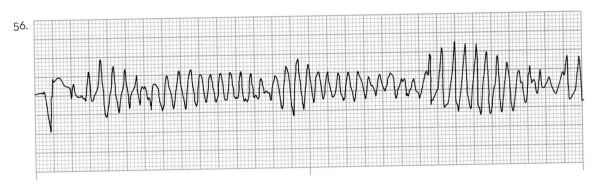

ECG Findings:

57.

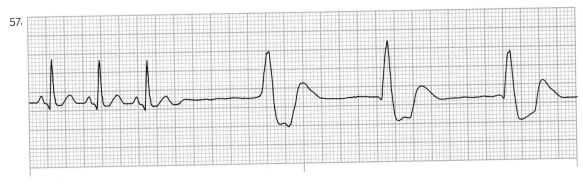

ECG Findings:

LO 15.1	• Ventricular dysrhythmias occur when the atria, AV junction, or both are unable to initiate an electrical impulse or when there is enhanced automaticity of the ventricular myocardium.
	• Ventricular dysrhythmias include PVC, ventricular escape complexes or rhythm, VT, VF, and asystole.
LO 15.2	• A key feature of ventricular dysrhythmias are wide (greater than 0.12 seconds in duration), bizarre-looking QRS complexes that have T waves in the opposite direction of the R wave and an absence of P waves.
LO 15.3	• PVCs are early ectopic beats that interrupt the normal rhythm and originate from an irritable focus in the ventricular conduction system or muscle tissue.
	• PVCs that look the same are called *uniform (unifocal)*, whereas PVCs that look different from each other are called *multiform (multifocal)*.
	• PVCs occurring every other beat are called *bigeminal PVCs*, whereas PVCs occurring every third beat are called *trigeminal PVCs*, and a PVC seen every fourth beat is called *ventricular quadrigeminy*.
	• Two PVCs in a row are called a *couplet* and indicate extremely irritable ventricles.
	• An interpolated PVC occurs but does not disrupt the normal cardiac cycle. It appears as a PVC squeezed between two regular complexes.
LO 15.4	• Ventricular escape beats occur when there is temporary cessation of the heartbeat, such as with sinus arrest or when the rate of the underlying rhythm falls to less than that the inherent rate of the ventricles. They are compensatory in nature.
LO 15.5	• Idioventricular rhythm is a slow dysrhythmia with wide QRS complexes that arise from the ventricles at a rate of 20 to 40 beats per minute.
LO 15.6	• Accelerated idioventricular rhythm is a regular rhythm at a rate of 40 to 100 beats per minute that features absent P waves and wide, bizarre-looking QRS complexes.
LO 15.7	• VT is a fast dysrhythmia, between 100 and 250 beats per minute, that arises from the ventricles. It is said to be present when there are three or more PVCs in a row. It may come in bursts of 6 to 10 complexes or may persist (sustained VT).
	• VT can occur with or without pulses, and the patient may be stable or unstable with this rhythm.
LO 15.8	• VT may be monomorphic, where the appearance of each QRS complex is similar, or polymorphic, where the appearance varies considerably from complex to complex.
	• Torsades de pointes, which means twisting about the points, is a form of VT that exhibits distinct characteristics on the ECG.
LO 15.9	• VF results from chaotic firing of multiple sites in the ventricles. This causes the heart muscle to quiver rather than contract efficiently, producing no effective muscular contraction and an absence of cardiac output.
	• On the cardiac monitor, VF appears like a wavy line, totally chaotic, without any logic.

LO 15.10	• Asystole is the absence of any cardiac activity. It appears as a flat (or nearly flat) line on the monitor screen and produces a complete cessation of cardiac output.
	• The presence of asystole should always be verified in two leads.
LO 15.11	• PEA is a condition in which there is an organized electrical rhythm on the ECG monitor (which should produce a pulse) but the patient is pulseless and apneic. Sinus rhythm, sinus tachycardia, idioventricular rhythm, or any rhythm that is generally expected to generate a pulse may be the electrical activity seen with PEA.

Assess Your Understanding

The following questions give you a chance to assess your understanding of the material discussed in this chapter. The answers can be found in Appendix A.

1. Ventricular dysrhythmias   (LO 15.1)
   a. originate above the bundle of His.
   b. are seldom life-threatening.
   c. have QRS complexes greater than 0.12 seconds in duration.
   d. have notched P waves that precede each QRS complex.

2. The presence of PVCs is an indication of which of the following?   (LO 15.3)
   a. Irritability of the myocardium
   b. Possible hypoxia
   c. Drug toxicity
   d. All of the above

3. Premature ventricular complexes (PVCs)   (LO 15.3)
   a. arise from the AV node.
   b. have a QRS complex that appears narrow and normal in configuration.
   c. have T waves in the same direction as the R waves of complexes.
   d. are followed by a compensatory pause.

4. Match the following types of PVCs with the correct description.   (LO 15.3)

Descriptions	Types of PVCs
a. PVCs that occur every other beat	_____ R-on-T phenomenon
b. PVCs that appear different from one another	_____ Multiform
c. Occurs when the PVC wave falls on the T wave	_____ Bigeminal
d. PVCs that occur every fourth beat	_____ Quadrigeminal

5. Two PVCs in a row are called   (LO 15.3)
   a. a run of PVCs.
   b. polymorphic.
   c. a couplet.
   d. interpolated.

6. Three PVCs in succession are referred to as a _____ of VT. (LO 15.3)
   a. couplet
   b. run
   c. trigeminy
   d. succession

7. Idioventricular rhythm has a rate of _____ beats per minute. (LO 15.5)
   a. 20 to 40
   b. 30 to 60
   c. 50 to 75
   d. 100 to 150

8. Idioventricular rhythm   (LO 15.5)
   a. is regularly irregular.
   b. has wide, bizarre-looking QRS complexes.
   c. has upright P waves preceding each QRS complex.
   d. arises from the AV junction.

9. VT has   (LO 15.7)
   a. inverted T waves.
   b. a rate of between 60 and 100 beats per minute.
   c. an irregular rhythm.
   d. QRS complexes greater than 0.12 seconds in duration.

10. Polymorphic VT   (LO 15.8)
    a. arises from one location in the ventricles.
    b. has narrow QRS complexes.
    c. has an appearance that varies considerably from complex to complex.
    d. results from chaotic firing of multiple sites in the ventricles.

11. VF appears on the ECG monitor as having   (LO 15.9)
    a. an overall pattern that appears irregularly shaped, chaotic, and lacks any regular repeating features.
    b. narrow QRS complexes.
    c. electrical signals that are the same height.
    d. P waves interspersed between bizarre-looking QRS complexes.

12. With asystole   (LO 15.10)
    a. the heart quivers.
    b. there is a flat line on the ECG monitor.
    c. there are QRS complexes but no P waves.
    d. there is an organized rhythm but no corresponding mechanical contraction of the heart.

13. Describe how the QRS complexes appear in a rhythm that originates from the ventricles. (LO 15.2)

14. Describe when ventricular escape beats will occur.   (LO 15.4)

Referring to the scenario at the beginning of this chapter answer the following questions.

15. There is no pulse with VF because the   (LO 15.9)
    a. heart is contracting too fast.
    b. heart is contracting too slow.
    c. heart is not contracting at all.
    d. stroke volume is too high.

16. The flat line rhythm that occurred after defibrillation is called   (LO 15.10)
    a. asystole.
    b. sinus arrest.
    c. ventricular arrest.
    d. agonal.

17. The goal of defibrillation is to create _____ and hope that a pace-maker will awaken and stimulate ventricular contraction.   (LO 15.9)
    a. sinus rhythm
    b. sinus tachycardia
    c. asystole
    d. VT

16 AV Heart Blocks

©rivetti/Getty Images

Chapter Outline

Courtesy Physio-Control

Learning Outcomes

LO 16.1   Describe the origin and key features of block of the AV node.

LO 16.2   Identify the causes, effects, and appearance of 1st-degree AV heart block.

LO 16.3   Identify the causes, effects, and appearance of 2nd-degree AV heart block, type I.

LO 16.4   Identify the causes, effects, and appearance of 2nd-degree AV heart block, type II.

LO 16.5   Identify the causes, effects, and appearance of 3rd-degree AV heart block.

LO 16.6   Identify the causes, effects, and appearance of atrioventricular dissociation.

Case History

A 65-year-old man presents to the emergency department complaining of palpitations. He denies chest pain, shortness of breath, or light headedness. He had a heart attack five years ago and has hypertension controlled by medication.

After obtaining his vital signs, normal except for a slow pulse, the charge nurse attaches the ECG leads and turns on the monitor. The emergency room (ER) physician enters the room and examines the tracing while the nurse starts an IV line. The physician calls the paramedic student into the exam room and hands her the ECG strip and asks her to interpret it. The student indicates that the rate is slow, but the rhythm is regular. She also notices that the PR interval is normal but that there are more P waves than QRS complexes.

16.1 Block of the Atrioventricular Node

AV heart blocks are partial delays or complete interruptions in the cardiac conduction pathway between the atria and ventricles (Figure 16-1). The block can occur at the AV node, the bundle of His, or the bundle branches.

With AV heart block, cardiac output can be negatively affected if the ventricular rate slows. This can occur either due to electrical impulses from the atria being

Figure 16-1
AV heart blocks.

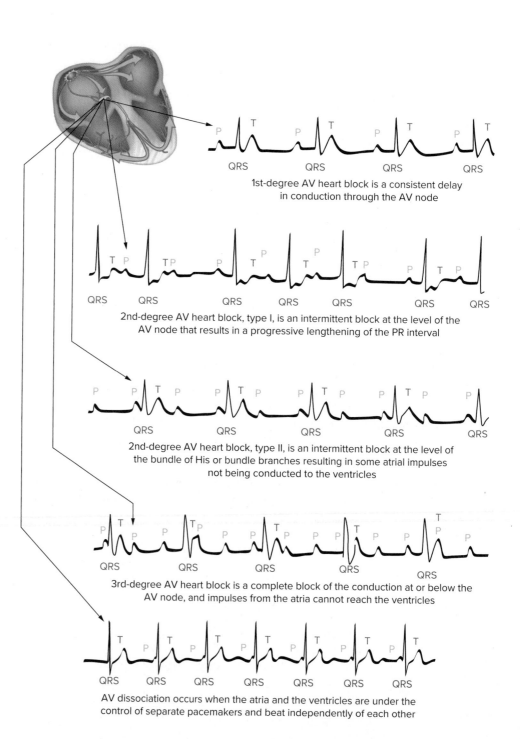

1st-degree AV heart block is a consistent delay in conduction through the AV node

2nd-degree AV heart block, type I, is an intermittent block at the level of the AV node that results in a progressive lengthening of the PR interval

2nd-degree AV heart block, type II, is an intermittent block at the level of the bundle of His or bundle branches resulting in some atrial impulses not being conducted to the ventricles

3rd-degree AV heart block is a complete block of the conduction at or below the AV node, and impulses from the atria cannot reach the ventricles

AV dissociation occurs when the atria and the ventricles are under the control of separate pacemakers and beat independently of each other

blocked from entering the ventricles, or, as what can occur with complete AV heart block, the ventricular escape beats arise from a site that is lower in the ventricles (producing a slower rate).

The most common causes of heart block are ischemia, MI, degenerative disease of the conduction system, congenital anomalies, and medications (especially digitalis preparations). Dysrhythmias resulting from AV heart blocks include the following:

- 1st-degree AV heart block
- 2nd-degree AV heart block, type I (Wenckebach)
- 2nd-degree AV heart block, type II
- 3rd-degree AV heart block

16.2 1st-Degree Atrioventricular Heart Block

Description

1st-degree AV heart block is not a true block because all impulses are conducted from the atria to the ventricles, but rather it is a consistent delay of conduction at the level of the AV node (Figure 16-2). It results in prolongation of the PR interval that remains the same duration from beat to beat.

In 1st-degree AV heart block, impulses arise from the SA node but their passage through the AV node is delayed.

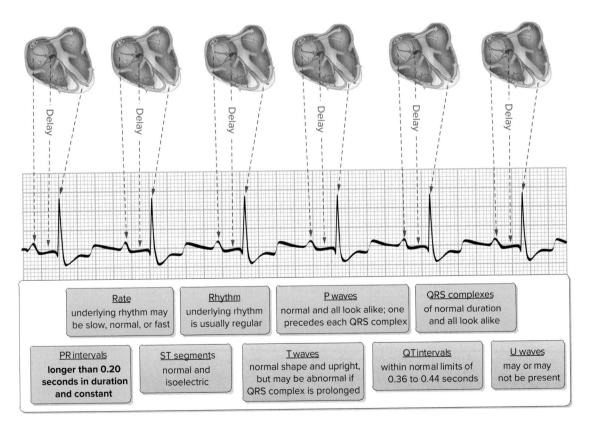

Rate	Rhythm	P waves	QRS complexes
underlying rhythm may be slow, normal, or fast	underlying rhythm is usually regular	normal and all look alike; one precedes each QRS complex	of normal duration and all look alike

PR intervals	ST segments	T waves	QT intervals	U waves
longer than 0.20 seconds in duration and constant	normal and isoelectric	normal shape and upright, but may be abnormal if QRS complex is prolonged	within normal limits of 0.36 to 0.44 seconds	may or may not be present

Figure 16-2
1st-degree AV heart block.

Causes

While 1st-degree AV heart block may occur in healthy persons for no apparent reason (particularly athletes), it is sometimes associated with other conditions such as myocardial ischemia, acute MI, increased vagal (parasympathetic) tone, and digitalis toxicity. Other causes of 1st-degree AV heart block are shown in the following box.

Causes of 1st-Degree Heart Block	
Cause	Examples
Cardiac diseases	Myocardial ischemia or infarction (often inferior wall MI), injury or ischemia to the AV node or junction, myocarditis, degenerative changes (age related) in the heart
Use of certain drugs	Beta-adrenergic blockers (i.e., propranolol), calcium channel blockers (i.e., verapamil), digoxin, antidysrhythmics (i.e., amiodarone, quinidine, procainamide)
Autonomic nervous system stimulation	Increased vagal tone
Noncardiac disorders	Hypokalemia and hyperkalemia, hypothermia, hypothyroidism

Effects

1st-degree AV heart block is of little or no clinical significance because all impulses are conducted to the ventricles. The patient experiencing this condition is usually asymptomatic. 1st-degree AV heart block may be temporary, particularly if it stems from ischemia early in the course of MI or certain medications. Conversely, it may also progress to higher degree block, especially in the presence of inferior wall MI.

ECG Appearance

The most obvious characteristic of 1st-degree AV heart block is PR intervals greater than 0.20 seconds in duration and constant. The underlying rhythm is usually regular whereas the rate is that of the underlying rhythm. It can occur in bradycardic, normal rate, and tachycardic rhythms. The P waves are normal; one precedes each QRS complex, and the QRS complexes are within normal limits.

1st-degree AV block is more of a condition than a dysrhythmia. For this reason, we typically describe it as part of the underlying rhythm (e.g., sinus rhythm, sinus tachycardia, or sinus bradycardia with 1st-degree AV block).

Treatment

As cardiac output is not affected, no specific treatment is indicated for 1st-degree AV heart block. However, efforts are directed toward identifying and treating the cause.

16.3 2nd-Degree Atrioventricular Heart Block, Type I

Description

2nd-degree AV heart block, type I, also called **Wenckebach** or **Mobitz Type I,** is an intermittent block at the level of the AV node (Figure 16-3). The PR interval (representing AV conduction time) increases until a QRS complex is not generated (dropped). By then, AV conduction recovers and the sequence repeats.

Some describe the pathophysiology of this dysrhythmia as a weakened AV junction that grows more tired with each heartbeat (thus producing a progressively longer PR interval following each P wave). Finally, the AV junction is too tired to carry the impulse, and a QRS complex is dropped (only a P wave appears). The lack of conduction through the AV junction allows it to rest; thus the next PR interval is shorter. Then as each subsequent impulse is generated and transmitted through the AV junction, there is a progressively longer PR interval until, again, a QRS complex is dropped. This cycle repeats.

The number of beats in which the P wave is followed by a QRS complex before a QRS complex is finally dropped can vary among individuals. For example, one patient may have a dropped beat after every third beat, and another patient may drop the QRS complex after every fifth complex.

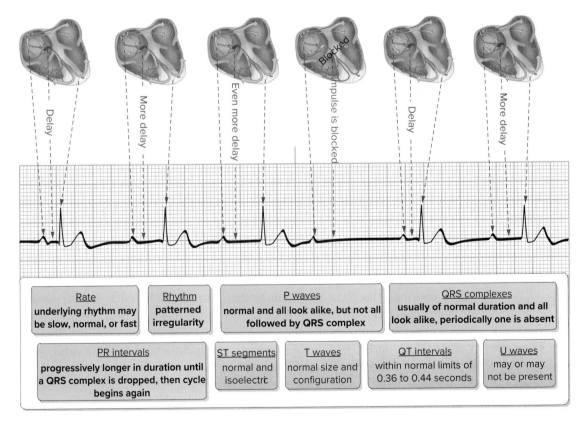

In 2nd-degree AV heart block, type I (Wenckebach), impulses arise from the SA node but their passage through the AV node is progressively delayed until the impulse is blocked.

Rate	Rhythm	P waves	QRS complexes
underlying rhythm may be slow, normal, or fast	patterned irregularity	normal and all look alike, but not all followed by QRS complex	usually of normal duration and all look alike, periodically one is absent

PR intervals	ST segments	T waves	QT intervals	U waves
progressively longer in duration until a QRS complex is dropped, then cycle begins again	normal and isoelectric	normal size and configuration	within normal limits of 0.36 to 0.44 seconds	may or may not be present

Figure 16-3
2nd-degree AV heart block, type I (Wenckebach).

Causes

2nd-degree AV heart block, type I, often occurs in acute MI or acute myocarditis. Common causes are shown in the following box.

Causes of 2nd-Degree AV Heart Block, Type I	
Cause	**Examples**
Cardiac diseases	AV nodal ischemia secondary to right coronary artery occlusion, myocardial ischemia or MI (inferior wall MI), cardiac surgery, myocarditis, rheumatic fever
Use of certain drugs	Beta-adrenergic blockers, calcium channel blockers, digitalis, quinidine, procainamide
Autonomic nervous system stimulation	Increased vagal tone
Noncardiac disorders	Electrolyte imbalance (hyperkalemia)

Effects

2nd-degree AV heart block, type I, may occur in otherwise healthy persons. By itself, it is usually transient and reversible, resolving when the underlying condition is corrected. However, particularly if it occurs early in MI, it may progress to more serious blocks. If dropped beats occur frequently, the patient may show signs and symptoms of decreased cardiac output. Those types of heart block that result in dropped QRS complexes or a slower ventricular rate can lead to a decrease in cardiac output (decreased heart rate × stroke volume = decreased cardiac output).

ECG Appearance

The characteristics that stand out in 2nd-degree AV heart block, type I, are a patterned, irregular rhythm; cycles of progressively longer PR intervals; and more P waves than QRS complexes. The irregularity appears as a pattern (the cycle seems to occur over and over); it is often described as *grouped beating*. The PR intervals become progressively longer until a P wave fails to conduct, resulting in a "dropped" QRS complex. After the blocked beat, the cycle starts again. The P waves are upright and uniform, but there are more P waves than QRS complexes as some of the QRS complexes are blocked.

The atrial rate is that of the underlying rhythm whereas the ventricular rate is slightly less than the atrial rate (slower than normal). The QRS complexes are within normal limits. Unless the block is infranodal, in which case the QRS complexes will be wide. Infranodal refers to a block caused by an abnormality below the AV node, either in the bundle of His or in both bundle branches. An infranodal block has more serious clinical implications than a block at the level of the AV node. It most often occurs in older persons. Symptoms include frequent episodes of fainting and a pulse rate of 20 to 40 bpm. The P-P interval is constant while the R-R interval increases until a QRS complex is dropped.

In 2nd-degree AV heart block, type II, impulses arise from the SA node but some are blocked in the AV node.

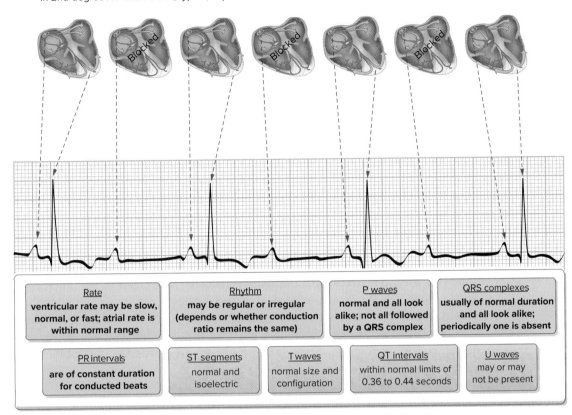

Rate	Rhythm	P waves	QRS complexes
ventricular rate may be slow, normal, or fast; atrial rate is within normal range	may be regular or irregular (depends or whether conduction ratio remains the same)	normal and all look alike; not all followed by a QRS complex	usually of normal duration and all look alike; periodically one is absent

PR intervals	ST segments	T waves	QT intervals	U waves
are of constant duration for conducted beats	normal and isoelectric	normal size and configuration	within normal limits of 0.36 to 0.44 seconds	may or may not be present

Figure 16-4
2nd-degree AV heart block, type II.

Treatment

If patient is asymptomatic, no specific treatment is needed. Patients who are symptomatic (e.g., chest pain, hypotension) should receive oxygen, an IV lifeline, and administration of atropine and transcutaneous pacing should be considered (see Figure 12-4).

16.4 2nd-Degree Atrioventricular Heart Block, Type II

Description

2nd-degree AV heart block, type II, is an intermittent block at the level of the AV node, bundle of His, or bundle branches resulting in atrial impulses that are not conducted to the ventricles (Figure 16-4). It is less common than type I, but it is more serious as it may progress to complete AV heart block. It differs from type I in that the PR interval is *constant* prior to a beat being "dropped."

Causes

Type II block is usually associated with anterior-wall MI, degenerative changes in the conduction system, or severe coronary artery disease. Common causes of type II block are shown in the box on the following page.

Causes of 2nd-Degree AV Heart Block, Type II	
Cause	**Examples**
Cardiac diseases	Anterior wall MI, severe coronary artery disease, organic heart disease, degenerative changes in the conduction system, cardiac surgery, myocarditis, rheumatic fever
Use of certain drugs	Beta-adrenergic blockers, calcium channel blockers, digitalis, quinidine, procainamide
Autonomic nervous system stimulation	Increased vagal tone
Noncardiac disorders	Electrolyte imbalance (hyperkalemia)

Effects

Type II is a serious dysrhythmia, usually considered malignant in the emergency setting when it is symptomatic. Slow ventricular rates result in decreased cardiac output and may produce signs and symptoms of hypoperfusion (low blood pressure, shortness of breath, congestive heart failure, pulmonary congestion, and decreased level of consciousness). It may progress to a more severe heart block and even to ventricular asystole.

ECG Appearance

The feature that stands out in 2nd-degree AV heart block, type II, is the presence of more P waves than QRS complexes. The number of P waves for each QRS complex (e.g., 2:1, 3:1, or 4:1), may be fixed or it may vary (e.g., 2:1, then 3:1, or 4:1, or vice versa). The PR interval may be normal or prolonged, but it is constant for each conducted complex. The QRS complexes are usually within normal limits, but they may also be wide if the block is infranodal. Infranodal refers to a block caused by an abnormality below the AV node, either in the bundle of His or in both bundle branches. An infranodal block has more serious clinical implications than a block at the level of the AV node. It most often occurs in older persons. Symptoms include frequent episodes of fainting and a pulse rate of 20 to 40 bpm. The rhythm is typically regular. It will be irregular if the conduction ratio (number of P waves to each QRS complex) varies. The atrial rate is that of underlying rhythm whereas the ventricular rate is less than the atrial rate.

The key characteristic that helps you differentiate between the two types of 2nd-degree AV block is whether the PR interval changes or remains constant.

Treatment

Patients who are symptomatic (e.g., chest pain, hypotension) should receive supplemental oxygen, an IV lifeline, and you should consider atropine, transcutaneous pacing or an infusion of epinephrine or dopamine (see Figure 12-4).

16.5 3rd-Degree Atrioventricular Heart Block

Description

3rd-degree AV heart block is a complete block of the conduction at or below the AV node so that impulses from the atria cannot reach the ventricles (Figure 16-5). It is also called **complete AV heart block.** The SA node serves as the pacemaker for the atria, typically maintaining a regular rate of 60 to 100 beats per minute. The pacemaker for the ventricles arises as an escape rhythm from the AV junction at a rate of 40 to 60 beats per minute or from the ventricles at a rate of 20 to 40 beats per minute. P waves and QRS complexes occur rhythmically, but the rhythms are unrelated to each other.

Causes

3rd-degree AV heart block occurring at the AV node is most commonly caused by a congenital condition. It may also occur in older adults because of chronic degenerative changes in the conduction system. Common causes are listed in the box on the following page.

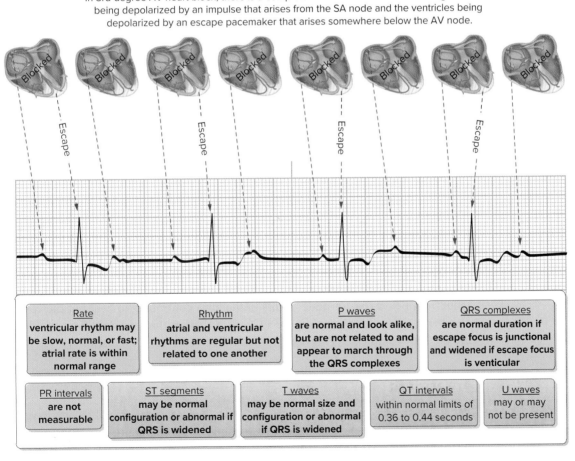

In 3rd-degree AV heart block, there is a complete block at the AV node resulting in the atria being depolarized by an impulse that arises from the SA node and the ventricles being depolarized by an escape pacemaker that arises somewhere below the AV node.

Rate	Rhythm	P waves	QRS complexes
ventricular rhythm may be slow, normal, or fast; atrial rate is within normal range	atrial and ventricular rhythms are regular but not related to one another	are normal and look alike, but are not related to and appear to march through the QRS complexes	are normal duration if escape focus is junctional and widened if escape focus is venticular

PR intervals	ST segments	T waves	QT intervals	U waves
are not measurable	may be normal configuration or abnormal if QRS is widened	may be normal size and configuration or abnormal if QRS is widened	within normal limits of 0.36 to 0.44 seconds	may or may not be present

Figure 16-5
3rd-degree AV heart block.

Causes of 3rd-Degree AV Heart Block	
Cause	**Examples**
Cardiac diseases	Anterior or inferior wall MI, AV node damage, cardiac surgery injury, severe coronary artery disease, degenerative changes in the conduction system, myocarditis, rheumatic fever, septal necrosis
Use of certain drugs	Beta-adrenergic blockers, calcium channel blockers, digitalis, quinidine, procainamide
Autonomic nervous system stimulation	Increased vagal tone
Noncardiac disorders	Hyperkalemia

Effects

Most occurrences of 3rd-degree AV heart block are well tolerated as long as the escape rhythm is fast enough to generate a sufficient cardiac output to maintain adequate perfusion. However, severe bradycardia and decreased cardiac output may occur with 3rd-degree AV heart block because of the slow ventricular rate and asynchronous action of the atria and ventricles. If it occurs with wide QRS complexes, it is considered an ominous sign. This type of heart block is potentially lethal, as patients with this dysrhythmia are often hemodynamically unstable.

ECG Appearance

The characteristic that stands out with 3rd-degree AV heart block is the presence of more P waves than QRS complexes. But further assessment of the dysrhythmia reveals that the P waves appear normal and seem to march right through the QRS complexes. This reveals that there is no relationship between the P waves and QRS complexes. As such, the PR intervals can be described as being "absent." The atrial rate is that of the underlying rhythm. The ventricular rate depends on the escape focus—40 to 60 beats per minute if it is junctional and 20 to 40 beats per minute if it is ventricular. The atrial and ventricular rhythms are regular but independent of each other. The QRS complexes are normal if the escape focus is junctional and widened if the escape focus is ventricular.

Going back to our discussion in Chapter 15, an escape pacemaker that fails to originate from below the AV node results in the atria continuing to beat but the ventricles being silent. It appears on the ECG as a rhythm that has only P waves; there are no QRS complexes. This condition is called ventricular standstill.

Treatment

Patients who are symptomatic (e.g., chest pain, hypotension) should receive oxygen, an IV lifeline, atropine, transcutaneous pacing, or an epinephrine or dopamine infusion (see Figure 12-4). Atropine may be used in 3rd-degree block with new wide QRS complexes while you are setting up the transcutaneous pacemaker.

Learning AV heart blocks can be challenging. Table 16-1 shows the key differences between the four types of AV heart block.

	Rhythm	P Waves	QRS Complexes	PR Intervals
1st-degree	Underlying rhythm is usually regular	Present and normal; all the P waves are followed by a QRS complex	Normal	Longer than 0.20 seconds and is constant
2nd-degree, type I	Patterned irregularity	Present and normal; not all the P waves are followed by a QRS complex	Normal or can be widened	Progressively longer until a QRS complex is dropped; the cycle then begins again
2nd-degree, type II	May be regular or irregular (depends on whether conduction ratio remains the same)	Present and normal; not all the P waves are followed by a QRS complex	Normal or can be widened	Constant for all conducted beats
3rd-degree	Atrial rhythm and ventricular rhythms are regular but not related to one another	Present and normal; not related to the QRS complexes; appear to march through the QRS complexes	Normal if escape QRS focus is junctional and widened if escape focus is ventricular	Not measurable

Table 16-1
Types of AV Heart Blocks

16.6 Atrioventricular Dissociation

Description

AV dissociation occurs when the atria and the ventricles are under the control of separate pacemakers and beat independently of each other. AV dissociation is not a dysrhythmia but instead is a result of some mechanism that causes independent beating of atria and ventricles. With AV dissociation, the ventricular rate is the same or faster than the atrial rate (Figure 16-6).

Causes

Four causes of AV dissociation include slowing of the primary pacemaker, acceleration of a pacemaker lower in the heart, AV block, and interference.

- With slowed or impaired sinus impulse formation or SA nodal conduction, such as sinus bradycardia or sinus arrest, a junctional or ventricular pacemaker emerges to take control of the ventricles, resulting in AV dissociation.

- Impulse formation in the AV junction or ventricles that is faster than the firing rate of the sinus node allows the lower pacemaker to take over control of the ventricles while the sinus node still controls the atria.

- Complete AV block results in the SA node and ventricles firing independently of each. However, AV block and AV dissociation are not necessarily the same thing: *All 3rd-degree block is AV dissociation but not all AV dissociation is 3rd-degree block.*

- Anything that interferes with the ability of an atrial impulse to conduct to the ventricles, such as pauses produced by premature beats, presents an opportunity for a pacemaker lower in the heart to emerge and take charge of the ventricles. Drug toxicity from digitalis, calcium channel blockers, disopyramide, and amiodarone also can result in interference.

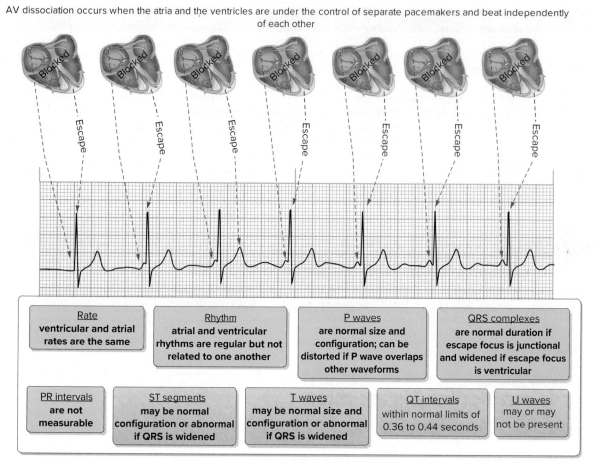

AV dissociation occurs when the atria and the ventricles are under the control of separate pacemakers and beat independently of each other

Rate	Rhythm	P waves	QRS complexes
ventricular and atrial rates are the same	atrial and ventricular rhythms are regular but not related to one another	are normal size and configuration; can be distorted if P wave overlaps other waveforms	are normal duration if escape focus is junctional and widened if escape focus is ventricular

PR intervals	ST segments	T waves	QT intervals	U waves
are not measurable	may be normal configuration or abnormal if QRS is widened	may be normal size and configuration or abnormal if QRS is widened	within normal limits of 0.36 to 0.44 seconds	may or may not be present

Figure 16-6
AV dissociation.

Effects

Signs and symptoms vary depending on the underlying cause. Signs of decreased cardiac output may be present if the heart rate is reduced.

ECG Appearance

With AV dissociation, the atrial and ventricular rhythms are regular. The atrial rate usually equals the ventricular rate. The P waves are of normal size and configuration. However, they can be distorted if the P wave overlaps other waveforms. The PR interval is absent or unidentifiable because the atria and ventricles are firing independently of each other. The QRS complex depends on the origin of the ventricular beat. A pacemaker high in the AV junction produces a narrow QRS complex whereas a pacemaker lower in the bundle of His produces a wide QRS complex. The T wave may be altered or inverted depending on the underlying cause.

Treatment

No treatment is needed if the condition causing AV dissociation is clinically insignificant. If the condition causing the AV dissociation reduces cardiac output, treatment is directed at managing the underlying problem. Specific measures include delivering atropine and other rate-accelerating agents and/or pacemaker insertion (see Figure 12-4). If drug toxicity caused the original disturbance, the drug should be discontinued.

Practice Makes Perfect

For each of the 53 ECG tracings on the following pages, practice the Nine-Step Process for analyzing ECGs. To achieve the greatest learning, you should practice assessing and interpreting the ECGs immediately after reading this chapter. Below are questions you should consider as you assess each tracing. Your answers can be written into the area above each ECG marked "ECG Findings." Your findings can then be compared to the answers provided in Appendix A. All tracings are six seconds in length.

1. Determine the heart rate. Is it slow? Normal? Fast? What is the ventricular rate? What is the atrial rate?

2. Determine if the rhythm is regular or irregular. If it is irregular, what type of irregularity is it? Occasional or frequent? Slight? Sudden acceleration or slowing in heart rate? Total? Patterned? Does it have a variable conduction ratio?

3. Determine if P waves are present. If so, how do they appear? Do they have normal height and duration? Are they tall? Notched? Wide? Biphasic? Of differing morphology? Inverted? One for each QRS complex? More than one preceding some or all the QRS complexes? Do they have a sawtooth appearance? An indiscernible chaotic baseline?

4. Determine if QRS complexes are present. If so, how do they appear? Narrow with proper amplitude? Tall? Low amplitude? Delta wave? Notched? Wide? Bizarre-looking? With chaotic waveforms?

5. Determine the presence of PR intervals. If present, how do they appear? Constant? Of normal duration? Shortened? Lengthened? Progressively longer? Varying?

6. Evaluate the ST segments. Do they have normal duration and position? Are they elevated? (If so, are they flat, concave, convex, arched?) Depressed? (If so, are they normal, flat, downsloping or upsloping?)

7. Determine if T waves are present. If so, how do they appear? Of normal height and duration? Tall? Wide? Notched? Inverted?

8. Determine the presence of QT intervals. If present, what is their duration? Normal? Shortened? Prolonged?

9. Determine if U waves are present. If present, how do they appear? Of normal height and duration? Inverted?

10. Identify the rhythm or dysrhythmia.

1.

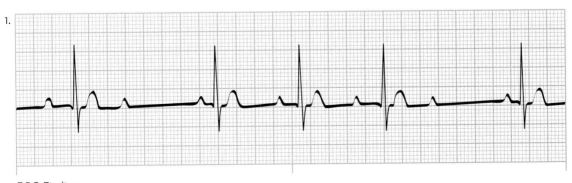

ECG Findings:

2.

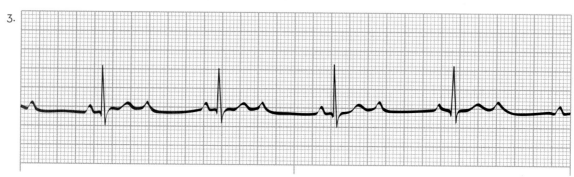

ECG Findings:

3.

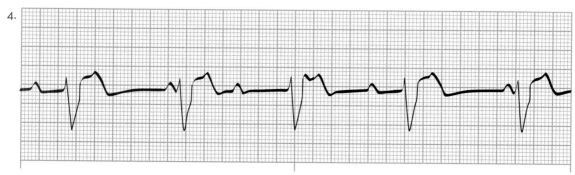

ECG Findings:

4.

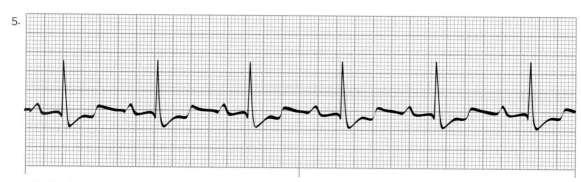

ECG Findings:

5.

ECG Findings:

6.

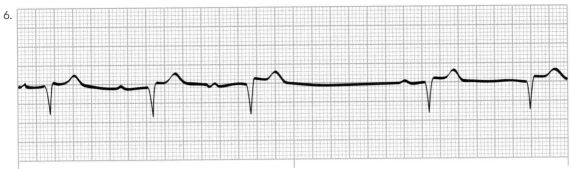

ECG Findings:

7.

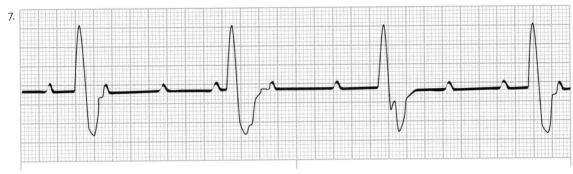

ECG Findings:

8.

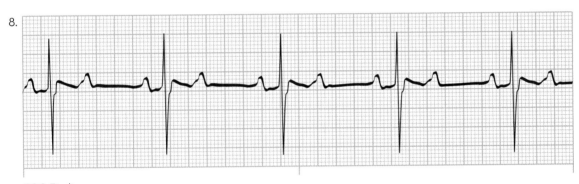

ECG Findings:

9.

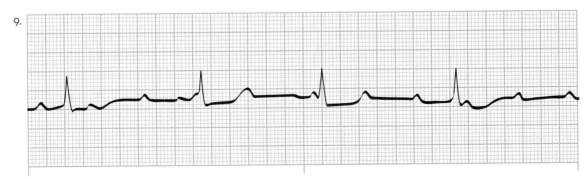

ECG Findings:

10.

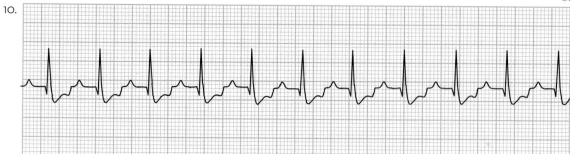

ECG Findings:

11.

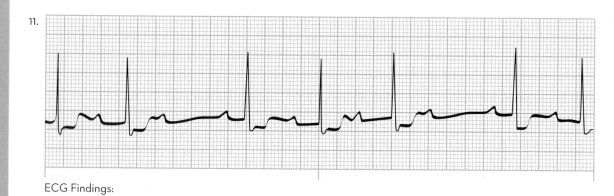

ECG Findings:

12.

ECG Findings:

13.

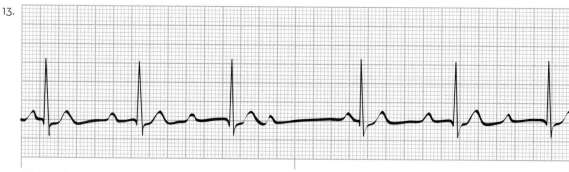

ECG Findings:

14.

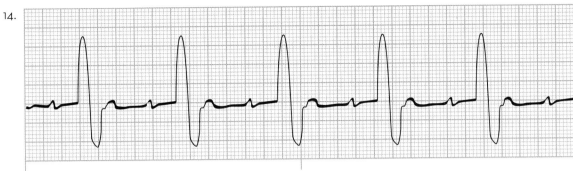

ECG Findings:

15.

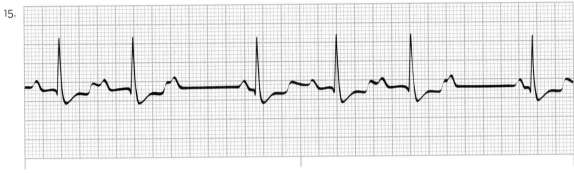

ECG Findings:

16.

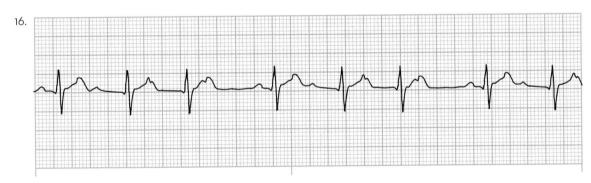

ECG Findings:

17.

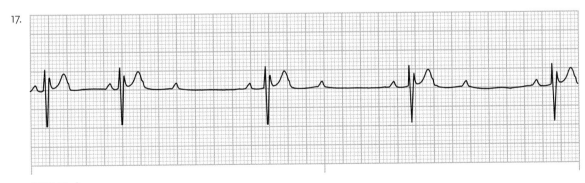

ECG Findings:

18.

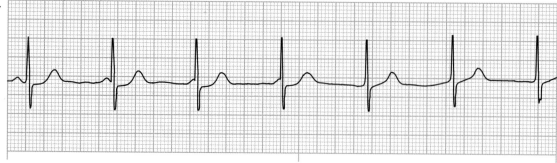

ECG Findings:

19.

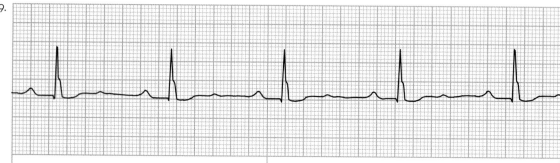

ECG Findings:

20.

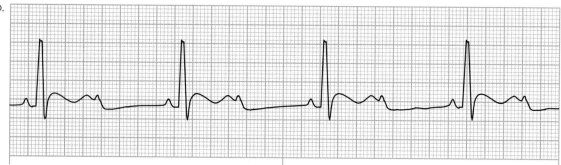

ECG Findings:

21.

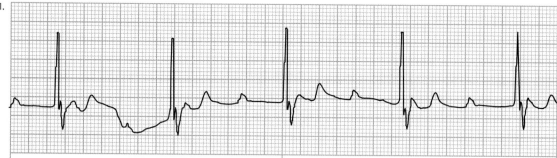

ECG Findings:

22.

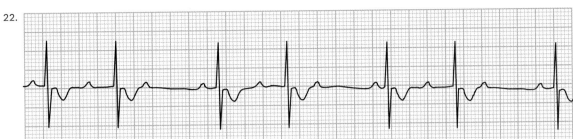

ECG Findings:

23.

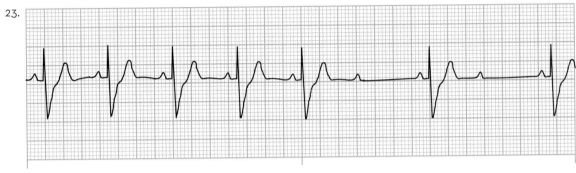

ECG Findings:

24.

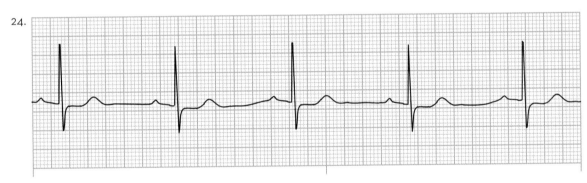

ECG Findings:

25.

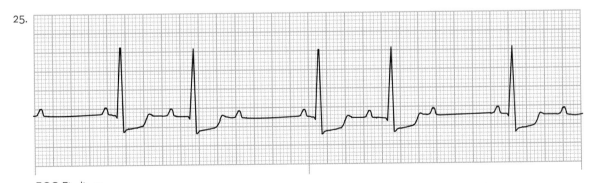

ECG Findings:

26.

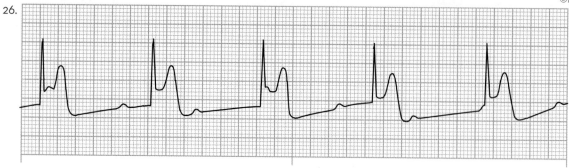

ECG Findings:

27.

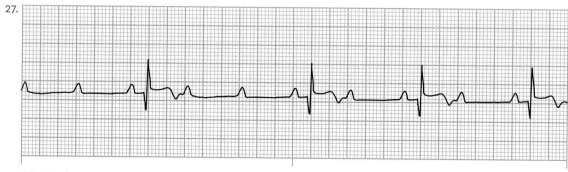

ECG Findings:

28.

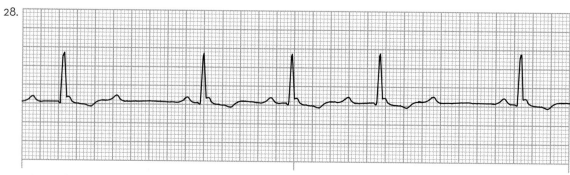

ECG Findings:

29.

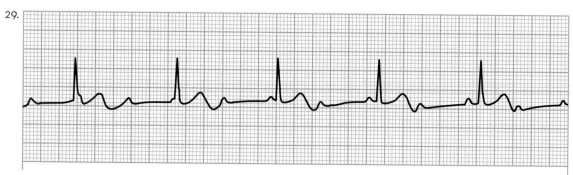

ECG Findings:

30.

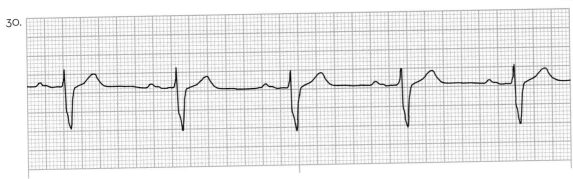

ECG Findings:

31.

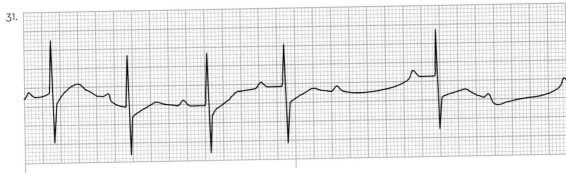

ECG Findings:

32.

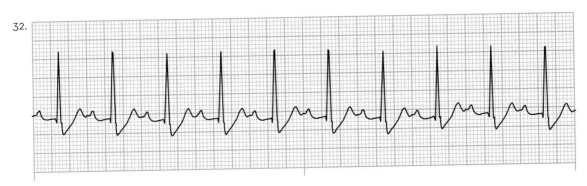

ECG Findings:

33.

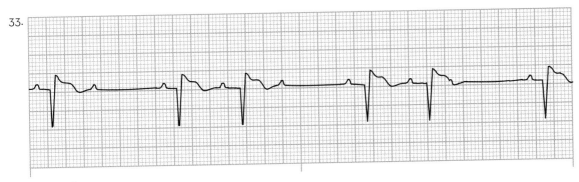

ECG Findings:

34.

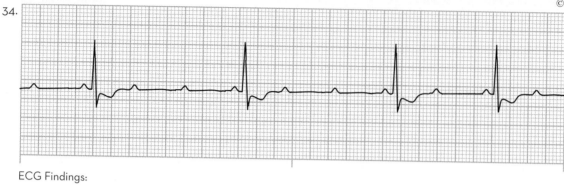

ECG Findings:

35.

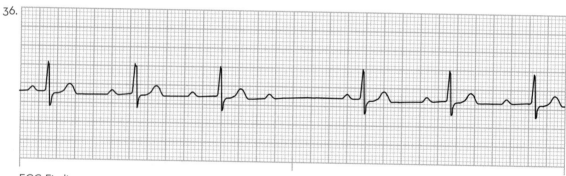

ECG Findings:

36.

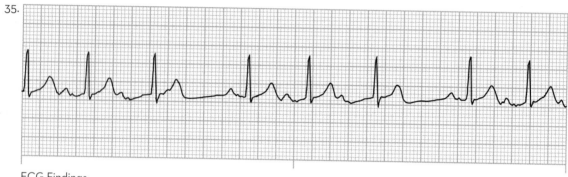

ECG Findings:

37.

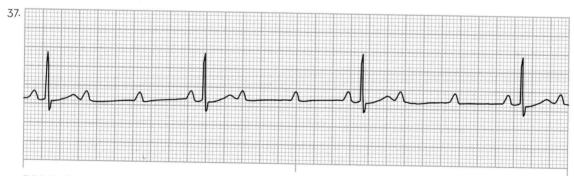

ECG Findings:

38.

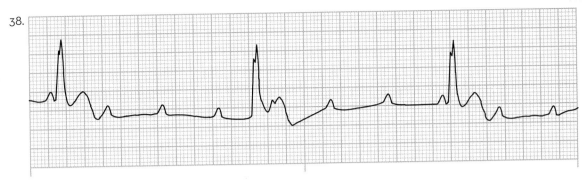

ECG Findings:

39.

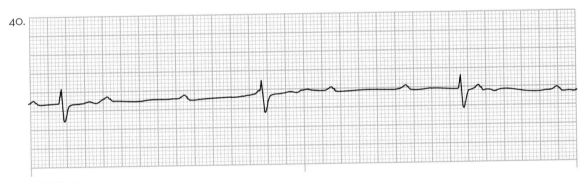

ECG Findings:

40.

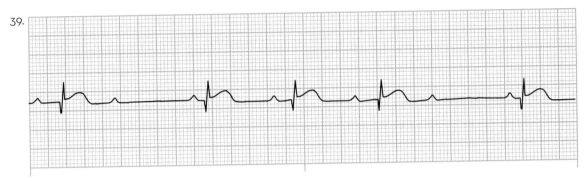

ECG Findings:

41.

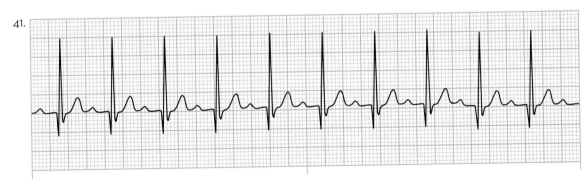

ECG Findings:

42.

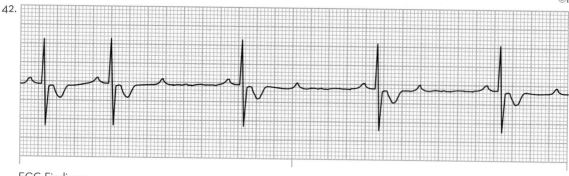

ECG Findings:

43.

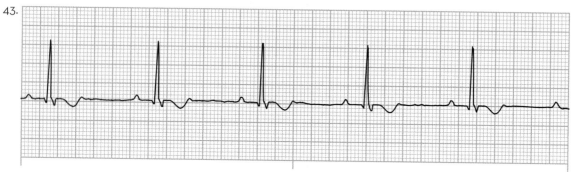

ECG Findings:

44.

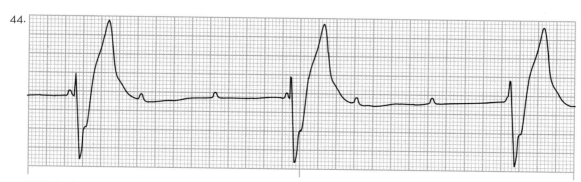

ECG Findings:

45.

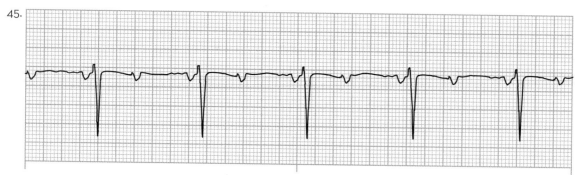

ECG Findings:

46.

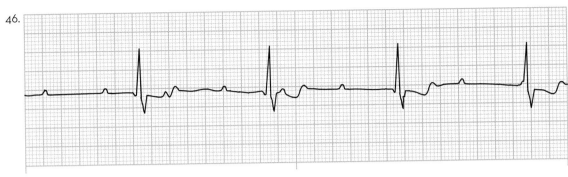

ECG Findings:

47.

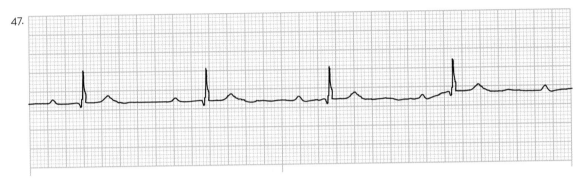

ECG Findings:

48.

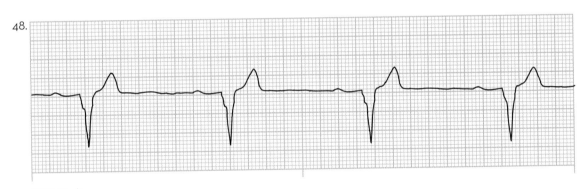

ECG Findings:

49.

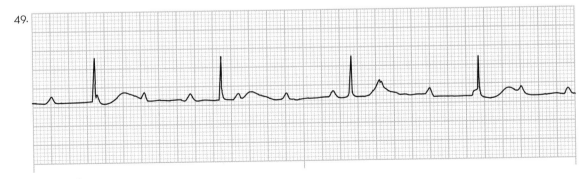

ECG Findings:

50.

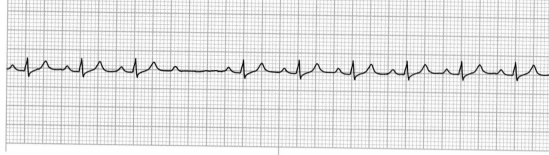

ECG Findings:

51.

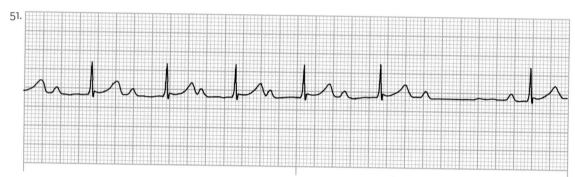

ECG Findings:

52.

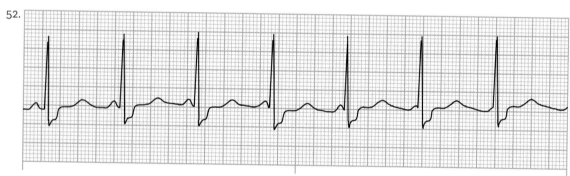

ECG Findings:

53.

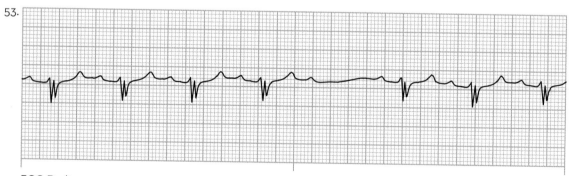

ECG Findings:

Key Points 🌡

LO 16.1	• AV heart blocks are partial delays or complete interruptions in the cardiac conduction pathway between the atria and ventricles. A block can occur at the AV node, the bundle of His, or the bundle branches.
LO 16.2	• 1st-degree AV heart block is not a true block. Instead, it is a consistent delay of conduction at the level of the AV node that results in a PR interval greater than 0.20 seconds in duration.
LO 16.3	• 2nd-degree AV heart block, type I, is also referred to as *Wenckebach* or *Mobitz Type I.* It is an intermittent block at the level of the AV junction.
	• With 2nd-degree AV heart block, type I, the PR interval increases until a QRS complex is dropped. After the dropped beat, the next PR interval is shorter. Then, as each subsequent impulse is generated and transmitted through the AV junction, there is a progressively longer PR interval until, again, a QRS complex is dropped. This cycle can repeat over and over.
	• With 2nd-degree AV heart block, type I, there are more P waves than QRS complexes, and the rhythm is regularly irregular. This is also called a *patterned irregularity.*
LO 16.4	• 2nd-degree AV heart block, type II, is an intermittent block at the level of the bundle of His or bundle branches resulting in atrial impulses that are not conducted to the ventricles.
	• With 2nd-degree AV heart block, type II, there are more P waves than QRS complexes, and the duration of PR interval of the conducted beats remains the same (is constant).
LO 16.5	• 3rd-degree AV heart block is a complete block of the conduction at or below the AV junction so that impulses from the atria cannot reach the ventricles.
	• In 3rd-degree AV heart block, the pacemaker for the atria arises from the SA node whereas the pacemaker for the ventricles arises as an escape rhythm from the AV junction at a rate of 40 to 60 beats per minute or from the ventricles at a rate of 20 to 40 beats per minute.
	• With 3rd-degree AV heart block, the upright and round P waves seem to march right through the QRS complexes. This reveals that there is no relationship between the P waves and QRS complexes.
	• 2nd- and 3rd-degree AV heart block can lead to decreased cardiac output.
LO 16.6	• In AV dissociation, the SA node loses control of the ventricular rate, leading to the atria and ventricles beating independently of each other, usually at about the same rate.

Assess Your Understanding ☑

The following questions give you a chance to assess your understanding of the material discussed in this chapter. The answers can be found in Appendix A.

1. AV heart blocks are   (LO 16.1)

   a. partial delays or complete interruptions in the cardiac conduction pathway between the atria and ventricles.

   b. early ectopic beats that originate outside the SA node.

c. a shifting of the pacemaker site between the SA node, atria, and/or AV junction.

d. the result of accessory conduction pathways between the atria and ventricles.

2. In 1st-degree AV heart block, (LO 16.2)

a. the PR intervals are greater than 0.20 seconds in duration.

b. not all the P waves are followed by QRS complexes.

c. the P waves are inverted.

d. the underlying rhythm is slow.

3. 1st-degree AV heart block is considered a/an (LO 16.2)

a. intermittent block.

b. complete block.

c. consistent delay in conduction.

d. partial block.

4. In 2nd-degree AV heart block, type I, (LO 16.3)

a. the PR intervals get progressively shorter.

b. not all the P waves are followed by a QRS complex.

c. the P waves are inverted.

d. the rhythm appears to be irregularly irregular.

5. In 2nd-degree AV heart block, type II, (LO 16.4)

a. the P waves are all the same.

b. a QRS complex follows each P wave.

c. the pacemaker site is in the AV junction.

d. the ventricular rate is less than 40 beats per minute.

6. Which of the following rhythms has a constant PR interval for all conducted beats? (LO 16.4)

a. 2nd-degree AV heart block, type I

b. 3rd-degree AV heart block

c. 2nd-degree AV heart block, type II

d. All of the above

7. The difference between 2nd-degree AV heart block, type I, and type II is that (LO 16.3, 16.4)

a. in type I, the PR intervals are progressively longer until a QRS complex is dropped and the cycle starts over again.

b. the ventricular rate in type II is usually greater than 60 beats per minute.

c. in type II, the P waves appear to march right through the QRS complexes.

d. in type II, there appears to be a pattern to the irregularity.

8. An ECG rhythm that has no correlation between P waves and QRS complexes is known as (LO 16.5)

a. 1st-degree AV heart block.

b. 2nd-degree AV heart block, type II.

c. 2nd-degree AV heart block, type I.

d. 3rd-degree AV heart block.

9. Which of the following dysrhythmias has P waves that seem to march through the QRS complexes? (LO 16.5)

    a. 2nd-degree AV heart block, type I

    b. 2nd-degree AV heart block, type II

    c. 3rd-degree AV heart block

    d. 1st-degree AV heart block

10. Which of the following heart blocks is irregular? (LO 16.3)

    a. 1st-degree AV heart block

    b. 2nd-degree AV heart block, type I

    c. 3rd-degree AV heart block

    d. All of the above

11. Match the following heart blocks with the correct characteristics.
(LO 16.2, 16.3, 16.4, 16.5)

Heart Blocks	Characteristics
a. 1st-degree AV heart block	_____ The PR interval gets progressively longer until a P wave fails to conduct, resulting in a dropped QRS complex. After the blocked beat, the cycle starts again.
b. 2nd-degree AV heart block, type I	_____ There is a complete block at or below the AV node; there is no relationship between the P waves and QRS complexes.
c. 2nd-degree AV heart block, type II	_____ Some beats are conducted while others are blocked.
d. 3rd-degree AV heart block	_____ This is not a true block; there is a delay at the AV node; each impulse is eventually conducted.

Referring to the scenario at the beginning of this chapter, answer the following questions.

12. What rhythm is the patient in? (LO 16.4)

    a. 1st-degree AV heart block

    b. 2nd-degree AV heart block, type I

    c. 2nd-degree AV heart block, type II

    d. 3rd-degree AV heart block

13. The most common cause of this condition is (LO 16.4)

    a. medication.

    b. MI.

    c. atherosclerotic heart disease.

    d. a malfunctioning artificial pacemaker.

14. The treatment of this condition usually requires which of the following? (LO 16.4)

    a. An artificial pacemaker

    b. An implanted defibrillator

    c. A heart transplant

    d. Cardiac stimulant medication

17 Pacemakers and Implanted Cardioverter-Defibrillators

©rivetti/Getty Images

Chapter Outline

Pacemakers and
Implantable
Defibrillators

Temporary Pacemakers

Permanent Pacemakers

Permanent Pacemaker
Components

Function of Permanent
Pacemakers

Appearance of the Paced
ECG

Pacemaker Failure and
Complications

Implantable Cardioverter-
Defibrillators

Courtesy Philips Healthcare

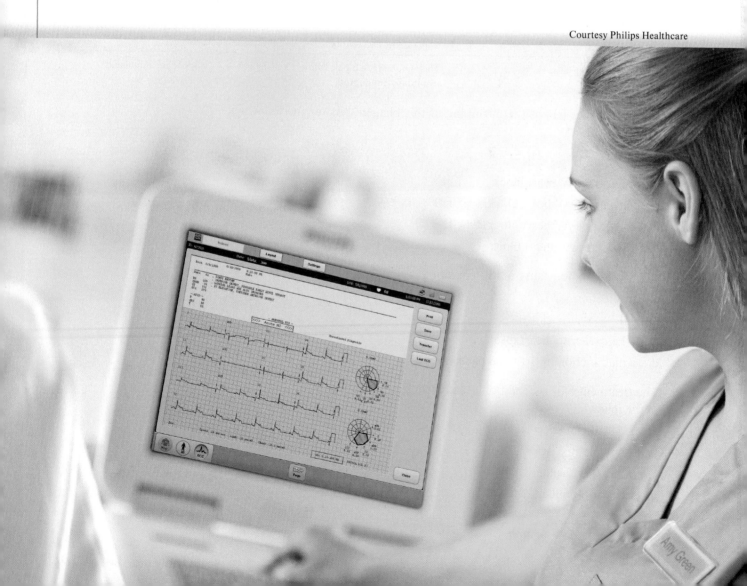

Learning Outcomes

LO 17.1 Describe the purpose and use of artificial pacemakers and implantable defibrillators.

LO 17.2 Describe three types of temporary pacemakers.

LO 17.3 Describe permanent pacemakers, where they are implanted, and their uses.

LO 17.4 List the different components of the permanent pacemaker and describe the function of each.

LO 17.5 Describe the function and different operating modes of the permanent pacemaker.

LO 17.6 Recall the characteristics associated with various paced rhythms seen on the ECG.

LO 17.7 Identify the different causes of pacemaker failure.

LO 17.8 Identify the uses for and operation of the implantable defibrillator.

Case History

A 45-year-old man experienced syncope, which prompted his family to call for an ambulance. The paramedics arrive to find the patient sitting upright but feeling lightheaded. Assessment reveals the patient is confused and has a blood pressure of 80/60 mmHg, a pulse rate of 40 beats per minute, and a respiratory rate of 14 breaths per minute. The ECG is attached to reveal the presence of 3rd-degree AV block with wide QRS complexes. The paramedics administer oxygen, place a IV line, position two electrode pads on the patient's chest, and begin pacing him. Within a few minutes, the patient's heart rate increases to 70 beats per minute, and he appears more alert. He agrees to be transported to the hospital, where he undergoes evaluation and treatment. The treating physician then consults with the on-call cardiologist, who orders implantation of an internal pacemaker. Following successful insertion of the pacemaker, the patient is discharged to recover at home.

17.1 Pacemakers and Implantable Defibrillators

In this chapter, we discuss artificial pacemakers and implanted defibrillators. Artificial pacemakers are medical devices that deliver electrical impulses to the heart to stimulate a normal heartbeat. They are used to maintain an adequate heart rate when the heart's own pacemaker cannot or when there is a block in its electrical conduction system that slows the heart rate. Some pacemakers are external to the body and used to provide temporary treatment, whereas others are permanently implanted in the chest. An implantable defibrillator is a device that delivers electrical shocks to the heart in the presence of life-threatening ventricular tachycardia or ventricular fibrillation. Some modern implantable devices have both a pacemaker and defibrillator combined into one unit.

17.2 Temporary Pacemakers

In Chapter 11, we discussed the use of the transcutaneous pacemaker, a temporary pacemaker in which the electrode pads are applied to the surface of the patient's chest. Two other temporary pacemakers can be used: the epicardial pacemaker and the transvenous pacemaker.

Epicardial Pacing

Epicardial pacing is employed during open heart surgery when the surgical procedure causes atrioventricular block and bradycardia. The electrode is placed in contact with the ventricular epicardium, and small electrical charges are delivered to maintain satisfactory cardiac output until a temporary transvenous electrode can be inserted.

Transvenous Pacing

With transvenous pacing, an electrode is fed through a vein and into either the right atrium or right ventricle. This is done using sterile technique. The pacing electrode is then connected to a pacemaker, the location of which is external to the body, to deliver emergency pacing. Transvenous pacing is used until a permanent pacemaker can be implanted or there is no longer a need for a pacemaker and the pacing electrode is removed.

17.3 Permanent Pacemakers

A permanent pacemaker is a medical device that is implanted in a surgically created pocket beneath the skin in the patient's chest wall just below the clavicle. It is often called an *internal pacemaker*. Pacemakers work because of two main things:

- Electrical current flow results in cardiac activity and can be identified.
- An intrinsic property of the heart is that an action potential, firing in a few cells, spreads as a self-propagating wave throughout the chamber.

First developed in the 1960s, pacemakers originally sent a steady flow of electrical impulses to the heart. Modern pacemakers can monitor the heart and activate only when necessary.

The basic function of pacemakers is to sense for normal heart activity, and if it does not occur, induce action potentials by passing depolarizing currents into a heart chamber, triggering a self-propagating wave. This restores electrical stimulation of the chamber and, ideally, mechanical pumping of the heart. The more complex permanent pacemakers have the ability to sense and/or stimulate both the atrial and ventricular chambers.

Uses

Pacemakers can be of tremendous clinical benefit in various circumstances including

- Symptomatic bradycardia
- Sick sinus syndrome
- Atrial fibrillation with bradycardia
- 3rd-degree (complete) AV heart block
- Symptomatic 2nd-degree AV heart block, particularly type II
- The sudden development of various combinations of AV heart block and bundle branch block in patients experiencing acute MI
- Recurrent tachycardias that can be overdriven and thereby terminated by pacemaker activity
- Synchronization of the heart beat in heart failure (cardiac resynchronization therapy)

Most often, implantation of a permanent pacemaker does not change a person's activities or lifestyle. Typically, only a small bump is seen in the skin over the site where the pacemaker has been placed. Whereas most people who receive pacemakers are 60 years or older, pacemakers can be implanted in persons of any age, even in children.

17.4 Permanent Pacemaker Components

The pacemaker consists of a generator and one or more lead wires that connect the device to the heart (Figure 17-1).

Generator

The pacemaker generator is a small device (about the size of a man's wristwatch), weighing less than 13 grams (under half an ounce). It consists of a power source, a sensing amplifier that processes the electrical activity of the heart as detected by the pacemaker electrodes, the computer logic for the pacemaker, and the output circuitry, which delivers the pacing impulse to the lead wire(s) to stimulate the heart to beat at the desired rate. The casing of the pacemaker is made of titanium or a titanium alloy and is hermetically sealed to prevent diffusion of water vapor from the body fluids into the electronic circuitry and rejection by the body's immune system (as titanium is very inert in the body). The power source, usually a lithium battery, is extremely durable. Depending on how many leads the pacemaker is configured with and how much energy the device uses, the battery can be expected to last an average of 7 to 15 years. Alternatively, cadmium/nickel oxide or nuclear batteries may be used. When a new battery is needed, the unit can be exchanged in a simple outpatient procedure.

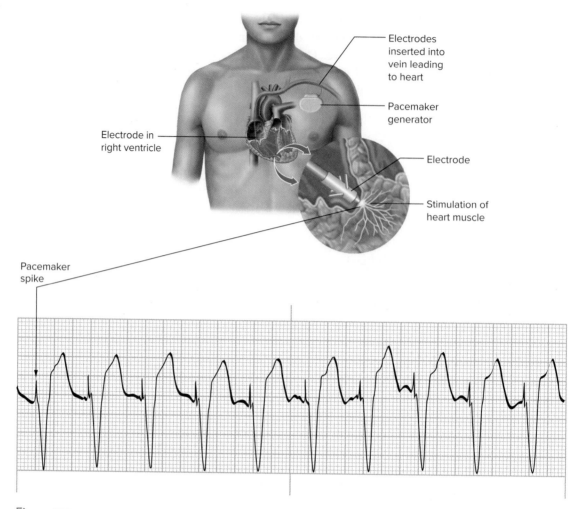

Figure 17-1

Pacemakers are used to provide electrical stimuli for hearts with an impaired ability to initiate or conduct an electrical impulse. The two primary components of the implanted pacemaker are the generator and the lead wire(s).

Lead Wire(s)

The lead wires are tiny, thin wires composed of a coiled metal conductor that is insulated with plastic. The only exposed part of the electrode is the distal end, which has a bare metal tip that is affixed against the wall of the heart chamber with soft plastic hooks, a short metal screw, or a screw-in tip. The tip has either one electrode (unipolar) or two electrodes (bipolar). The proximal end of the lead wire is attached to the pacemaker. Because the heartbeat is constantly flexing the lead wire, it must be resistant to fracture.

To insert the lead wire, an incision is made in a suitable vein and the lead wire(s) is fed along, until it eventually passes into the atrial chamber or through the heart valve and into the ventricular chamber. The procedure is facilitated by fluoroscopy, which enables the viewing of the passage of the lead wire lead. Once in place, the lead wire(s) pass electricity from the spontaneous cardiac activity into the generator and from the pacemaker into the heart.

17.5 Function of Permanent Pacemakers

Today's pacemakers can be externally programmed, receiving and transmitting data/programming through the skin via electromagnetic waves. They can be adjusted according to pacing mode, output, sensitivity, refractory period, and rate adaption. This allows the cardiologist to select the optimum pacing modes for individual patients. Further more, the ECG can be used to find if a patient's pacemaker is properly working. You need to be familiar with the pacemaker modes, capabilities, and coding in order to compare the intended versus actual pacemaker function. Lastly, in this section, we discuss the use of another type of pacemaker, the biventricular pacemaker.

Pacing Modes

Permanent pacemakers are categorized by how many chambers they stimulate and how they operate. They include the following:

- *Single-chamber.* One pacing lead is inserted into either the right atrium or right ventricle but not both.

- *Dual-chamber.* Electrodes are placed into two chambers of the heart. One lead paces the atrium, while the other paces the ventricle. By assisting the heart in coordinating the function between the atria and ventricles, this type of pacemaker acts similarly to how the heart naturally paces itself and may also be referred to as an AV sequential pacemaker. Most dual-chamber systems can be programmed to a single-chamber mode, which can be useful if the atrial lead wire fails.

- *Fixed-rate.* This pacemaker paces the heart at a single, preset rate.

- *Demand.* This pacemaker senses the heart's native electrical impulses and, if no electrical activity is detected within a certain period of time, delivers a short low-voltage pulse to the chamber to stimulate the heartbeat. This sensing and stimulating activity continues on a beat-by-beat basis. It is the most widely used pacemaker.

- *Rate-responsive.* This pacemaker has sensors that identify increases or decreases in the patient's physical activity and automatically adjusts the base pacing rate to meet the body's metabolic needs. It can boost the heart rate in response to motion or increased respirations for those patients whose body, because of either disease of the sinus node or the effects of medications, cannot appropriately increase the heart rate during activity.

Output

The energy delivered to the lead wires can be adjusted. Using a lower energy level conserves power and increases battery life, whereas using a higher energy level makes certain pacing occurs when necessary. The voltage output is usually set to 50% above the capture threshold. The threshold is the pacemaker output that just captures the chamber. The pulse duration can also be altered.

Sensitivity

A pacemaker should detect the presence of appropriate cardiac electrical activity (and not pace), whereas at the same time it needs to ignore nonrelevant electrical activity. For atrial pacemakers, the relevant activity is the P wave; for ventricular systems, it is the R wave. Pacemaker sensitivity is the level of current the generator recognizes or ignores as coming from a cardiac chamber. The pectoral muscle generates 2 to 3 mV currents while the heart generates 5 to15 mV currents. Most often, sensitivity is set at around 4 to 5 mV, thus excluding pectoral muscle as well as electrical activity occurring in distant chambers. If the lead wire starts to fail, the amount of electricity passed up the lead wire may decrease, leading to failure to sense and inappropriate pacing. Altering the sensitivity can deal with this problem.

Refractory Period

It is important that pacemakers not misinterpret electrical activity from one chamber as coming from another. As an example, an atrial lead wire can sense ventricular activity, label it as atrial, and inappropriately inhibit pacing when it might be needed. Programming the generator not to detect electrical activity at certain times of the cardiac cycle (e.g., inhibiting the atrial lead wire from detecting electrical activity during the time period when normal ventricular activity occurs) can prevent this.

Rate Adaptation

Rate-responsive pacemakers detect physical activity (using motion detectors or QT interval sensors), and react by increasing the pacing rate. These are useful in mimicking the physiological effects of exercise on the heart rate.

Coding System

In order to be standardized, the pacemaker mode and function are described by a coding system that uses five letters. In practice however, only three to four are commonly used.

The first letter (Position I) represents the heart chamber being paced. This letter may be

- O = none
- A = atrium
- V = ventricle
- D = dual (ventricle and atrium)

The second letter (Position II) represents the chamber of the heart being sensed by the pacemaker. This letter may be

- O = none
- A = atrium
- V = ventricle
- D = dual (ventricle and atrium)

The third letter (Position III) indicates how the pacemaker generator responds to sensing. This letter may be

- O = none
- T = triggers pacing
- I = inhibits pacing
- D = dual (triggers and inhibits pacing)

The fourth letter (Position IV) has to do with adjustment of the pacing rate in response to exercise. If the pacemaker is rate responsive, it is denoted with the letter R. These pacemakers have extra sensors that can detect a patient's physical activity and respond by increasing heart rate to a set limit. If there is none, it is denoted as "O." A common pacemaker is a VVI-R. This has a single lead that senses and paces the ventricle, and if the patient exercises, increases the paced heart rate.

The fifth letter indicates multisite pacing. This letter may be

- O = none
- A = atrium
- V = ventricle
- D = dual (ventricle and atrium)

Thus, the coding system for the pacemaker might be

- VOO. In this mode, the ventricle is paced and there is no sensing function. This is a temporary mode used to check pacing threshold.
- AAI. In this mode, the pacemaker paces and senses in the atrium. When it senses atrial activity, pacing is inhibited.
- VVI. In this mode, the ventricle is paced and sensed. If spontaneous cardiac output is detected, then the device is inhibited.
- VDD. Here the pacemaker paces the ventricle and senses both the atrium and ventricle. On sensing intrinsic atrial activity, the pacemaker triggers ventricular pacing; on sensing ventricular activity, the pacemaker inhibits pacing. It is also known as a *P-synchronous pacer.*
- DVI. In this mode, the pacemaker can pace in the atrium, the ventricle, or both. Sensing takes place only in the ventricle. When the pacemaker senses intrinsic ventricular activity, it inhibits pacing.
- DDD. In this mode, the pacemaker paces and senses in the atrium, the ventricle, or both. On sensing activity in either chamber, the pacemaker inhibits pacing in that chamber. Or, on sensing atrial activity, the pacemaker may trigger ventricular pacing.

Cardiac Resynchronization Therapy

Another type of pacemaker is the biventricular pacemaker. Also called *cardiac resynchronization therapy* (CRT), this pacemaker paces both the right and left ventricles. It is used to resynchronize a heart that does not beat in synchrony, a common problem in patients with heart failure.

Asynchronous contractions lead to blood being moved within, rather than out of the ventricle. This results in reduced cardiac output. Further, these patients experience greater mitral regurgitation and reduced diastolic filling time.

The biventricular pacemaker employs three leads: one is placed in the right atrium, one is located in the right ventricle, and the last one is inserted through the coronary sinus to pace the free wall of the left ventricle. These three wires are connected to a CRT generator and programmed so that the two ventricular wires are activated simultaneously.

Simultaneous contraction of the right and left ventricles results in less blood being shunted within the ventricle and more blood being propelled into the aorta.

These devices relieve symptoms of heart failure, improve prognosis, and lessen the chance of ventricular dysrhythmias.

17.6 Appearance of the Paced ECG

On the ECG, the firing of the pacemaker and its movement down the lead wire appears as a narrow vertical line referred to as a *pacing spike* or *pacing artifact*. What the waveforms of the chamber being depolarized look like depends on which chamber(s) is (are) being paced (Figure 17-2).

A lead wire that is positioned in the atria produces a pacing spike and a P wave of normal duration and shape. The reason for this is that the atrial lead wire is close to the sinus node, resulting in the wave of depolarization spreading across the atria in a normal manner.

With a ventricular pacemaker, the pacing spike is followed by a broad QRS complex. This occurs because the electrode sits at the apex of the right ventricle. After the

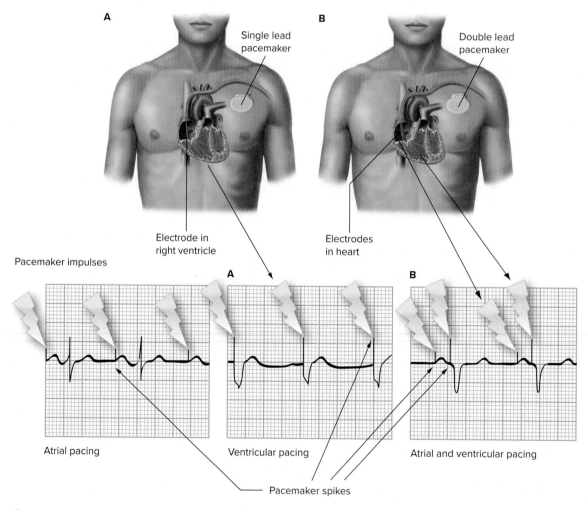

Figure 17-2
Pacing spikes.

pacemaker fires, the wave of depolarization slowly spreads from myocyte to myocyte across the right ventricle. The left ventricle is then depolarized. Because the left ventricle is the last to depolarize, it produces a pattern identical to left bundle branch block. A retrograde P wave may or may not be seen.

With a dual-chamber pacemaker, there will be an initial pacing spike followed by a P wave and a second pacing spike followed by a broad QRS complex.

Simultaneous stimulation of both ventricles (biventricular pacing) produces a QRS that is not as wide as is typically associated with right ventricular pacing.

It is important to note that pacemaker spikes do not appear in every lead. For this reason, you should check more than one lead to identify their presence. Because lead aV_R produces small QRS complexes, the pacemaker spikes are less likely to be obstructed and may be easier to see.

Unipolar vs. Bipolar Systems

In a unipolar system, the positive electrode is positioned in the heart tissue and the negative electrode is connected to the pulse generator. Because the two poles are so far apart, a unipolar system produces tall pacing spikes on the ECG. In a bipolar system, the electrodes are only millimeters apart in the cardiac tissue. This produces short pacemaker spikes.

Another issue to be aware of is that the ECG tracing may include indicators such as arrows below the pacemaker spikes to help you see their presence. The ECG machine will begin the typical analysis of the tracing but then may defer to a default that shows on the printout, for example, "pacemaker rhythm, no further analysis."

In some patients, pacemaker spikes are not easy to see on an ECG because their amplitude is less than 1 mV. If the patient is able to communicate and is appropriately oriented, asking the question "Do you have a pacemaker?" may be of use when broad QRS complexes and left axis deviation are seen on the ECG.

17.7 Pacemaker Failure and Complications

Pacemakers may not work properly for a number of reasons, including the following: failure to capture, failure to pace, failure to sense, oversensing, and pacemaker-mediated tachycardia. Regardless of why a pacemaker does not work, you should always focus on treating the patient first and not the device. We also discuss some complications associated with pacemakers.

Failure to Capture

Failure to capture occurs when the pacemaker fires but does not trigger an atrial or ventricular depolarization.

Identifying It

Failure to capture can be recognized by the presence of pacemaker spikes which are not followed by P waves (if an atrial electrode is used) or by broad QRS complexes

Pacemaker fires but does fails to capture ventricular depolarization

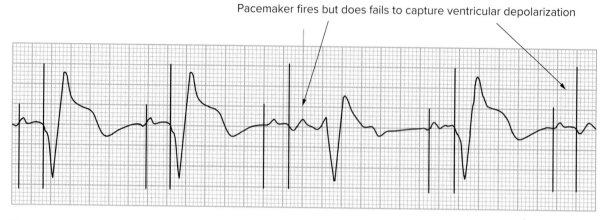

Figure 17-3
Example of failure to capture.

(if a ventricular electrode is used) (Figure 17-3). Other signs and symptoms that may be seen include bradycardia, fatigue, and hypotension.

Causes

Failure of the pacemaker to capture can be caused by an electrode tip that is out of place, the pacemaker voltage being set too low, a dead battery, a broken lead wire, the presence of edema or scar tissue at the electrode tip, perforation of the myocardium by the lead, or MI at the electrode tip.

Interventions

Correcting the problem depends on the cause. Interventions used to correct failure to capture include repositioning the electrode tip, increasing the voltage, replacing the battery, repositioning or replacing the lead wire, and/or surgical intervention.

Failure to Pace

Failure to pace occurs when the pacemaker fails to send a signal to depolarize one or both chambers.

Identifying It

The absence of apparent pacemaker activity on the ECG is the primary way of identifying a failure to pace (Figure 17-4). Depending on the underlying heart rate, the patient may present with bradycardia and hypotension. Another indicator has to do with a magnet being applied over the pacemaker. Normally, a magnet will cause a synchronous pacemaker to be converted to an asynchronous pacemaker. With failure to pace, the application of a magnet produces no response.

Causes

Possible causes of failure to pace include a battery or circuit failure, a displaced or broken lead, or a loose lead-generator connection.

Interventions

Interventions used to correct failure to pace include replacing the battery, generator, or lead; repairing the connection; or tightening the terminal (if it is a temporary pacemaker).

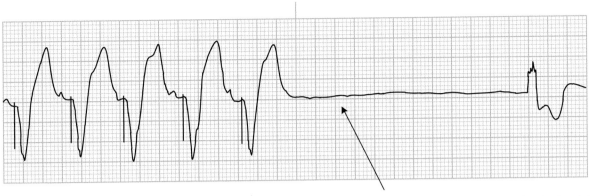

Failure to pace results in an absence of pacemaker spikes and electrical activity

Figure 17-4
Failure to pace is seen as an absence of pacemaker spikes when they should be present.

Failure to Sense

With correct sensing, when an atrial or ventricular beat occurs, there should be no paced beat within the time period determined by the minimum paced rate. If there is, then the pacemaker has failed to sense.

Identifying It

Failure of the pacemaker to sense is seen as ECG pacemaker spikes that fall where they should not. For example, the pacemaker spikes may be seen falling on T waves (Figure 17-5). Also, the patient may complain of palpitations or skipped beats and may even develop ventricular tachycardia.

Causes

Possible causes of a failure to sense include a dead battery, an electrode tip that is out of position, a broken lead wire, the presence of edema or fibrosis at the electrode tip causing decreased R or P wave amplitude, or the "sensitivity" being set too high (this is a programmable function, and can be adjusted).

Pacemaker fails to sense impulse and fires unnecessarily

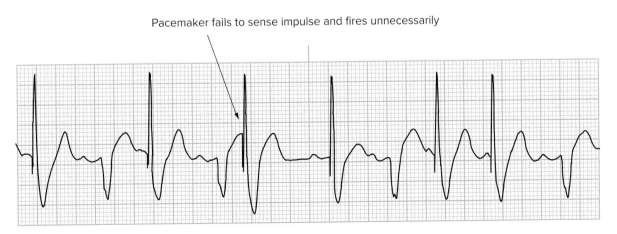

Figure 17-5
Failure to sense is seen as pacemaker spikes occurring when they should not be present.

Interventions

Interventions used to resolve failure to sense include replacing the battery or lead, repositioning the electrode, positioning the patient on his or her left side, or making sure the sensitivity setting is correct.

Oversensing

Oversensing occurs when the pacemaker misinterprets certain types of electrical activity as being cardiac activity and does not fire when it should. This leaves the patient vulnerable to a slow heart rate.

Identifying It

The main identifying feature of oversensing is an absence of pacemaker spikes in the presence of a heart rate that is slower than the rate set for the pacemaker (which is normally faster than the patient's spontaneous rate). The patient may also be symptomatic (i.e., fatigue and hypotension) due to decreased cardiac output that may coincide with the slow heart rate.

Causes

One cause of oversensing is the pacemaker's sensitivity being set too low. Another is the pacemaker's detection of nonrelevant electrical activity and its perception that the atria have depolarized when they haven't. This results in the pacemaker not firing when it should. Examples include skeletal muscle contractions (in unipolar lead systems) or distant cardiac chamber electrical activity (T waves or atrial impulses). Electromagnetic interference can also result in pacemaker oversensing.

Interventions

Interventions used to correct oversensing include inserting a bipolar lead, testing with a magnet, increasing the sensitivity setting (if the pacemaker is temporary), and, lastly, reprogramming the internal pacemaker.

Pacemaker-Mediated Tachycardia

In dual-chamber pacemakers, the electrodes and generator can sometimes act as an accessory pathway leading to tachydysrhythmias.

Identifying It

Pacemaker-mediated tachycardia is seen as a fast heart rate with a pacemaker spike preceding each QRS complex on EGG (Figure 17-6).

Causes

Pacemaker-mediated tachycardia occurs due to retrograde conduction through the AV node (from dual-chamber pacing), which depolarizes the atria and triggers the rapid heart rate.

Interventions

Due to the rapid heart rate, the patient must be monitored for ventricular tachycardia and ventricular fibrillation. Interventions include reprogramming the pacemaker to an atrial nonsensing mode or prolonging the postventricular atrial refractory period of the pacemaker so that it will no longer sense retrograde P waves. In an

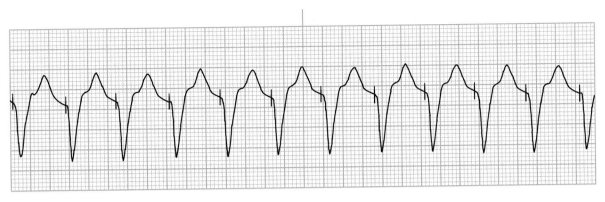

Figure 17-6
Pacemaker-mediated tachycardia is seen as a tachycardia with a pacemaker spike preceding each QRS complex on the EGG.

emergency, the heart rate can be decreased by holding a magnet over the pulse generator. This acts to convert the pacemaker from synchronous to asynchronous until it can be reprogrammed.

Complications of Pacemakers

Aside from the pacemaker's failing to properly function there are two complications that deserve discussion: infection and pacemaker syndrome.

Infection

Infection should be suspected in anyone with an implanted pacemaker presenting with an inflammatory illness. Signs and symptoms might include fever, weight loss, anemia, raised erythrocyte sedimentation rate (ESR), and C-reactive protein (CRP), and/or positive blood cultures. Treatment includes antibiotic administration, complete pacemaker system removal, and, when sterility is achieved, inserting a new system.

Pacemaker Syndrome

Patients who have a VVI pacemaker (ventricle paced and sensed device inhibited if spontaneous cardiac output detected) and intact AV conduction can develop retrograde ventricular atrial conduction. As such, shortly after ventricular systole, the atria fire, contracting on a closed AV valve. This results in atrial distension, brain natriuretic peptide release, and reflexes that lower blood pressure, causing faintness or syncope. This problem can be corrected by lowering the rate at which pacing starts (less pacing occurs) or upgrading to a DDD system.

17.8 Implantable Cardioverter-Defibrillators

An implantable cardioverter-defibrillator (ICD) is an electronic device that is positioned in the left pectoral region similarly to how pacemakers are implanted. It can perform several functions including pacing to treat bradycardia, overdrive pacing, or cardioversion to treat ventricular tachycardia and delivering internal defibrillation to the ventricle to treat ventricular fibrillation. Generally, ICDs are used in cases where drug therapy, surgery, or catheter ablation have failed to prevent the dysrhythmia(s) for which the ICDs are being implanted. The ICD consists of a programmable pulse generator and one or more lead wires.

Pulse Generator

The pulse generator is a small (weight of only 70 grams and thickness of about 12.9 mm), fully programmable, battery-powered computer (Figure 17-7). It monitors the heart's electrical activity and delivers electrical therapy when an abnormal rhythm is identified. The device has a solid-state memory that allows it to store information about the heart's activity before, during, and after the dysrhythmia, along with tracking delivered treatments and outcomes. Many devices also store electrograms (which are similar to ECGs). Using an interrogation device, this information can be retrieved and used to evaluate ICD function and battery status as well as to adjust the ICD settings. Because of all the functions performed by the ICD, it is equipped with a battery that stores substantial energy and charges the capacitor rapidly.

Electrode Wires

The ICD includes an electrode wire that is passed through a vein to the right chambers of the heart and lodged in the right ventricular apex. When the ICD discharges, energy is delivered to the heart via this electrode wire.

Recognition of Ventricular Dysrhythmias

ICDs continually monitor the heart rate and rhythm and can deliver electrical therapy when the heart's electrical activity exceeds the preset number. They can distinguish between ventricular tachycardia and ventricular fibrillation. These devices use a number of methods to determine if a fast rhythm is normal, ventricular

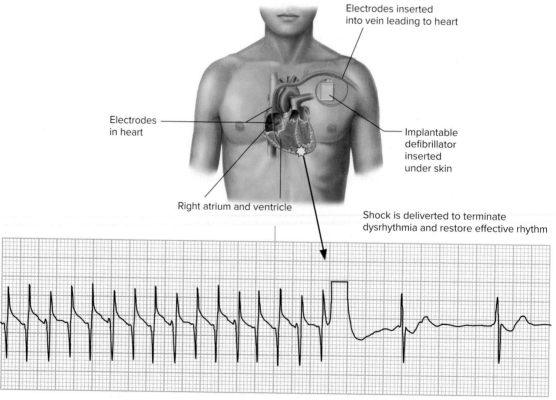

Electrodes inserted into vein leading to heart

Electrodes in heart

Implantable defibrillator inserted under skin

Right atrium and ventricle

Shock is deliverted to terminate dysrhythmia and restore effective rhythm

Figure 17-7
The ICD implanted in a patient's chest.

tachycardia, or ventricular fibrillation. This includes rate discrimination, rhythm discrimination, and morphology discrimination.

Rate Discrimination

The ICD evaluates the ventricular rate and compares it with the atrial rate. If the atrial rate is faster than or equal to the ventricular rate, then the rhythm is not likely of ventricular origin. Instead it is usually something benign; in this case, the ICD will not deliver any therapy. The following are problems that can occur when using the heart rate to diagnose ventricular tachycardia or fibrillation.

- If the heart rate increases from supraventricular dysrhythmias such as atrial fibrillation, the rhythm may be interpreted as not having P waves.
- Occasionally, prominent T waves are counted as R waves, leading to double counting.

Incorrect diagnosis can result in the ICD identifying ventricular tachycardia as being present, thus leading to the inappropriate delivery of antitachycardia therapy.

Rhythm Discrimination

The ICD determines the regularity of ventricular tachycardia. Generally, ventricular tachycardia is regular. If the rhythm is irregular, it is usually due to the presence of an irregular rhythm that originates in the atria, such as atrial fibrillation.

Morphology Discrimination

The ICD assesses the morphology of every ventricular beat and compares it with what the ICD believes to be a normally conducted ventricular impulse for the patient. This normal ventricular impulse is often an average of a multiple of beats of the patient taken in the recent past.

Therapies Provided by the ICD

The ICD can treat several conditions including bradycardia (pacing), ventricular tachycardia (overdrive pacing or cardioversion), and ventricular fibrillation (internal defibrillation). Each of these therapies is described in Table 17-1.

Table 17-1
Types of ICD Therapy

Therapy	Description
Antibradycardia pacing	As described earlier, electrical pacing is used to increase the heart rate when it is too slow. Most ICDs can pace one chamber at a preset rate. Some devices sense and pace both chambers.
Antitachydysrhythmia pacing	This mode generates a series of small, rapid electrical pacing pulses to interrupt ventricular tachycardia (by "capturing" the ventricle and breaking the reentry circuit) and return the heart to its normal rhythm.
Cardioversion	This mode uses a low- or high-energy shock (up to 35 joules), which is timed to the R wave (which is detected by the right ventricular electrode) to terminate VT and return the heart to its normal rhythm.
Defibrillation	This mode delivers a high-energy shock (up to 35 joules) to the heart to terminate ventricular fibrillation and return the heart to its normal rhythm.

Antitachydysrhythmia pacing is also known as *fast pacing* or *overdrive pacing*. It is only effective if the underlying rhythm is monomorphic ventricular tachycardia; it is never effective if the rhythm is ventricular fibrillation or polymorphic ventricular tachycardia. Occasionally, this type of pacing results in the ventricular tachycardia speeding up and becoming hemodynamically unstable or progressing to ventricular fibrillation, in which case defibrillation is delivered. This treatment is not used for all patients; instead it may be ordered after appropriate evaluation of electrophysiologic studies.

Dysrhythmia definitions are programmed into the ICD and altered according to the clinical situation. For example, the device may be programmed to identify the following heart rates and deliver an appropriate therapy:

- Normal heart rhythm: 60 to 150 beats per minute; no action taken
- Slow ventricular tachycardia: 150 to 200 beats per minute; leads to antitachydysrhythmia pacing
- Fast ventricular tachycardia: 200 to 250 beats per minute; leads to several attempts of antitachydysrhythmia pacing followed by cardioversion/ defibrillation
- Ventricular fibrillation: greater than 250 beats per minute; leads to immediate defibrillation.

Provider Safety

A patient who has an ICD should be treated as if they did not have a device. If the ICD discharges while you are in contact with the patient, you may feel the shock; however, it is not likely to be dangerous. Providers shocked by an ICD report sensations similar to contact with an electrical current. If the patient has an ICD that is delivering shocks as demonstrated by the patient's muscles contracting in a manner similar to that observed during external defibrillation, you should allow 30 to 60 seconds for the ICD to complete the treatment cycle before attaching an AED or defibrillator pads. Occasionally, the analysis and shock cycles of automatic ICDs and AEDs will conflict. There is the potential for pacemaker or ICD malfunction after defibrillation when the pads are in close proximity to the device. For this reason, avoid placing the pads or paddles over the device. Pacemaker spikes with unipolar pacing may confuse AED software and may prevent VF detection. The anteroposterior and anterolateral locations are acceptable in patients with these devices. In patients with ICDs or pacemakers, pad/paddle placement should not delay defibrillation.

Practice Makes Perfect

For the following tracings, identify if a pacemaker is present and what type it is. Then determine if there is a pacemaker failure. Record your answers on a separate sheet of paper or directly on these pages. Your findings can then be compared to the answers provided in Appendix A.

1.

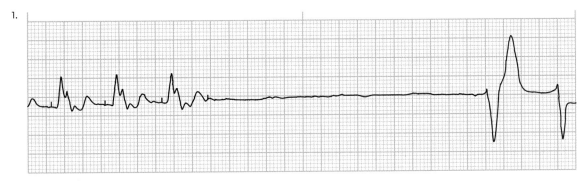

ECG Findings:

2.

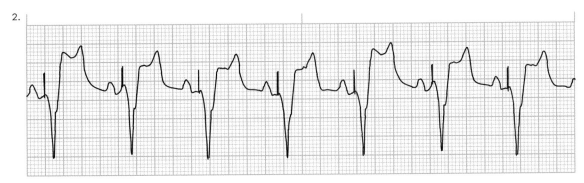

ECG Findings:

3.

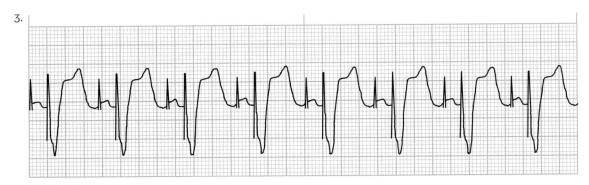

ECG Findings:

4.

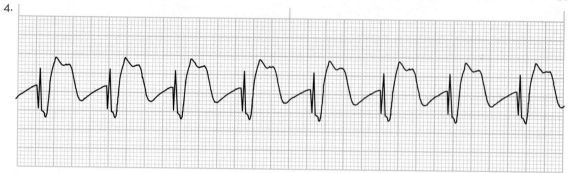

ECG Findings:

5.

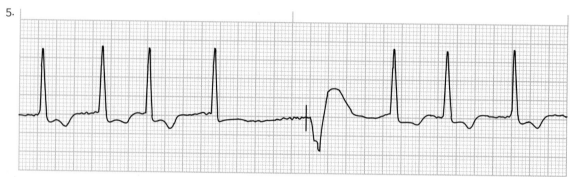

ECG Findings:

6.

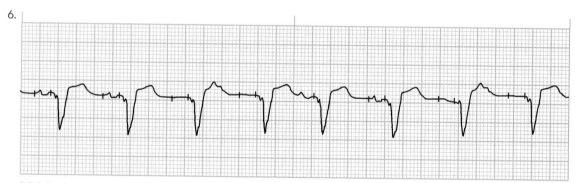

ECG Findings:

7.

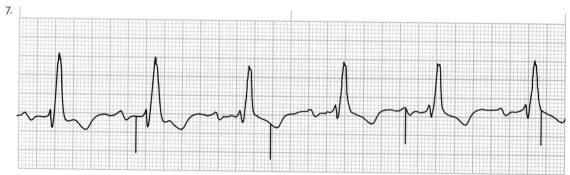

ECG Findings:

8.

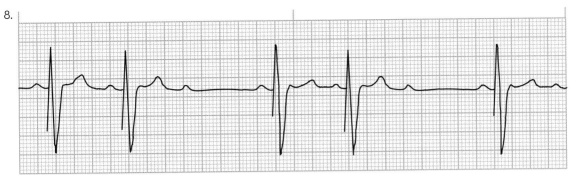

ECG Findings:

9.

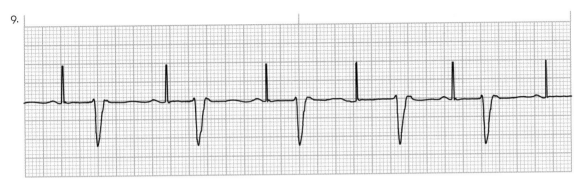

ECG Findings:

10.

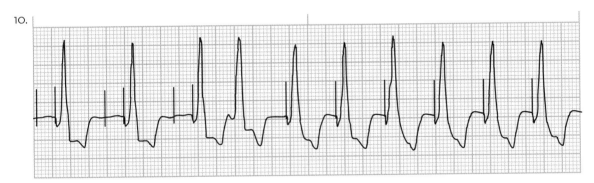

ECG Findings:

11.

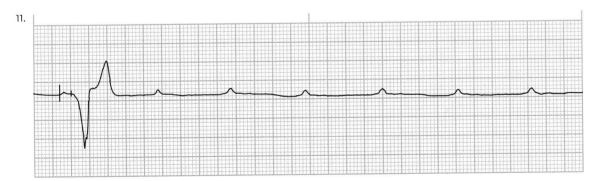

ECG Findings:

12.

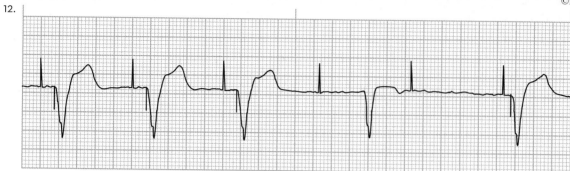

ECG Findings:

13.

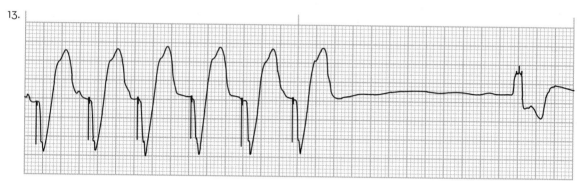

ECG Findings:

14.

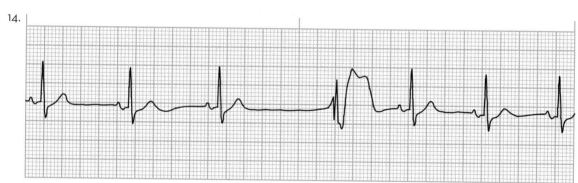

ECG Findings:

15.

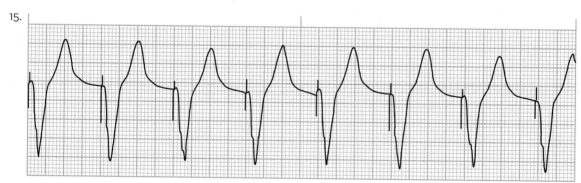

ECG Findings:

16.

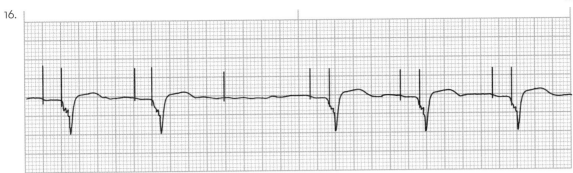

ECG Findings:

17.

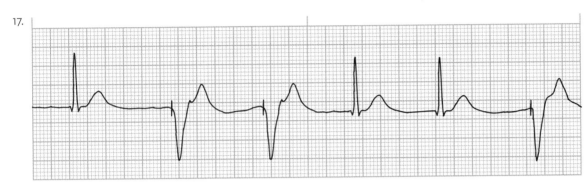

ECG Findings:

18.

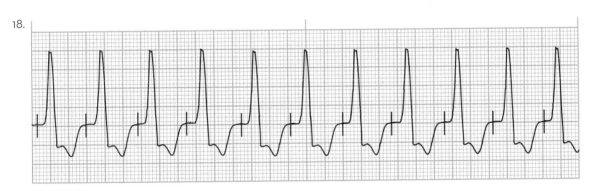

ECG Findings:

19.

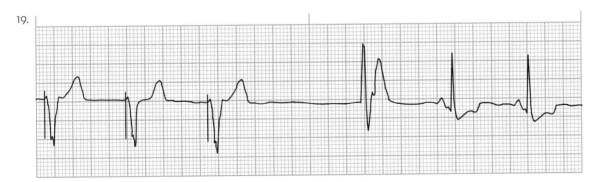

ECG Findings:

Key Points

LO 17.1	• Artificial pacemakers are medical devices used to deliver electrical impulses to the heart to stimulate a normal heartbeat. Some pacemakers are external to the body and provide temporary treatment whereas others are permanently implanted in the chest.
LO 17.2	• One type of temporary pacemaker is the transcutaneous pacemaker. It delivers electrical impulses through lead wires to electrode pads that are applied to the surface of the patient's chest. Epicardial pacing is employed during open heart surgery with the electrode placed in contact with the ventricular epicardium. With transvenous pacing, an electrode is fed through a vein into either the right atrium or right ventricle.
LO 17.3	• Permanent pacemakers are implanted in a surgically created pocket beneath the skin in the patient's chest wall just below the clavicle. They may be used to increase the heart rate when it is slow (antibradycardia devices), slow the heart rate when it is fast (antitachycardia devices), and synchronize the heart beat in heart failure (CRT devices).
LO 17.4	• A pacemaker consists of a generator and one or more lead wires. The generator has a power source (often a lithium battery) and logic circuits that detect cardiac electrical activity and determine the appropriate response.
	• Pacemaker electrodes are either positioned in the atrium or ventricle alone (single-chamber pacemakers) or, more often, in both chambers (dual-chamber pacemakers or AV sequential pacemakers).
LO 17.5	• Pacemakers are programmable; they receive and transmit data/programming instructions through the skin using electromagnetic waves.
	• The demand pacemaker is most commonly used. It fires only when the patient's intrinsic heart rate falls below a given threshold level. For example, if the pacemaker is set at 60 beats per minute, it remains inactive until there is a pause between beats that translates into a rate below 60. The pacemaker then fires.
LO 17.6	• Depending on how many chambers are paced, the firing of a pacemaker produces one or two narrow pacemaker spikes on the ECG.
	• A paced ECG complex shows two features: (a) a narrow "pacing spike," which reflects the impulse depolarizing the paced chamber, and (b) a P wave or QRS complex that immediately follows the pacing spike.
LO 17.7	• Pacemakers may not work properly for several reasons, including a failure to capture, a failure to pace, a failure to sense, oversensing, and pacemaker-mediated tachycardia.
	• Failure to capture is the presence of pacemaker spikes that are not followed by a P wave or broad QRS complex. Failure of the pacemaker to sense is the presence of ECG pacemaker spikes that fall where they shouldn't. Oversensing is an absence of pacemaker spikes in the presence of a heart rate that is slower than the rate set for the pacemaker. Pacemaker-mediated tachycardia is a fast heart rate with a pacemaker spike preceding each QRS complex on EGG.
LO 17.8	• An ICD is implanted in patients who are at risk of sudden cardiac death due to ventricular fibrillation and ventricular tachycardia. The device is programmed to detect cardiac dysrhythmias and correct them by delivering paced beats, cardioversion, or defibrillation.

Assess Your Understanding

The following questions give you a chance to assess your understanding of the material discussed in this chapter. The answers can be found in Appendix A.

1. A cardiac pacemaker is used to deliver electrical: (LO 17.1)
   a. impulses to the heart to stimulate a normal heartbeat.
   b. shocks to the heart in the presence of life-threatening ventricular fibrillation.
   c. currents to the autonomic nervous system to stimulate myocardial contraction.
   d. signals to the SA node to initiate the heartbeat through the normal conductive pathways.

2. A pacemaker that is temporarily employed during open heart surgery, with the electrode being placed in contact with the ventricular epicardium, is called a _____ pacemaker. (LO 17.2)
   a. transcutaneous
   b. transvenous
   c. internal
   d. epicardial

3. A permanent pacemaker is implanted in a surgically created pocket beneath the skin in the patient's (LO 17.3)
   a. abdomen.
   b. chest.
   c. neck.
   d. upper thigh.

4. Permanent pacemakers can be used to treat all of the following conditions EXCEPT (LO 17.3)
   a. symptomatic bradycardia.
   b. 3rd-degree AV heart block.
   c. recurrent tachycardia.
   d. premature ventricular complexes.

5. The pacemaker generator (LO 17.4)
   a. consists of a power source, sensing amplifier, computer logic for the pacemaker, and output circuitry.
   b. is powered by two or three AAA rechargeable batteries.
   c. is susceptible to damage from body water vapor diffusing into the computer circuitry and rejection by the immune system.
   d. is typically replaced after every 5 years of use.

6. The lead wires of the permanent pacemaker (LO 17.4)

    a. are inflexible to prevent dislodgement of the electrode from the chamber wall.

    b. are composed of a coiled metal conductor that is insulated with a fine linen mesh.

    c. consist of two electrodes implanted in the outside myocardial wall (epicardium).

    d. are passed through a suitable vein into one or both of the heart's chambers.

7. A pacemaker that senses the heart's native electrical impulses and, if no electrical activity is detected within a certain period of time, delivers a short low voltage pulse to the chamber to stimulate the heartbeat is referred to as a _____ pacemaker. (LO 17.5)

    a. fixed-rate

    b. demand

    c. rate-responsive

    d. dual-chamber

8. In the following mode the ventricle is paced and sensed. If spontaneous cardiac output is detected, then the device is inhibited. (LO 17.5)

    a. VDD

    b. DVI

    c. VVI

    d. DDD

9. On the ECG, the firing of the pacemaker and its movement down the lead wire appears as a (LO 17.6)

    a. narrow vertical line.

    b. square box.

    c. jagged biphasic wave.

    d. flat line.

10. With a ventricular pacemaker, the pacing spike is followed by a (LO 17.7)

    a. normal P wave.

    b. broad P wave and normal QRS complex.

    c. broad QRS complex.

    d. narrow QRS complex.

11. In a unipolar system, the positive electrode (LO 17.6)

    a. produces short pacemaker spikes.

    b. is positioned in the heart tissue while the negative electrode is connected to the pulse generator.

    c. has two electrodes that are only millimeters apart in the cardiac tissue.

    d. is attached to myocardial epicardium.

12. Describe how firing of a dual-chamber pacemaker appears
on the ECG. (LO 17.6)

13. Failure to capture can be recognized by (LO 17.7)

a. the presence of pacemaker spikes that are not followed by
P waves or by broad QRS complexes.

b. an absence of pacemaker spikes in the presence of a heart rate that is
slower than the rate set for the pacemaker.

c. ECG pacemaker spikes that fall where they shouldn't.

d. a tachycardia with a pacemaker spike preceding each QRS complex
on the EGG.

14. An ICD can perform all of the following functions EXCEPT (LO 17.8)

a. pacing to treat bradycardia.

b. overdrive pacing or cardioversion to treat ventricular tachycardia.

c. delivering internal defibrillation to the ventricle to treat ventricular
fibrillation.

d. pacing both the right and left ventricles to resynchronize an
asynchronous heartbeat.

These questions refer to the case history at the beginning of the chapter.

15. The paramedics used what type of pacemaker to increase
the patient's heart rate? (LO 17.1)

a. Transcutaneous

b. Transvenous

c. Internal

d. Epicardial

16. The cardiologist has ordered the implantation of a(n) (LO 17.2)

a. implantable cardioverter-defibrillator.

b. biventricular pacemaker.

c. transvenous pacemaker.

d. permanent pacemaker.

©rivetti/Getty Images

18 Overview of 12-Lead ECGs and Electrical Axis

Chapter Outline

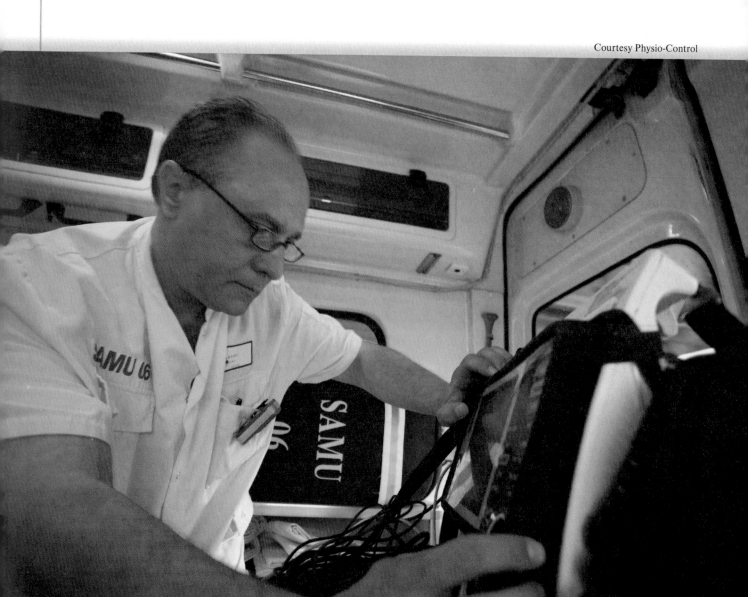

Learning Outcomes

LO 18.1 Discuss the benefits and use of a 12-lead ECG.

LO 18.2 Recall the ECG electrodes that make up the limb leads, where they are placed, and the deflection of waveforms seen with each.

LO 18.3 Recall the ECG electrodes that view the horizontal plane, make up the precordial leads and what direction the waveforms deflect in each lead.

LO 18.4 Identify the ECG leads that provide the inferior, lateral, septal, and anterior ECGs leads and define the term contiguous leads.

LO 18.5 Describe how to analyze the printout of a 12-lead ECG.

LO 18.6 Define the terms *instantaneous vectors, electrical axis,* and *mean QRS axis.*

LO 18.7 Identify the plane and ECG leads used to determine the electrical axis, and recall how to use the four-quadrant and degree methods.

LO 18.8 Define the terms *right axis deviation, left axis deviation,* and *extreme axis deviation* and describe when each is seen.

LO 18.9 Identify causes of altered electrical axis.

Case History

Paramedics are called to the home of an 83-year-old man complaining of shortness of breath. After obtaining a set of vital signs that reveal the patient to be hypertensive, one of the paramedics auscultates the lungs and discovers rales throughout both lung fields. They place the patient on supplemental oxygen, administer furosemide, and take a 12-lead ECG. The 12-lead ECG indicates that the patient has an enlarged heart but does not show evidence of a heart attack.

18.1 The 12-Lead ECG

The 12-lead ECG is used to identify various cardiac conditions as well as to differentiate between dysrhythmias where the origin is uncertain (i.e., supraventricular vs. ventricular).

Views

As mentioned before, a 12-lead ECG is just that; instead of looking at just one view of the heart, you can see 12 different views. Each lead provides a view that the other leads cannot. This gives you a more complete picture of the heart's electrical activity as well as specific diagnostic information.

These views are obtained by placing electrodes on the patient's extremities and chest (or upper torso). The location or sites where the electrodes are placed vary depending on which view of the heart's activity is being assessed. Although we say there are 12 leads, there are actually only 10 lead wires.

Depending on which lead is used and where the positive electrode is placed, the waveforms generated by the electrical activity of the heart can be seen as upright deflections or downward deflections, or they can be biphasic.

While this was covered in great depth in Chapter 2, we will provide a brief review. With the 12-lead ECG, electrodes are placed at specific spots on the patient's extremities and/or torso and chest wall to view the heart's electrical activity from two distinct planes, the frontal and horizontal (see Figure 2-9). These planes provide a cross-sectional view of the heart. The location where the electrodes are placed determines the view of the heart or lead.

18.2 Limb Leads

The limb leads view the frontal plane of the heart. There are six limb leads: I, II, III, aV_R, aV_L, and aV_F. The four electrodes that make up these leads are positioned as follows on the body (Figure 18-1):

- The right arm electrode, labeled RA, is positioned on the right arm.
- The left arm electrode, labeled LA, is positioned on the left arm.
- The left leg electrode, labeled LL, is positioned on the left leg.
- The right leg electrode, labeled RL, is positioned on the left leg.

These electrodes can be placed far down on the limbs or close to the hips and shoulders, but they must be even (right vs. left). Alternatively, these electrodes can be placed on the chest or upper torso as shown.

Leads I, II, and III

Leads I, II, and III are referred to as the *standard* limb leads (based on the use of the first ECG). All three are bipolar leads.

With lead I, the LA lead is the positive electrode, the RA lead is the negative electrode, and the LL and RL leads are the ground and there to complete the circuit.

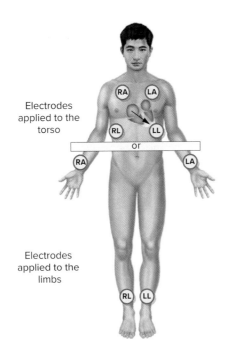

Figure 18-1
Position of limb-lead
electrodes.

Electrodes
applied to the
torso

Electrodes
applied to the
limbs

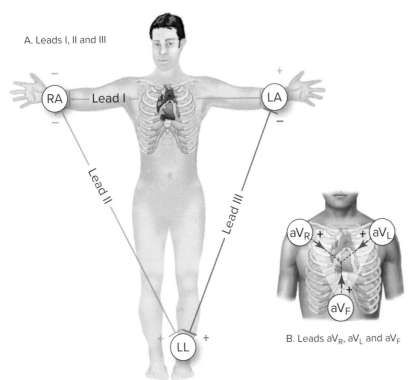

A. Leads I, II and III

Figure 18-2
Limb leads. (A) Leads I. II and
III are bipolar, (B) Leads aV_R,
aV_L and aV_F are unipolar.

B. Leads aV_R, aV_L and aV_F

The waveforms in lead I are mostly upright because the wave of depolarization is moving toward the positive electrode (Figure 18-2A). This includes the P wave, the R wave, and the T wave. The Q wave (if seen) will take an opposite direction as it represents the brief depolarization of the interventricular septum, which is moving away from the positive lead.

With lead II, the LL lead is the positive electrode, the RA lead is the negative electrode, and the LA and RL leads are ground and there to complete the circuit. The waveforms in lead II are mostly upright because the wave of depolarization is moving toward the positive electrode (Figure 18-2A). This includes the P wave, R wave, and T wave. The Q wave (if seen) will take an opposite direction, as it represents the brief depolarization of the interventricular septum, which is moving away from the positive lead.

With lead III, the LL lead is the positive electrode, the LA lead is the negative electrode, and the RA and RL leads are ground. The P and T waves should be upright while the QRS complex may be mostly positive, although the R wave is not as tall as in lead II. Alternatively, the QRS complex may be biphasic as depolarization of the ventricles intersects the negative to positive layout of the ECG electrodes (Figure 18-2A). The Q wave is typically not seen in this lead.

Leads aV$_R$, aV$_L$, and aV$_F$

The other three leads that view the frontal plane are the augmented limb leads, aV$_R$, aV$_L$, and aV$_F$. The ECG waveforms produced by these leads are small, and for this reason the ECG machine enhances or augments them by 50%. This results in their amplitude being comparable to other leads.

In lead aV$_R$ the RA lead is the positive electrode. Because the heart's electrical activity moves away from the positive electrode, the waveforms take a negative deflection (Figure 18-2B). This includes the P wave, R wave, and T wave.

In lead aV$_L$ the LA lead serves as the positive electrode. Because the heart's electrical activity moves toward the positive electrode (or perpendicular to it), the waveforms take a positive (or biphasic) deflection (Figure 18-2B). This includes the P wave, R wave, and T wave. The Q wave (if seen) will take an opposite direction because it represents the brief depolarization of the interventricular septum, which is moving away from the positive lead.

In lead aV$_F$ the LL lead is the positive electrode. The waveforms take a positive deflection because the heart's electrical activity moves toward the positive electrode (Figure 18-2B). This includes the P wave, R wave, and T wave. The Q wave is typically not seen in this lead.

18.3 Precordial Leads

The leads arranged along the horizontal plane give us an anterior, lateral, and inferior view of the heart and are called *precordial leads, V leads,* or *chest leads.*

The precordial leads are unipolar, requiring only a single positive electrode. The opposing pole of those leads is the center of the heart as calculated by the ECG. The six precordial leads are positioned in order across the chest and include V$_1$, V$_2$, V$_3$, V$_4$, V$_5$, and V$_6$. Because of their close proximity to the heart, they do not require augmentation.

As discussed in Chapter 2, the QRS complexes start out in a downward direction, then go through a transitional zone where they become half upright and half downward, and then become upright (Figure 18-3). If the transition occurs in V$_1$ or V$_2$, it is considered early transition, and if it occurs in V$_5$ or V$_6$, it is considered late transition. Both early and late transition can indicate the presence of various cardiac conditions.

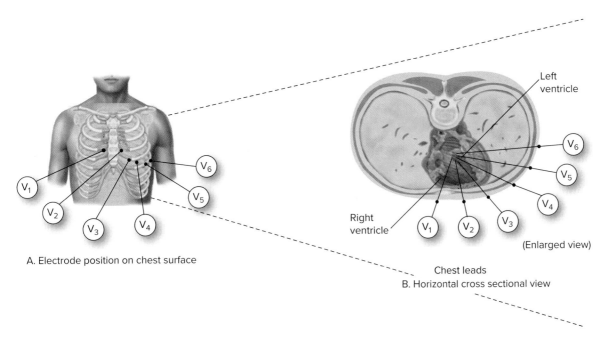

Left
ventricle

V6

V5

V4

Right
ventricle

V1

V2

V3

(Enlarged view)

A. Electrode position on chest surface

Chest leads
B. Horizontal cross sectional view

Figure 18-3
Precordial leads, (A) electrodes are placed across the horizontal plane from right to left, (B) View
of the heart for each precordial lead

Lead V$_1$

For lead V$_1$, the positive electrode is positioned in the 4th intercostal space (between
the fourth and fifth ribs) just to the right of the sternum. Since septal and right
ventricular depolarization send the current toward the positive electrode, there is a
small initial upright deflection (the R wave). Then, as the current travels toward the
left ventricle and away from the positive electrode, it produces a negative deflection,
a deep S wave (Figure 18-3).

Lead V$_1$ is particularly effective at showing the P wave, QRS complex, and the ST
segment. It is helpful in identifying ventricular dysrhythmias, and bundle branch
blocks.

Lead V$_2$

In lead V$_2$, the positive electrode is positioned in the 4th intercostal space (between
the fourth and fifth ribs) just to the left of the sternum. Horizontally, it is at the
same level as lead V$_1$ but on the opposite side of the sternum (Figure 18-3). Similar
to what we see with lead V$_1$, in lead V$_2$ the depolarization initially moves toward the
positive electrode, producing an R wave. Then, as depolarization moves away from
the positive electrode, it produces a less deep S wave.

Lead V$_3$

Lead V$_3$ is located midway between leads V$_2$ and V$_4$. For this reason you need to
locate the position for lead V$_4$ before you apply the electrode for lead V$_3$. Depolarization of the septum and right ventricle produces such a small electrical force that
it is not seen, whereas the current depolarization of the left ventricle moves perpendicular to the positive electrode, resulting in a biphasic waveform that has an
R wave and S wave that are relatively the same amplitude in lead V$_3$ (Figure 18-3).

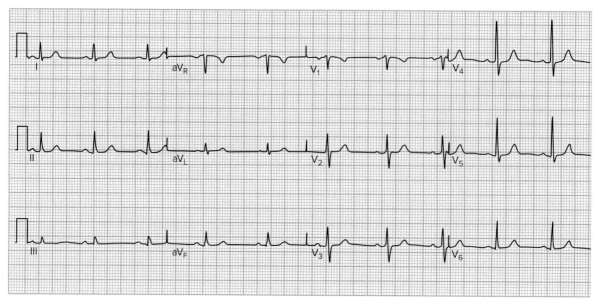

Figure 18-4
Normal appearance of ECG complexes in 12-lead ECG.

Lead V₄

Lead V$_4$ is placed at the 5th intercostal space (between the fifth and sixth ribs) in the midclavicular line (the imaginary line that extends down from the midpoint of the clavicle). Depolarization of the septum and right ventricle produces such a small electrical force that it is not seen while the current depolarization of the left ventricle moves perpendicular to the positive electrode, resulting in a biphasic waveform that has an R wave larger than the S wave in lead V$_4$ (Figure 18-3).

Lead V₅

Lead V$_5$ is placed in the 5th intercostal space at the anterior axillary line. Horizontally, it is even with V$_4$ but in the anterior axillary line. As the impulse conducts through the septum and right ventricle, it moves away from the positive lead, producing a small Q wave in lead V$_5$. Then, as the impulse conducts through the left ventricle, it moves toward the positive electrode, producing a tall R wave (Figure 18-3). Because of its location closest to the left ventricle, it will have the tallest R waves of the precordial leads. Then depolarization of the posterobasal right and left ventricular free walls and basal right septal mass produces a small S wave.

Lead V₆

Lead V$_6$ is located horizontally level with V$_4$ and V$_5$ at the midaxillary line (middle of the armpit). As the impulse conducts through the septum and right ventricle, it moves away from the positive lead, producing a small Q wave in lead V$_5$. Then, as the impulse conducts through the left ventricle, it moves toward the positive electrode, producing a tall R wave (Figure 18-3). Because it is not as close to the left ventricle, the R wave in lead V$_6$ is smaller than that found in lead V$_5$. Lastly, depolarization of the posterobasal right and left ventricular free walls and basal right septal mass produces a small S wave.

An important thing to remember is that if you have placed leads V$_4$ through V$_6$ correctly, they should line up horizontally.

Each view provides different information. When assessing the 12-lead ECG, we look for characteristic normalcy and changes in all leads. Figure 18-4 shows the normal appearance of the 12-lead ECG. Memorizing what view(s) of the heart each lead provides will help you decide which lead to use to gain the information you need to make a proper assessment or diagnosis.

18.4 ECG Views of the Heart

As discussed in Chapter 1, the 12 ECG leads each record the electrical activity of the heart from a different view, which also correlates to different anatomical areas of the heart.

Leads II, III, and aV_F have a view of the inferior (or diaphragmatic) surface of the heart. For this reason they are called the *inferior leads*.

Leads I, aV_L, V_5, and V_6 have a view of the lateral wall. For this reason they are called the *lateral leads*. Because the positive electrode for leads I and aV_L is located on the left arm or upper torso, these leads are sometimes referred to as the *high lateral leads*, and due to the positive electrodes for leads V_5 and V_6 being on the patient's chest, they are sometimes referred to as the *low lateral leads*.

Leads V_1 and V_2 have a view of the septal wall of the ventricles and are called the *septal leads*.

Leads V_3 and V_4 have a view of the anterior wall and are called the *anterior leads*. They also have an apical view of the heart. Lead V_2 can also be considered an anterior lead as it, too, has a view of the anterior wall. The combination of leads V_1, V_2, V_3, and V_4 are referred to as *anteroseptal leads*.

Contiguous Leads

Two leads that look at neighboring anatomical areas of the heart are said to be contiguous. As an example, V_4 and V_5 are contiguous (as they are next to each other on the patient's chest), even though V_4 is an anterior lead and V_5 is a lateral lead. The following lists the contiguous leads:

Contiguous Inferior Leads

- Lead II is contiguous with lead III.
- Lead III is contiguous with leads II and aV_F.
- Lead aV_F is contiguous with lead III.

Contiguous Septal, Anterior, and Lateral Leads

- V_1 is contiguous with V_2.
- V_2 is contiguous with V_1 and V_3.
- V_3 is contiguous with V_2 and V_4.
- V_4 is contiguous with V_3 and V_5.
- V_5 is contiguous with V_4 and V_6.
- V_6 is contiguous with V_5 and lead I.
- Lead I is contiguous with V_6 and aV_L.
- aV_L is contiguous with lead I.

Identifying changes in contiguous leads helps in determining whether an abnormality on the ECG is likely to represent true disease such as acute coronary ischemia or injury or just a false finding.

18.5 Analyzing the 12-Lead ECG

To analyze a 12-lead ECG printout, we begin by looking at the left column from top to bottom, the middle from top to bottom, and the right from top to bottom. That doesn't mean you can't do it differently; it is just one approach (Figure 18-5). Now, much the same as what we do when analyzing an ECG tracing for characteristics

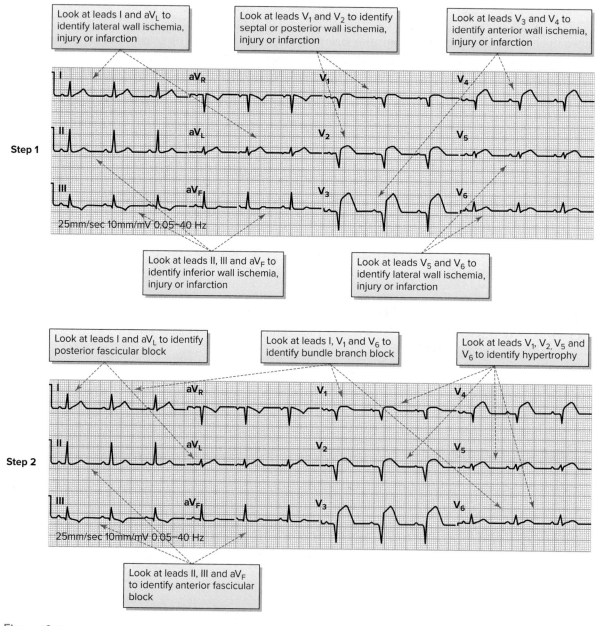

Figure 18-5

Another approach that can be used to analyze the 12-lead ECG tracing is to look for characteristics of given conditions in specific lead groupings. Often the first step is to look for characteristics of myocardial ischemia, injury, and/or infarction. The second step may be to look for characteristics of bundle branch block, fascicular block, and/or hypertrophy.

that help to identify the presence of dysrhythmias, we do the same by looking for certain characteristics that tell us the presence of given cardiac conditions.

In Table 18-1 that follows, we list the conditions we cover in this text. Although more can be identified through the use of a 12-lead ECG, the ones listed are the more common conditions we see.

Condition	Description
Hypertrophy	Hypertrophy is a condition in which the muscular wall of the ventricle(s) becomes thicker than normal. It results from the ventricle having to pump against increased resistance within the cardiovascular system. Conditions that cause hypertrophy include systemic hypertension and aortic stenosis.
Chamber Enlargement	An enlarged chamber occurs as the result of volume overload—the chamber dilates to accommodate the increased blood volume. It is therefore bigger and can hold more blood than normal. Unlike hypertrophy, the muscular wall of the dilated chamber typically does not become thicker. Enlargement is most often seen with certain types of valvular disease. For example, left atrial enlargement can result from mitral valve insufficiency.
Bundle Branch Block	Bundle branch block is a disorder that leads to one or both of the bundle branches failing to properly conduct impulses. This produces a delay in the depolarization of the ventricle it supplies. Right bundle branch block occurs with anterior wall MI, coronary artery disease, and pulmonary embolism. It may also be caused, as can any type of heart block, by drug toxicity. Left bundle branch block can be caused by anterior wall MI, hypertensive heart disease, aortic stenosis, or degenerative changes of the conduction system.
Myocardial Ischemia	Myocardial ischemia is a deprivation of oxygen and nutrients to the myocardium. It typically occurs when the heart has a greater need for oxygen than the narrowed coronary arteries can deliver or when there is a loss of blood supply to the myocardium. Causes of myocardial ischemia include atherosclerosis, vasospasm, thrombosis, embolism, decreased ventricular filling time (such as that produced by tachycardia), and decreased filling pressure in the coronary arteries (such as that caused by severe hypotension or aortic valve disease).
Myocardial Injury	If the myocardial oxygen demand fails to lower, the coronary artery blockage worsens, or if the ischemia is allowed to progress untreated, myocardial injury will occur. Myocardial injury reflects a degree of cellular damage beyond ischemia. It occurs if the blood flow is not restored (and ischemia reversed) within a few minutes. If blood flow is not restored to the affected area, tissue death will occur.
Myocardial Infarction	MI is the death of injured myocardial cells. It occurs when there is a sudden decrease or total cessation of blood flow through a coronary artery to an area of the myocardium. MI commonly occurs when the intima of a coronary artery ruptures, exposing the atherosclerotic plaque to the blood within the artery. This initiates the abrupt development of a clot (thrombus). The vessel, already narrowed by the plaque, becomes completely blocked by the thrombus. The area of the heart normally supplied by the blocked artery goes through a characteristic sequence of events described as zones of ischemia, injury, and infarction.
Pericarditis	Pericarditis is inflammation of the pericardium. The most common causes of pericarditis are viral and bacterial infections. Other causes include uremia, renal failure, rheumatic fever, posttraumatic pericarditis, connective tissue disease, and cancer. Sometimes pericarditis can accompany an acute MI. When pericarditis occurs in conjunction with MI, it develops several days after the infarction.
Pericardial Effusion	Pericardial effusion is a buildup of an abnormal amount of fluid and/or a change in the character of the fluid in the pericardial space. The pericardial space is the space between the heart and the pericardial sac. Formation of a substantial pericardial effusion dampens the electrical output of the heart, resulting in low voltage QRS complex in all leads.

Continued

Pulmonary Embolism	A pulmonary embolism is an acute blockage of one of the pulmonary arteries by a blood clot or other foreign matter. This leads to obstruction of blood flow to the lung segment supplied by the artery. The larger the artery occluded, the more massive the pulmonary embolus and, therefore, the larger the effect the embolus has on the heart. Due to the increased pressure in the pulmonary artery caused by the embolus, the right atrium and ventricle become distended and unable to function properly, leading to right-sided heart failure.
Electrolyte Imbalances	Changes in potassium and calcium serum levels can profoundly affect the ECG. Those involving potassium are the most immediately life-threatening. Levels that are too high (hyperkalemia) or too low (hypokalemia) can quickly result in serious cardiac dysrhythmias. The presence of ECG changes may be a better measure of clinically significant potassium toxicity than is the serum potassium level.
Medication Use and Toxicity	Some medications can have striking effects on ECG waveforms by altering the way the cells depolarize, repolarize, and innervate surrounding tissues. These changes may indicate expected and, therefore, benign effects of certain medications, as well as provide evidence of medication toxicity.

Table 18-1
Common Cardiac Conditions identified through the use of the 12 lead ECG.

Because we have 12 views, and each sees different areas of the heart, we can use them to help us identify certain conditions. So, in addition to remembering characteristics of given conditions, you also need to remember in which lead you can expect to find those characteristics. For example, the better view of the ventricles is seen in leads V_1, V_2 (right ventricle), V_5, and V_6 (left ventricle) as well as in leads I and aV_F. Thus, to identify abnormal conduction through the ventricles such as with bundle branch block or hypertrophy of one or both ventricles, we need to look at those leads.

Seems complicated, doesn't it? Well, once you start doing it, it really isn't all that bad.

18.6 Electrical Axis and the ECG

Changes in the size or condition of the heart muscle and/or conduction system will affect how electrical impulses are conducted through the tissue and, subsequently, the appearance of the ECG waveforms. What we discuss next will help you to understand why given waveforms change in direction, duration, or appearance. Furthermore, it will help you identify the presence of certain cardiac conditions when using the 12-lead ECG.

Vectors

The wave of depolarization and repolarization of the cardiac cells produces many small electrical currents or impulses called *instantaneous vectors*. Remember, depolarization is the advancing wave of ionic exchange. If you think about the millions of cells in the ventricles alone, that is a lot of electrical current. These currents vary in intensity and direction and can be visually represented by arrows (Figure 18-6). While these tiny swirls of current are moving simultaneously in many different directions, the ECG only records the sum or average of the current flow or instantaneous vectors at any given moment. The sum of these vectors is called the *mean*

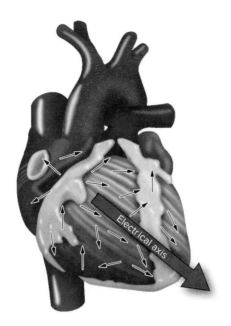

Figure 18-6
Depolarization and repolarization of the heart produces many small electrical currents or impulses. The overall direction and force of the current flow through the heart is called the heart's *electrical axis*.

instantaneous vector. The instantaneous vectors are represented by each part of the QRS complex. The mean vector or electrical axis is derived by looking at the QRS complex as a whole and deciding whether it is more positive or negative. This is what clinically determines the axis.

Mean Instantaneous Vector

Whereas the sum of the small vectors is called the *mean instantaneous vector,* its direction and magnitude as it travels through the heart is called the heart's *electrical axis* (See Figure 18-6). The heart's electrical axis reveals important information about the movement of electrical impulses through the heart. The axis is depicted as a single large arrow. The angle of orientation represents the direction of current flow, whereas its length represents its voltage (amperage). Axis is defined in the frontal plane only. The mean instantaneous vector generated during the depolarization of the atria and its direction and amplitude is called the *P axis*. It is rarely determined. The mean instantaneous vector generated during the depolarization of the ventricles and its direction and amplitude is referred to as the *mean QRS axis*. The mean QRS axis is the most clinically important and therefore the axis we measure. Normally, the QRS axis points to the left and downward, reflecting the dominance of the left ventricle over the right ventricle.

Waveform Direction

Remember from our earlier discussion that when the electrical current traveling through the heart is moving toward a positive electrode, the ECG machine records it as a positive or upright waveform. When the impulse is traveling away from a positive electrode and toward a negative electrode, the ECG machine records it as a negative or downward deflection (Figure 18-7).

Ventricular Depolarization and Mean QRS Axis

In a healthy heart, waves of depolarization originate in the SA node and travel through the atria, the AV node, and on to the ventricles. The ventricular conduction system quickly conducts the stimulus to the ventricles (Figure 18-8).

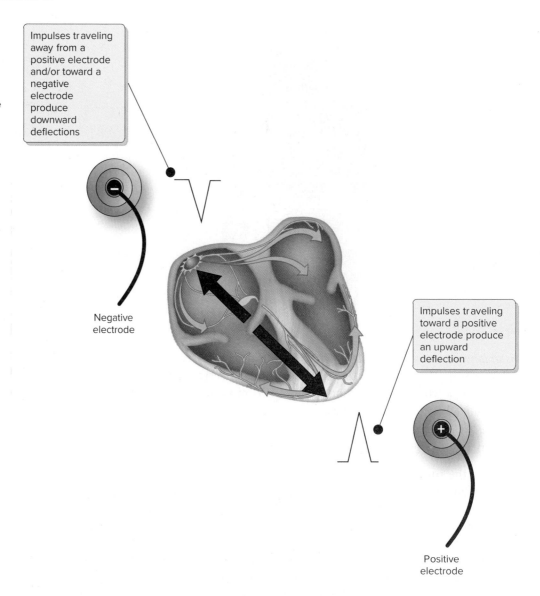

Figure 18-7
Direction of ECG waveforms when the electrical current is traveling toward a positive ECG electrode or away from it.

Impulses traveling away from a positive electrode and/or toward a negative electrode produce downward deflections

Negative electrode

Impulses traveling toward a positive electrode produce an upward deflection

Positive electrode

The Purkinje fibers carry the impulse from the endocardial lining of the right and left ventricles in the apical region of the heart near the septum through the full thickness of the ventricular wall toward the outside surface (or epicardium). The impulse continues through the ventricles in all areas at once and ends in the lateral and posterior aspect of the left ventricle near its base.

Even though the impulse arrives at the subendocardial lining of both ventricles at about the same time, completion of right ventricle depolarization occurs first. This is because the thinner wall of the right ventricle transmits the impulse in a fraction of the time that it takes the impulse to travel through the thick lateral wall of the left ventricle.

As identified earlier, the QRS complex represents the simultaneous depolarization of both ventricles (Figure 18-9).

The vectors arising in the right ventricle are directed mostly to the right when viewed in the frontal plane; those in the left ventricle are directed mostly to the left (Figure 18-10). The left ventricular vectors are larger and persist longer than those

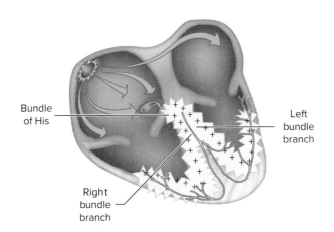

Bundle
of His

Left
bundle
branch

Right
bundle
branch

Figure 18-8
The ventricular conduction
system quickly conducts the
impulse to the ventricles.

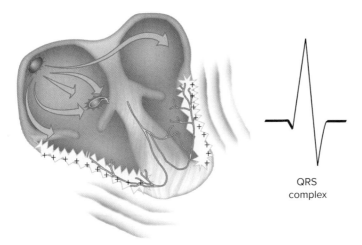

QRS
complex

Figure 18-9
Ventricular depolarization and
the QRS complex.

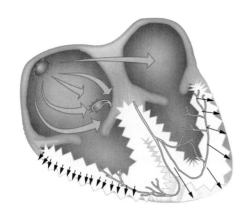

Figure 18-10
The vectors of the left
ventricle are larger and persist
longer than those of the right
ventricle.

of the smaller right ventricle, primarily because of the greater thickness of the left
ventricular wall.

Again, we call the sum of all the small vectors of ventricular depolarization the
mean QRS vector. The mean QRS axis represents both its direction and magnitude.
The small depolarization vectors of the thicker left ventricle are larger, resulting in
greater magnitude, so the mean QRS axis points more to the left (Figure 18-11).
The ventricles are in the left side of the chest and angle downward to the left. For
this reason the mean QRS axis points downward and toward the patient's left side.

Figure 18-11
The mean QRS axis points
downward and toward the
patient's left side.

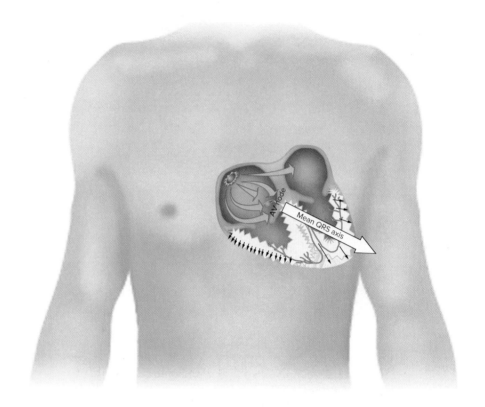

18.7 Determining the Electrical Axis

As described earlier, the electrical axis is the average direction of the heart's electrical activity during ventricular depolarization. ECG leads positioned on the body sense the sum of the heart's electrical activity and record it as waveforms. The patient's electrical axis can be identified by assessing the waveforms recorded by the six leads that view the frontal plane. This includes leads I, II, III, aV_R, aV_L, and aV_F. Imaginary lines drawn from the location of each of the leads intersect in the center of the heart and form a diagram referred to as the *hexaxial reference system*.

Hexaxial Reference System

To better explain the hexaxial reference system, think of an imaginary circle drawn over the patient's chest (Figure 18-12). This circle lies in the frontal plane. The center of the circle is the AV node. The limb leads view electrical forces moving up and down and left and right through this circle. Within the circle are six bisecting lines, each representing one of the six limb leads. The intersection of all lines divides the circle into equal, 30-degree segments. Lead I starts at 0 degrees and is located at the 3 o'clock position. Lead aV_F starts at 90 degrees and is located at the 6 o'clock position. Remember though, that the starting point of each lead is the position of the positive electrode.

As you can see moving from the 3 o'clock position counterclockwise, the degrees are more and more negative, until you reach the 9 o'clock position, where it is ±180 degrees. Moving clockwise from the 3 o'clock position, each segment represents positive-degree positions; however, it doesn't necessarily mean that pole is positive.

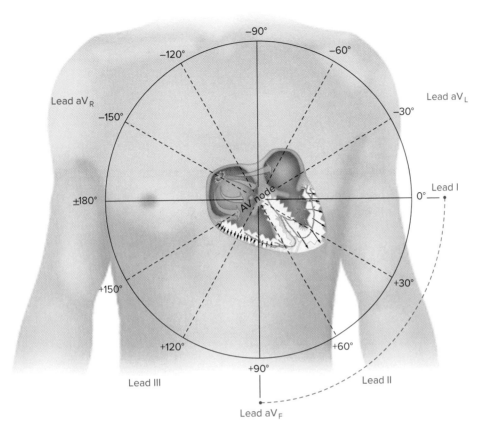

Figure 18-12
The imaginary circle used to determine the heart's axis is divided into equal 30-degree segments with lead I at the 3 o'clock position and lead aV$_F$ located at the 6 o'clock position.

A normal axis is one that falls between 0 and +90 degrees. An axis that falls between +90 and +180 degrees is considered right axis deviation. One that falls between 0–90 degrees indicates left axis deviation, one that falls between −90 and ±180 degrees is considered extreme axis deviation (also referred to as *indeterminate axis*).

To determine the electrical axis, we can use one of two methods: the four-quadrant method and the degree method.

Four-Quadrant Method

With the four-quadrant method, the axis is determined by examining the QRS complex in leads I and aV$_F$.

The mean QRS axis normally points downward and to the patient's left, between 0 and 90 degrees (Figure 18-13). As long as it remains within this range, it is considered normal. If it is outside this range, it is considered abnormal.

Any lead records a positive deflection if the wave of depolarization is moving toward it. Lead I is oriented at 0 degrees. If the mean QRS axis is directed anywhere between −90 and +90 degrees, the right half of the circle, you can expect lead I to record a positive QRS complex (Figure 18-14). Lead I is created by making the left arm positive and the right arm negative, so the left half of the circle is positive, while the right half is negative.

Lead aV$_F$ is oriented at +90 degrees. If the mean QRS axis is directed anywhere between 0 and ±180 degrees, the bottom half of our circle, you can expect lead aV$_F$

Figure 18-13
The mean QRS axis normally points downward and to the patient's left between 0 and +90 degrees.

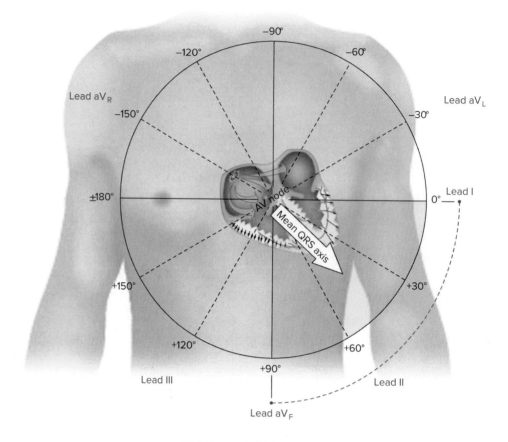

Figure 18-14
If the mean QRS axis is directed anywhere between −90 and +90 degrees (the left half of our circle), a positive QRS complex is seen in lead I.

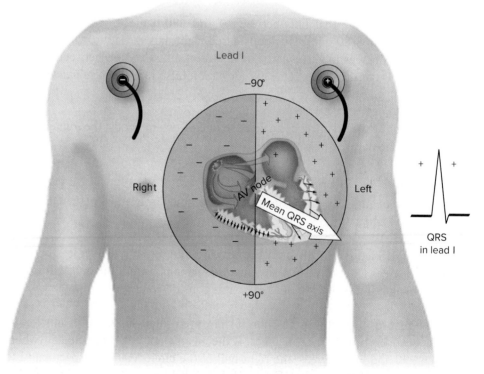

to record a positive QRS complex (Figure 18-15). Lead aV$_F$ is created by making the legs positive and the other limbs negative, so the bottom half of the circle is positive while the upper half is negative.

With lead I we can see whether the impulses are moving to the right or to the left, whereas lead aV$_F$ tells us whether they are moving up or down.

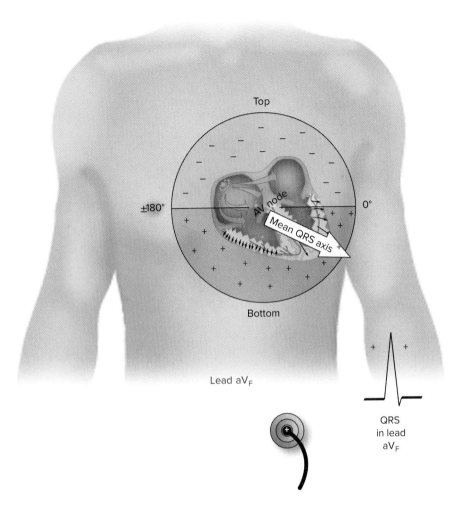

Figure 18-15
If the mean QRS axis is directed anywhere between 0 and ±180 degrees (the bottom half of our circle), a positive QRS complex is seen in lead aV$_F$.

If the main deflection of the QRS complex points up (positive) in lead I, then the electrical impulses are moving from right to left in a normal manner. Conversely, if the deflection points down (negative), then the impulses are moving abnormally from left to right (Figure 18-16).

If the deflection of the QRS complex is positive in lead aV$_F$, the electrical impulses are traveling downward in a normal manner. If the deflection is negative, the impulses are traveling upward; this would be considered abnormal (Figure 18-17).

An easy way to remember the direction of the deflection of the QRS complexes in leads I and aV$_F$ is to think of your thumbs pointing up or down. Two thumbs up indicates normal axis; anything else is considered abnormal.

To determine whether the QRS complex is positive or negative, you examine it and decide whether the average deflection is positive or negative. Often, this is easy to estimate by merely looking at the QRS complex and deciding what the overall complex looks like. More accuracy can be obtained by measuring the number of small squares on the ECG paper grid of negative and positive deflection. If there are more small squares on the negative side, the QRS complex is referred to as negative. If there are more small squares on the positive side, the QRS complex is referred to as positive.

Degree Method

Although the quadrant method is easier, a more accurate axis calculation is the degree method. Aside from being more accurate, it also allows you to determine the axis even when the QRS complex is not clearly positive or negative in leads I and aV$_F$.

Figure 18-16
If the mean QRS axis is directed anywhere between −90 and +90 degrees (the right half of our circle), a negative QRS complex is seen in lead I.

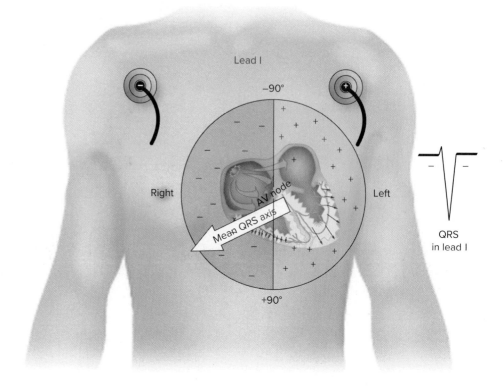

Figure 18-17
If the mean QRS axis is directed anywhere between 0 and 180 degrees (the top half of our circle), a negative QRS complex is seen in lead aV$_F$.

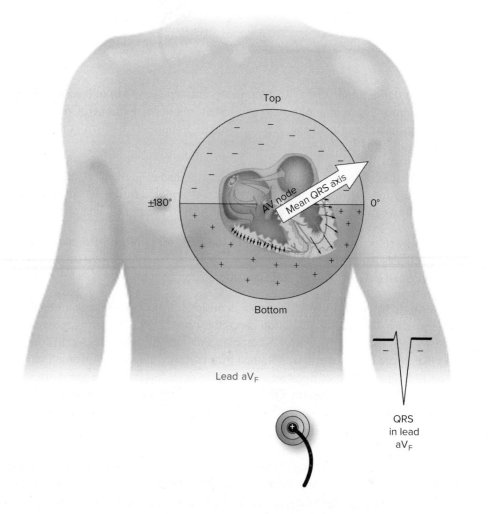

There are three easy ways to employ the degree method. In the six limb leads of the ECG tracing:

1. Identify the lead with tallest (most positive deflection) QRS. It points directly toward the ECG axis.

2. Alternatively, identify the lead with the most negative QRS. It points directly away from the ECG axis.

3. Alternatively, identify the lead with an equiphasic QRS (which has an equal deflection above and below the baseline) or the smallest QRS complex. This lead is at right angles (or perpendicular) to the QRS axis. For example, if lead III has the smallest QRS complex, then the lead perpendicular to the line representing lead III would be lead aV$_R$. After you have identified the perpendicular lead, examine its QRS complex. If the electrical activity is moving toward the positive pole of a lead, the QRS complex deflects upward. If it is moving away from the positive pole of a lead, the QRS complex deflects downward. Plot this information on the hexaxial diagram to determine the direction of the electrical axis.

Let's look at some examples to see these three ways to determine axis in use (Figure 18-18).

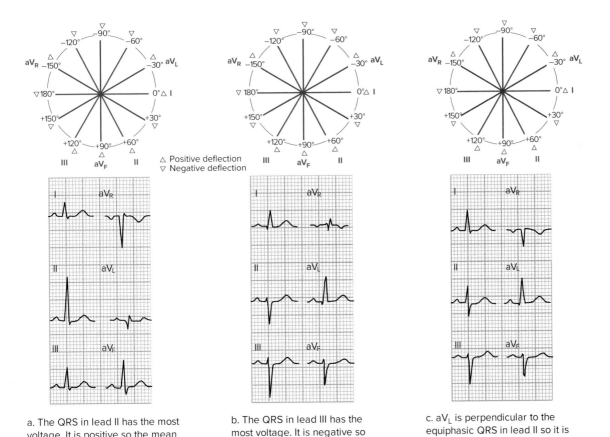

a. The QRS in lead II has the most voltage. It is positive so the mean QRS vector is pointing directly at lead II or +60 degrees. For this example, the axis is +60 degrees.

b. The QRS in lead III has the most voltage. It is negative so the mean QRS vector is pointing directly away from lead III. For this example the axis is −60 degrees.

c. aV$_L$ is perpendicular to the equiphasic QRS in lead II so it is either ±150 or −30 degrees. Since the deflection in lead aV$_L$ is positive, it indicates the mean QRS vector is traveling toward the electrode. The axis for this example is −30 degrees.

Figure 18-18
Shown above are three examples using the degree method to determine QRS axis.

18.8 Altered QRS Axis

As was mentioned earlier, the mean QRS axis normally points downward and to the patient's left, between 0 and +90 degrees. An axis between +90 and ±180 degrees indicates right axis deviation, and one between 0 and −90 degrees indicates left axis deviation. An axis deviation between ±180 and −90 degrees indicates extreme axis deviation and is called an *indeterminate axis* (Figure 18-19).

Again, if the QRS complex deflection is positive in both leads I and aV$_F$, the electrical axis is normal. If the QRS complex in lead I is upright and the QRS complex in lead aV$_F$ has a negative deflection, then left axis deviation exists. If the QRS complex in lead I is negative and lead aV$_F$ has a QRS complex with a positive deflection, then right axis deviation exists. If the QRS complex is negative in both leads, extreme axis deviation exists (Figure 18-20).

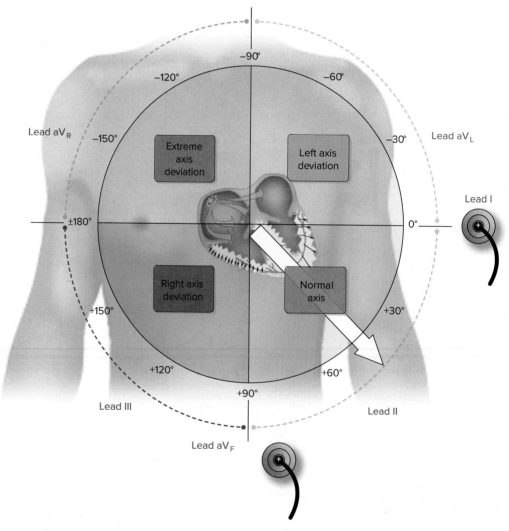

Figure 18-19

Normal axis, right axis deviation, extreme axis deviation, and left axis deviation.

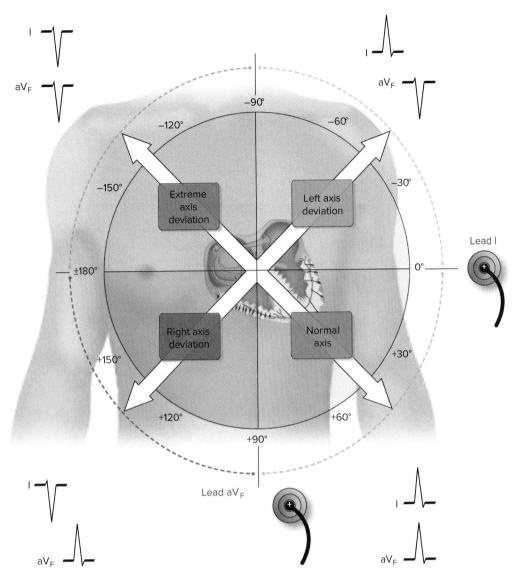

Figure 18-20
Positive QRS complexes in leads I and aV$_F$ indicate a normal QRS axis. A negative QRS complex in lead I and an upright QRS complex in lead aV$_F$ indicate right axis deviation. Negative QRS complexes in both lead I and lead aV$_F$ indicate extreme axis deviation, and an upright QRS complex in lead I and a negative QRS complex in lead aV$_F$ indicate left axis deviation.

18.9 Causes of Altered Electrical Axis

If the heart is displaced, the mean QRS axis is also displaced in the same direction. For example, in tall, slender individuals the heart may be rotated toward the patient's right side. This is referred to as a *vertical heart*. This causes the mean QRS axis to shift toward the right. With obesity and in pregnancy, the diaphragm (and the heart) is pushed up by the increased abdominal pressure so that the mean QRS axis will likely point directly to the patient's left. This is referred to as a *horizontal heart* (Figure 18-21).

Another condition that causes the axis to shift is enlargement or *hypertrophy*, of one or both of the heart's chambers (Figure 18-22). The resultant greater depolarization activity displaces the mean P or QRS axis toward the enlarged, or hypertrophied, region. This is because an enlarged, or hypertrophied, chamber has more (and larger) vectors, which draws the mean P or QRS axis in that direction. As such, right axis

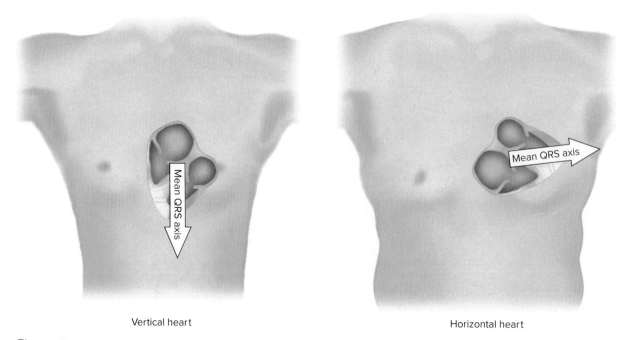

Vertical heart

Horizontal heart

Figure 18-21
In thin individuals the apex of the heart can be directed vertically, whereas with very obese individuals and in pregnancy the apex can be directed horizontally. As the apex of the heart shifts, so does the mean QRS axis.

Figure 18-22
With enlargement, or hypertrophy, of one or both of the heart's chambers, the mean QRS axis shifts toward the enlarged or hypertrophied side.

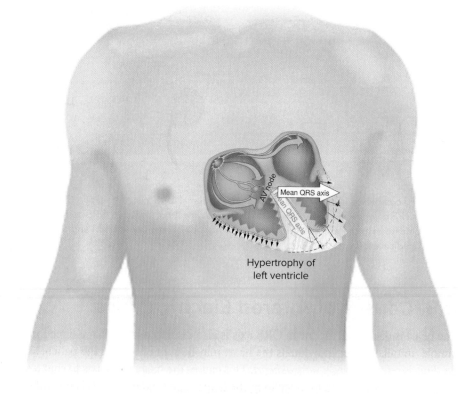

Hypertrophy of left ventricle

deviation (lead I deflects negatively and lead aV$_F$ deflects positively) is usually caused by right ventricular hypertrophy, and left axis deviation (lead I deflects positively and lead aV$_F$ deflects negatively) is usually caused by left ventricular hypertrophy.

MI is another condition that causes the mean QRS axis to point in a different direction than normal (Figure 18-23). With MI, a branch of the coronary arteries supplying an area of the heart with blood becomes occluded and the tissue becomes necrotic. This infarcted tissue cannot depolarize and therefore has no vectors. The vectors from the

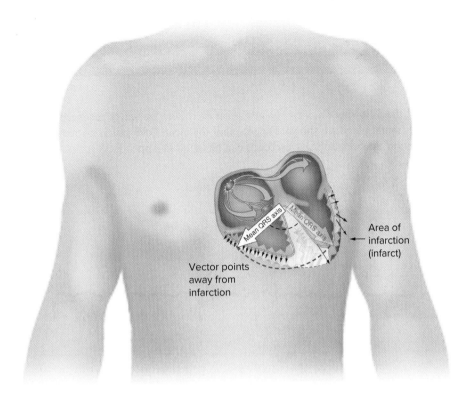

Figure 18-23
The infarcted tissue seen with MI cannot depolarize and therefore has no vectors. The vectors from the other side are unopposed vectors because of this, so the mean QRS vector tends to point away from the infarct.

other side are unopposed vectors because of this, so the mean QRS vector tends to point away from the infarct. As an example, left axis deviation may be seen in inferior wall MI and right axis deviation may be seen in lateral wall infarction.

Axis deviation isn't always a cause for alarm, and it isn't always cardiac in origin. For example, infants and children normally have right axis deviation, whereas pregnant women normally have left axis deviation. Other causes of axis deviation are shown in Table 18-2.

Finding a patient's electrical axis can help confirm a diagnosis or narrow a range of possible diagnoses. As mentioned above, factors that influence the location of the axis include the heart's position in the chest, the heart's size, the patient's body size or type, the conduction pathways, and the force of the electrical impulses being generated. Remember that electrical activity in the heart swings away from areas of damage or necrosis, so the damaged part of the heart will be the last area to be depolarized.

Condition	Description
Left axis deviation	Left ventricular hypertrophy, inferior wall MI, left bundle branch block, left anterior fascicular block, pacemaker rhythm, Wolff-Parkinson-White syndrome (preexcitation syndromes), ventricular pacing/ectopy, congenital heart disease (atrial septal defect), horizontally oriented heart (short, squat person)
Right axis deviation	Right ventricular hypertrophy, lateral wall MI, right bundle branch block, left posterior fascicular block, right ventricular load (i.e., Cor Pulmonale as in COPD), hyperkalemia, sodium channel blockade (i.e., TCA poisoning), Wolff-Parkinson-White syndrome (preexcitation syndromes), dexocardia, ventricular ectopic rhythms, vertically oriented heart (tall, thin person)
Extreme axis deviation	Ventricular rhythms, hyperkalaemia, severe right ventricular hypertrophy

Table 18-2
Factors that cause axis deviation.

Practice Makes Perfect

For each of the following *Practice Makes Perfect* exercises, the waves recorded by the six leads of the frontal plane are shown. Determine if the mean QRS axis is normal or if there is axis deviation. Answers to these exercises can be found in Appendix A.

1.

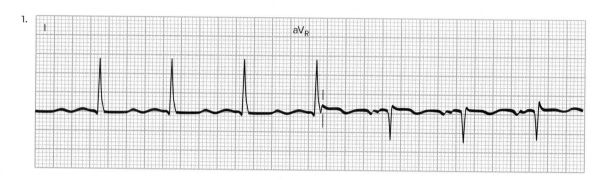

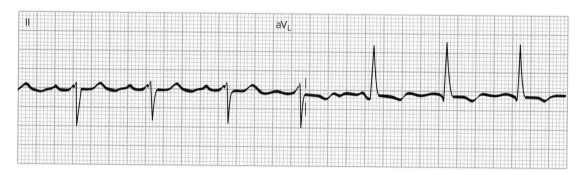

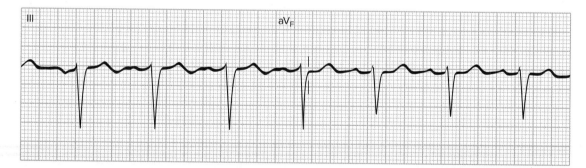

2.

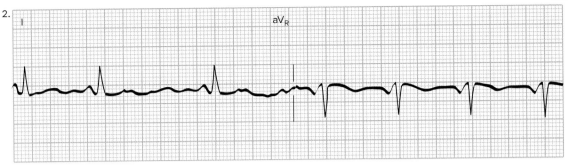

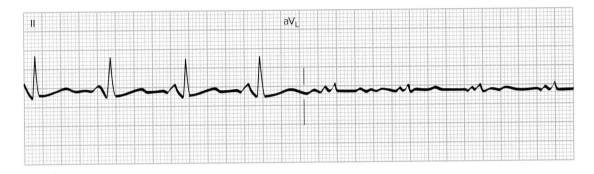

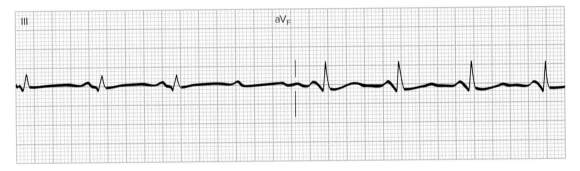

3.

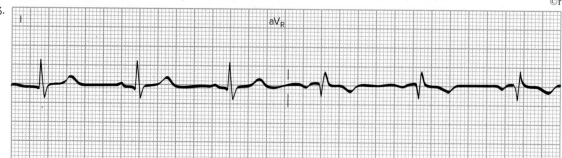

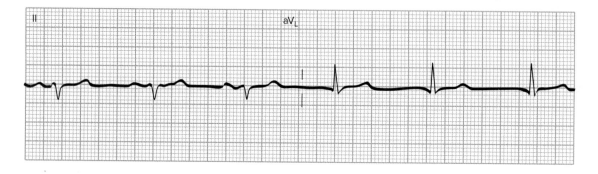

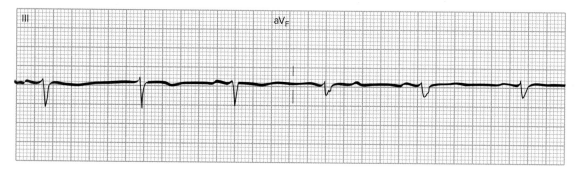

4.

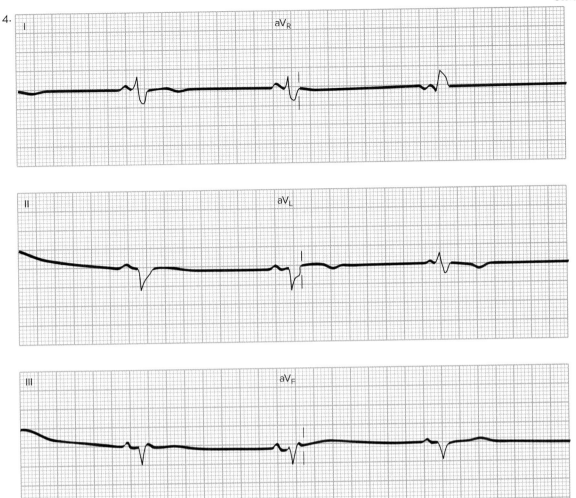

Key Points

LO 18.1	• Because the electrodes positioned on the patient's skin detect the heart's electrical activity, having them placed in many locations gives us many views of the heart.
LO 18.2	• The shape of the waveform is described from the perspective of the positive electrode of the selected lead.
	• The limb leads are produced by placing electrodes on the right arm (RA), left arm (LA), left leg (LL), and right leg (RL). The limb leads include leads I, II, and III; augmented voltage right (aV_R); augmented voltage left (aV_L); and augmented voltage foot (aV_F). They provide a view of the electrical activity along the heart's frontal plane.
LO 18.3	• The precordial leads include leads V_1, V_2, V_3, V_4, V_5, and V_6. They provide information about the electrical activity along the horizontal plane of the heart.
LO 18.4	• To analyze a 12-lead ECG printout, we can begin by looking at the left column from top to bottom, the middle from top to bottom, and the right from top to bottom.
LO 18.5	• Inferior leads include II, III, and aV_F; septal leads include V_1 and V_2; anterior leads include V_3 and V_4; and lateral leads include V_5, V_6, I, and aV_L.
	• Two leads that look at neighboring anatomical areas of the heart are said to be contiguous.
LO 18.6	• Depolarization and repolarization of the cardiac cells produce many small electrical currents or impulses that vary in intensity and direction. These currents are called *instantaneous vectors*.
	• The mean or average of all the instantaneous vectors the ECG detects is called the *mean vector*. The direction of the mean vector is called the *electrical axis*.
	• When the electrical current traveling through the heart is moving toward a positive electrode, the ECG machine records it as a positive or upright waveform.
	• When the impulse is traveling away from a positive electrode and toward a negative electrode, the ECG machine records it as a negative or downward deflection.
	• The mean of all vectors that result from ventricular depolarization is called the *QRS axis*. It is the most important and also the most frequently determined axis.
	• A sequence of vectors is produced as the Purkinje fibers carry the impulse from the endocardial lining of the right and left ventricles through the full thickness of the ventricular wall toward the epicardium.
	• Completion of right ventricular depolarization occurs first because the thinner wall of the right ventricle transmits the impulse in a fraction of the time it takes the impulse to travel through the thick lateral wall of the left ventricle.
	• The sum of all the small vectors of ventricular depolarization is called the *mean QRS vector*. The mean QRS axis points more to the left because the instantaneous vectors of the thicker left ventricle are larger.

LO 18.7	• The limb leads provide information about the frontal plane and are used to determine the position of the mean QRS axis, described in degrees within an imaginary circle drawn over the patient's chest. The center of the circle is the AV node.
	• If the QRS complex is positive in leads I and aV_F, the QRS axis is normal.
	• Leads I and aV_F can be used to quickly determine whether the mean QRS axis on any ECG is normal.
	• The intersection of all lines divides the circle into equal 30-degree segments. Lead I starts at 0 degrees and is located at the 3 o'clock position. The mean QRS axis normally points downward and to the patient's left, between 0 and +90 degrees.
	• An axis between −90 and ±180 degrees indicates right axis deviation, and one between 0 and −90 degrees indicates left axis deviation. An axis deviation between ±180 and −90 degrees indicates extreme axis deviation and is called an *indeterminate axis*.
LO 18.8	• If the QRS complex is upright in lead I and negative in lead aV_F, then left axis deviation exists. If the QRS complex is negative in lead I and positive in lead aV_F, then right axis deviation exists. If the QRS complex is negative in both leads, extreme right axis deviation exists.
LO 18.9	• Factors that influence the location of the axis include the heart's position in the chest, the heart's size, the patient's body size or type, the conduction pathways, and the force of the electrical impulses being generated. Specific conditions that cause axis deviation include hypertrophy, MI, bundle branch block, fascicular block, and congenital heart disease among others.

Assess Your Understanding

The following questions give you a chance to assess your understanding of the material discussed in this chapter. The answers can be found in Appendix A.

1. Describe the uses for the 12-lead ECG. (LO 18.1)

2. With lead I, the _____ lead wire is the positive electrode. (LO 18.2)
   a. LL
   b. RA
   c. LA
   d. RL

3. With the aV_F lead, the positive electrode is positioned (LO 18.2)
   a. on the right arm.
   b. on the left leg.
   c. in the 4th intercostal space in the right sternal border.
   d. in the 5th intercostal space in the left midaxillary border.

4. Which leads view the heart on the horizontal plane? (LO 18.3)
   a. V_1, V_2, V_3, V_4, V_5, and V_6
   b. aV_R, aV_L, and aV_F

    c. MCL$_1$ and MCL$_6$

    d. I, II, and III

5. The limb leads are obtained by placing electrodes on the   (LO 18.2)

6. Leads _____ are referred to as the standard limb leads.   (LO 18.2)

    a. I, II, and III

    b. aV$_R$, aV$_L$, and aV$_F$

    c. V$_1$, V$_2$, V$_3$, V$_4$, V$_5$, and V$_6$

    d. MCL$_1$ and MCL$_6$

7. Leads _____ are referred to as the precordial leads.   (LO 18.3)

    a. MCL$_1$ and MCL$_6$

    b. aV$_R$, aV$_L$, and aV$_F$

    c. I and II

    d. V$_1$, V$_2$, V$_3$, V$_4$, V$_5$, and V$_6$

8. Match the following ECG leads with the correct description.   (LO 18.5)

Leads		Referred to as
a.	V$_1$ and V$_2$	_____ inferior leads
b.	V$_3$ and V$_4$	_____ lateral leads
c.	II, III, and aV$_F$	_____ septal leads
d.	V$_1$, V$_2$, V$_3$, and V$_4$	_____ anterior leads
e.	I, aV$_L$, V$_5$, and V$_6$	_____ anteroseptal leads

9. Contiguous ECG leads   (LO 18.5)

    a. include any two standard limb leads.

    b. include any two leads that are anatomically next to one another.

    c. are also called anteroseptal leads.

    d. are those that view the heart from a superior direction.

10. Describe how to analyze a 12-lead ECG tracing.   (LO 18.5)

11. The currents produced by depolarization and repolarization of the cardiac cells are called   (LO 18.6)

    a. mean vectors.

    b. instantaneous vectors.

    c. the mean electrical axis.

    d. positive deflections.

12. The electrical axis is depicted as a   (LO 18.6)

    a. large circle.

    b. negative deflection.

    c. positive deflection.

    d. single large arrow.

13. Axis is defined   (LO 18.6)

    a. in the frontal plane only.

    b. immediately upon attaching the ECG leads.

    c. as a waveform that has a negative deflection.

    d. as the amperage of a waveform.

14. The QRS axis   (LO 18.6)

    a. is rarely determined.

    b. is the mean of all vectors generated during the depolarization of the atria.

    c. is the most important and also the most frequently determined axis.

    d. normally points up and to the right.

15. Completion of right ventricular depolarization occurs _____ completion of left ventricular depolarization.   (LO 18.6)

    a. before

    b. after

    c. during

    d. both a and b

16. Describe why _____ ventricular vectors are larger and persist longer than those of the _____ ventricle.   (LO 18.6)

17. The sum of all the small vectors of ventricular depolarization is called the mean   (LO 18.6)

    a. P wave vector.

    b. QRS axis.

    c. T axis.

    d. ST axis.

18. The center(s) of the circle used to determine the axis of the mean QRS vector is (are) the   (LO 18.6)

    a. AV node.

    b. SA node.

    c. Purkinje fibers.

    d. apical region of the ventricle.

19. The circle used to determine the axis of the mean QRS vector is divided into equal _____-degree segments.   (LO 18.7)

    a. 10   b. 15

    c. 30   d. 60

20. On the circle used to determine the axis of the mean QRS vector, lead I starts at _____ degrees and is located at the _____ o'clock position.   (LO 18.7)

    a. −90, 6

    b. ±180, 9

    c. −120, 11

    d. 0, 3

21. The mean QRS axis normally points _____ and to the patient's
    _____, between _____ degrees. (LO 18.7)
    a. downward, right, −90 and ±180
    b. upward, left, 0 and −90
    c. downward, left, 0 and +90
    d. upward, right, ±180 and −90

22. Match the following axis deviations with the correct degrees. (LO 18.8)

Deviation Degrees		
a.	Right axis deviation	_____	0 and 90 degrees
b.	Left axis deviation	_____	180 and 90 degrees
c.	Extreme axis deviation	_____	90 and 180 degrees

23. Fill in the following table of QRS deflections (LO 18.8)

	Direction of QRS in Lead I	Direction of QRS in Lead aV$_F$	Electrical Axis
a.	Positive	Positive	
b.	Positive	Negative	
c.	Negative	Positive	
d.	Negative	Negative	

24. What direction will the ECG waveform take (negative, positive, or biphasic) in this example? (LO 18.6)

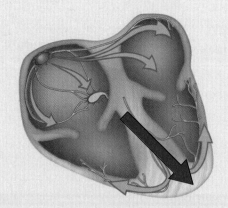

Positive electrode

25. In this illustration showing lead I, identify with + or – signs which half of the circle will be positive and which half will be negative. (LO 18.8)

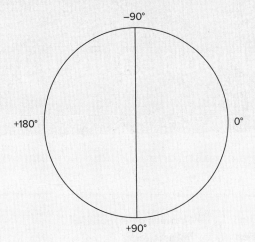

26. In this illustration showing lead aV$_F$, identify with + or – signs which half of the circle will be positive and which half will be negative. (LO 18.8)

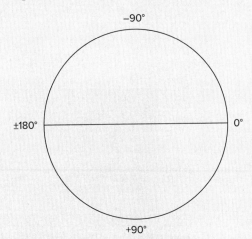

Referring to the scenario at the beginning of this chapter, answer the following questions.

27. When the left ventricle is enlarged, the axis of the heart moves (LO 18.9)
    a. leftward.
    b. rightward.
    c. upward.
    d. lower.

28. The axis shifts because the sum of the electrical impulses in the enlarged ventricle is (LO 18.9)
    a. equal on both sides of the heart.
    b. greater on the left than the right.
    c. lesser on the left than the right.
    d. greater on the bottom than the top.

29. Describe whether or not infarcted tissue depolarizes and if electrical vectors are produced. (LO 18.9)

19 Myocardial Ischemia, Injury, and Infarction

©rivetti/Getty Images

Courtesy Physio-Control

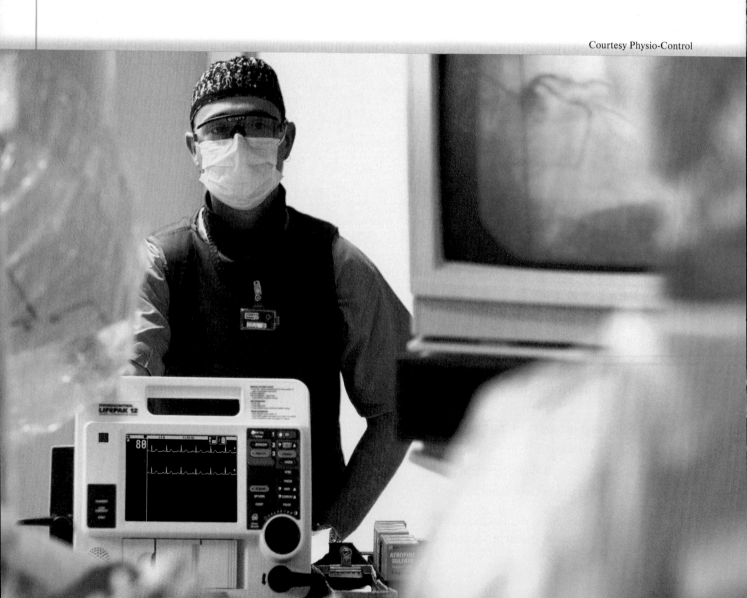

Learning Outcomes

LO 19.1 Identify the arteries of the coronary vasculature.

LO 19.2 Recall the normal morphology of the Q wave, ST segment, and T wave.

LO 19.3 Describe the ECG changes that occur with myocardial ischemia, injury, and infarction.

LO 19.4 List the criteria for diagnosing MI.

LO 19.5 Recall the ECG leads that can be used to identify myocardial ischemia, injury, and infarction in various areas of the heart.

LO 19.6 List the treatments for myocardial infarction.

Case History

EMS is called to a local office building to evaluate a 46-year-old man complaining of chest pain. Upon entering the man's office, the paramedics see an ashen, grey, and diaphoretic middle-aged man sitting at his desk. They notice an overflowing ashtray on the desk as they move to his side.

The paramedics perform a rapid assessment and find the patient to be hypertensive with labored respirations. Oxygen is applied and an IV established. Following their cardiac protocol, the paramedics give the patient an aspirin, start him on nitroglycerin, and obtain a 12-lead ECG.

The 12-lead ECG shows 5 mm of ST elevation in leads II, III, and aV_F, indicating that the chest pain may be from a heart attack. The patient gives the paramedics permission to transport him to the nearest cardiac center.

The paramedics call the cardiac center and speak to the emergency medicine physician on duty. They have a diagnostic 12-lead ECG, so the physician mobilizes the cardiac team, and the patient is taken immediately for emergent angioplasty. Three days later, the cardiologist tells the patient that because of the rapid care of the paramedics, combined with their ability to obtain and interpret a 12-lead ECG, he did not sustain any permanent damage to his heart and should fully recover.

19.1 Coronary Circulation

In Chapter 1, we discussed the coronary circulation. Because understanding the coronary circulation helps us identify where myocardial ischemia, injury, and/or infarction is occurring within the heart, we will review it in more depth here.

The heart constantly pumps blood to the body. Because of this, its oxygen consumption is proportionately greater than that of any other single organ. With the oxygen demand in the myocardial cells being so high, the heart must have its own blood supply. The coronary arteries provide a continuous supply of oxygen and nutrients to the myocardial cells.

Arising from the aorta just above the aortic valve, the right and left coronary arteries lie on the heart's surface and have many branches that penetrate all parts of the heart. The terminal branches of the arteries have many interconnections, forming an extensive vascular network—each perfusing a particular portion of the myocardium. The three main coronary arteries we will discuss in more depth are the right coronary artery (RCA), the left anterior descending artery (LAD), and the left circumflex artery (LCx). The LAD and LCx are branches of the left coronary artery.

Right Coronary Artery (RCA)

The RCA (Figure 19-1) originates from the right anterior aortic sinus. It travels along the right atrioventricular (AV) groove on its way to the crux of the heart (where posterior interventricular and atrioventricular grooves meet).

The first branch of the RCA is often the conus artery, which supplies the right ventricular outflow tract. However, in 50% of hearts, the conus artery originates from the aorta. The second branch of the RCA is often the SA nodal artery. In most people, the RCA supplies the SA nodal artery. In the remaining hearts, the SA

Figure 19-1
Anatomy of right coronary artery (RCA).
©McGraw-Hill Education

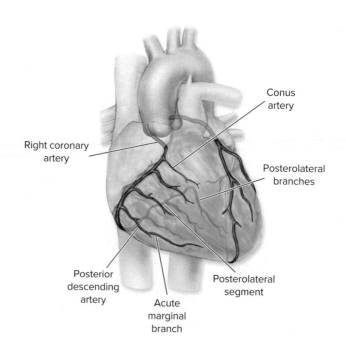

nodal artery arises from either a continuation of an anterior branch of the LCx or from both the RCA and LCx.

At its midportion, as the RCA travels around the right ventricle, it gives off, at right angles, one or more acute marginal (AM) branches that supply the anterior wall of the right ventricle. On occasion, the AM branches provide collateral circulation when the left anterior descending artery (LAD) is occluded. Proximal occlusion of the RCA will compromise blood flow to the right ventricle. For this reason, approximately one-third of inferior STEMI (ST-Elevation Myocardial Infarction) patients have some degree of right ventricular infarction.

As the RCA turns at the crux to become the posterior descending artery (PDA), the AV nodal artery arises and passes superiorly to supply the AV node, proximal parts of the bundles (branches) of His, and the parts of the posterior interventricular septum that surround the bundle branches. This artery can occasionally arise from the LCx.

In approximately 65% of hearts, the RCA continues past the crux and after giving off the PDA, continues as the posterolateral segment, giving off one or more posterolateral branches, which supply the inferior (or posterior) wall of the left ventricle. In hearts with this anatomy, acute occlusion of the RCA will result in both inferior and posterior wall MI.

In most hearts, the PDA is a branch of the RCA and the person is said to have "right dominance." In some people, the RCA is a smaller vessel, terminating before it reaches the crux. To compensate, the LCx is larger and gives rise to the PDA. These persons are said to have "left dominance." The PDA can also be supplied by an anastomosis of the left and right coronary arteries (known as co-dominance). Knowing this information is helpful as the left-dominant anatomic variant accounts for infero-posterolateral infarction pattern of simultaneous ST segment elevation in inferior and anterior lead areas. Understandably, such infarctions are often quite large. Such patients may also manifest, type I acute LCx occlusion because the AV nodal artery is usually supplied by the LCx when this vessel is dominant.

The PDA continues along the undersurface of the heart. In 50% of hearts, the PDA terminates at the left ventricular apex; in the other 50%, it terminates before the apex. The PDA supplies the posterior and inferior walls of the left ventricle. It also gives rise to several small inferior septal branches that travel upward to supply the inferior third of the septum (connecting with septal branches traveling down from the LAD).

What It Supplies

In addition to supplying blood to the right ventricle, the RCA supplies blood to 25% to 35% of the left ventricle. In patients with right dominance, the RCA gives off the PDA (remember, in those who are left dominant, the PDA is given off by the left circumflex artery). The PDA supplies the inferior wall, ventricular septum, and the posteromedial papillary muscle. The RCA supplies the SA nodal artery in 60% of patients. The other 40% of the time, the SA nodal artery is supplied by the left circumflex artery. The RCA also supplies the AV node and bundle of His.

Left Coronary Artery (LCA)

The LCA arises from the left aortic sinus. It usually runs a short course (between 0.5 and 2.5 cm) as the left main coronary artery (LMain) before bifurcating into the LAD and the LCx in approximately two-thirds of people. Acute occlusion of the LCA can result in rapid demise of the patient (since it usually leads to infarction of almost the entire left ventricle). These patients generally die before reaching the hospital. In approximately one-third of people, the LCA trifurcates and there is a third branch called the ramus intermedius artery, which supplies the anterolateral left ventricular wall.

Left Anterior Descending Artery (LAD)

The LAD (Figure 19-2) runs along the anterior epicardial surface of the heart in the interventricular groove on its path toward the cardiac apex. The major branches of the LAD are the perforating intramural septal branches and diagonal branches. There may be one or many septal branches depending on individual anatomy. The first septal branch is typically the largest; its origin is generally just after the takeoff of the first diagonal branch. These perforating septal branches supply the anterior two-thirds of the interventricular septum, as well as the bundle of His and bundle branches of the conduction system. A septal branch may supply the anterior muscle of the right ventricle. The interventricular septum is the most densely vascularized area of the heart because it has an integral role in providing blood supply to the heart's conduction system. Septal perforators normally run a vertical path downward following their takeoff from the proximal LAD.

Downward penetrating septal branches from the LAD typically connect with upward penetrating septal branches from the PDA branch of the RCA. In this way, there is usually a network of collaterals from both LCA and RCA systems in the event of disease in one system. How adequately collaterals from one system compensate for disease in the other is subject to individual variation (as well as to how rapidly occlusive disease develops).

Figure 19-2
Anatomy of left anterior descending artery (LAD).
©McGraw-Hill Education

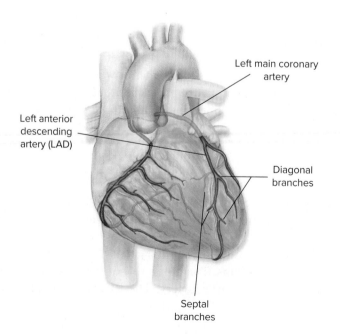

Left main coronary artery

Left anterior descending artery (LAD)

Diagonal branches

Septal branches

Very proximal LAD lesions have been known as "willow-makers." Especially if they are proximal to the first septal perforator (and first diagonal), these lesions are essentially "left-main equivalents" because of the extent of injury and conduction system damage they produce.

Diagonal branch anatomy is also highly variable. There may be one, two, or three diagonals supplying the anterolateral wall. Occasionally there is no diagonal branch, but rather the ramus intermedius (mentioned earlier) arises from between the LAD and LCx to supply the anterolateral surface. Typically the first diagonal branch (D-1) is the largest. D-1 supplies the high lateral wall (assessed on the ECG by lead aV_L). Suspect D-1 involvement when there is ST elevation in lead aV_L in a patient with anterior MI. The LAD typically supplies the anterior wall of the heart and the cardiac apex. However, in some patients the LAD may terminate prior to the apex. In such cases, the apex is supplied by a larger and longer-than-usual PDA arising from either a very dominant RCA or a dominant LCx. In other patients, the LAD is especially long and "wraps around" the apex to supply the undersurface of the heart. At times, it may even serve the function of the PDA.

Assessing acute apical infarction by ECG is often difficult since the area is not optimally viewed by the standard 12 leads. Appreciation of the potential for LAD anatomic variation provides insight to the various ECG patterns that may be seen.

What It Supplies   In general, the LAD artery and its branches supply most of the interventricular septum; the anterior, lateral, and apical wall of the left ventricle; most of the right and left bundle branches; and the anterior papillary muscle of the bicuspid valve (left ventricle). It also provides collateral circulation to the anterior right ventricle, the posterior part of the interventricular septum, and the posterior descending artery.

Left Circumflex Artery (LCx)

The LCx (Figure 19-3) is the other main branch of the LCA. After arising from the LCA, the LCx wraps around the lateral wall of the heart, and then runs in the left posterior AV groove in its path to the infero-postero-interventricular groove.

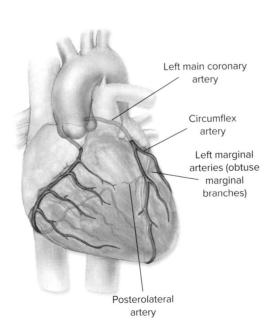

Left main coronary artery

Circumflex artery

Left marginal arteries (obtuse marginal branches)

Posterolateral artery

Figure 19-3
Anatomy of left circumflex artery (LCx).
©McGraw-Hill Education

The LCx gives rise to one or more left marginal arteries, also called obtuse marginal branches (OM), as it curves toward the posterior surface of the heart. It helps form the posterior left *ventricular branch* or posterolateral artery. The LCx becomes relatively small after giving rise to its marginal branches.

What It Supplies   The LCx supplies the posterolateral left ventricle and the anterolateral papillary muscle. It also supplies the sinoatrial nodal artery in 38% of people and supplies 15% to 25% of the left ventricle in right-dominant systems. If the coronary anatomy is left-dominant, the LCx supplies 40% to 50% of the left ventricle.

19.2 The ECG Waveforms

As discussed in earlier chapters, the electrical impulse that initiates the heartbeat normally arises from the SA node. From there it travels through the atria to the AV node, where there is a delay in the conduction of the impulse before it is sent to the ventricles. This delay in conduction is represented by a flat line, called the *PR segment* (Figure 19-4). In the past, the PR segment was used as a baseline from which to evaluate depression or elevation of the ST segment. Currently, the flat line between the T wave and the P wave of the following beat (called the *TP segment*) is considered to be a more accurate baseline for evaluation of depression or elevation of the ST segment. The reason for this change is that sometimes the PR segment itself is depressed and can falsely make the ST segment appear elevated when it is not. However, if the TP segment cannot be identified because of a rapid heart rate or artifact, then the PR segment should be used as the baseline.

Figure 19-4
The TP or PR segment can be used as a baseline to evaluate normal, depressed or elevated ST segments.

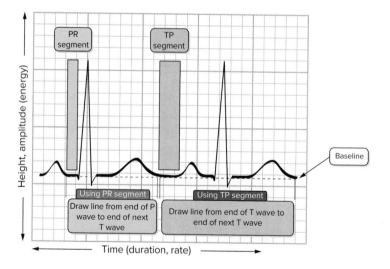

The impulse then travels through the bundle of His, the right and left bundle branches, and the Purkinje fibers. The spread of impulse through the ventricles produces the QRS complex.

Q Waves

Remember, the interventricular septum is the wall of muscle separating the right and left ventricles. A tiny septal branch (fascicle) of the left bundle branch conducts the wave of depolarization to the interventricular septum. It happens so quickly that septal depolarization is not always visible on the ECG, but when it is, this small left-to-right depolarization produces a tiny, negative deflection from the baseline called the

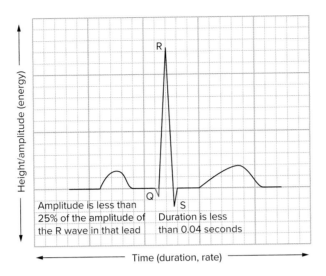

Figure 19-5
The Q wave is the first part of the QRS complex; it is a tiny negative deflection from the baseline.

Q wave (Figure 19-5). It may be seen in leads I, aV$_L$, V$_5$, V$_6$, and sometimes in the inferior leads and in V$_3$ and V$_4$. When it is present, the normal, and clinically insignificant, Q wave should have a duration of less than 0.04 seconds (1 mm or 1 small square) and be less than 25% the amplitude of the R wave in that lead.

Characteristics of a Normal Q Wave

- Small negative deflection from the baseline

- Duration of less than 0.04 seconds (1 mm or 1 small square)

- Less than 25% the amplitude of the R wave in that lead

Normal ST Segments

The ST segment is the line that follows the QRS complex and connects it to the T wave (Figure 19-6). The term *ST segment* is used regardless of whether the final wave of the QRS complex is an R or an S wave. It begins at the isoelectric line extending from the S wave until it gradually curves upward to the T wave. During the period of the ST segment, the ventricles are beginning to repolarize. Under normal circumstances, the ST segment appears as a flat line (neither positive nor negative), although

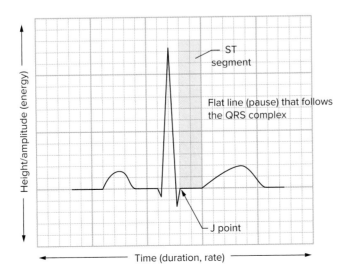

Figure 19-6
The ST segment is the line that follows the QRS complex and connects it to the T wave.

Figure 19-7
Measuring ST segment
displacement from the
baseline to a horizontal line
that extends across one small
box (0.04 seconds) to the
right of the J point.

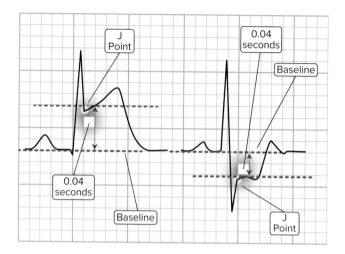

it may vary by 0.5 to 1.0 mm in some precordial leads. The point that marks the end
of the QRS and the beginning of the ST segment is known as the *J point.*

Again, we can use the TP segment, which extends from the end of the T wave to the
beginning of the following P wave, as our baseline. We then draw a line extending
from the baseline through the end of the following T wave. From the J point of the
QRS complex, we draw a vertical line to the baseline and count the number of small
squares in between. Any elevation of depression of the ST segment from the baseline of more than one small square is considered abnormal (Figure 19-7).

Characteristics of a Normal ST Segment

- Flat line

- Follows the QRS complex and connects it to the T wave

- Can vary by 0.5 to 1.0 mm in some precordial leads

Normal T Waves

Next, we should see the T wave (Figure 19-8). The T wave represents the completion
of ventricular repolarization. It is larger than the P wave and slightly asymmetrical. The
peak of the T wave is closer to the end than the beginning, and the first half has a more
gradual slope than the second half. Normally, the T wave is not more than 5 to 6 mm

Figure 19-8
The T wave follows the ST
segment and represents
ventricular repolarization.

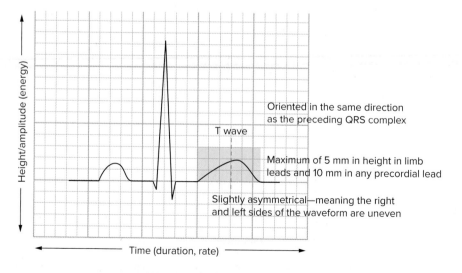

in height in the limb leads or 10 to 12 mm in height in any precordial lead. The T wave is normally oriented in the same direction as its associated QRS complex. T waves are normally positive in leads I and II and V_2 to V_6 and negative in lead aV_R. They are also positive in aV_L and aV_F but may be negative if the QRS complex is less than 6 mm in height. In leads III and V_1, the T wave may be positive or negative.

Characteristics of a Normal T Wave
• Slightly asymmetrical
• Maximum of 5 mm in height in the limb leads
• Maximum of 10 mm in height in any precordial lead
• Oriented in the same direction as the preceding QRS complex

19.3 ECG Indicators of Ischemia, Injury, and Infarction

Chapter 10 provided an overview of cardiovascular disease and reviewed the development and different stages of coronary artery disease. Over the past several decades, the ECG machine has become an effective tool for early identification of persons experiencing myocardial ischemia, injury, and infarction. The ECG can also help identify which coronary artery or branch is occluded. Having this information allows timely intervention and treatment, thus serving to improve patient outcomes and reduce mortality.

One important thing to understand though is that myocardial infarction can occur without significant ECG changes. For this reason, be sure to include all the other assessment tools described later in this chapter in your overall assessment of the patient experiencing a possible acute coronary event.

Key ECG indicators of ischemia, injury, and/or MI are (Figure 19-9):

1. Changes in the T wave (peaking or inversion)
2. Changes in the ST segment (depression or elevation)
3. Enlarged Q waves or appearance of new Q waves
4. A new or presumably new left bundle branch block
5. Reciprocal changes in leads facing the affected area from opposing angles

Although the ECG usually progresses through these three indicators during an MI, any one of these changes may be present without any of the others. For this reason, it is not unusual to see ST segment depression without T wave inversion. As we discuss ischemia, remember that it can be transient, such as in angina, or it can be related to MI.

T Wave Changes

Hyperacute T Waves

Sometimes hyperacute T waves may be the first sign of MI. These are abnormally tall and narrow (Figure 19-10) and can also be referred to as *peaked T waves*. Hyperacute T waves are often not seen because they are present only for 2 to 30 minutes following the onset of infarction. After a period of time, these T waves

Figure 19-9
Key ECG changes with
myocardial ischemia, injury,
or infarction.

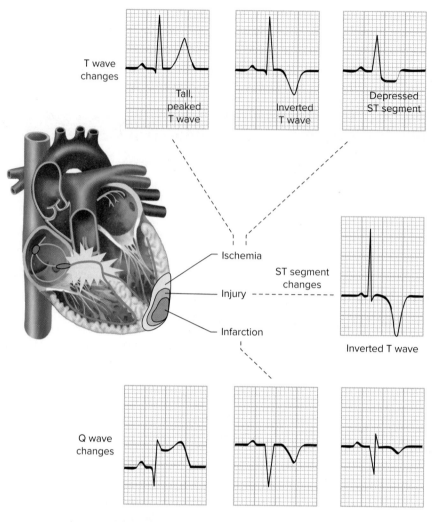

Figure 19-10
Abnormally tall and narrow
T waves are an early sign
of MI.

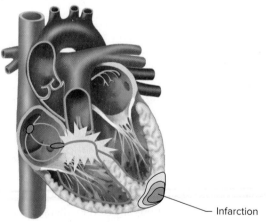

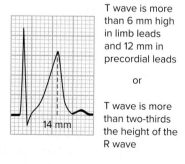

T wave is more
than 6 mm high
in limb leads
and 12 mm in
precordial leads

or

T wave is more
than two-thirds
the height of the
R wave

invert. In general, the T wave should not be more than 6 mm high in the limb leads and 12 mm high in the precordial leads. A helpful rule to follow is this: If the T wave is more than two-thirds the height of the R wave, it is abnormal.

Peaked T waves can also be seen with hyperkalemia, so be sure to distinguish between the two conditions (early onset MI and hyperkalemia).

Inverted T Waves

Inverted T waves are a characteristic sign of myocardial ischemia (Figure 19-11) and indicate that ischemia is present through the full thickness of the myocardium. T waves invert because the lack of oxygen results in leaking of potassium from the ischemic tissue. This delays the process of repolarization.

T wave inversion may be present with angina but may also be seen during an MI due to the ischemic tissue surrounding the infarct. In transient ischemia, the ECG must be obtained as closely as possible to the onset of chest pain; otherwise, the ECG changes may disappear. The inverted T wave from ischemia is symmetrical, meaning that the right and left sides of the T wave are mirror images. T waves inverted in the leads where they should be positive are referred to as *"flipped" T waves.*

Inverted T waves will be present in the leads facing the affected area of the myocardium. The chest leads (V_1 through V_6) are closest to the ventricles, so they are the best for checking for T wave inversion (Figure 19-11). Also, be sure to check the limb leads.

Aside from myocardial ischemia, there are many other causes of T wave inversion, including pericarditis, ventricular hypertrophy, bundle branch block, shock, electrolyte

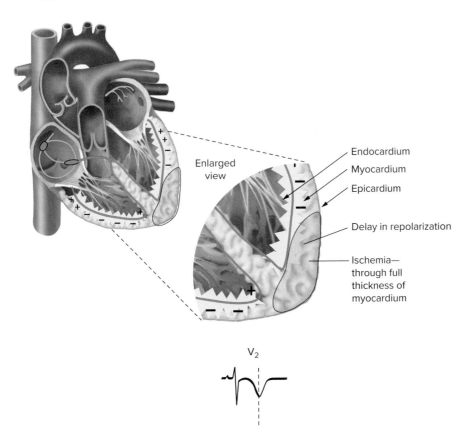

Figure 19-11
Inverted T waves are a characteristic of myocardial ischemia.

Enlarged view

Endocardium
Myocardium
Epicardium

Delay in repolarization

Ischemia— through full thickness of myocardium

V_2

T wave is symmetrical—meaning the right and left sides of the waveform are the same size

Figure 19-12
Examples of T wave inversion
due to other causes.

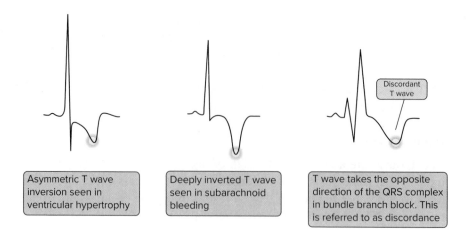

Asymmetric T wave inversion seen in ventricular hypertrophy

Deeply inverted T wave seen in subarachnoid bleeding

T wave takes the opposite direction of the QRS complex in bundle branch block. This is referred to as discordance

Discordant T wave

imbalance, and subarachnoid hemorrhage (Figure 19-12). Asymmetrical T waves with a gentle downslope and rapid upslope are an indicator of inverted T waves associated with other causes rather than inverted T waves associated with ischemia.

Inverted T waves will revert to normal if the ischemia subsides, as with angina. T wave inversion may persist from months to years after an MI because of the damage done to the myocardium, which affects its ability to repolarize normally. Therefore, T wave inversion by itself is only indicative of ischemia if it is consistent with the patient's clinical findings and is not diagnostic of MI.

ST Segment Changes

As with differences in the T wave appearance, changes we see in the ST segment can be due to injured myocardium delaying the process of repolarization.

ST Segment Depression

ST segment depression is another characteristic sign of myocardial ischemia (Figure 19-13). ST segment changes are considered significant when the ST segment falls more than 1 mm below the baseline at a point one small square (0.04 seconds) to the right of the J point and is seen in two or more leads facing the same anatomic area of the heart.

The slope of the ST segment can help differentiate between changes that are significant and those that are likely due to the heart rate.

The slope of the ST segment may be described as "horizontal," or "downsloping," (both of which are significant) or "upsloping," which is not significant and possibly influenced by heart rate. Depressed but upsloping ST segment generally rules out ischemia as a cause.

Two other things to consider are: 1) ST depression seen in leads V_1 and V_2 are indicative of possible posterior wall injury; and 2) ST segment seen in leads opposite of the injured area of the heart are called reciprocal and further support the presence of myocardial injury. Both are discussed later in this chapter.

ST segments may be depressed and may or may not include T wave inversion.

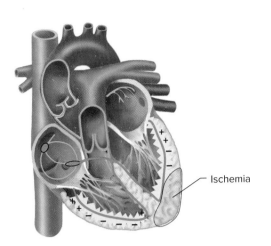

Figure 19-13
ST segment depression is
another sign of myocardial
ischemia.

Ischemia

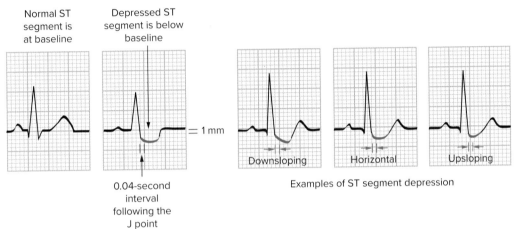

Normal ST segment is at baseline

Depressed ST segment is below baseline

= 1 mm

0.04-second interval following the J point

Downsloping Horizontal Upsloping

Examples of ST segment depression

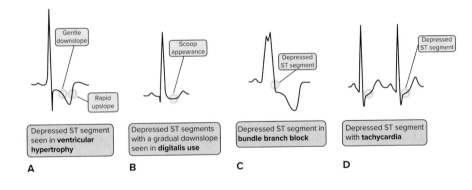

Gentle downslope

Rapid upslope

Depressed ST segment seen in **ventricular hypertrophy**

A

Scoop appearance

Depressed ST segments with a gradual downslope seen in **digitalis use**

B

Depressed ST segment

Depressed ST segment in **bundle branch block**

C

Depressed ST segment

Depressed ST segment with **tachycardia**

D

Figure 19-14
Some causes of ST segment
depression other than MI.

ST Segment Depression from Other Causes   ST segment depression is not always indicative of myocardial ischemia. Instead, it may be caused by digoxin effect, right or left ventricular hypertrophy, pulmonary embolus, right or left bundle branch block, hypokalemia, hypothermia, tachycardia, mitral valve prolapse, and central nervous system disease. The appearance of the ST segment depression may help in identifying the cause (Figure 19-14). Also, the patient's chief complaint, history, and physical exam should help in identifying the cause of the ST segment depression. We discuss a number of these causes in later chapters.

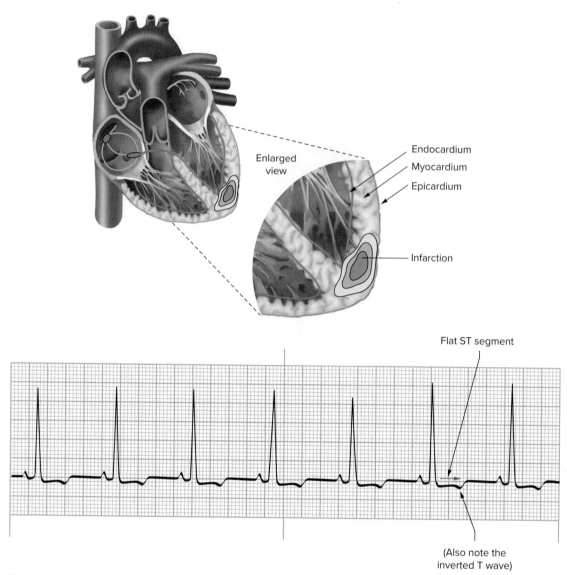

Figure 19-15
The ST segment is flat (or downsloping) in subendocardial infarction.

Flat ST Segment Depression

A flat depression of the ST segment occurs with subendocardial infarction (often referred to as *subendocardial injury*) (Figure 19-15). It may be either horizontal or downsloping in leads where the QRS complex is upright. Subendocardial infarction is an infarct that does not extend through the full surface of the ventricular wall; it involves only a small area of myocardium just below the endocardial lining.

ST Segment Elevation

As described earlier, myocardial injury reflects a degree of cellular damage beyond ischemia. Myocardial injury produces ST segment elevation in the leads facing the affected area (Figure 19-16). This occurs because depolarization of the injured myocardium is incomplete, with the tissue remaining electrically more positive than the uninjured areas surrounding it.

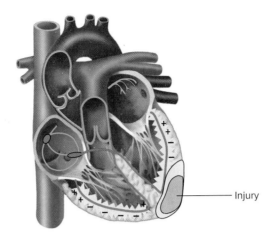

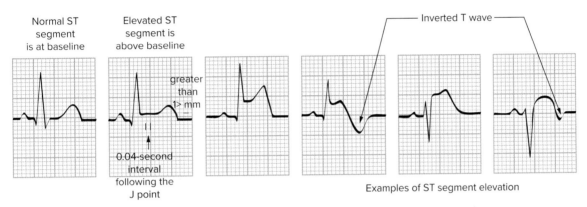

Figure 19-16
ST segment elevation is a sign of myocardial injury.

Myocardial injury produces a distinctive ST segment that bows upward and tends to merge imperceptibly with the T wave. The ST segment may be only slightly elevated or it may be elevated as much as 10 mm above the baseline. It is said to be significant when the ST segment is raised more than 1 mm above the baseline at a point one small square (or 0.04 seconds) to the right of the J point in the limb leads or more than 2 mm in the precordial leads, with these changes being seen in two or more leads facing the same anatomic area of the heart. ST segment depression may be seen in the leads opposite the affected area.

ST segment elevation with an upward concavity, sometimes described as a "smiley" configuration (Figure 19-17A), is usually benign, especially when seen in someone who is otherwise healthy and asymptomatic. If seen with notching of the J point in

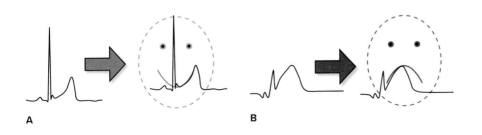

Figure 19-17
Draw a smiley face over the ST segment to determine ST segment concavity. (A) Upward concavity (concave). (B) Downward concavity (convex).

one or more leads, it is a benign normal variant known as early repolarization. It may also be seen with patients experiencing pericarditis. Again, draw a couple of "eyes" above it, it looks like a smiling face, it may be benign.

The shape of ST segment elevation can also help identify acute MI. Look for ST segment elevation with coving or a downward concavity. It is often described as a "frowny" configuration; in other words if you draw a couple of "eyes" above it, it looks like a frowning face (Figure 19-17B).

Another way to identify if the ST segment elevation is either downward concavity or upward concavity is to draw a straight line from the J point to the tip of the T wave (Figure 19-18). If the T wave projects above the straight line, it is downward concavity. Conversely, if the T wave lies below the straight line, it is upward concavity.

Figure 19-18
Use a straight line to determine ST segment concavity. (A) Upward concavity (also called con cave). (B) Downward concavity (also called convex).

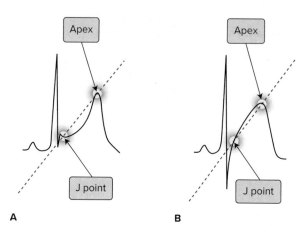

However, these are only a guide; exceptions can and do occur. History is always important. Although ST elevation with a "smiley" configuration and J point notching often reflects a normal variant, this is only true in asymptomatic patients. An identical ST pattern seen in a patient with chest pain must be assumed to be an acute MI until proven otherwise. If possible, compare the tracing to old ECGs to see if the same ST segment elevation appearance is present. Repeat the ECG to identify evolving changes.

ST segment elevation is the earliest reliable sign that MI has occurred. It tells us that the MI is acute.

STEMI and NSTEMI

The ECG findings allow us to distinguish between ST elevation MI (STEMI) and non-ST elevation MI (NSTEMI). The reason this is important is that most cases of STEMI are treated emergently with reperfusion therapy such as thrombolysis (the use of thrombolytic drugs) or, if possible, with percutaneous coronary intervention (PCI) that includes angioplasty and stent insertion, in the hospitals that have facilities for coronary angiography.

With NSTEMI, the patient has symptoms such as chest pain but there is an absence of ST segment elevation. Instead, there may ST segment depression (Figure 19-19) and other findings. Often, the cause is significant narrowing of one or more coronary vessels.

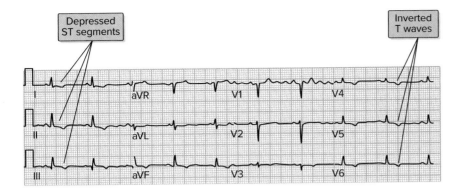

NSTEMI is managed through the use of medication, although PCI is often performed in the course of hospital admission if there are elevated serum cardiac markers, persistent or reoccurring chest pain, and/or changes in the ECG suggestive of compromised myocardial blood flow. In patients suffering from multiple coronary arteries with significant narrowing and whose conditions are relatively stable, or in some extraordinary emergency cases, bypass surgery may be used as a treatment for the largely narrowed coronary artery(s).

Other Causes of ST Segment Elevation

ST segment elevation may be caused by conditions other than myocardial injury. They are sometimes referred to as STEMI impostors and can pose a challenge when it comes to differentiating them from a STEMI. ST segment elevation may be a normal variant (so-called male pattern, which is seen in approximately 90% healthy young men). It may also be:

- Benign early repolarization (BER) (Figure 19-20A): a normal variant of the ST segment, seen in 1 to 2% of patients, especially young men. It is characterized by elevation of the J point and the ST segment. The ST segment may be concave. There are also tall R waves, large symmetrical T waves, and an absence of reciprocal ST depression. These findings are most often present in the middle chest leads V_2 to V_5. In another form, there is a notch or slur at the end of the QRS complex called a J wave. Recently, a different form of early repolarization has been linked to idiopathic ventricular fibrillation. It is most often seen in lead II and consists of terminal QRS slurring that looks like a "hump," without ST elevation.

- Ventricular aneurysm (Figure 19-20B): a weakening and bulging of the ventricular wall, should be suspected if ST segment elevation following an MI persists.

- Brugada syndrome (Figure 19-20C): discussed in more detail in Chapter 20.

- Bundle branch block (Figure 19-20A): discussed in more detail in Chapter 20.

- Pericarditis: an inflammation of the heart. With this condition, the ST segment elevation is usually flat or concave (the middle sags downward) and is seen in all leads except aV_R (Figure 19-21). This resolves with time. It also seems to elevate the entire T wave off the baseline, with the baseline gradually slanting back downward all the way to the next QRS complex. The P wave is often a part of this downward slanting of the baseline. It is discussed in more detail in Chapter 21.

Figure 19-20
Other causes of ST segment elevation (A) benign early repolarization (BEN), (B) ventricular aneurysm, (C) Brugada, (D) Left bundle branch block.

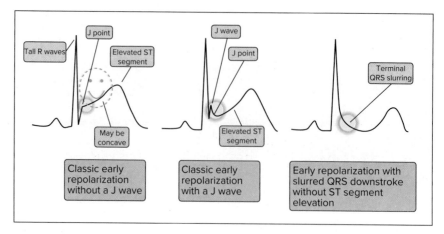

A, Benign early repolarization (BEN)

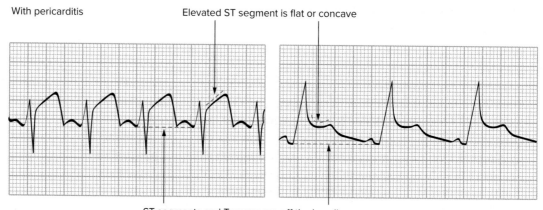

Ventricular aneurysm   Brugada   Left bundle branch block

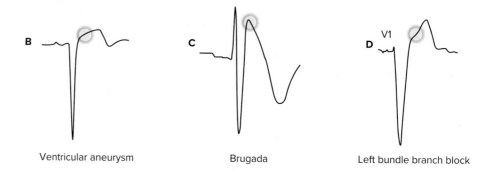

With pericarditis   Elevated ST segment is flat or concave

ST segments and T waves are off the baseline,
gradually angling back down to the next QRS complex

Figure 19-21
ST segment elevation may be caused by a variety of other conditions, including pericarditis.

Q Wave Changes

The presence of pathologic (also called *significant*) Q waves generally indicates that irreversible myocardial damage has occurred (Figure 19-22). As the Q wave develops, the R wave disappears. Although we can expect to see the Q waves within several hours of the onset of infarction, in some patients they may take several days to appear. Further, by the time the Q waves appear, the ST segment usually has returned to the baseline. Q waves often persist for the lifetime of the patient.

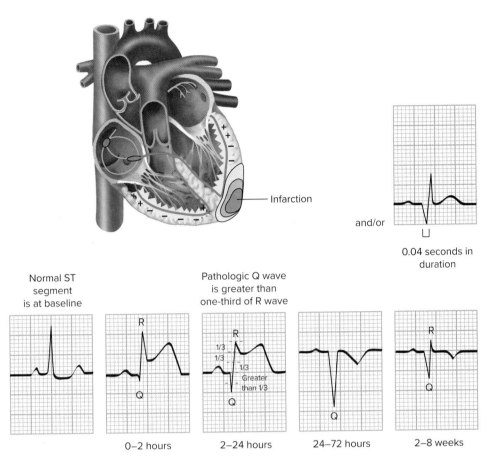

Figure 19-22
Pathological Q waves are a
sign of MI.

Infarction

and/or

0.04 seconds in
duration

Normal ST
segment
is at baseline

Pathologic Q wave
is greater than
one-third of R wave

0–2 hours   2–24 hours   24–72 hours   2–8 weeks

For the Q waves to be considered pathologic, they must

- Be greater than 0.04 seconds in duration or;
- Have a depth at least one-third the height of the R wave in the same QRS complex and;
- Be present in two or more contiguous leads (leads that look at the same area of the heart)

Q waves develop because the infarcted areas of the heart fail to depolarize, producing an electrical void (Figure 19-23). As a result, the electrical forces of the heart are directed away from the area of infarction. An electrode overlying the infarction therefore records a deep negative deflection, a Q wave.

In the past, Q wave infarcts were referred to as *transmural,* meaning that the infarct extended completely through the myocardial wall. However, recent discoveries indicate that this is not the case and confirms that the development of Q waves has more to do with the resulting electrical void than it does the amount of tissue injured.

MI can occur without Q waves developing. The only changes in the ECG with non-Q wave infarctions are T wave inversion and ST segment elevation.

Left Bundle Branch Block

A new or presumably new left bundle branch block can be another indicator of MI. However, you must have access to the patient's old ECGs to confirm this. A challenge with left bundle branch block (as well as pacing) is that distortion of the QRS

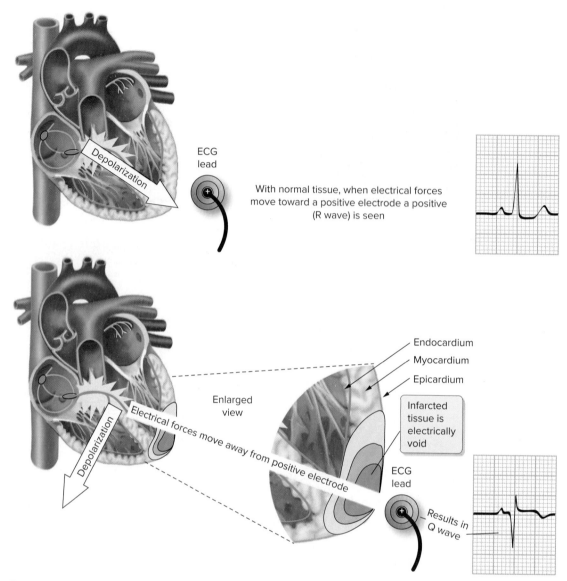

Figure 19-23
Q waves occur because infarcted tissue is electrically silent, allowing the wave of depolarization moving away from the positive electrode to be seen.

complex interferes with your ability to identify acute MI by making it difficult to accurately assess the ST segment. We discuss how to identify bundle branch block in the presence of MI in Chapter 20.

Reciprocal Changes

Reciprocal changes seen on the 12-lead ECG may assist you in distinguishing between MI and conditions that mimic it. Further, closely scrutinizing the contour of the ST segment may also be helpful. With MI the ST segment tends to be straight or upwardly convex (nonconcave).

The simplest explanation of reciprocal changes is that they are a mirror image that occurs when you have two leads viewing the same MI from opposite angles (Figure 19-24).

An example is when you have a lead located over the infarcted area and a lead on the direct opposite side of the heart. The infarcted zone is electrically neutral,

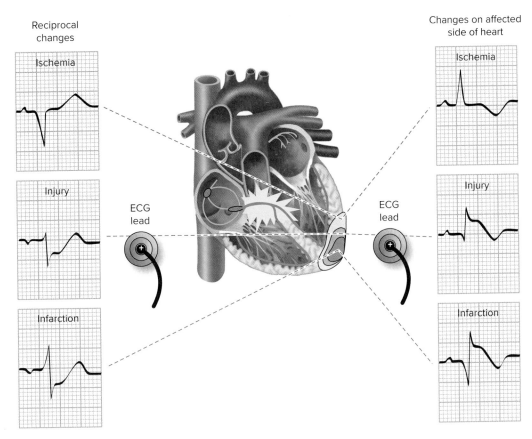

Figure 19-24
Reciprocal changes are a mirror image (opposite image) that occurs when two leads view the same infarction from opposite angles.

so the lead directly over the heart registers only an unopposed vector moving away from it. This causes the waveform to be negative, thus producing a Q wave. Next, the lead registers the other vectors contributing to the QRS and the injured tissue that causes ST elevation. The T wave is flipped because of repolarization abnormalities generated by the areas of ischemia and injury. The electrode opposite of the infarction sees an unopposed vector coming at it, giving rise to a high R wave. This lead registers the zones of ischemia as a pattern of ST depression and the injury as an upright T wave. The second lead essentially sees the mirror image of what the first electrode saw. The recording made by the lead on the wall opposite the MI is registering the reciprocal change of that MI. This concept is particularly important as it relates to looking for ECG changes in areas of the heart (posterior and the right ventricle) where there is not a lead directly over that area.

ECG Evolution during Myocardial Infarction

During MI the ECG often evolves through three stages (Figure 19-25). First, with the acute onset of ischemia, there is the appearance of hyperacute T waves (peaking) followed by T wave inversion. In the presence of actual MI, the T wave can remain inverted for months to years. Next, as the infarction progresses from ischemia to actual injury of the myocardium, ST segment elevation occurs. Acutely, the ST segment elevates and merges with the T wave. If myocardial injury progresses to infarction, the ST segment often returns to baseline within a few hours. Later, if perfusion is not restored, there may be loss of R wave height, and new Q waves

Figure 19-25
ECG changes associated with
evolution of MI.

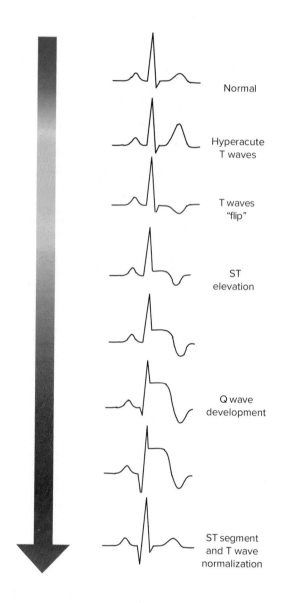

appear, usually within hours to days. Q waves signify MI and the formation of scar tissue. Usually, the T wave recovers, leaving a pathologic Q wave as the only lasting sign that an MI has occurred.

Whereas during MI the ECG characteristically evolves through these three stages, any of these changes may be present without the others. For example, it is not unusual to see ST segment elevation without T wave inversion. In addition, as mentioned earlier, MI does not always generate Q waves.

MI—Age Indeterminate

Sometimes, the analysis by the ECG machine includes the wording "MI—age indeterminate" or similar phrasing. Typically it occurs when Q waves are present, representing that an injury has occurred and that tissue has died. It is important to assess Q waves for height, width, leads present in, and the clinical scenario. A Q wave MI is

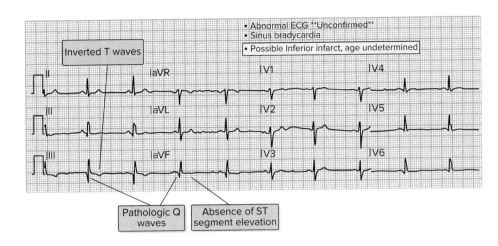

Figure 19-26
In this example, the ECG machine identified pathological Q waves and absence of current STEMI and recorded it as "Possible inferior infarct, age undetermined." This represents that myocardial infarction occurred sometime in the past.

called an old (indeterminate age) MI because all the changes, both on the ECG and in the myocardium, that would allow you to time or date the damage have already occurred (Figure 19-26). The exception occurs when ST and T wave changes are still present and Q waves appear; then you can determine that the infarct is more recent.

Also, this result is sometimes seen in normal hearts when the ECG machine fails to identify the presence of small R waves or when there is incorrect placement of ECG leads on the chest. For this reason, a second ECG should be performed. If there remains some question, an echocardiogram can be used to distinguish between an old MI and a normal heart.

19.4 Criteria for Diagnosing Myocardial Infarction

The diagnosis of MI is based on the presence of at least two of the following three criteria:

1. Clinical history of ischemic-type chest discomfort/pain

2. Rise and fall in serum cardiac markers

3. Changes on serially obtained ECG tracings

Clinical History

Symptoms characteristic of MI includes gradual (over several minutes) onset chest pain, shortness of breath, nausea, vomiting, palpitations, diaphoresis, and anxiety (often expressed as a sense of impending doom). The patient may describe the pain as tightness, pressure, squeezing, crushing, or vice-like. Some patients tell of it feeling like there is an elephant standing on their chest. The pain often radiates to the left arm or left side of the neck but may also radiate to the lower jaw, neck, right arm, back, and epigastrium, where it may mimic heartburn. Women often have fewer typical symptoms than men, most commonly shortness of breath, weakness, a feeling of indigestion, and fatigue. Approximately 25 percent of MIs are silent, in other words, occurring without characteristic chest pain or other symptoms. This is often seen in the elderly, women and diabetics.

Serum Cardiac Markers

The cell membranes of dying myocardial cells break and leak substances into the blood. Blood tests can be used to determine the presence of these substances in the blood. These substances (called *cardiac markers* or *serum cardiac markers*) include creatine kinase MB isoforms, troponin, and myoglobin. Troponin-T and troponin-I are two tests that may be ordered for a patient with a suspected MI. Elevated levels (positive test) indicate that MI has occurred. The troponin-I test may have better specificity than troponin-T. Troponin-T and troponin-I start to rise at 4 to 6 hours and remain high for up to two weeks. CK-MB starts to rise at 4 to 6 hours and falls to normal within 48 to 72 hours. Cardiac markers should rise above two times the upper limit of the reference range.

Also, a full blood count can provide additional information. Elevation of the white blood cell (WBC) count is usual in MI, and the erythrocyte sedimentation rate (ESR) and C-reactive protein (CRP) may also elevate.

ECG Findings

Often, characteristic ECG changes accompany MI. The earliest changes occur almost immediately following the onset of compromised myocardial perfusion. As a matter of practice, a 12-lead ECG should be immediately performed on anyone even remotely suspected of experiencing MI. However, early ECGs do not always reveal MI because the progression of electrocardiographic changes varies between individuals and because unstable ischemic syndromes have rapidly changing supply versus demand characteristics. For this reason, it is important to obtain serial 12-lead ECGs throughout patient assessment and treatment. This is particularly true if the first ECG is obtained during a pain-free episode.

19.5 Identifying the Myocardial Infarction Location

As we discussed earlier, initial (and serial) ECGs can help identify the presence of acute (or impending) occlusion but also the probable culprit coronary artery or branch and the location (and extent) of evolving injury. Leads II, III, and aV_F provide a view of the tissue supplied by the right coronary artery, whereas leads I, $aV_L, V_1, V_2, V_3, V_4, V_5,$ and V_6 view the tissue supplied by the left coronary artery. However, it is important to remember that coronary artery anatomy can vary between given individuals, so the precise vessel involved is not always completely predictable from the ECG.

When evaluating the extent of infarction produced by a left coronary artery occlusion, it is important to determine how many of these leads show changes consistent with an acute infarction. The more of these eight leads demonstrating acute changes, the larger the infarction.

When assessing the 12-lead tracing, it is easiest to use lead groupings to identify ischemia, injury, and/or infarction. Remember from our discussion in Chapters 2 and 18:

- leads II, III, and aV_F view the inferior wall of the heart
- leads V_1 and V_2 view the septal area
- leads V_3 and V_4 view the anterior wall of the left ventricle
- leads V_5, V_6, I, and aV_L view the lateral wall of the left ventricle

Another point to be aware of is there can be overlap in certain lead areas. Two examples are: 1) lead V_2 is between septal and anterior areas of the heart, and 2) lead V_4 is between anterior and lateral areas of the heart.

Lastly, not all areas of the heart are well visualized on a standard 12-lead ECG. Among the more difficult areas to see on ECG are the high lateral wall, the right ventricle, the posterior wall of the ventricle, and the cardiac apex. Special leads such as V_4R or use of the "mirror test" may be needed to detect acute changes in these areas.

Septal Wall Infarction

Leads V_1 and V_2 overlie the ventricular septum, so ischemic changes seen in these leads, and possibly in the adjacent precordial leads, are considered to be septal wall infarctions (Figure 19-27). To identify septal wall MI, look for ST segment elevation in leads V_1 and V_2. Commonly, the ST segment will have coving or a downward concavity. T wave inversion and diagnostic Q waves may also be present. Also, look for disappearance of the septal Q waves in leads V_5 and V_6. The suspected culprit artery in septal wall infarctions is the left anterior descending artery—more specifically, one or more of the septal branches. Reciprocal changes are not seen in the other leads.

Because the left coronary artery also supplies blood to the anterior wall of the left ventricle, it is common for a septal wall MI to accompany an anterior wall MI.

Figure 19-27
Leads V_1 and V_2 are used to identify septal wall MI.

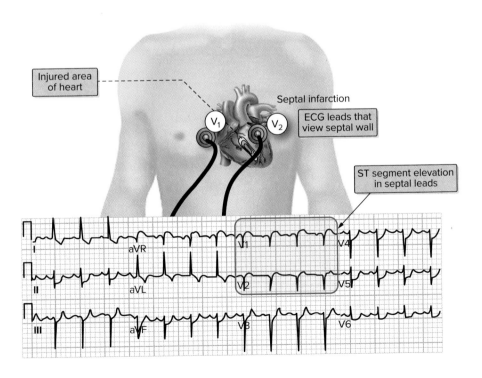

Injured area of heart

Septal infarction

ECG leads that view septal wall

ST segment elevation in septal leads

Anterior Wall Infarction

Anterior infarction involves the anterior surface of the left ventricle (Figure 19-28). It occurs as a result of left anterior descending artery occlusion. Remember, this coronary artery also supplies the ventricular septum and portions of the right and left bundle branches. Leads V_3 and V_4 are immediately over the anterior surface of the heart; therefore, they provide the best view for identifying anterior MI. ST segment elevation, T wave inversion, and the development of significant Q waves in these leads indicate MI involving the anterior surface of the heart. The ST segment elevation typically appears concave downward. In some cases, it seems to overwhelm the T wave. This is called "tombstoning" because it has a similar shape to a tombstone (Figure 19-29). Also, watch for the disappearance of the r (or R) waves as the Q waves develop. Reciprocal ST segment depression may be seen in leads II, III, and aV_F (the inferior leads). Axis is normal at +60 degrees.

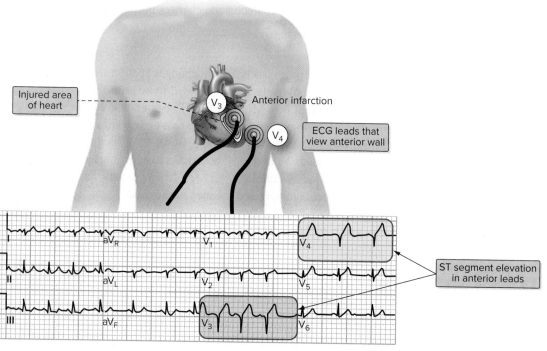

Figure 19-28
Leads V_3 and V_4 are used to identify anterior MI.

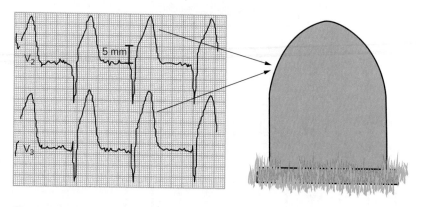

Figure 19-29
ST segment seems to overwhelm T wave giving it a "tombstone like" appearance.

Complications of anterior wall MI include heart failure, ventricular dysrhythmias, AV heart block and/or BBB, and sinus tachycardia.

- *Heart failure:* as a result of extensive anterior MI, contractility of the left ventricle can become compromised, leading to cardiogenic shock.
- *Ventricular dysrhythmias*: ventricular fibrillation or tachycardia can occur in a significant minority of patients with anterior MI.
- *AV and BBB block*: 1st-degree AV block may be seen due bundle of His involvement. 2nd-degree AV block with wide QRS complexes may occur due to a resulting blockage of the conduction system below the AV node. New RBBB may be seen and is a sign of severe conduction system damage. RBBB may also present as a bifascicular block, usually with left anterior fascicular block but if there is extensive damage, left posterior fascicular block.
- *Sinus tachycardia*: may be seen due to increased sympathetic tone and/or associated heart failure.

If the ECG changes are seen in leads V_1, V_2, V_3, and V_4, we call it an *anteroseptal wall infarction*.

Anteroseptal MI is due to occlusion of the proximal left anterior descending artery, distal to the first digital branch and proximal to the first septal branch.

Lateral Wall Infarction

Infarction of the left lateral wall of the heart is referred to as a *lateral wall infarction*. It is often due to occlusion of the smaller branch arteries that supply the lateral wall, such as the first diagonal branch of the left anterior descending artery, the obtuse marginal branch of the left circumflex artery, or the ramus intermedius. With lateral wall infarction, you should see ECG changes such as ST segment elevation, T wave inversion, and the development of significant Q waves in leads I, aV_L, V_5, and V_6 (Figure 19-30). Reciprocal changes are seen in leads II, III, and aV_F. But are only present if there is ST segment elevation in leads I and aV_L.

ST elevation primarily localized to leads I and aV_L is referred to as a high lateral STEMI.

Lateral wall MI often causes premature ventricular complexes and varying degrees of AV heart block.

Lateral wall MIs are often associated with a larger area such as an anterolateral or inferolateral wall MI.

Anterolateral Wall Infarction

If the ECG changes are seen in leads V_3, V_4, V_5, V_6, and sometimes in I, and aV_L, we call it an *anterolateral wall infarction* (Figure 19-31). Reciprocal changes are seen in leads II, III, and aV_F. The suspected culprit artery in anterolateral wall infarctions is the proximal left anterior descending artery or left main coronary artery. When the ST segment in aV_R is elevated to a greater extent than in V_1, it is indicative of left main coronary artery occlusion.

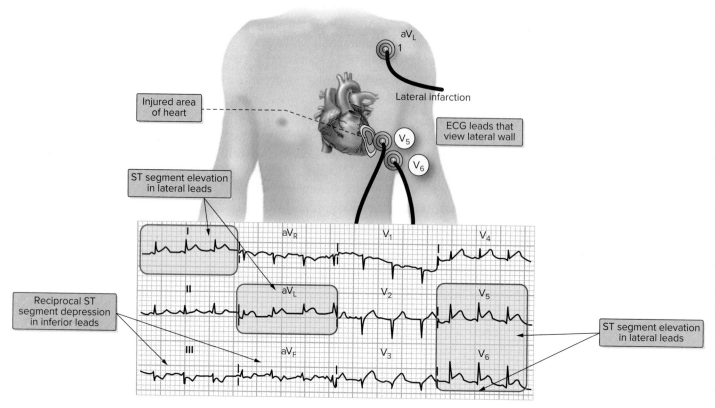

Figure 19-30
Leads I, aV_L, V_5, and V_6 are used to identify lateral MI.

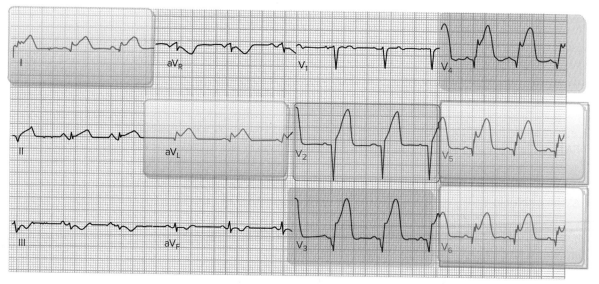

Figure 19-31
Leads V_3, V_4, V_5, V_6, I, and aV_L are used to identify anterolateral MI.

Extensive Anterior Wall Infarction

The term *extensive anterior wall infarction* (sometimes called *anteroseptal with lateral extension*) is used if we see ECG changes in leads V_1, V_2, V_3, V_4, V_5, V_6, I, and aV_L (Figure 19-32). Reciprocal changes are seen in leads II, III, and aV_F. The suspected

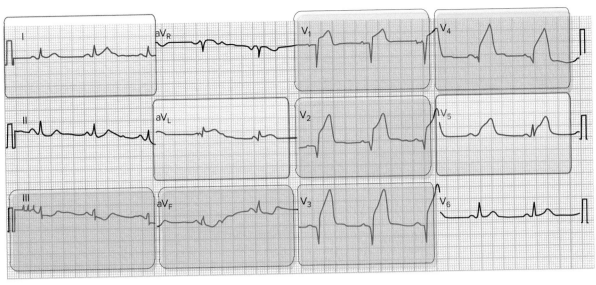

Figure 19-32
Leads V_1, V_2, V_3, V_4, V_5, V_6, I, and aV_L are used to identify extensive anterior MI. Note in this example, lead V_6 is not highlighted as changes are not seen but leads III and aV_F are highlighted as reciprocal changes are present.

culprit artery in extensive anterior wall infarction is the proximal left main coronary artery (proximal to the first diagonal branch).

Inferior Infarction

Inferior infarction involves the inferior or diaphragmatic surface of the heart. It is frequently caused by occlusion of the right coronary artery, although in the 10 to 20% of patients with left-dominate circulation, it may be due to occlusion of the circumflex artery. ST segment elevation, T wave inversion, and the development of significant Q waves in leads II, III, and aV_F indicate MI involving the inferior surface of the heart (Figure 19-33). Reciprocal changes may be seen in leads I and aV_L. *It is important to know that sinus bradycardia is common in inferior-wall MI.* Also, 1st-degree AV heart block, 2nd-degree AV heart block, and 3rd-degree AV heart block may be seen. If these dysrhythmias occur within the first six hours, increased vagal tone is the likely cause. For this reason, atropine may be effective for treating unstable bradycardia.

3rd-degree AV block seen with inferior MI is usually at the level of the AV node and thus the escape rhythm has narrow QRS complexes and a heart rate that is generally in the range of 40 to 60 beats per minute. For this reason, the patient may have an adequate blood despite the presence of complete AV heart block.

Right ventricular infarction should be suspected in any patient with inferior wall MI.

Posterior Wall Infarction

Posterior infarctions are usually caused by occlusion of the posterior descending branch of the right coronary artery (RCA) or circumflex artery and involve the posterior surface of the heart. They are usually associated with inferior or lateral wall infarctions but can be isolated. Since there are no ECG leads over the posterior surface of the heart (unless posterior leads are used), posterior infarctions can be

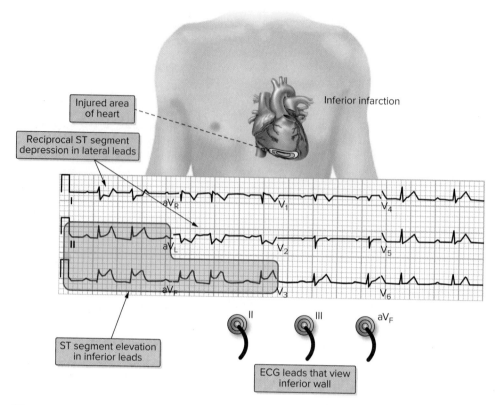

Figure 19-33
Leads II, III, and aV$_F$ are used to identify inferior wall MI.

identified by looking for reciprocal changes in leads V$_1$ and V$_2$ (the anterior leads). Normally, the R waves in leads V$_1$ and V$_2$ are mostly negative. An unusually large R wave in leads V$_1$, V$_2$, and sometimes V$_3$ can be the reciprocal of a posterior Q wave (Figure 19-34). In other words, there is a positive "mirror test." This is where there are taller R waves in V$_1$, V$_2$, and/or V$_3$ with ST depression that looks like a Q with ST elevation if the tracing is flipped over.

Right Ventricular Infarction

About 50% of patients experiencing inferior wall (of the left ventricle) MI also have right ventricular involvement. For this reason, a patient showing signs of inferior wall MI should receive a right-sided ECG analysis in addition to application of the standard 12 leads. Right ventricular MI (Figure 19-35) usually occurs due to an obstruction of the proximal right coronary artery proximal to the first acute marginal branch. The marginal branch is critical in supplying blood to the free wall of the right ventricle. The proximal segment of this vessel supplies the SA node and the right atrial wall; the middle segment supplies the lateral and inferior right ventricle; and the distal segment supplies the posterior portion of the left ventricle, the inferior septum, inferior left ventricular wall and AV node. An occlusion of just the right ventricular marginal branch results in an isolated right ventricular infarction. It is important to recognize the signs of a right ventricular MI because treatment differs significantly between patients with an right ventricular MI and those with other infarction sites.

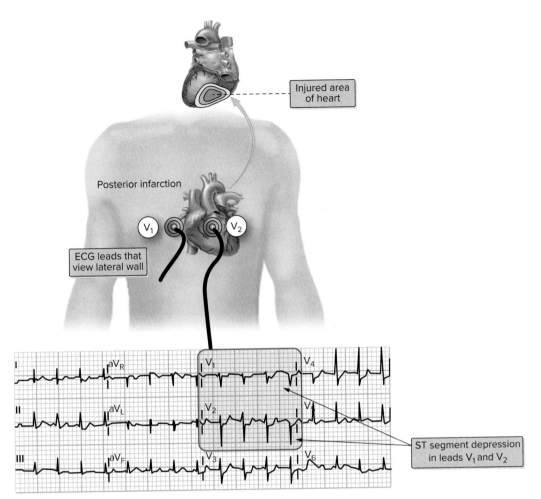

Figure 19-34
Leads V_1 and V_2 are used to identify posterior MI.

Life-threatening complications from right ventricular MI can occur early in the event.

Associated findings may include atrial infarction (PR segment displacement, elevation or depression in leads II, III, and aV_F), symptomatic sinus bradycardia, atrioventricular node block, and atrial fibrillation. Hemodynamic effects of right ventricular dysfunction may include failure of the right ventricle to pump sufficient blood through the pulmonary circuit to the left ventricle, with consequent systemic hypotension.

A 12-lead tracing that shows ST segment elevation in leads II, III or aV_F (inferior leads) should immediately trigger acquisition of a right-sided 12-lead. With right ventricular infarction, the ECG may also show an acute anterior Q wave pattern in leads V_1 through V_3. The right-sided ECG should be done just the same as a left-sided ECG with the exception of where the electrodes are placed. Alternatively, just the lead V_4 electrode can be moved to the same location on the opposite side of the chest. This has been shown to be approximately 80% reliable in comparison to the 90% reliability we see with moving all six leads. ST segment elevation in V_4R is diagnostic of right ventricular infarction (Figure 19-36); however, any ST elevation in the leads V_2R through V_6R should

Figure 19-35
Right ventricular STEMI and 12-lead labeled to show that V4 was moved to the right side of the chest.

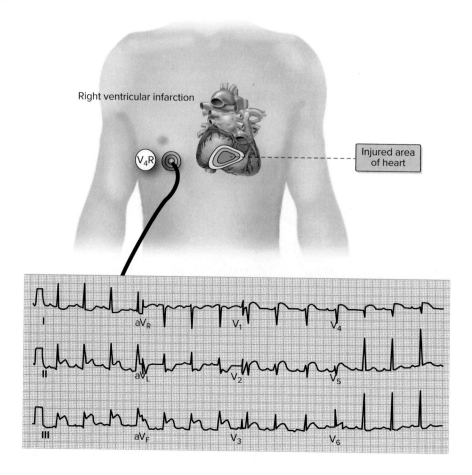

Figure 19-36
Leads V$_1$R through V$_6$R can be used to identify right ventricular infarction.

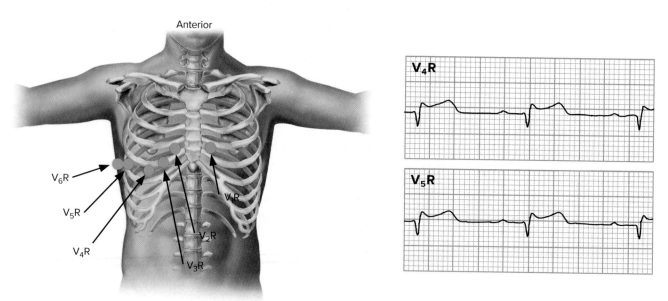

raise suspicion of a right ventricular infarction. Further, in right ventricular infarction the T wave in V$_4$R usually has a convex or "domed" shape when injury is occurring. Pathological Q waves reflect a completed MI. When the tracing is printed, write "R" beside the V leads to mark that they are right leads, not left.

Although this is useful information, remember that these guidelines are generalized as each person's ECG tracings will vary due to variances in body structure, underlying heart and lung disease, and other factors. Thus, abnormal findings may overlap somewhat in the various leads.

Use of Additional ECG Leads

The use of additional ECG leads such as the posterior leads V_7, V_8, and V_9 may improve your ability to detect posterior MI (Figure 19-37). The reason for this is that the standard 12-lead ECG does not directly view the posterior basal and lateral walls of the left ventricle. In particular, MI resulting from blockage of the circumflex artery may produce a nondiagnostic ECG. For these reasons, the use of these additional leads should be considered. An additional point of information is inferior wall MI may reveal the presence of right ventricular MI. For this reason, the VR leads should be applied.

Use of the aV_R Lead

In recent years, a number of studies have demonstrated the value of using lead aV_R during the analysis of the 12-lead ECG in patients who are likely experiencing acute coronary syndrome. These studies have shown that lead aV_R is a strong predictor of left main coronary artery occlusion when used in isolation or in conjunction with other leads. The presence of simultaneous ST segment elevation in leads $aV_R + aV_L$ or the presence of ST segment elevation in aV_R that exceeds the amount of ST segment elevation in lead V_1 is highly specific for left main coronary artery occlusion in patients with acute coronary syndrome (Figure 19-38). Other studies have

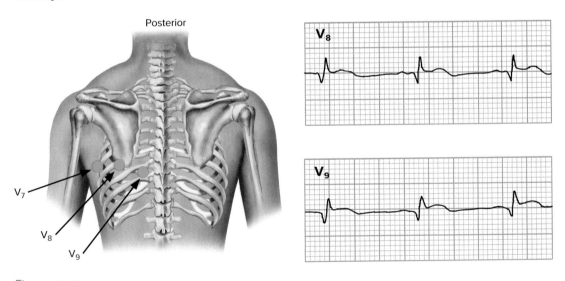

Figure 19-37
Posterior leads V_7, V_8, and V_9 can be used to identify posterior MI.

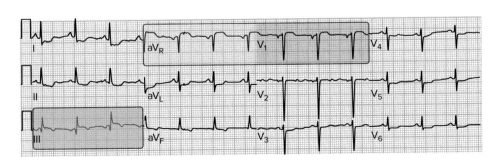

Figure 19-38
ST Segment elevation in leads aV_R, I and III along with widespread ST segment depression.

discussed ST segment elevation in lead aV$_R$ in less specific terms, simply citing that this finding is indicative of either left main coronary artery occlusion or left anterior artery occlusion, or indicative of either left main coronary artery occlusion or triple-vessel disease. Regardless, the literature continues to show with increasing consistency that ST segment elevation in lead aV$_R$ in patients with acute coronary syndromes is associated with more ominous coronary occlusions. Patients with left main coronary artery occlusions, left anterior artery occlusions, or triple-vessel occlusions have a worse prognosis, requiring more aggressive immediate therapy and often bypass surgery. Thus, lead aV$_R$ is a potentially useful tool in electrocardiography and should be considered when analyzing the 12-lead ECG in patients with suspected acute coronary syndrome.

19.6 Treatment of Myocardial Infarction

Initial treatment of MI includes obtaining a 12-lead ECG, the administration of aspirin, oxygen, nitroglycerin, and/or opiates (such as morphine, fentanyl, etc.). The initial 12-lead ECG should be followed by serial 12-lead ECGs in order to identify an evolving MI. Also, an IV line should be placed to allow for the delivery of medications if needed. Furthermore, many recommend placing defibrillator pads on the patient's chest to allow for rapid defibrillation in the event of sudden cardiac arrest.

Unless there is an untoward delay, consideration should be given to obtaining the 12-lead ECG before administration of aspirin and/or nitroglycerin as the STEMI may disappear after their administration. This could result in a loss of evidence of the myocardial injury and cause a delay in the delivery of otherwise beneficial care. Often, the symptoms and STEMI will reoccur after a while but time to definite treatment has been increased.

All patients presenting with acute coronary syndromes should receive aspirin in a dose of at least 162 to 325 mg, unless there is a clear history of aspirin allergy. Patients with aspirin intolerance still should receive aspirin at presentation. Chewed aspirin is preferred, as this promotes rapid absorption into the bloodstream to achieve faster therapeutic levels. Aspirin works by blocking platelet aggregation, in other words, it keeps them from clumping together. This can help reduce the risk of or the progression of clotting in both everyday life or in the setting of acute myocardial infarction.

Nitrates are usually given as a 0.4 mg dose in a sublingual tablet, followed by close observation of the effect on chest pain and the hemodynamic response. If the initial dose is well tolerated, further nitrates can be administered. The most common side effects of nitrates are hypotension and headache.

The initial dose of morphine of 2 to 4 mg as a slow IV bolus can be given, with increments of 2 to 4 mg repeated.

Early recognition of an evolving MI is essential. The reason is that therapy delivered within the first few hours after the onset of MI can, in effect, prevent the completion of the infarct and improve chances for patient survival. Among these treatments are thrombolytic agents and direct plasminogen activators that act to lyse (disintegrate) clots that are occluding the coronary arteries and return blood flow to the heart muscle before actual myocardial death occurs.

Other drugs used in the treatment of myocardial infarction in addition to those already mentioned include: unfractionated or low-molecular weight heparin, bivalirudin, P2Y^{12} receptor inhibitors, IV nitroglycerin, beta-blockers, and Glycoprotein (GP) IIb/IIIa receptor inhibitors.

Further, many hospitals are equipped to deliver catheterization and angioplasty. Emergency angioplasty, performed within the initial 6 hours of infarction onset, offers greater survival than thrombolysis alone. Upon successful completion of angioplasty, stents coated with cytotoxic drugs are placed to prevent reocclusion at the site of the original vascular injury. To further improve patient outcome, oral and intravenous platelet inhibiting agents are administered.

Regardless of which treatment is employed, the key to success is timing. You must act quickly to prevent lasting myocardial damage. Being able to recognize the acute changes of an MI on the ECG is a critical diagnostic skill that can make a difference between life and death.

Right Ventricular Infarction Treatment

Treatment for right ventricular MI differs from treatment used for other infarction sites. Since the right ventricle is dependent on preload for adequate stroke volume, left ventricular stroke volume and cardiac output, treatments that produce vasodilation, can have a deleterious effect on the heart. In right ventricular MI, or inferior wall MI with right ventricular involvement, the administration of nitroglycerin and morphine could cause an abrupt drop in blood pressure and thus should be avoided. For these patients, management is directed toward recognition of right ventricular infarction, reperfusion, volume loading, rate and rhythm control, and inotropic support.

Careful fluid infusion is the mainstay of treatment. If adequate blood pressure and cardiac output cannot be maintained with fluids alone, dobutamine infusion is the preferred agent in right ventricular MI patients with systolic BP of 70 to 100 mmHg and no signs of shock. Other inotropic agents that can be employed include milrinone, norepinephrine, and possibly low-dose vasopressin. Symptomatic bradycardias can be treated with atropine and AV blocks may require external pacing.

Further, these patients presenting within 6 hours of the onset of inferior wall infarction with right ventricular involvement benefit from thrombolytic therapy or coronary angioplasty.

For the following 12-lead ECGs, identify if myocardial ischemia, injury and/or infarction are present and the location.

1.

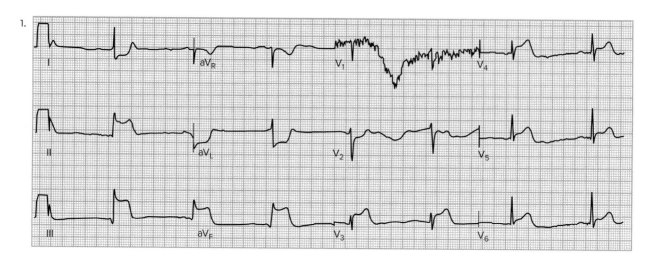

2.

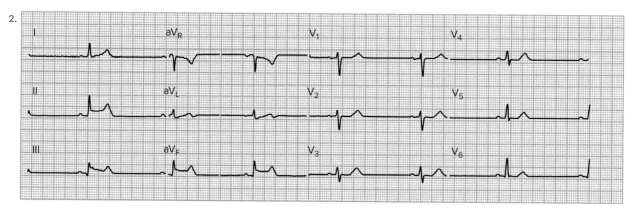

3.

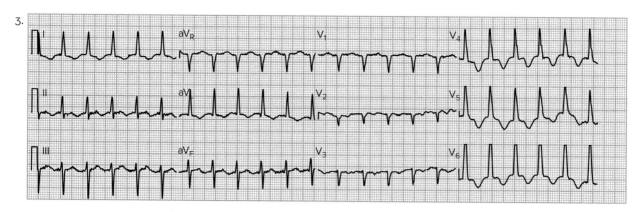

4.

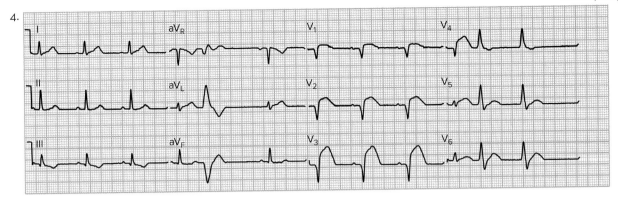

5.

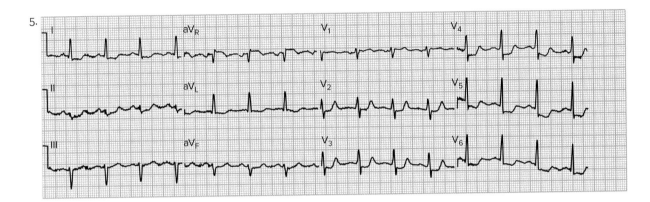

6.

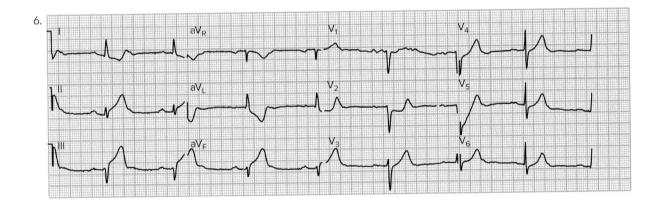

7.

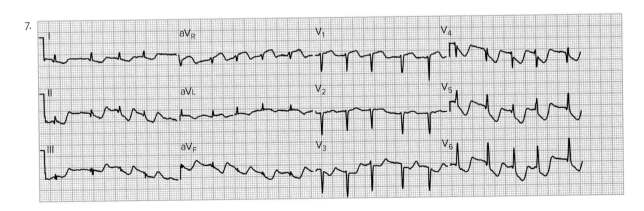

8.

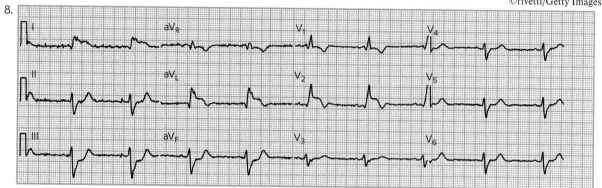

9.

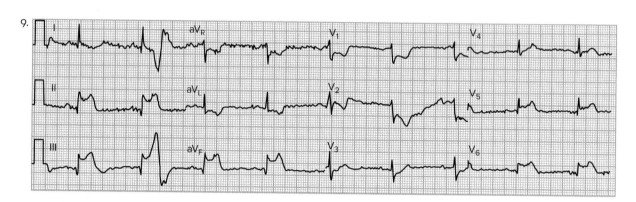

10.

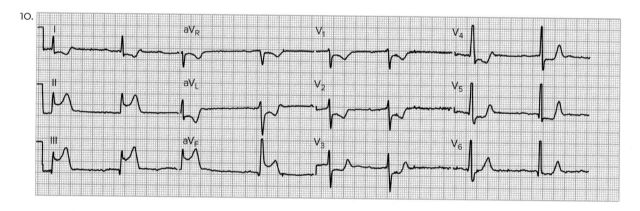

11.

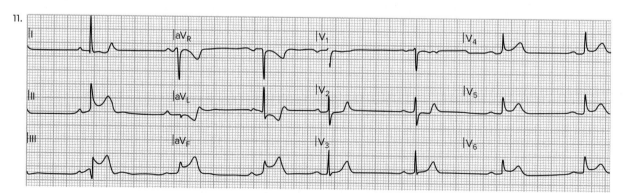

12.

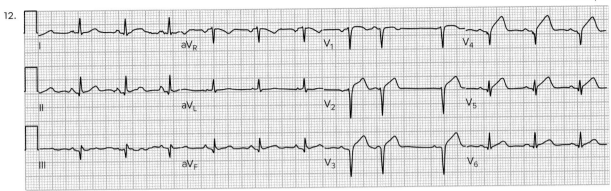

13.

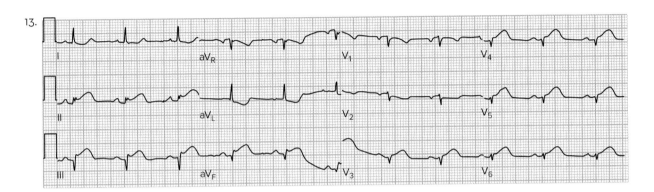

14.

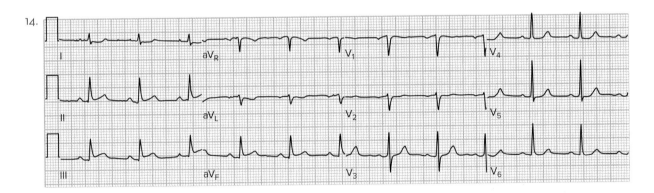

15.

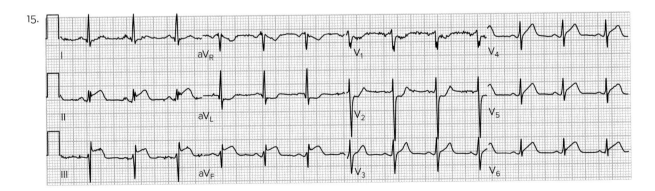

16.

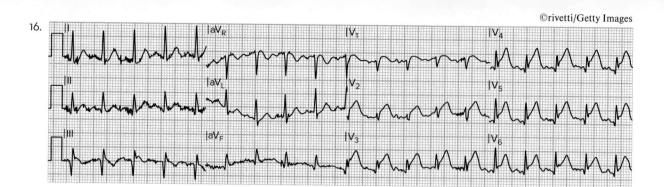

17.

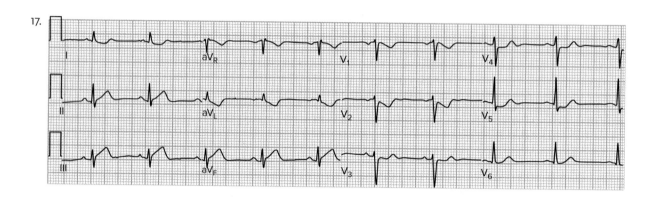

18.

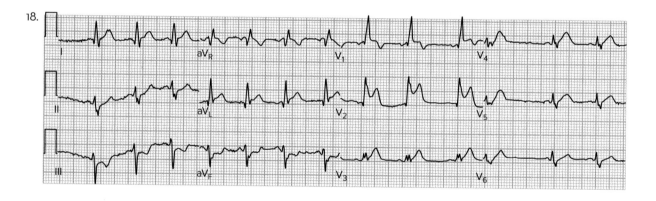

19.

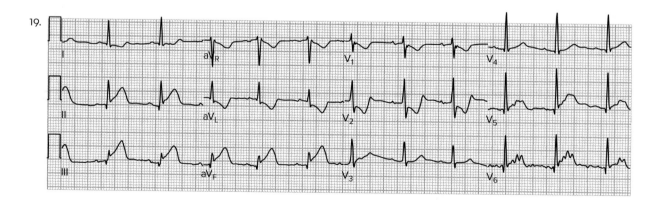

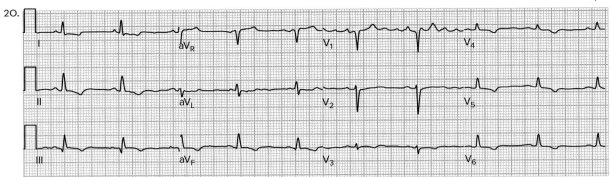

20.

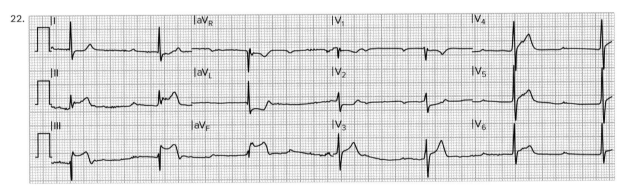

21.

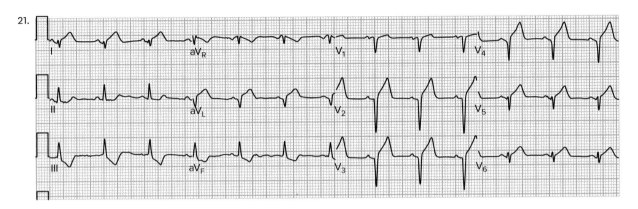

22.

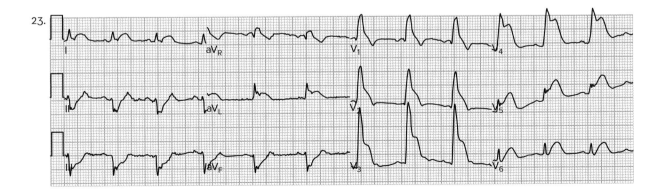

23.

24.

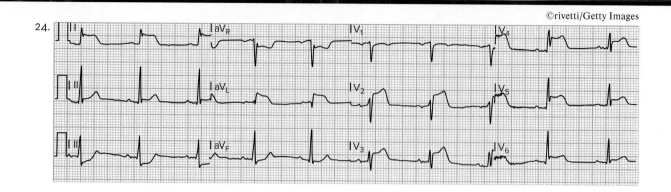

25.

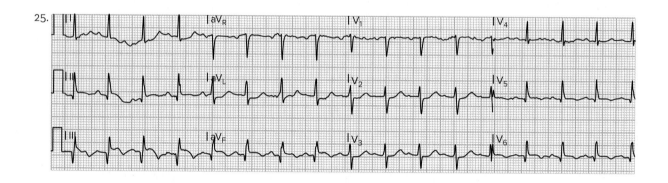

Key Points

LO 19.1	• The oxygen and nutrient demand of the heart is extremely high; therefore, it requires its own continuous blood supply. The coronary arteries deliver the needed blood supply to the myocardial cells. • The two main coronary arteries are the right coronary artery and the left coronary artery. The left coronary artery has two primary branches; the left anterior descending artery and the circumflex artery.
LO 19.2	• The ST segment is the flat line that follows the QRS complex and connects it to the T wave. • The ST segment can be compared with the TP segment to evaluate ST segment depression or elevation. Alternatively, PR segment can be used instead. • The T wave is slightly asymmetrical and oriented in the same direction as the preceding QRS complex. • The Q wave is the first part of the QRS complex. It is the first downward deflection from the baseline. It is not always present.
LO 19.3	• The ECG can help identify the presence of ischemia, injury, and/or infarction of the heart muscle. • Peaked T waves, inverted T waves, and ST segment depression are considered three characteristic signs of myocardial ischemia. • ST segment elevation occurs with myocardial injury. It is the earliest reliable sign that MI has occurred and tells us that the MI is acute. • Subendocardial infarction results in a flat depression of the ST segment. • Pathologic Q waves indicate the presence of irreversible myocardial damage or a past MI. • MI can occur without the development of Q waves.
LO 19.4	• Criteria for diagnosing myocardial infarction include: 1. Clinical history of ischemic-type chest discomfort/pain; 2. Rise and fall in serum cardiac markers; and 3. Changes on serially obtained ECG tracings.
LO 19.5	• Septal wall MI can be identified by ECG changes seen in leads V_1 and V_2. • Leads V_3 and V_4 provide the best view for identifying anterior MI. Leads V_5 and V_6, while located more laterally, may also help identify anterior infarction. • Lateral infarction is identified by ECG changes such as ST segment elevation; T wave inversion; and the development of significant Q waves in leads I, aV_L, V_5, and V_6. • Inferior infarction is determined by ECG changes such as ST segment elevation; T wave inversion; and the development of significant Q waves in leads II, III, and aV_F. • Posterior infarctions can be diagnosed by looking for reciprocal changes in leads V_1 and V_2. • Use of lead VR4 can help identify right ventricular MI. Alternatively, all six chest leads can be moved to the opposite side of the chest to help identify it. The right sided chest lead(s) should be employed whenever an inferior wall STEMI is seen on the 12-lead ECG.
LO 19.6	• Initial treatment of MI includes the administration of aspirin, oxygen, nitroglycerin. • Additional treatments for STEMI include thrombolytic agents and direct plasminogen activators that act to lyse (disintegrate) clots that are occluding the coronary arteries and return blood flow to the heart muscle before actual myocardial death occurs.

> • Emergency angioplasty, performed within the initial six hours of infarction onset, offers greater survival than thrombolysis alone. Upon successful completion of angioplasty, stents coated with cytotoxic drugs are placed to prevent reocclusion at the site of the original vascular injury.
>
> • Drugs that cause vasodilation are contraindicated in right ventricular MI as they can produced profound hypotension. Instead, careful fluid infusion is the mainstay of treatment in these cases.

Assess Your Understanding

The following questions give you a chance to assess your understanding of the material discussed in this chapter. The answers can be found in Appendix A.

1. The blood vessels that supply the myocardial cells with a continuous supply of oxygen and nutrients are referred to as the   (LO 19.1)
   a. coronary veins.
   b. cerebral arteries.
   c. coronary arteries.
   d. carotid arteries.

2. Describe why the heart requires its own blood supply.   (LO 19.1)

3. ST segment depression or elevation can be evaluated by comparing the ST segment with the   (LO 19.2)
   a. Q wave.
   b. TP segment.
   c. position of the T wave.
   d. QT interval.

4. Which is true regarding the normal Q wave?   (LO 19.2)
   a. It is always present.
   b. It is a small positive deflection from the baseline.
   c. It has a duration of less than 0.04 seconds.
   d. It is greater than 25% the amplitude of the R wave in that lead.

5. The ST segment   (LO 19.2)
   a. contains at least two waves.
   b. begins at the peak of the R wave and gradually curves downward to the T wave.
   c. represents right and left ventricular depolarization.
   d. is the line that follows the QRS complex and connects it to the T wave.

6. The normal T wave is   (LO 19.2)
   a. perfectly round.
   b. not more than 5 mm tall in the limb leads.
   c. oriented in the opposite direction of the preceding QRS complex.
   d. normally inverted in lead II.

7. Which of the following is supplied by the right coronary artery in most people? (LO 19.1)

   a. Most of the left ventricle

   b. Circumflex artery

   c. SA nodal artery

   d. Perforating intramural septal branches

8. Myocardial _____ is the most severe result of an interruption of coronary artery blood flow. (LO 19.3)

   a. ischemia

   b. infarction

   c. injury

   d. interruption

9. Describe how the ECG can help identify the presence of ischemia, injury, and/or infarction of the heart muscle. (LO 19.3)

10. All of the following are key ECG indicators of myocardial ischemia, injury, and/or infarction EXCEPT (LO 19.3)

    a. inversion of the T wave.

    b. absence of Q waves in the limb leads.

    c. elevation of the ST segment.

    d. peaking of the T wave.

11. Describe how myocardial injury is identified in comparison to myocardial ischemia and myocardial infarction. (LO 19.3)

12. Match the following ECG changes with the cardiac condition in which they are seen. (LO 19.3)

ECG Change	Myocardial Condition
a. Enlarged Q waves or presence of new Q waves _____	_____ ischemia
b. ST segment depression _____	_____ injury
c. ST segment elevation _____	_____ infarction

13. ST segment depression is considered significant when the (LO 19.3)

    a. ST segment falls more than 1 mm below the baseline at a point one small square to the right of the J point.

    b. changes are seen in at least one or more leads facing the opposite anatomic area of the heart.

    c. associated T wave is inverted.

    d. All of the above.

14. On the following illustrations, draw examples of each of the items listed beneath the figures. (LO 19.3)

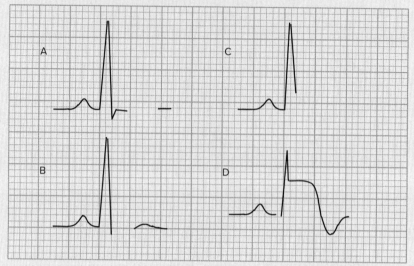

a. Inverted T wave

b. Depressed ST segment

c. Elevated ST segment

d. Enlarged Q wave

15. Inverted T waves associated with myocardial ischemia (LO 19.3)

a. indicate the ischemia is present through the full thickness of the myocardium.

b. occur just at the start of the ischemia.

c. are asymmetrical.

d. are observed best in the limb leads.

16. Of the following three examples of inverted T waves, circle those that are symmetrical. (LO 19.3)

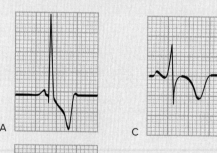

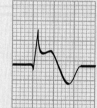

17. List three characteristic signs of myocardial ischemia.

18. ST segment elevation associated with myocardial injury   (LO 19.3)

    a. is usually only slight.

    b. produces a distinctive ST segment that has a sharp inclination upward and tends to merge abruptly with the T wave.

    c. occurs because there is rapid depolarization of the injured myocardium.

    d. is seen in the leads facing the affected area.

19. List three criteria used to identify myocardial infarction.   (LO 19.4)

20. The appearance of Q waves   (LO 19.3)

    a. indicates that irreversible myocardial damage has occurred.

    b. usually disappears after the MI ends.

    c. always occurs in conjunction with ST segment elevation.

    d. is considered pathologic if they are less than 0.04 seconds in duration.

21. Match the following leads where ECG changes will be seen with type of infarction.   (LO 19.4)

Infarction Location Affected Leads	
a. Anterior _____	_____ V_1 and V_2 (for reciprocal changes)
b. Inferior _____	_____ II, III, and aV_F
c. Lateral _____	_____ V_3 and V_4
d. Posterior _____	_____ I, aV_L, V_5, and V_6

Referring to the scenario at the beginning of this chapter, answer the following questions.

22. What part of the heart was affected by the MI?   (LO 19.4)

    a. Anterior

    b. Anteroseptal

    c. Posterior

    d. Inferior

23. What is the minimum amount of ST elevation required to make the diagnosis of myocardial injury   (LO 19.3)

    a. 1 mm

    b. 1.5–2 mm

    c. 2.5–3 mm

    d. 3.5–5 mm

24. The most important intervention the paramedics can perform to improve the patient's outcome is to   (LO 19.5)

    a. apply oxygen.

    b. administer nitroglycerin.

    c. administer aspirin.

    d. rapidly transport to an appropriate facility.

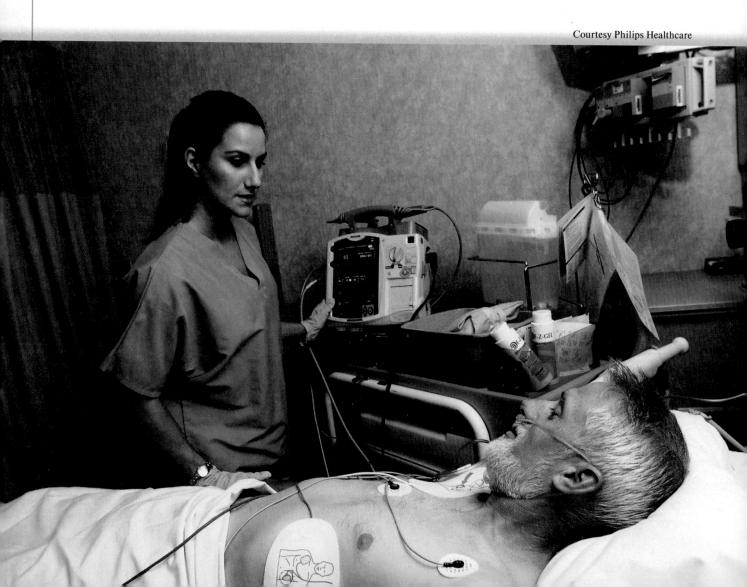

20 Bundle Branch Block

©rivetti/Getty Images

Chapter Outline

Ventricular Conduction
Disturbances

The Bundle Branches

Bundle Branch Block

Fascicular Block

Identifying MI and/or
Hypertrophy in the
Presence of BBB

Treatments of Bundle
Branch Block

Inherited Conditions That
Mimic Bundle Branch
Block

Courtesy Philips Healthcare

Learning Outcomes

LO 20.1 Define the term *ventricular conduction disturbance*.

LO 20.2 Recall how impulses are conducted through right and left bundle branches.

LO 20.3 Describe how bundle branch block occurs and list the possible sites of block within the ventricular conduction system and the characteristics seen with right and left bundle branch blocks, and nonspecific intraventricular conduction defect.

LO 20.4 Define the term *fascicular block* and describe the mechanism by which it occurs as well as the characteristic ECG changes seen.

LO 20.5 Describe how to identify acute MI in the presence of bundle branch block.

LO 20.6 Recall treatments for bundle branch block.

LO 20.7 Recall inherited conditions that mimic bundle branch block, including Brugada syndrome and Arrhythmogenic Right Ventricular Cardiomyopathy.

Case History

Your patient is a 60-year-old male who is complaining of chest pain. He was brought into the emergency department by the local fire department. They tell you he was picked up at work after a brisk walk at lunch time. You find a pale, sweating man lying on a cot. He is awake and alert but is holding his chest. After introducing yourself, you ask him how he feels. The patient indicates that he experienced a sudden onset of chest pain and shortness of breath. The paramedic hands you a card with the patient's medical history. It indicates that he has a history of hypertension and high cholesterol.

Oxygen is being administered to the patient and an intravenous line has been established. After obtaining a 12-lead ECG, you hand it to the doctor who has just entered the treatment room. He comments on the pattern of the QRS complex in V_1. You note that it looks like rabbit ears. The QRS complex is also much wider than normal, indicating that the patient's heart has an intraventricular conduction disturbance.

20.1 Ventricular Conduction Disturbances

Ventricular conduction disturbances involve delays in electrical conduction through either the bundle branches (right or left), the fascicles (anterior, posterior), between the Purkinje fibers and the adjacent myocardium or a combination of these.

20.2 The Bundle Branches

The normal electrical impulse begins in the SA node and quickly moves across the atria, depolarizing them. It then slows as it passes through the AV node and progresses rapidly into the bundle of His. From there it moves into the right and left **bundle branches** (Figure 20-1). As it progresses, it activates the intraventricular septum from left to right. The right bundle branch continues in the direction of the apex, spreading throughout the right ventricle. The left bundle branch divides further into three divisions—the septal fascicle, the anterior fascicle, and the posterior fascicle. The septal fascicle carries the impulse to the interventricular septum in a right-to-left direction. The anterior fascicle carries the impulse to the anterior (superior) portions of the left ventricle. The posterior fascicle carries the impulse to the posterior (inferior) portions of the left ventricle. The very small Purkinje fibers take the current from the bundle branches to the individual myocardial cells. The right bundle branch does not divide into separate fascicles (or at least they have not yet been identified anatomically).

The normal QRS begins with activation of the septum from left to right, which produces a small R wave in leads V_1 and V_2 (as the impulse travels toward the positive electrode) and a small q wave in leads V_5 and V_6 (as the impulse travels away from the positive electrode). Then, as the impulse moves through the left ventricle, it produces a deep S wave in V_1 and V_2 (as it is moving away from the positive electrode) and a tall R wave in V_5 and V_6 (as the impulse is traveling toward the positive electrode). The deep S wave and tall R wave represent the electrical dominance of the left ventricle.

The right and left ventricles are quickly and simultaneously stimulated so that the QRS complex is narrow—0.10 seconds in duration or less (Figure 20-1). Although depolarization of both ventricles is simultaneous, the electrical forces of the left ventricle dominate those of the right because the mass of the left ventricle is so much larger than that of the right. This results in the electrical axis being drawn leftward, lying between 0 and +90 degrees. Thus, with normal ventricular depolarization, the QRS complex is narrow and the electrical axis is between 0 and +90 degrees (Figure 20-1).

20.3 Bundle Branch Block

A **bundle branch block** leads to one or both of the bundle branches failing to conduct impulses (Figure 20-2). This produces a delay in the depolarization of the ventricle it supplies because the impulse must travel through the surrounding muscle (which conducts more slowly than the specialized tissue of the bundle branch) to stimulate the ventricle below the block. Depolarization of the individual ventricles is still of normal duration; however, the unblocked ventricle will depolarize out of synch with the blocked ventricle, thus producing wider than normal QRS complexes.

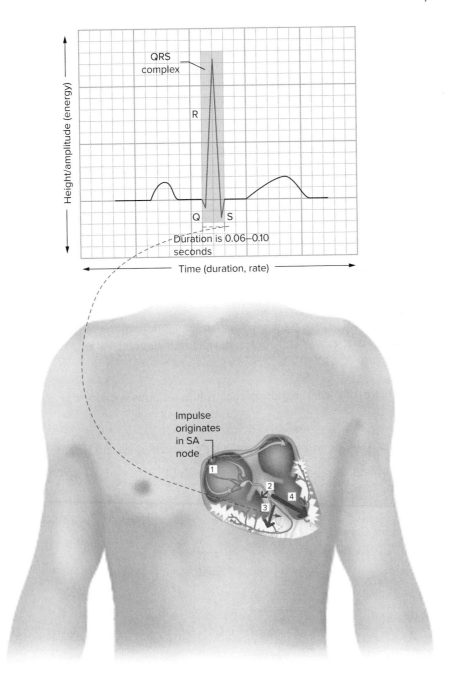

Figure 20-1
The QRS complex represents ventricular depolarization from beginning to end and is narrow—0.10 seconds or less in duration.

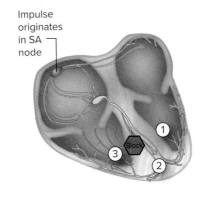

Figure 20-2
In bundle branch block, conduction of the electrical impulse is partially or completely blocked in one of the bundle branches while conduction (1) continues uninterrupted through the unaffected bundle branch. Then the impulse activates the intraventricular septum (2). Finally, the impulse activates the other ventricle (3).

Bundle branch block occurs only in supraventricular rhythms (sinus, atrial, and junctional) because the impulse originates above the ventricles and travels down through the bundle branches in these rhythms. Conversely, ventricular rhythms originate from the ventricular tissue and therefore do not conduct through the bundle branches to stimulate the ventricles.

The most telling sign of bundle branch block is a wide QRS complex, at least 0.12 seconds (three small squares) or more in duration. As such, bundle branch block is diagnosed by first analyzing the width of the QRS complexes.

Then we look at the morphology. Because of the nonsimultaneous depolarization of the ventricles (one depolarizes normally while the other is delayed), it can produce what looks like two QRS complexes (Figure 20-3) superimposed on one another. The ECG records this combined electrical activity as notched or slurred QRS complexes or QRS complexes having two peaks (sometimes referred to as an *M like* or a *rabbit ears* appearance).

When we see two R waves, we name them in sequential order, the R and R′. The R′ is referred to as *R prime* and represents delayed depolarization of the blocked ventricle (Figure 20-3).

If only a portion of the left bundle branch is blocked, this is called a *fascicular block* or *hemiblock.*

Possible sites of block within the ventricular conduction system include

- The right bundle branch—right bundle branch block
- The left bundle branch—left bundle branch block
- The left anterior fascicle—left anterior fascicular block (LAFB), also called *left anterior hemiblock* (LAHB)

Figure 20-3
QRS complex changes in bundle branch block are due to depolarization of the ventricles being out of sync with one another.

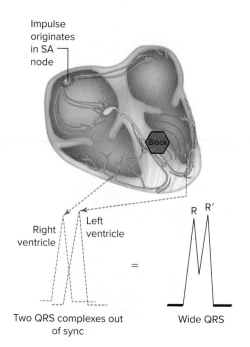

- The left posterior fascicle–left posterior fascicular block (LPFB), also called *left posterior hemiblock* (LPHB)
- Any combination of these or along with prolongation of the PR interval (1st-degree AV block)

ECG Leads Used to Identify Ventricular Conduction Disturbances

The best ECG leads in which to identify bundle branch block are leads V_1, V_2, I, V_5, and V_6. Also, leads V_3 and aV_L often help in revealing bundle branch block (Figure 20-4).

Fascicular blocks are identified using the limb leads I, II, III, aV_L, and aV_F (Figure 20-5).

Right Bundle Branch Block

With **right bundle branch block (RBBB)**, conduction through the right bundle branch is blocked, causing depolarization of the right ventricle to be delayed; conduction does not start until the left ventricle is almost fully depolarized (Figure 20-6).

The impulse starts off normally. It is initiated in the SA node and travels through the atria, depolarizing them. It then transverses the AV node and passes through the bundle of His. It then moves down the left bundle, initiates septal

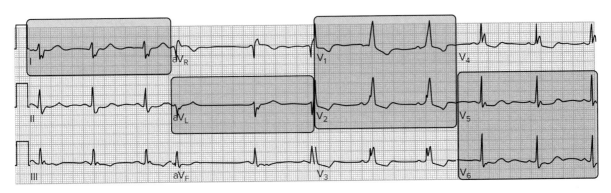

Figure 20-4
To identify bundle branch block, look at leads V_1, V_2, I, aV_L, V_5, and V_6.

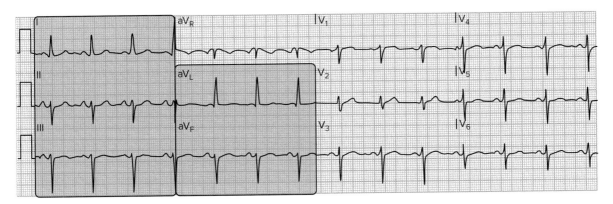

Figure 20-5
To identify fascicular block, look at leads I, aV_L, II, III, and aV_F.

Figure 20-6
In RBBB, conduction through the right bundle is blocked, causing depolarization of the right ventricle to be delayed; conduction does not start until the left ventricle is almost fully depolarized.

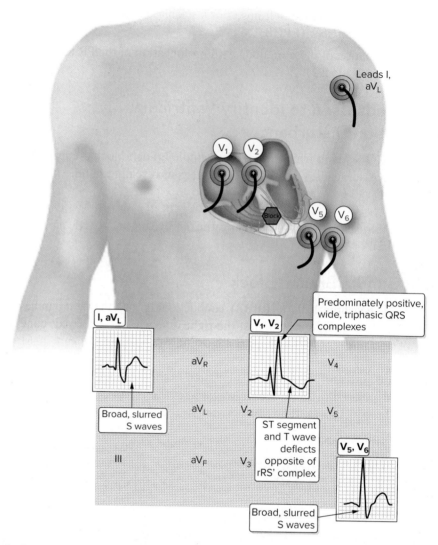

I, aV$_L$

Broad, slurred S waves

V$_1$, V$_2$

Predominately positive, wide, triphasic QRS complexes

ST segment and T wave deflects opposite of rRS' complex

V$_5$, V$_6$

Broad, slurred S waves

depolarization, and at the same time attempts to travel down the right bundle branch. Because the right bundle is blocked, the impulse travels through the anterior and posterior fascicles and into the Purkinje fibers of the left ventricle. The impulse then travels from the left ventricle across the septum and depolarizes the right ventricle. Its movement across the ventricles is slower than normal because it is moving cell by cell instead of via the faster conduction pathways.

Very distinct characteristics appear on the ECG in RBBB. The QRS begins normally with activation of the septum from left to right, which produces a small R wave in leads V$_1$ and V$_2$ (as the impulse travels toward the positive electrode) and a small q wave in leads V$_5$ and V$_6$ (as the impulse travels away from the positive electrode). Then, as the impulse moves through the left ventricle, it produces an S wave in V$_1$ and V$_2$ (as it is moving away from the positive electrode) and an R wave in V$_5$ and V$_6$ (as it is traveling toward the positive electrode). Then, the delayed depolarization of the right ventricle produces a positive deflection, an R' wave in leads V$_1$ and V$_2$ (as it is moving toward the positive electrode), and a broad, wide terminal S wave in leads V$_5$ and V$_6$ (as it is moving away from the positive electrode), as well as the other leads overlying the left ventricle (I and aV$_L$).

RBBB causes the QRS complex to have a unique shape (see Figure 20-6). It occurs because of late depolarization of the right ventricle. In leads V$_1$ and V$_2$, the QRS

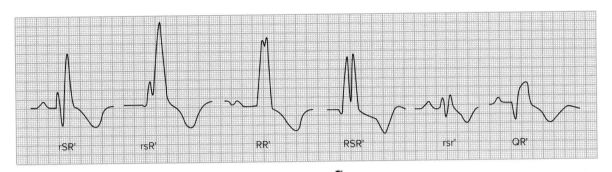

Figure 20-7
Examples of right bundle branch block in lead V₁.

complexes deflect positively and are wide and triphasic. This does not distort the QRS complex but instead adds a terminal deflection in the septal leads. The secondary R wave is usually taller than the initial R wave. They may appear as rSR′, rsR′, RR′, RSR′, or rsr′ complexes (as examples; see Figure 20-7). In some cases, the waveforms may have an M-shaped appearance. Some liken its appearance to rabbit ears.

The ST segments and T waves of these complexes deflect opposite that of the terminal portion (see Figure 20-6). This is referred to as *discordance*. These changes represent repolarization abnormalities secondary to the abnormal depolarization of RBBB. Suspect abnormal pathology such as ischemia if the ST segments and T waves are concordant (take the same direction) as the terminal portion of the QRS complex.

In RBBB, the late right ventricular depolarization results in the terminal S waves in leads V_6 and I (and often V_5 and aV_L) being different from normal. They are broad (sometimes referred to as slurred) and of a greater duration than the associated R waves (see Figure 20-6).

The axis may be normal, right or left. If left axis is seen consider the presence of left anterior fascicular block.

Characteristics of RBBB

- In lead V_1 (and often V_2 and V_3) the QRS complexes differ from normal:
  Deflection: the QRS complexes deflect positively rather than negatively.
  Width: the QRS complexes are 0.12 seconds or greater in duration.
  Appearance: the QRS complexes have an initial R wave and secondary R wave (i.e., an rSR′, rsR′, RR′, RSR′, rsr′ complex).
  Discordance: the ST segments and T waves deflect opposite that of the terminal portion of the QRS complex.

- In leads V_6 and I (and often V_5 and aV_L), the terminal S waves are different from normal:
  Width: the S waves are of greater duration than the R waves.
  Appearance: the S waves are broad and sometimes referred to as slurred.

RBBB can be a normal finding in adults of all ages. It can also occur with anterior wall MI, coronary artery disease, hypertension, scar tissue that develops after heart surgery, myocarditis, and pulmonary embolism. It may also be caused, as can any type of heart block, by drug toxicity. Further, it can also be due to a congenital heart abnormality such as atrial septal defect.

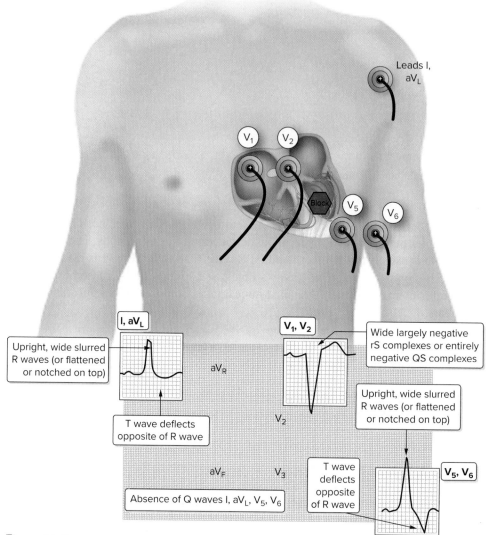

Leads I, aVL

V₁ V₂

Block V₅ V₆

I, aVL

Upright, wide slurred R waves (or flattened or notched on top)

aVR

V₁, V₂

Wide largely negative rS complexes or entirely negative QS complexes

V₂

Upright, wide slurred R waves (or flattened or notched on top)

T wave deflects opposite of R wave

aVF   V₃

T wave deflects opposite of R wave

V₅, V₆

Absence of Q waves I, aVL, V₅, V₆

Figure 20-8

In LBBB, conduction through the left bundle is blocked, causing depolarization of the left ventricle to be delayed; it does not start until the right ventricle is almost fully depolarized.

Left Bundle Branch Block

In **left bundle branch block (LBBB),** depolarization of the left ventricle is delayed (Figure 20-8). Again, the impulse starts off normally. It is initiated in the SA node and travels through the atria, depolarizing them. It then transverses the AV node, passes through the bundle of His, and attempts to move down the right and left bundle branches. It passes through the right bundle normally but is blocked from passing through the left bundle. Following right ventricular depolarization, the impulse travels across the septum and depolarizes the left ventricle. Its movement across the ventricle is slower than normal because depolarization is moving cell by cell instead of via the faster conduction pathways of the anterior and posterior fascicles and Purkinje fibers.

Activation of the ventricles in the left bundle branch is greatly different than what occurs in the normal heart. This leads to a much different-looking QRS complex.

In leads V₁ and V₂ (which overlie the right ventricle), the initial movement of the impulse from right to left instead of left to right through the septum can produce a small q

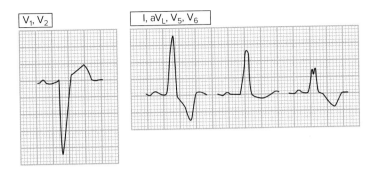

Figure 20-9
Examples of left bundle branch block in lead V_1, I, aV_L, V_5, and V_6.

wave (which is often not seen) as the impulse is traveling away from the positive electrode. Then as the right ventricle depolarizes, it can appear as a very small R wave (or it may even be absent) as the impulse is now moving toward the positive electrode. The abnormal depolarization of the left ventricle produces a wide, largely negative S wave as the current is moving away from the positive electrode. This is referred to as a wide rS complex. If the R wave is absent, it is an entirely negative and wide QS complex.

Normally, in leads V_5, V_6, I, and aV_L (which overlie the left ventricle), the QRS complexes are seen as a Q wave and tall R wave. In LBBB, because the impulse moves from right to left through the septum by way of the right ventricle (instead of from left to right), there is an absence of a Q wave in these leads. Then, as the impulse is traveling toward the positive electrode, the slow left ventricular depolarization produces a wide, tall, slurred R wave that can vary in appearance (see Figure 20-8). The R waves can also be flattened on top or notched (with two tiny points). However, the M-shaped appearance is less likely to be seen than in RBBB. Sometimes, there is an RS (biphasic) pattern in leads V_5 and V_6 (attributed to displaced transition of the QRS complex).

Also, like with RBBB, the ST segment and T wave are discordant (in the opposite direction of) the terminal portion of the QRS complex. These changes represent repolarization abnormalities secondary to abnormal depolarization. Suspect abnormal pathology such as ischemia if ST segments and T waves are concordant with the terminal portion of the QRS complex.

Further, left axis deviation may be present in LBBB.

Characteristics of LBBB

- To quickly identify LBBB, look for the following:
  In lead V_1 (and often V_2 and V_3) the QRS complexes are different from normal:
  Deflection: The complexes deflect negatively.
  Width: The complexes are 0.12 seconds or greater in duration.
  Appearance: The complexes have a QS or rS configuration
  Discordance: The ST segments and T waves deflect opposite that of the terminal portion of the QRS deflection.

- In leads V_6 and I (and often V_5 and aV_L) the R waves are different from normal:
  Width: the complexes are 0.12 seconds or greater in duration.
  Appearance: the R waves are monophasic, slurred, or notched.

- Left axis deviation may be present

LBBB is rarely seen in normal hearts. Its presence almost always represents significant underlying cardiac disease. It can be caused by anterior wall MI; hypertensive heart disease; aortic stenosis; degenerative changes of the conduction system; or thickened, stiffened, or weakened heart muscle (cardiomyopathy).

Incomplete Bundle Branch Block

At times either the right or the left bundle branch conducts the electrical impulse more slowly than normal, but is not totally blocked. When this happens on the side of the slow conduction, the electrical impulse reaches the ventricle slightly later than normal. As a result, the QRS complex may have a similar appearance to bundle branch block and is slightly wider than normal (Figure 20-10), but it is not as wide as with complete bundle branch block (having a duration of no greater than 0.11 seconds). This slight widening of the QRS is often referred to as **incomplete bundle branch block.**

Whereas incomplete bundle branch block may indicate underlying heart disease, incomplete bundle branch block is often of no significance, especially when it occurs on the right side (i.e., incomplete RBBB).

Nonspecific Intraventricular Conduction Defect

Nonspecific intraventricular conduction defect is a delay or block in the Purkinje system, and not a blockage of the right or left bundle branch. It is characterized by QRS complexes that are 0.12 second or greater (since the ventricular muscle must be activated through ordinary myocardium instead of the specialized conductive

Figure 20-10
Incomplete bundle branch block.

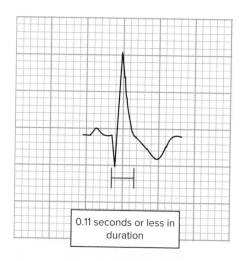

0.11 seconds or less in duration

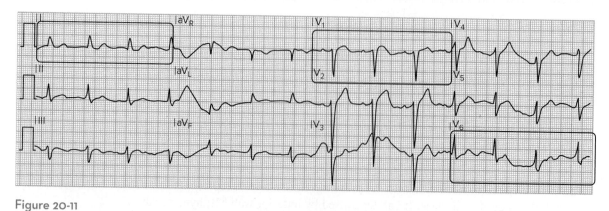

Figure 20-11
The QRS complexes on lead I have the characteristics of left bundle branch block but the QRS complexes in lead V₆ do not. They appear more like what is seen in right bundle branch block.

tissues) but do not conform to all the characteristics found in either left or right bundle branch block. An example would be the presence of rSR′ complexes in V_1 and V_2 and broad S waves in lead I but not in V_6 (Figure 20-11).

20.4 Fascicular Block

As previously described, the left bundle branch consists of three separate branches or fascicles. A block in conduction through one of the fascicles is called a **fascicular block.** There are many fascicles in the heart's conduction system, but the anterior and posterior fascicles of the left bundle branch are the most clinically significant. Whereas the ECG appearance of anterior and posterior fascicular block differs from that of bundle branch blocks, the mechanism is essentially the same—the normal conduction pathway is blocked, and the affected myocardium receives its electrical stimulation via cell-to-cell depolarization and retrograde conduction from the remaining intact parts of the conduction system. Because there is minimal prolongation with a fascicular block, it is not enough to widen the QRS complex to any appreciable degree. However, the morphology of the QRS complex does change.

As mentioned earlier, the limb leads are used to identify fascicular block.

Axis deviation is the key ECG characteristic of fascicular blocks. It occurs because, when one fascicle is blocked, the electrical current travels down the other to stimulate the heart. This causes the axis to shift accordingly. When diagnosing fascicular block, be sure to rule out other causes of axis deviation, such as ventricular hypertrophy and myocardial infarction.

The anterior fascicle is longer and thinner and has a more fragile blood supply than the posterior fascicle, so LAFB is far more common than LPFB. While LAFB can be seen in both normal and diseased hearts, LPFB almost always is associated with heart disease.

Left Anterior Fascicular Block

With LAFB, conduction down the left anterior fascicle is blocked (Figure 20-12). For this reason, the electrical impulse rushes down the left posterior fascicle to the inferior surface of the heart toward leads II, III, and aV_F. This produces a small r wave. As the impulse moves away from the lateral leads, I and aV_L, a small q wave is recorded. Depolarization of the left ventricle occurs, progressing in an inferior-to-superior and right-to-left direction and producing a prominent S wave in leads II, III, and aV_F, and an R wave in leads I and aV_L. The axis of ventricular depolarization is therefore redirected upward and slightly to the left. This results in left axis deviation.

Characteristics of Left Anterior Fascicular Block
Normal QRS duration and no ST segment or T wave changes
Small r waves and deep S waves in leads II, III, and aV_F
Small q waves and tall R waves in leads I and aV_L
Left axis deviation between −30 and −90 degrees
No other cause of left axis deviation is present

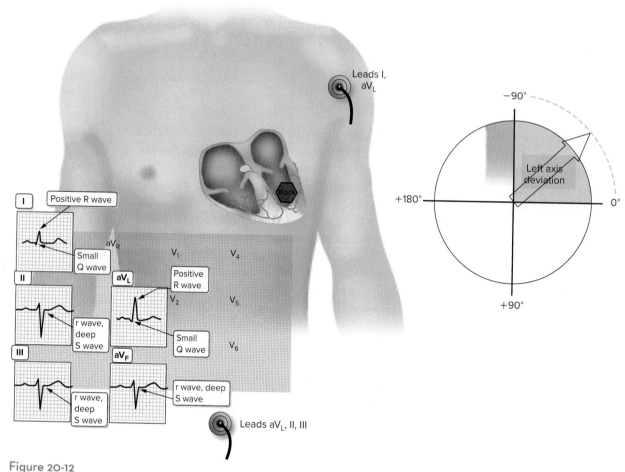

Figure 20-12

With LAFB, conduction down the left anterior fascicle is blocked, resulting in all the current rushing down the left posterior fascicle to the inferior surface of the heart. There are small q waves and R waves in leads I and aV_L and small r waves and deep S waves in leads II, III, and aV_F and left axis deviation.

Left Posterior Fascicular Block

In LPFB, the electrical impulse rushes down the left anterior fascicle, resulting in ventricular myocardial depolarization occurring in a superior-to-inferior and left-to-right direction. Initial conduction is directed away from leads II, III, and aV_F producing a small q wave and toward leads I and aV_L producing a small r wave. Then as conduction is directed toward leads II, III, and aV_F, R waves are produced. As it moves away from leads I and aV_L, S waves are produced (Figure 20-13). Therefore, the main electrical axis is directed downward and to the right. This results in right axis deviation.

In contrast to complete LBBB and RBBB, the QRS complex in fascicular block is not prolonged (wide).

Characteristics of Left Posterior Fascicular Block
Normal QRS duration and no ST segment or T wave changes
Small r waves and deep S waves in leads I and aV_L
Small q and tall R waves in leads II, III, and aV_F
Right axis deviation
No other cause of right axis deviation is present

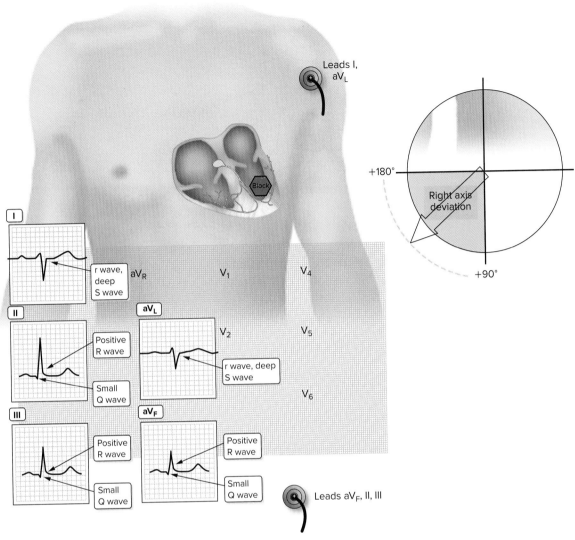

Figure 20-13
With LPFB, conduction down the left posterior fascicle is blocked, resulting in ventricular myocardial depolarization occurring in a superior-to-inferior and left-to-right direction. There are S waves in the left lateral leads (lead I), R waves inferiorly (lead aV$_F$), and right axis deviation.

Patients with any type of ventricular conduction block and especially those with a combination of blocks are at high risk of developing complete heart block.

Bifascicular and Trifascicular Blocks

Although a 2009 American Heart Association Electrocardiography and Arrhythmias Committee, Council on Clinical Cardiology; the American College of Cardiology Foundation; and the Heart Rhythm Society (AHA/ACCF/HRS) scientific statement on the standardization and interpretation of the electrocardiogram recommended against using the terms *bifascicular* and *trifascicular block* because these patterns lack unique anatomic and pathologic substrates, we have included coverage of these terms because they are still widely used.

Bifascicular block is a conduction disturbance in which two of the three main fascicles of the His/Purkinje system are blocked. Most often, it refers to a combination of RBBB and either LAFB (more commonly) or LPFB.

Figure 20-14
Trifascicular block.

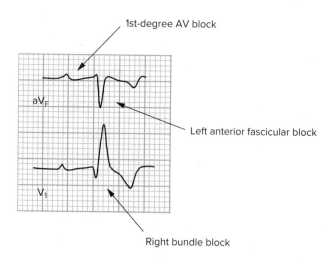

1st-degree AV block

aV$_F$

Left anterior fascicular block

V$_1$

Right bundle block

LBBB may be included in the definition of bifascicular block because the block occurs above the bifurcation of the left anterior and left posterior fascicles of the left bundle branch.

Trifascicular block is a conduction disturbance in which there are three features seen on the ECG (Figure 20-14):

- Prolongation of the PR interval (1st-degree AV block)
- RBBB
- Either LAFB or LPFB

Trifascicular block is uncommon.

20.5 Identifying MI and/or Hypertrophy in the Presence of Bundle Branch Block

While not impossible, identifying acute or old MI in the presence of bundle branch block is always more difficult. Remember our discussion in Chapter 18; new LBBB in patients complaining of chest pain is an indication of MI—much the same as ST segment elevation.

The first thing to look at to find acute events in both RBBB and LBBB are the ST segments and T waves. If they are concordant with the terminal portion of the QRS complexes, MI may exist (Figure 20-15).

Because the same artery that feeds the right bundle branch also feeds the septum, RBBB and anteroseptal MI commonly occur together. Look for ST segment elevation of greater than 1 mm and concordant ST segments and T waves in the ECG leads overlying the septum.

In LBBB, because the QRS complexes are distorted, and ST segment elevation, poor R wave progression in leads V$_1$ through V$_3$, and increased voltage are commonly seen, identification of acute or old infarction is difficult. Criteria for identifying myocardial infarction in the presence of LBBB include the presence of excessive discordance. This is where the ST segment elevation is greater than would be expected in LBBB alone. A way to measure for this is where a clear J point can

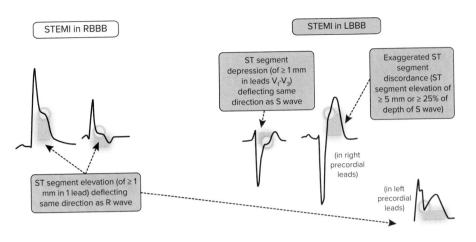

Figure 20-15
The first three QRS complexes are examples in which the ST segments are concordant with the terminal position of the QRS complexes. The fourth QRS represents depressed ST segment, in which the ST segment is concordant with the terminal portion of the QRS complex. The fifth QRS complex represents exaggerated ST segment discordance. All suggest the possibility of MI in the presence of bundle branch block.

be seen and it is up or down equal to or greater than 5 mm in any lead. An alternative that can be employed is to compare the ST segment change to the S wave. If the ST segment elevation is greater than 0.25 (with at least 1 mm of ST elevation) of depth of the S wave, STEMI is likely present. (Figure 20-15). Additionally, the presence of Q waves in leads I, aV_L, V_5, and V_6 LBBB suggest myocardial infarction. Normally in LBBB, Q waves are never present in these leads, so finding them indicates that myocardial infarction occurred at some point.

LVH should be suspected in the presence of RBBB if the R wave in aV_L is equal to or greater than 12 mm or the R wave in V_5 or V_6 is equal to or greater than 25 mm (Figure 20-16). Because LBBB distorts the normal vector forces, left ventricular hypertrophy cannot usually be identified in the presence of LBBB.

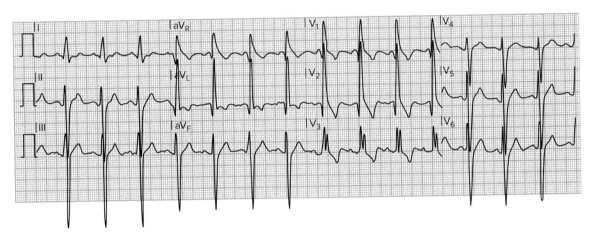

Figure 20-16
Example of right bundle branch block with left ventricular hypertrophy.

20.6 Treatments of Bundle Branch Block

Bundle branch block is relatively common and occurs in a variety of medical conditions. It is seen in medical conditions that affect the right side of the heart or the lungs, so its presence on the ECG should prompt a screening exam for such conditions. However, RBBB is common in normal, healthy individuals, and the screening exam often turns up no medical problems. In such cases, RBBB has no apparent medical significance and can be written off as a "normal variant" and safely ignored.

In contrast, LBBB usually indicates underlying cardiac pathology. Whereas, occasionally, LBBB occurs in apparently healthy people, its appearance should prompt a thorough assessment for underlying cardiac problems.

Fortunately, in general it is quite uncommon for stable bundle branch block to progress to complete heart block. For this reason, despite the fact that bundle branch block is a common ECG finding, it does not require pacemaker implantation. Often patients can be monitored to determine if they progress to a more complete block.

However, a few conditions in which people with bundle branch block require implantation of a pacemaker include:

1. RBBB along with LAHB, which is associated with acute MI. Here, the conduction system disease tends to be unstable and can progress to complete heart block. These patients often need pacemakers.

2. When bundle branch block is associated with syncope. Generally speaking, an electrophysiology study should be considered to test for impending complete heart block. A permanent pacemaker can then be used to eliminate the problem.

3. In certain patients with dilated cardiomyopathy and either complete or incomplete bundle branch block. Here a new form of pacing—called *CRT*—has been shown to improve symptoms and prolong life. CRT can be considered in any patient with heart failure and bundle branch block.

20.7 Inherited Conditions That Mimic Right Bundle Branch Block

Several other conditions can have similar appearance to right bundle branch block, including Brugada syndrome and Arrhythmogenic Right Ventricular Cardiomyopathy.

Brugada Syndrome

This is a rare congenital disorder that restricts sodium transport across the ion channels in the right ventricle but does not affect the structure of the heart. It is caused by genetic defect that often exists in other family members (autosomal dominant inherited disease). It more commonly affects young men of southeast Asian descent. It is not a common condition in the western world, but those affected are mainly young to middle-aged men and some women. Patients who have this condition are susceptible to developing episodic syncope brought on by polymorphic ventricular tachycardia or, worse, cardiac arrest due to ventricular fibrillation.

Brugada syndrome has a similar appearance to RBBB and is characterized by persistent ST segment elevation in leads V_1, V_2, and V_3. However, unlike RBBB, the rSR' is not more than 0.12 seconds wide and there are no broad S waves in leads I and V_6. Three main patterns of Bruguda syndrome have been described—Type 1, Type 2, and Type 3 (Figure 20-17):

• *Type 1*: Presents as prominent high takeoff J point elevation with "coved-type" ST segment elevation with an amplitude of 2 mm or greater and T wave inversion. The terminal portion of the ST segment gradually descends.

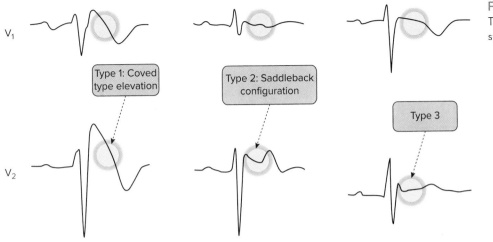

Figure 20-17
Three types of Brugada syndrome.

- *Type 2*: Presents as prominent high takeoff J point elevation of 2 mm or greater with gradually descending ST segment elevation that creates a 1 mm or greater "trough" leading to a positive or biphasic T wave that gives it a "saddleback configuration."

- *Type 3*: Appears as ST segment elevation of greater than 1 mm with either a coved-type or saddleback configuration.

The only available treatment for these patients is an implantable cardioverter defibrillator (ICD).

Arrhythmogenic Right Ventricular Dysplasia (AVRD) (ARVC)

Arrhythmogenic right ventricular dysplasia, also referred to as *arrhythmogenic right ventricular cardiomyopathy* (ARVC) was initially discussed in Chapter 10. It is another rare condition that produces an atypical RBBB pattern. It is often an inherited cardiomyopathy characterized by fatty or fibro-fatty invasion of the right ventricular myocardium, and often the left ventricle as well. It typically occurs in young adults (less than 35 years of age). This change in myocardial tissue disrupts the heart's electrical signals and causes dysrhythmias such as PVCs, paroxysmal ventricular tachycardia, and ventricular fibrillation. Often, sudden cardiac arrest is the first presenting sign in young athletes afflicted with this condition. ARVC is more common in men than women and in people of Italian or Greek descent. There is often a family history of sudden cardiac death. Over time, surviving patients develop features of right ventricular failure, which may progress to severe biventricular failure and dilated cardiomyopathy.

The most recognizable feature of ARVD is that it looks like an incomplete RBBB pattern with a post-depolarization epsilon wave (Figure 20-18) at the end of the QRS complex. The epsilon wave is described as a terminal notch in the QRS complex and is seen in approximately 40% of cases. It is due to slowed intraventricular conduction and is best seen in leads V_1 and V_2. There is also T wave inversion in leads V_1, V_2, and V_3 and prolonged S wave upstroke. There may also be localized QRS widening (0.08 to 0.11 seconds) in those same leads. This may be due to

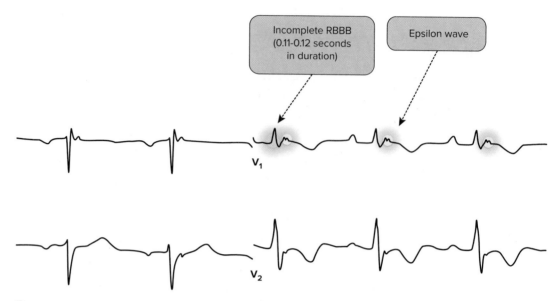

Figure 20-18

In this figure, ARVC appears as incomplete RBBB pattern with a post-depolarization epsilon wave at the end of the QRS complex.

delayed activation of the right ventricle, rather than any intrinsic abnormality in the right bundle branch.

Electrophysiology studies in ARVD demonstrate inducible polymorphic ventricular tachycardia. Echocardiography reveals a mildly dilated right ventricle with akinetic (poorly contracting) regions due to fibrotic adipose tissue replacing areas of the right ventricular musculature. A definitive diagnosis of ARVC comes from magnetic resonance imaging (MRI).

Treatment for patients living with ARVC is life-long dysrhythmia prophylaxis with beta-blockers or an implantable cardioverter defibrillator (ICD).

Practice Makes Perfect

For the following 12-lead ECGs, identify if bundle branch block or fascicular block are present and the types.

1.

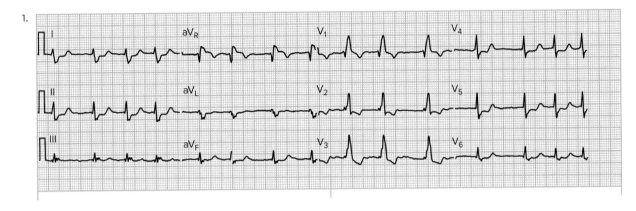

2.

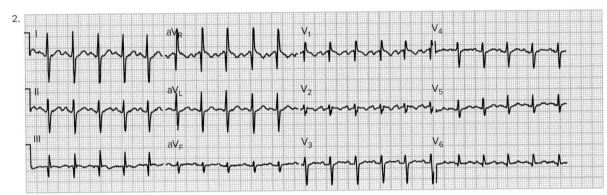

3.

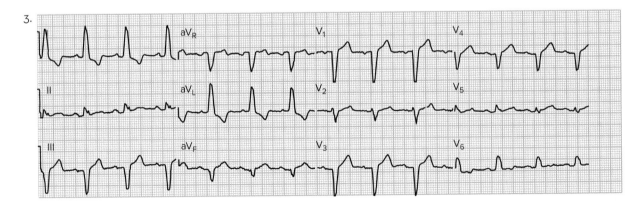

4.

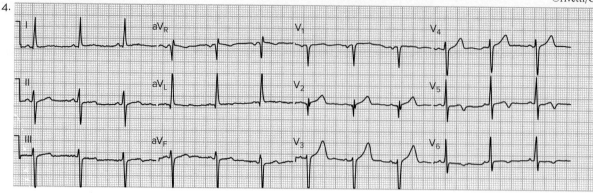

5.

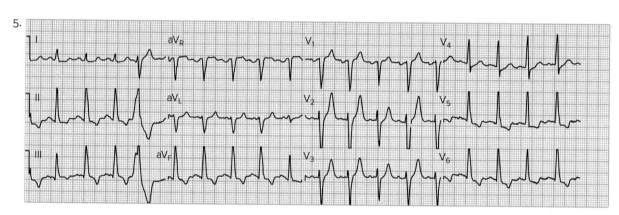

6.

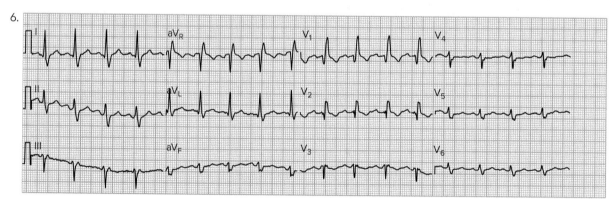

7.

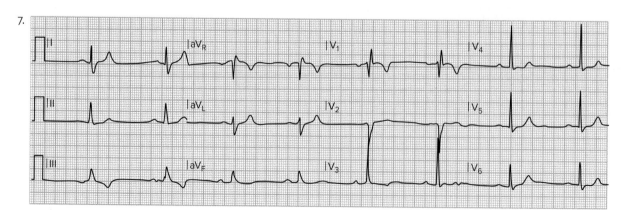

8.

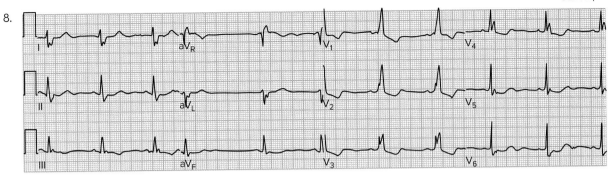

9.

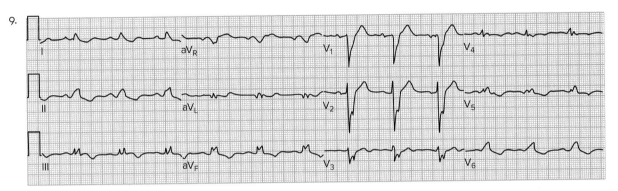

10.

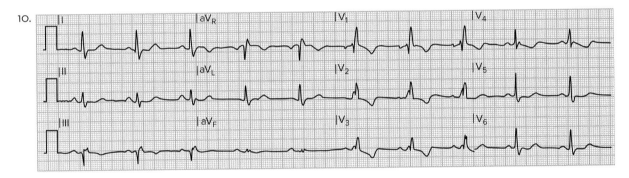

11.

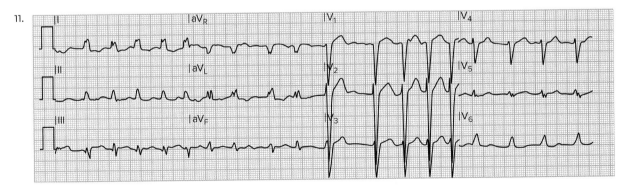

12.

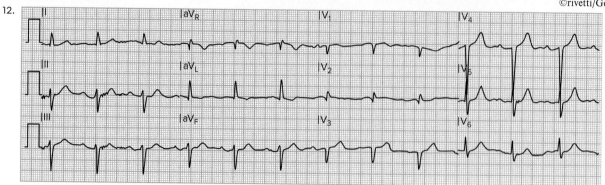

13.

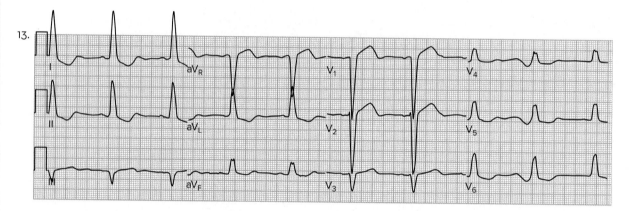

14.

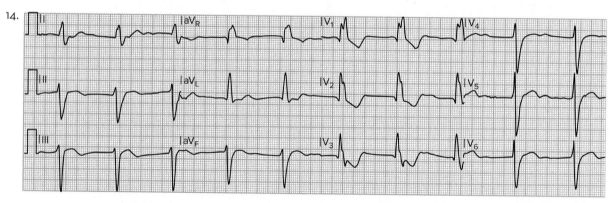

15.

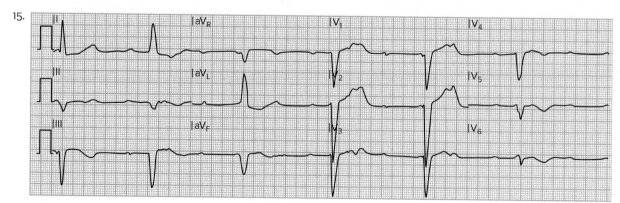

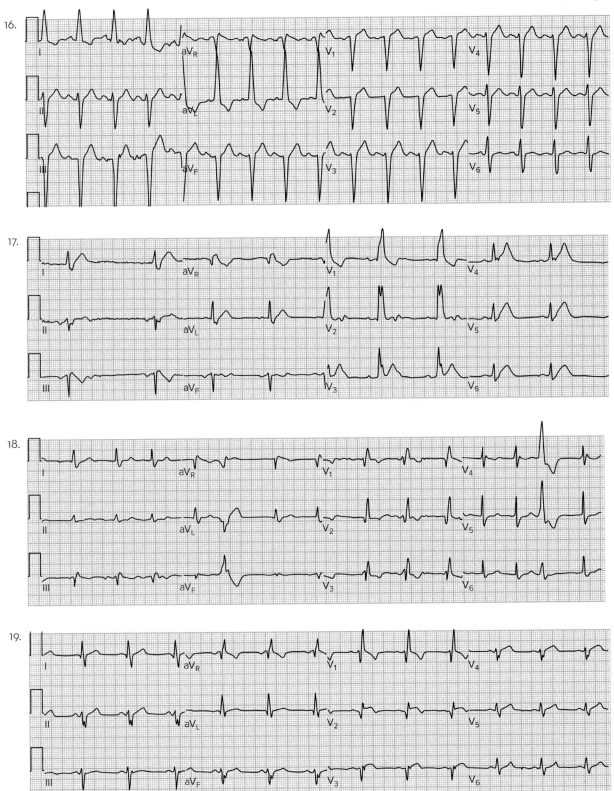

16.

17.

18.

19.

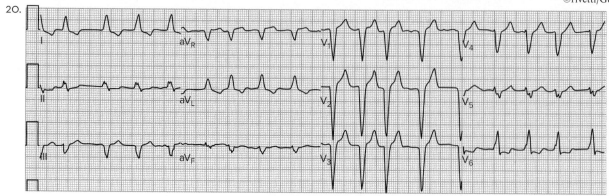

20.

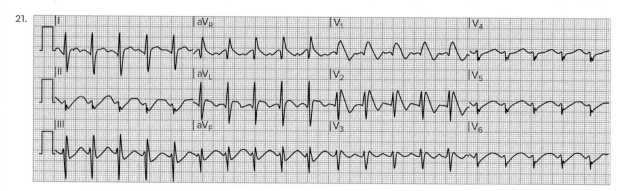

21.

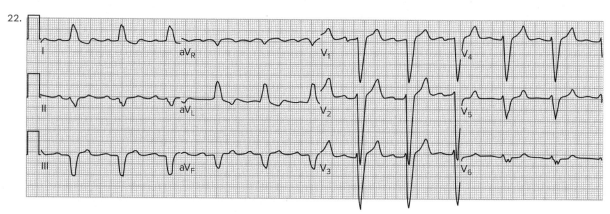

22.

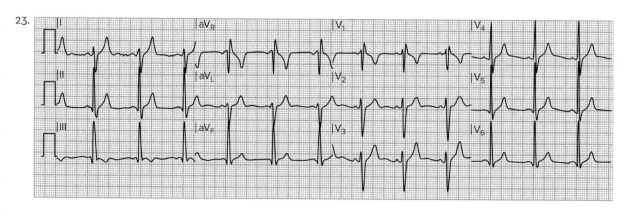

23.

24.

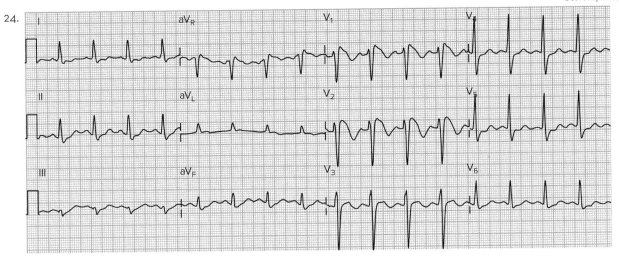

25.

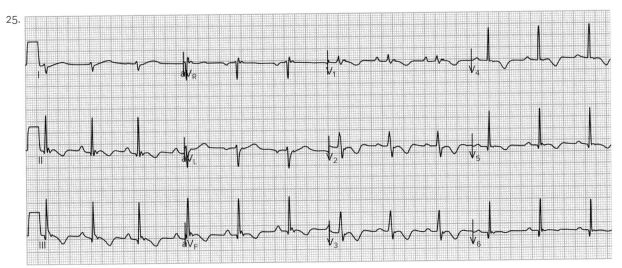

Key Points

LO 20.1	• Ventricular conduction disturbances involve delays in electrical conduction through either the AV node, the bundle branches (right or left), the fascicles (anterior, posterior), or a combination of these.
LO 20.2	• The bundle of His divides to form the right and left bundle branches. The left bundle branch then divides to form the septal, anterior, and posterior fascicles. • The QRS complex is normally narrow—less than 0.10 seconds in duration—and the electrical axis lies between 0 and +90 degrees.
LO 20.3	• Bundle branch block is a disorder that leads to one or both of the bundle branches failing to conduct impulses. This produces a delay in the depolarization of the ventricle it supplies. • In bundle branch block you will see wide, bizarre-looking QRS complexes (equal to or greater than 0.12 seconds in duration) and discordant ST segments and T waves in the leads that view the right and left ventricles. • To diagnose RBBB, check for wide, positively deflecting, triphasic QRS complexes (with some variation of an rSR′ configuration) with discordant ST segments and T waves in leads V_1 and V_2 and broad (wide) terminal S waves in leads I, V_5, and V_6. • To diagnose LBBB, check for positively deflecting, wide, and slurred or notched QRS complexes with discordant ST segments and T waves in leads V_5 and V_6 and wide, slurred, or notched negatively deflecting QRS complexes and discordant ST segments and T waves in leads V_1 and V_2. • Sometimes the right or the left bundle branch conducts the electrical impulse more slowly than normal, but it is not totally blocked. This produces QRS complexes that appear similar to RBBB or LBBB but the width of the complex is 0.11 or less. This is called *incomplete bundle branch block*. • Nonspecific intraventricular conduction defect is a conduction abnormality located in the ventricles and not the right or left bundle branch. It is characterized by QRS complexes that are 0.12 second or greater but do not conform to all the characteristics found in either left or right bundle branch block.
LO 20.4	• If only a portion of the left bundle branch is blocked, it is called a *fascicular block*. • Fascicular blocks cause axis deviation. LAFB results in left axis deviation. LPFB results in right axis deviation. • Characteristics of LAFB are small r waves and S waves in leads II, III, and aV_F and small q waves and R waves in leads I and aV_L. Characteristics of LPFB are small r waves and S waves in leads I and aV_L and small q and R waves in leads II, III, and aV_F. • Bifascicular block is a conduction disturbance in which two of the three main fascicles of the His/Purkinje system are blocked. Trifascicular block is a conduction disturbance in which there is 1) prolongation of the PR interval, 2) RBBB, and 3) either LAFB or LPFB.
LO 20.5	• To identify an acute MI in the presence of RBBB or LBBB, first look at the ST segments and T waves. If they are concordant with the terminal portion of the QRS complexes, an acute MI may exist. • LVH should be suspected in the presence of RBBB if the R wave in lead aV_L is equal to or greater than 12 mm or the R wave in lead V_5 or V_6 is equal to or greater than 25 mm. Because LBBB distorts the normal vector forces, left ventricular hypertrophy cannot usually be identified in the presence of LBBB.

LO 20.6	• Bundle branch block often does not require treatment.
LO 20.7	• Brugada syndrome has a similar appearance to RBBB and is characterized by persistent ST segment elevation in leads V, V_2, and V_3. However, unlike RBBB, the rSR′ is not more than 0.12 seconds wide and there are no broad S waves in leads I and V_6.
	• Brugada syndrome presents as prominent high takeoff J point elevation with "coved-type" ST segment elevation with an amplitude of 2 mm or greater and T wave inversion or a gradually descending ST segment elevation that creates a 1 mm or greater "trough" leading to a positive or biphasic T wave that gives it a "saddleback" configuration.
	• Arrhythmogenic right ventricular dysplasia (ARVD) is another rare condition that produces an atypical RBBB pattern. The most recognizable feature of ARVD is that it looks like an incomplete RBBB pattern with a post-depolarization epsilon wave at the end of the QRS complex.

Assess Your Understanding

The following questions give you a chance to assess your understanding of the material discussed in this chapter. The answers can be found in Appendix A.

1. Describe what is meant by the term *ventricular conduction disturbance.*

    (LO 20.1)

2. The QRS complex   (LO 20.2)
    a. is normally narrow.
    b. is greater than 0.5 seconds in direction.
    c. begins just as the impulse enters the AV node.
    d. represents ventricular repolarization.

3. With normal ventricular depolarization, the electrical axis   (LO 20.2)
   is drawn
    a. rightward, between 0 and 90 degrees.
    b. leftward, between 0 and 90 degrees.
    c. upward, at 45 degrees.
    d. downward, between 0 and 30 degrees.

4. Bundle branch block   (LO 20.3)
    a. causes narrowed QRS complexes to appear on the ECG.
    b. is a disorder in which one or both of the bundle branches fails to conduct impulses.
    c. accelerates depolarization of the ventricle it supplies.
    d. is diagnosed by analyzing only the width of the QRS complexes.

5. To diagnose RBBB, you check for an RR′ in leads   (LO 20.3)
   a. aV_F and aV_L.   b. I and II.
   c. V_1 or V_2.   d. V_3 or V_4.

6. The width of the QRS complexes in LBBB is _____ seconds in duration.   (LO 20.3)
   a. 0.08–0.10   b. 0.10–0.12
   c. equal to or greater than 0.12   d. less than 0.12

7. QRS complex configuration in LBBB in leads V_5 and V_6 is   (LO 20.3)
   a. wide and slurred or notched.
   b. triphasic.
   c. wide and deflecting negatively.
   d. narrow and deflecting positively.

8. Describe the expected direction the ST segment and T waves take in BBB.   (LO 20.3)

9. Fascicular blocks   (LO 20.4)
   a. cause axis deviation.
   b. result from accessory pathways between the atria and ventricles.
   c. result in R waves being superimposed on the other.
   d. lead to a complete absence in conduction below the level of block.

10. Describe the appearance of QRS complexes and ST segments/T waves for RBBB and LBBB in each of the leads listed below.   (LO 20.3)

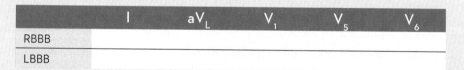

	I	aV_L	V_1	V_5	V_6
RBBB					
LBBB					

11. Describe the appearance of q waves, R waves and S waves in left anterior fascicular block and left posterior fascicular block in each of the leads listed below.   (LO 20.4)

	I	aV_L	II	III	aV_F
Left anterior fascicular block					
Left posterior fascicular block					

12. Describe the ECG characteristics of Brugada syndrome.   (LO 20.7)

13. Describe the ECG characteristics of ARVD.   (LO 20.7)

14. Describe the treatment of bundle branch block.   (LO 20.6)

Referring to the scenario at the beginning of this chapter, answer the following questions.

15. The appearance of the QRS complex in this case indicates the presence of a(n) (LO 20.3)
    a. complete AV heart block.
    b. RBBB.
    c. incomplete heart block.
    d. 1st-degree AV heart block.

16. The rabbit ear appearance of the QRS complex is technically referred to as (LO 20.3)
    a. QR'S.
    b. Q'R'S'.
    c. RR'.
    d. R'SR.

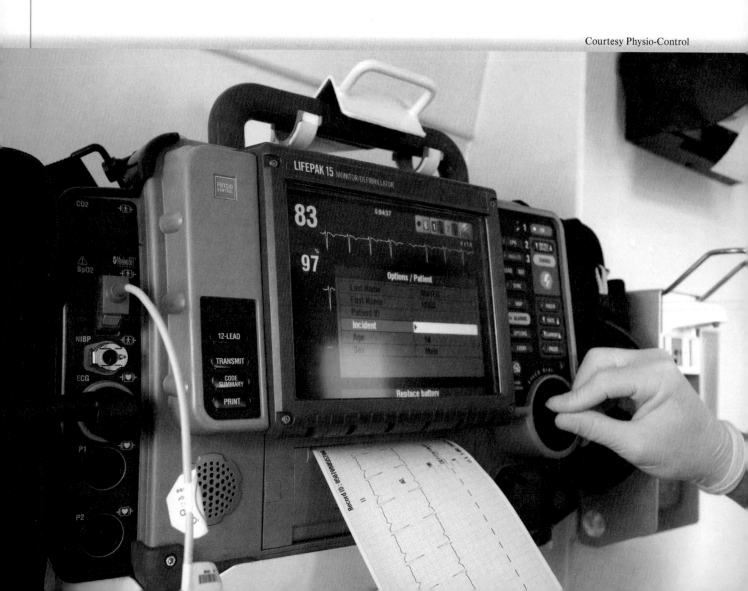

21 Atrial Enlargement and Ventricular Hypertrophy

©rivetti/Getty Images

Chapter Outline

ECG Changes Used to Identify Enlargement and Hypertrophy

Atrial Enlargement

Ventricular Hypertrophy

ST Segment and T Wave Changes

Courtesy Physio-Control

Learning Outcomes

LO 21.1 Recall the electrical waveforms that can change with enlargement and hypertrophy.

LO 21.2 Describe the ECG changes that can be used to identify right and left atrial enlargement.

LO 21.3 Describe the ECG changes that can be used to identify right and left ventricular hypertrophy.

LO 21.4 Identify the ST segment and T wave changes that may be seen with severe hypertrophy.

Case History

You are a member of the EMS squad who responds to the local high school for a 16-year-old boy complaining of chest pain. You enter the nurse's station where the school nurse is kneeling over a thin, pale young man lying on a cot. He is awake and alert but is holding his chest. After introducing yourself, you ask him how he feels. The patient indicates that he gets chest pain when he overexerts himself in physical education class. The nurse hands you a card with the patient's medical history. It indicates that he has an enlarged heart as a result of a viral infection that attacked his heart, resulting in cardiomyopathy.

You apply oxygen to the patient and establish an IV line. After obtaining a 12-lead ECG, you hand it to your partner who comments on the unusually tall R waves in V_1 and V_2. You also note a down-sloping ST segment depression and T-wave inversion in those leads.

21.1 ECG Changes Used to Identify Enlargement and Hypertrophy

Changes that we see in the ECG that indicate the presence of enlargement or hypertrophy include

- An increase in the amplitude of the waveform
- An increase in duration of the waveform
- Axis deviation
- T wave changes

Increases in the amplitude of the R or S waves represent the voltage criteria used to identify ventricular hypertrophy. The other criteria such as increased QRS complex width, ST segment depression, and T-wave inversion are referred to as *non-voltage criteria*. As a point of information, the ECG is not all that effective at identifying hypertrophy. In about 50% of cases where hypertrophy is shown to exist, the ECG will be normal. Other diagnostic tools, such as x-ray and echocardiography, are much more effective at revealing hypertrophy. However, when the electrocardiographic indicators are present, the ECG is (about 90%) accurate.

21.2 Atrial Enlargement

The duration of a normal P wave is no more than 0.10 seconds, whereas its amplitude, whether a positive or negative deflection, should not exceed 2.5 mm. The first portion of the P wave represents right atrial depolarization, whereas the terminal part represents left atrial depolarization (Figure 21-1).

Leads II and V_1 provide the necessary information to assess atrial enlargement. Lead II is nearly parallel to the flow of current through the atria (the mean P wave vector). For this reason, it records the largest deflections and can reveal abnormalities in atrial depolarization. Lead V_1 is placed just to the right of the sternum in the 4th intercostal space. This positions it directly over the atria and gives the most accurate information about atrial enlargement, which allows easy separation of the right and left atrial components of the P wave. A P wave that has both positive and negative deflections is termed *biphasic*. A biphasic P wave in lead V_1 is normal. A biphasic P wave will have two components, the initial component from the right atrium and the terminal component from the left atrium. Note that a biphasic P wave has deflections above and below the baseline.

Right Atrial Enlargement

Right atrial enlargement results in the right atrium's electrical dominance over the left atrium. The diagnosis of right atrial enlargement is made when you see an increase in the amplitude or peaking of the first part of the P wave (Figure 21-2). Remember, the first part of the P wave represents depolarization of the right atrium.

The width of the P wave, however, stays within normal limits because its terminal part originates from the left atrium, which depolarizes normally if left atrial enlargement is absent.

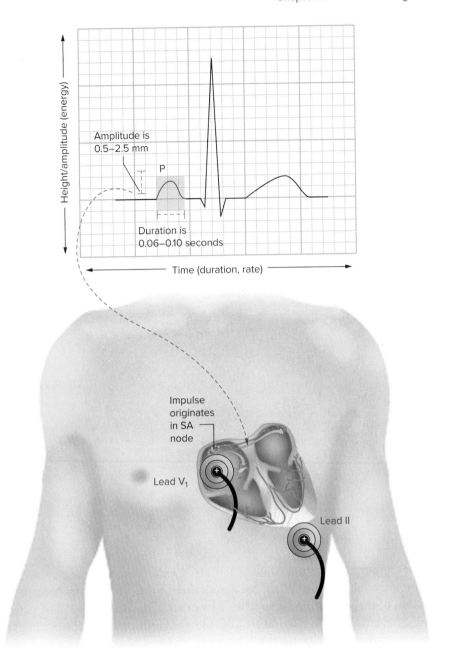

Figure 21-1
Atrial conduction and the
normal P wave. Leads II and V₁
provide the best information
regarding atrial conduction.

Characteristics of Right Atrial Enlargement

- The P wave is taller than 2.5 mm in lead II (as well as in the other inferior leads III and aV$_F$)
- The initial component of the P wave is taller than the terminal component (if it is biphasic)

Right atrial enlargement is often secondary to pulmonary hypertension related to emphysema. Therefore, the P wave changes in right atrial enlargement are termed *P pulmonale.*

Left Atrial Enlargement

With left atrial enlargement, the amplitude of the terminal portion of the P wave may increase in V₁ (Figure 21-3). Diagnostically we say left atrial enlargement is

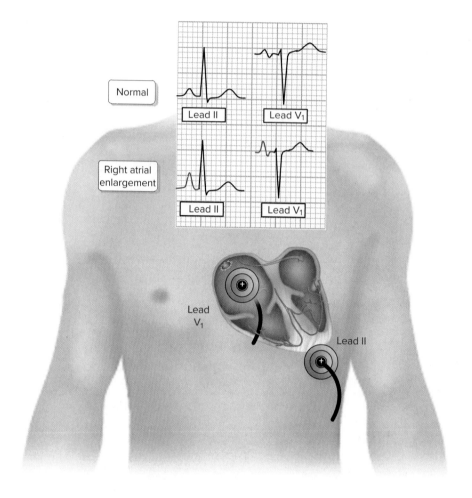

present when the terminal (left atrial) portion of the P wave drops more than 1 mm below the isoelectric line (in lead V_1).

However, a more telling indicator of left atrial enlargement is an increase in the duration or width of the terminal portion of the P wave. This occurs because depolarization of the left atrium is the latter part of atrial depolarization. The diagnosis of left atrial enlargement requires that the terminal portion of the P wave be greater than one small square (0.04 seconds) in duration.

Often the presence of ECG evidence of left atrial enlargement reflects only a nonspecific conduction irregularity. For this reason, you must carefully weigh the interpretation of atrial enlargement on the ECG with the patient's clinical presentation. However, the presence of ECG evidence of left atrial enlargement may also be the result of mitral valve stenosis causing the left atrium to enlarge to force blood across the stenotic (tight) mitral valve. The presence of left atrial enlargement from mitral stenosis is termed *P mitrale*.

Characteristics of Left Atrial Enlargement

- The amplitude of the terminal (negative) portion of the P wave increases, descending more than 1 mm below the isoelectric line in lead V_1

- The duration of the P wave is increased to greater than 0.12 seconds in duration, or the terminal (negative) portion of the P wave is greater than 1 small square (0.04 seconds) in width

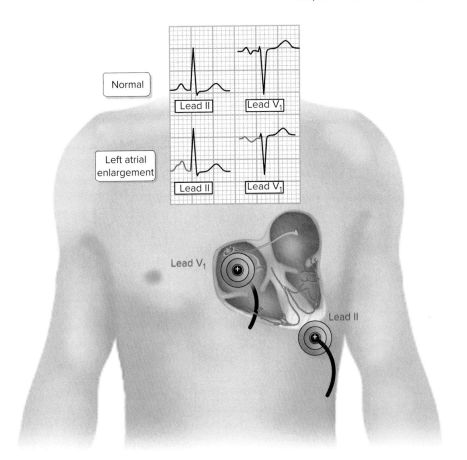

Figure 21-3
Left atrial enlargement leads to an increase in the amplitude and width of the terminal portion of the P wave.

21.3 Ventricular Hypertrophy

Using the ECG to diagnose ventricular hypertrophy requires careful assessment of the QRS complex using voltage criteria in many leads. Voltage criteria are based on measurement of the R waves and S waves of the QRS complexes. These criteria have various degrees of **sensitivity** and **specificity**. Also, axis deviation can aid in identifying certain types of hypertrophy.

Let's begin by quickly reviewing how the QRS complex should appear in lead V_1. Ventricular depolarization will move downward to the left side and posteriorly because the thicker left ventricle is located mostly posteriorly. The V_1 electrode is positive; therefore, the wave of depolarization moving through the left ventricle will be mostly moving away from it, producing a mainly negative QRS complex. For this reason, the R wave is usually very short, whereas the S wave is much larger than the R wave in this lead.

Right Ventricular Hypertrophy

Right ventricular hypertrophy is far less common than left ventricular hypertrophy. It is often the result of chronic lung disease, valvular heart disease, congenital heart disease, or primary pulmonary hypertension.

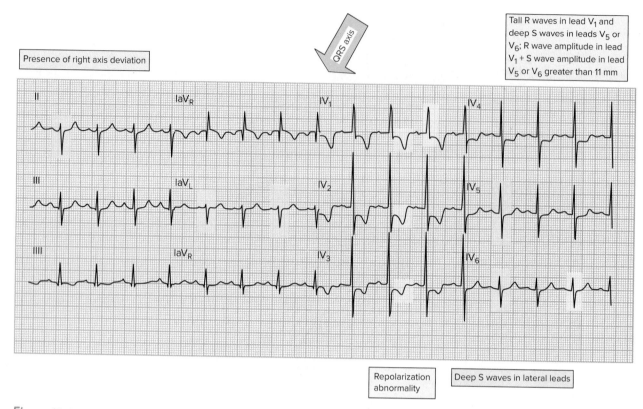

Presence of right axis deviation

QRS axis

Tall R waves in lead V_1 and deep S waves in leads V_5 or V_6; R wave amplitude in lead V_1 + S wave amplitude in lead V_5 or V_6 greater than 11 mm

Repolarization abnormality

Deep S waves in lateral leads

Figure 21-4

Lead V_1 normally has a short R wave and a deeper S wave. With right ventricular hypertrophy, the thick wall of the enlarged right ventricle causes tall R waves in lead V_1 (and leads that lie close to it) and deep S waves in V_6.

Given the opposing forces of the thicker left ventricle, the sensitivity of electrocardiographic criteria for right ventricular hypertrophy is relatively low. The mass of the left ventricle is still greater than that of the right ventricle in persons with right ventricular hypertrophy. However, the presence of certain ECG findings can help identify right ventricular hypertrophy.

Right axis deviation is the most consistent characteristic seen with right ventricular hypertrophy (Figure 21-4). Instead of falling within the normal range of 0 to +90 degrees, the QRS axis now falls between +90 and ±180 degrees. This causes the QRS complex to be more negative than positive in lead I and positive in lead aV_F.

Voltage Criteria

The precordial leads can also be helpful in identifying right ventricular hypertrophy (Figure 21-5). The thick wall of the enlarged right ventricle causes the R waves to be more positive in the leads that lie closer to lead V_1 (the electrode located directly over the right ventricle). Thus, moving from the right to the left chest leads (from V_1 to V_6), the R waves go from being more positive to being smaller. This is a progressive but gradual change. Also, the S waves are smaller in lead V_1 and become larger as you move from V_1 to V_6. This produces the following:

- Lead V_1—the R waves are equal to or larger than the S waves or the R wave is greater than 7 mm

- Leads V_5, V_6—the S waves are larger than the R waves or the S wave is greater than 7 mm deep

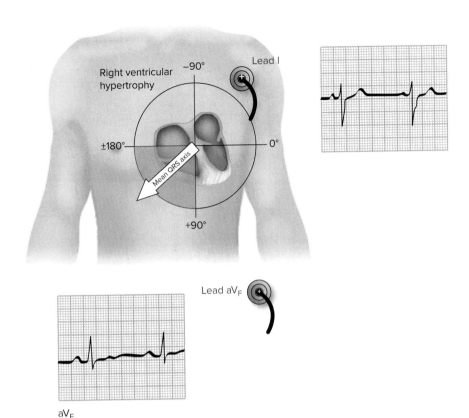

Figure 21-5
In right ventricular hypertrophy the QRS axis moves to between +90 and ±180 degrees. The QRS complexes in right ventricular hypertrophy are more negative in lead I and positive in lead aV$_F$.

Additionally, the QRS complex in lead V$_1$ is wider (but rarely beyond 0.1 seconds). Also, look for the following supporting criteria for right ventricular hypertrophy: dominant S waves in leads I, II, and III (referred to as S1, S2, S3 pattern), right atrial enlargement (P pulmonale) and ST depression/T-wave inversion (repolarization abnormality) in the right precordial (V$_1$ to V$_4$) and inferior (II, III, and aV$_F$) leads.

Characteristics of Right Ventricular Hypertrophy

- The presence of right axis deviation (with the QRS axis exceeding +110 degrees)
- The R waves are larger than or equal to S waves in leads V$_1$ and the S waves are larger or equal to the R waves in leads V$_5$, V$_6$
- The R waves in V$_1$ and S waves in lead V$_6$ are equal to or greater than 7 mm
- QRS complex in V$_1$ is greater than 120 milliseconds (but not due to right bundle branch block)
- S1, S2, S3 pattern (dominant S waves in leads I, II, and III)
- right atrial enlargement (P pulmonale)
- ST depression/T-wave inversion in the right precordial (V$_1$ to V$_4$) and inferior (II, III, and aV$_F$) leads

Left Ventricular Hypertrophy

Frequent causes of left ventricular hypertrophy are systemic hypertension and valvular heart disease. The thickened left ventricular wall leads to prolonged depolarization (increased R wave peak time) and delayed repolarization (ST and T-wave abnormalities) in the lateral leads.

The most commonly used criteria for left ventricular hypertrophy are based on QRS voltage criteria. Voltage criteria must be accompanied by non-voltage criteria

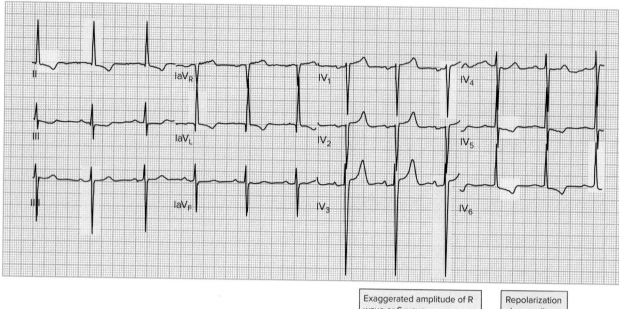

R wave amplitude in lead I + the S wave in lead III greater than 25 mm; R wave amplitude in lead aV$_L$ is greater than 11 mm

S wave amplitude in lead V$_1$ + R wave amplitude in lead V$_6$ is greater than 35 mm

Exaggerated amplitude of R wave or S wave

Repolarization abnormality

Figure 21-6
The thick wall of the enlarged left ventricle causes the R waves to be more positive in the leads that lie closer to leads V$_5$, V$_6$ and the S waves to be deeper in the leads that are closer to V$_1$, V$_2$.

to be considered diagnostic of this condition. Through many studies, multiple criteria have been developed to diagnose left ventricular hypertrophy (some of which are summarized below).

Voltage Criteria

The precordial leads are more sensitive than the limb leads in the diagnosis of left ventricular hypertrophy. Therefore, increased R wave amplitude in those leads overlying the left ventricle forms the voltage basis for the ECG diagnosis of left ventricular hypertrophy (Figure 21-6). Also, the S waves are smaller in the leads overlying the left ventricle (leads V$_5$, V$_6$) but deeper in the leads overlying the right ventricle (leads V$_1$, V$_2$).

Specific voltage criteria for identifying left ventricular hypertrophy in the precordial leads include any of the following:

- The amplitude of S waves in V$_1$ + the amplitude of the tallest R wave in leads V$_5$ or V$_6$ exceed 35 mm
- R wave amplitude in lead V$_4$, V$_5$, or V$_6$ exceeds 26 mm
- R wave amplitude in lead V$_6$ exceeds the R wave amplitude in lead V$_5$
- Largest R wave plus largest S wave in precordial leads exceeds 45 mm

Characteristics of Left Ventricular Hypertrophy

- There is increased R wave amplitude over those leads overlying the left ventricle

- The S waves are smaller in the leads overlying the left ventricle but deeper in the leads overlying the right ventricle

The more criteria present, the more likely the patient has left ventricular hypertrophy.

The limb leads can also provide evidence of left ventricular hypertrophy. The limb-lead voltage criteria for left ventricular hypertrophy include one or more of the following

- R wave amplitude in lead I + the S wave amplitude in lead III exceeds 25 mm
- R wave amplitude in lead aV_L exceeds 11 mm
- R wave amplitude in lead aV_F exceeds 20 mm
- S wave amplitude in lead aV_R exceeds 14 mm
- R wave in aV_L + the S wave in V_3; sum is greater than 28 mm in males or greater than 20 mm in females

Non-Voltage Criteria

Non-voltage criteria include those findings that are not related to measuring the amplitude of the R waves and S waves of the QRS complex and include one or more of the following:

- QRS complex becomes slightly widened and the time to the peak of the R wave is increased (but rarely greater than 0.1 seconds) in leads V_5 or V_6
- ST segment depression and T-wave inversion in the left-sided leads (repolarization abnormality)
- left atrial enlargement (P mitrale)
- Some ST segment elevation in leads with deep S waves (represents appropriate proportional discordance)

Left axis deviation of in excess of –15 degrees is often seen but is not necessary for the diagnosis of left ventricular hypertrophy.

The criteria listed are of little value in persons younger than 35 years of age because they frequently have increased voltage due to a relatively thin chest wall.

It is possible for both the right and left ventricles to be hypertrophied. This produces a combination of features such as left ventricular hypertrophy in the precordial leads and right axis deviation in the limb leads. However, most often the indicators of left ventricular hypertrophy hide those of right ventricular hypertrophy.

21.4 ST Segment and T Wave Changes

In ventricular hypertrophy, you may also see changes in the ST segment and T waves. This is called *secondary repolarization abnormalities* and includes the following (Figure 21-7):

1. Down-sloping ST segment depression
2. T-wave inversion (which takes an opposite direction to the R wave)

The depressed ST segment and the inverted T-wave appear to blend together, forming a single asymmetric wave. The downward slope is gradual, whereas the upslope is abrupt.

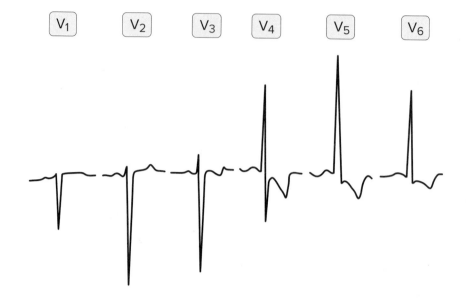

Repolarization abnormalities are not uncommon. They are easier to see in those leads with tall R waves because the electrical forces of the hypertrophied ventricle are greater. For this reason, right ventricular repolarization abnormalities are seen in leads V_1 to V_4 and the inferior leads, I, II, and aV_F. Left ventricular repolarization abnormalities are seen in leads I, aV_L, V_5, and V_6.

Hypertrophic Cardiomyopathy

As described in Chapter 10, hypertrophic cardiomyopathy is a pathologic cardiac condition in which the intraventricular septum is abnormally thickened. It may also result in thickening of the ventricular wall.

The classic ECG finding in hypertrophic cardiomyopathy are deep, narrow ("dagger-like") Q waves in the lateral (V_5 to V_6, I, aV_L)—and sometimes inferior (II, III, aV_F)—leads due to the abnormally hypertrophied intraventricular septum. These Q waves may mimic prior myocardial infarction, although the Q-wave morphology is different: infarction Q waves are typically greater than 40 milliseconds in duration while septal Q waves in this condition are less than 40 milliseconds in duration. Lateral Q waves are more common than inferior Q waves in hypertrophic cardiomyopathy.

Left ventricular diastolic dysfunction may lead to compensatory left atrial hypertrophy, with signs of left atrial enlargement (P mitrale) on the ECG.

Criteria for left ventricular hypertrophy, as described above, is usually present.

Wolff-Parkinson-White, or WPW, syndrome can be associated with this condition as well. Atrial fibrillation and supraventricular tachycardia are common. Ventricular dysrhythmias such as ventricular tachycardia can also occur and may be a cause of sudden death.

If there is localized hypertrophy of the left ventricular apex, the classic ECG finding is giant T-wave inversion in the precordial leads.

For the following 12-lead ECGs, identify if atrial enlargement or hypertrophy are present and the type.

1.

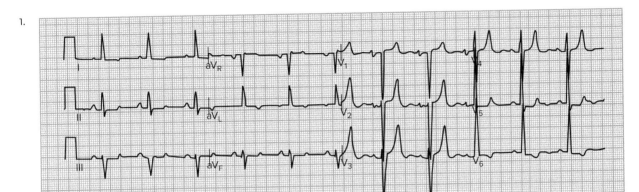

2.

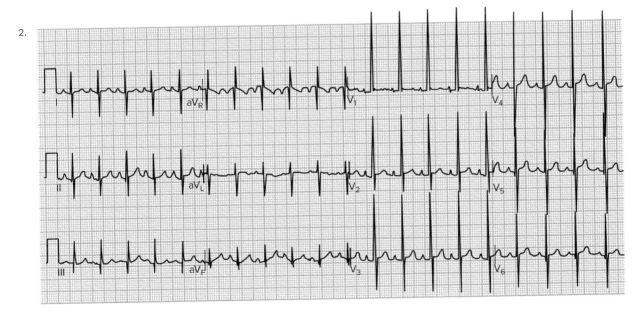

3.

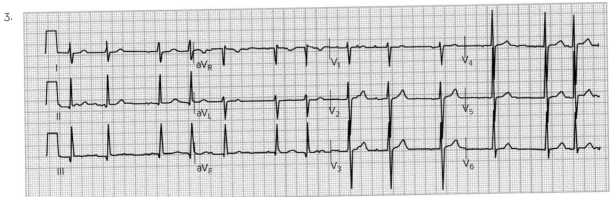

4.

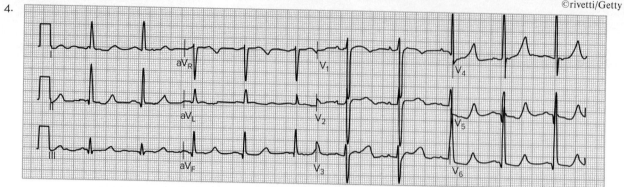

5.

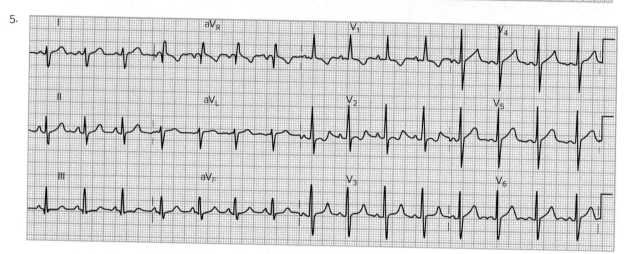

6.

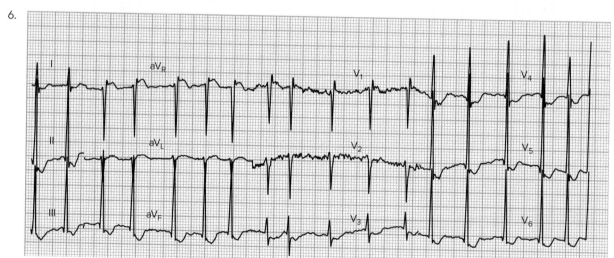

Key Points

LO 21.1	• We look at the P wave to assess for atrial enlargement. Similarly, we examine the QRS complex to help us identify ventricular hypertrophy. • Indicators of enlargement or hypertrophy include an increase in the duration of the waveform, an increase in the amplitude of the waveform, and axis deviation.
LO 21.2	• The duration of the normal P wave is less than 0.10 seconds, whereas its amplitude should not exceed 2.5 mm. • The first portion of the P wave represents right atrial depolarization, whereas the terminal portion represents left atrial depolarization. • Leads II and V_1 provide the necessary information to assess atrial enlargement. • Key characteristics of right atrial enlargement include the following: The first part of the P wave will be taller than normal in lead V_1, P waves have an amplitude that exceeds 2.5 mm in the inferior leads, and there is no change in the duration of the P wave. The duration of the P wave does not appear longer in right atrial enlargement because it is hidden by the portion of the P wave that represents left atrial depolarization. • Two indicators of left atrial enlargement are increased amplitude in the terminal portion of the P wave in V_1 and increased duration or width of the P wave.
LO 21.3	• Normally, since the V_1 electrode is positive, the wave of depolarization moving through the left ventricle will be moving away from it, producing QRS complexes that are negative (very short R waves with much larger S waves). • Right axis deviation is the most common characteristic seen with right ventricular hypertrophy. • In the precordial leads, right ventricular hypertrophy causes the R waves to be more positive in the leads that lie closer to lead V_1 and the S waves to be deeper in the leads than lie closer to V6. • Left ventricular hypertrophy is identified by increased R wave amplitude in the precordial leads overlying the left ventricle (V_5, V_6) and S waves that are smaller in those same leads. Also, the S waves should be deeper in the leads (lead V_1) overlying the right ventricle. • The duration of the QRS complex may be slightly prolonged in both right and left ventricular hypertrophy. However, it is rarely greater than 0.12 seconds.
LO 21.4	• In severe hypertrophy, you may also see changes in the ST segment and T waves. This is called *secondary repolarization abnormalities* and includes ST segment depression and T-wave inversion.

Assess Your Understanding

The following questions give you a chance to assess your understanding of the material discussed in this chapter. The answers can be found in Appendix A.

1. List the ECG changes that can indicate the presence of enlargement and hypertrophy.

    (LO 21.1)

2. Hypertrophy is identified by changes in the   (LO 21.3)
   a. P waves.
   b. QRS complexes.
   c. T waves.
   d. PR intervals.

3. Which of the following are changes seen in the ECG that can indicate the presence of enlargement or hypertrophy?   (LO 21.1)
   a. An increase in duration of the waveform
   b. A biphasic waveform
   c. Axis deviation
   d. All of the above

4. The P wave   (LO 21.2)
   a. is normally 5 mm in amplitude.
   b. normally has a duration of less than 0.10 seconds.
   c. represents left atrial depolarization.
   d. is normally upright in lead V_1.

5. Leads used to assess atrial enlargement include   (LO 21.2)
   a. V_1 through V_6
   b. All the limb leads
   c. I and aV_F
   d. II and V_1

6. Characteristics of right atrial enlargement include   (LO 21.2)
   a. the P wave appearing tallest in lead II instead of lead aV_F.
   b. an increase in the amplitude of the first part of the P wave in V_1.
   c. increased amplitude of the terminal portion of the P in V_1.
   d. the terminal portion of the P wave being at least 0.04 seconds in width.

7. With left atrial enlargement,   (LO 21.2)
   a. the P wave vector swings to the right.
   b. the initial part of the P wave is enlarged.
   c. there is an increase in the duration of the terminal portion of the P wave.
   d. the P wave is taller than 2.5 mm in lead V_1.

8. The ECG evidence of atrial enlargement   (LO 21.2)
   a. almost always correlates with pathological changes in the atrium.
   b. is seen in only 10% to 20% of the patients diagnosed using other procedures.
   c. should be tempered with the patient's clinical presentation.
   d. requires careful assessment of the QRS complex in many leads.

9. To diagnose ventricular hypertrophy, we assess the   (LO 21.3)
    a. P waves.
    b. T waves.
    c. QRS complexes.
    d. P wave axis.

10. Normally, in lead V_1, the QRS complex will appear   (LO 21.3)
    a. mainly negative.
    b. positive, with a large R wave and a short S wave.
    c. biphasic, with an R wave and an S wave of equal height.
    d. mainly positive.

11. Characteristic(s) seen with right ventricular hypertrophy include   (LO 21.4)
    a. R waves that are more positive in the leads that lie closer to lead V_6.
    b. S waves that are larger in lead V_1.
    c. QRS complexes that are slightly more positive than negative in lead I and negative in lead aV_F.
    d. right axis deviation.

12. Common characteristic(s) seen with left ventricular hypertrophy   (LO 21.3)
    include
    a. R waves that are more positive in the leads that lie closer to lead V_6.
    b. S waves that are smaller in lead V_1.
    c. R waves in lead V_2 that have an amplitude that exceeds 26 mm.
    d. right axis deviation.

13. Left ventricular repolarization abnormalities can be identified   (LO 21.4)
    by down-sloping ST segment depression and T-wave inversion
    in leads
    a. I, II, and III.
    b. V_1 and V_2.
    c. V_5 and V_6.
    d. I, aV_L, V_5, V_6.

14. The patient in this case is likely experiencing: (LO 21.3)
    a. right atrial enlargement.
    b. left atrial enlargement.
    c. left ventricular hypertrophy.
    d. right ventricular hypertrophy.

22 Other Cardiac Conditions and the ECG

©rivetti/Getty Images

Chapter Outline

Pericarditis

Pericardial Effusion

Pulmonary Embolism

Electrolyte Imbalance

Drug Effects and Toxicity

Courtesy Physio-Control

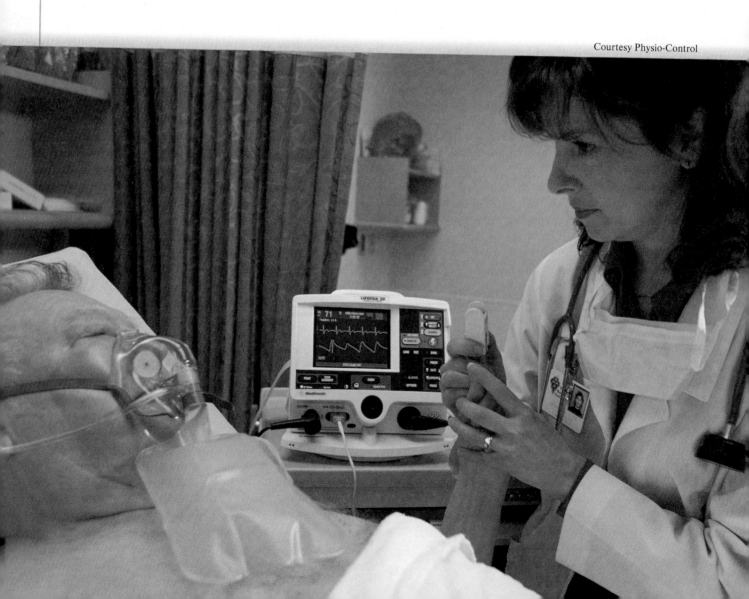

Learning Outcomes

Case History

A 72-year-old man with renal failure presents to the emergency department by ambulance for evaluation of weakness and palpitations. The physician enters the exam room and introduces himself to the patient. While doing so, he notices that the patient is in sinus rhythm on the monitor, with frequent PVCs.

The patient indicates he is feeling very weak. He missed his last two dialysis appointments, and he can't seem to catch his breath. The physician examines the patient and determines that he has frothy pulmonary edema. The ECG shows runs of ventricular tachycardia and tall, peaked T-waves with a prolonged QT interval. The physician immediately summons the nurse and tells her that the patient is volume overloaded and he suspects the patient's potassium is critically elevated because he missed his dialysis appointments.

The nurse places the patient on supplemental oxygen, and the physician orders blood work and one ampule of calcium gluconate, two ampules of D_{50}, 20 units of insulin, and a dose of sodium bicarbonate to be administered in fairly rapid succession.

Follow-up ECGs indicate normal T-waves and diminished ectopy. The patient is taken to dialysis, where the excess fluid in his system is removed.

22.1 Pericarditis

Pericarditis is another condition that can lead to changes in the ECG (Figure 22-1). It is inflammation of the pericardium that can produce sharp, substernal chest pain.

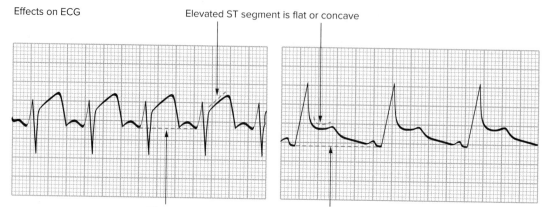

Effects on ECG

Elevated ST segment is flat or concave

ST segments and T waves are off the baseline,
gradually angling back down to the next QRS complex

Figure 22-1
Pericarditis and ST segment elevation.

ECG Changes

Initially with pericarditis, the T-wave is upright and may be elevated. During the recovery phase it inverts. This may take several weeks to develop. The ST segments demonstrate a diffuse, upward-concave elevation (not convex as in MI) in all leads except aV_R and V_1. It is sometimes described as a saggy appearance. If the ST segment is elevated, it appears off the baseline, gradually sloping back down to the next QRS complex (and may include the P wave). Given that the signs and symptoms are similar, it is easy to mistake these changes for a developing MI. Certain features of the ECG can be helpful in differentiating pericarditis from MI:

Characteristics of Pericarditis

• Because pericarditis involves the whole heart, the ST segment and T-wave changes in pericarditis are diffuse. Therefore, the ECG changes will be present in all leads, while those of MI are localized to the affected area.

• In pericarditis, T-wave inversion usually occurs only after the ST segments have returned to baseline. In MI, T-wave inversion is usually seen before ST segment normalization.

• In pericarditis, development of a Q wave does not occur.

• In pericarditis, there is sometimes depression of the PR segment in virtually any lead except aV_R.

22.2 Pericardial Effusion

The inflammatory process associated with pericarditis can stimulate the body's immune response, resulting in white cells or serous, fibrous, purulent, and hemorrhagic exudates being sent to the injured area. This can lead to pericardial effusion (Figure 22-2). Pericardial effusion is a buildup of an abnormal amount of fluid and/or a change in the character of the fluid in the pericardial space. The pericardial space is the space between the heart and the pericardial sac.

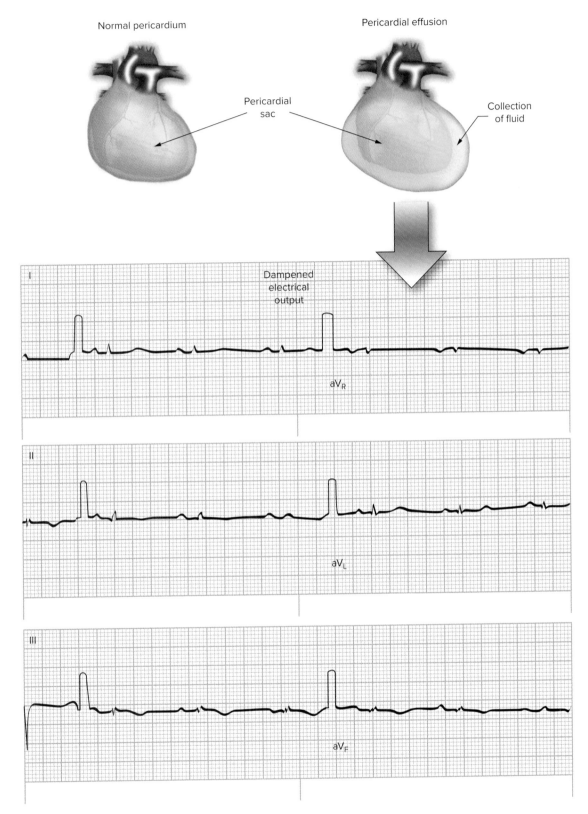

Figure 22-2
Large pericardial effusion with low-voltage QRS complexes.

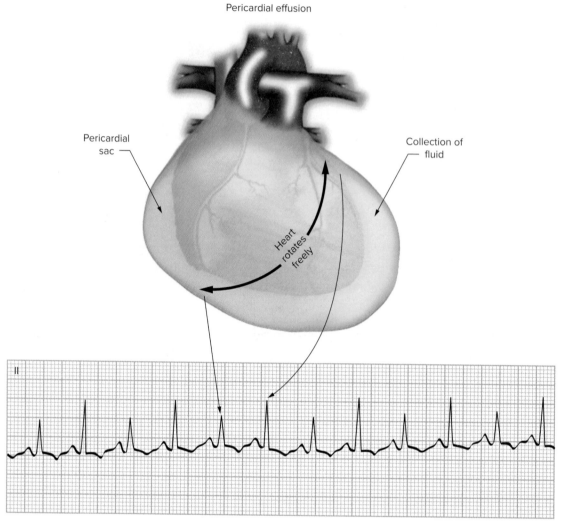

Figure 22-3
Large pericardial effusion with electrical alternans.

ECG Changes

Formation of a substantial pericardial effusion dampens the electrical output of the heart, resulting in low-voltage QRS complexes in all leads. However, the ST segment and T-wave changes of pericarditis may still be seen.

If an effusion is large enough, the heart may rotate freely within the fluid-filled sac. This can cause electrical alternans, a condition in which the electrical axis of the heart varies with each beat (Figure 22-3). A varying axis is most easily recognized on the ECG by the presence of QRS complexes that change in height with each successive beat. This condition can also affect the P and T-waves.

The accumulation of fluid or exudate in the pericardial sac can lead to the development of cardiac tamponade—a condition in which the heart is compressed. Cardiac tamponade can lead to compromised cardiac output.

22.3 Pulmonary Embolism

A pulmonary embolism is an acute blockage of one of the pulmonary arteries by a blood clot or other foreign matter (Figure 22-4). This leads to obstruction of blood flow to the lung segment supplied by the artery. The larger the artery occluded, the

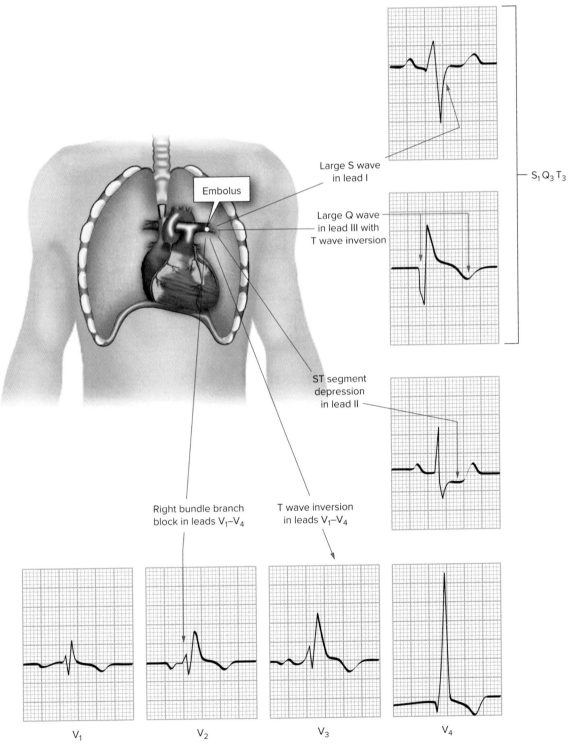

Figure 22-4
ECG changes seen with pulmonary embolism.

more massive the pulmonary embolus and therefore the larger the effect the embolus has on the lungs and heart. Due to the increased pressure in the pulmonary artery caused by the embolus, the right atrium and ventricle become distended and unable to function properly, leading to right heart failure. This condition is referred to as *acute cor pulmonale.*

Massive pulmonary embolism impairs oxygenation of the blood, and death may result. The most common source of the clot is in one of the large pelvic or leg veins. The pain that accompanies a pulmonary embolus is pleuritic, and shortness of breath is often present.

ECG Changes

ECG changes that suggest the development of a massive pulmonary embolus are shown in the box below.

Characteristics of Pulmonary Embolism
• Tall, symmetrically peaked P waves in leads II, III, and aV_F and sharply peaked biphasic P waves in leads V_1 and V_2. These changes are indicative of right atrial enlargement.
• A large S wave in lead I, a deep Q wave in lead III, and an inverted T-wave in lead III. This is called the *S1 Q3 T3 pattern* (see Figure 22-4).
• ST segment depression in lead II.
• Right bundle branch block (which usually subsides after the patient improves).
• A QRS axis greater than 90 degrees (right axis deviation).
• Inversion of the T-waves in leads V_1 to V_4.
• Q waves generally limited to lead III.

A number of dysrhythmias may be seen with massive pulmonary embolism, most commonly sinus tachycardia and atrial fibrillation. In the case of a minimal or small pulmonary embolism, the ECG is usually normal, or it may show a sinus tachycardia.

22.4 Electrolyte Imbalance

Changes in potassium and calcium serum levels can profoundly affect the ECG. Those involving potassium are the most immediately life-threatening. Levels that are too high (hyperkalemia) or too low (hypokalemia) can quickly result in serious cardiac dysrhythmias. The presence of ECG changes may be a better measure of clinically significant potassium toxicity than the serum potassium level.

Hyperkalemia

Hyperkalemia is an abnormally elevated level of potassium in the blood. The normal potassium level in the blood is 3.5 to 5.0 milliequivalents per liter (mEq/L). Potassium levels between 5.1 mEq/L and 6.0 mEq/L reflect mild hyperkalemia. Potassium levels of 6.1 mEq/L to 7.0 mEq/L reflect moderate hyperkalemia, and levels above 7 mEq/L reflect severe hyperkalemia. Signs and symptoms of hyperkalemia include weakness, paralysis, and respiratory failure.

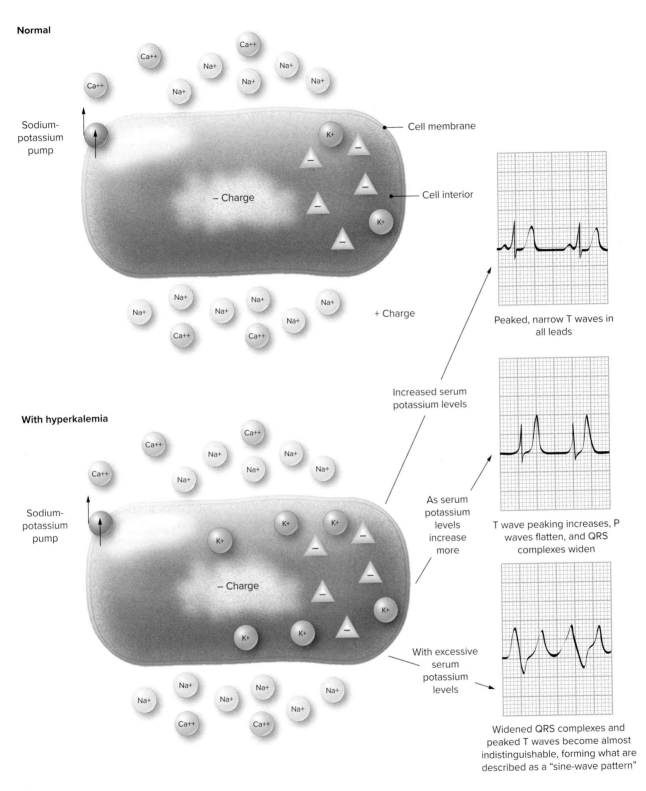

Normal

Ca++ Ca++ Na+ Na+ Na+ Na+ Na+

Sodium-potassium pump

Ca++

K+

Cell membrane

– Charge

– – – –

Cell interior

K+

Na+ Na+ Na+ Na+ Na+ Na+

Ca++ Ca++

+ Charge

Peaked, narrow T waves in all leads

Increased serum potassium levels

With hyperkalemia

Ca++ Ca++ Na+ Na+ Na+ Na+

Sodium-potassium pump

Ca++

K+ K+ K+

– Charge

– – – –

K+

K+ K+

Na+ Na+ Na+ Na+ Na+

Ca++ Ca++

As serum potassium levels increase more

T wave peaking increases, P waves flatten, and QRS complexes widen

With excessive serum potassium levels

Widened QRS complexes and peaked T waves become almost indistinguishable, forming what are described as a "sine-wave pattern"

Figure 22-5
ECG changes seen with hyperkalemia.

Hyperkalemia can generate a rapid progression of changes in the ECG that can end in ventricular fibrillation and death (Figure 22-5). For this reason you should take immediate corrective action with any changes you see on the ECG that are due to hyperkalemia.

T-wave peaking begins as the potassium level starts to rise. You may remember, from Chapter 19, peaked T-waves can also be seen in MI. The distinction between the two is that the changes in MI are limited to those leads overlying the infarcted area, whereas, with hyperkalemia, the changes are seen in all leads. Another distinctive feature of the hyperkalemic T-wave is that both the up and down slopes are concave, giving the appearance of a tent. As the serum potassium increases more, the PR interval becomes prolonged, and the P wave eventually flattens and then disappears. Finally, the QRS complex widens until it blends with the T-wave, forming what is called a *sine-wave pattern.* Idioventricular rhythm or ventricular fibrillation may ultimately develop.

ECG changes seen with hyperkalemia are shown in the box below.

Characteristics of Hyperkalemia
• Peaked T-waves (tenting)
• Flattened P waves
• Prolonged PR interval (1st-degree AV heart block)
• Widened QRS complex
• Deepened S waves and merging of S and T-waves
• Concave up and down slopes of the T-wave
• Sine-wave pattern

Hypokalemia

Hypokalemia is a lower-than-normal amount of potassium in the blood. Symptoms of hypokalemia include weakness, fatigue, paralysis, respiratory difficulty, constipation, and leg cramps. Severe hypokalemia can lead to dysrhythmias and pulseless electrical activity (PEA) or asystole.

The ECG is also an effective means of determining the presence of hypokalemia (Figure 22-6). The four key changes in the ECG seen with serious hypokalemia are shown in the box below.

Characteristics of Hypokalemia
• ST segment depression
• Flattening of the T-wave
• Appearance of U waves
• Prolongation of the QT interval

A wave that occurs after the T-wave in the cardiac cycle is called a *U wave.* Although U waves are the most distinguishing characteristic of hypokalemia, by themselves they are not diagnostic. U waves are seen in other conditions and can sometimes be seen in patients with normal hearts and normal serum potassium levels.

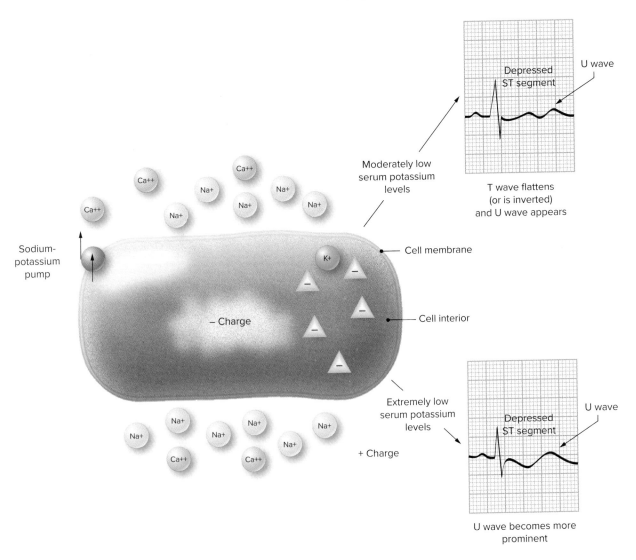

Figure 22-6
ECG changes seen with hypokalemia.

Calcium Disorders

Calcium is needed by the myocardial cells to contract. Alterations in serum calcium levels mainly affect the QT interval (Figure 22-7). Hypocalcemia prolongs the QT interval, while hypercalcemia shortens it. Torsades de pointes, a variant of ventricular tachycardia, is seen in patients with prolonged QT intervals. Hypocalcemia also results in decreased cardiac contraction.

22.5 Drug Effects and Toxicity

Some medications can have striking effects on ECG waveforms by altering the way the cells depolarize, repolarize, and innervate surrounding tissues. These changes may indicate expected, and, therefore, benign effects of certain medications as well as provide evidence of medication toxicity.

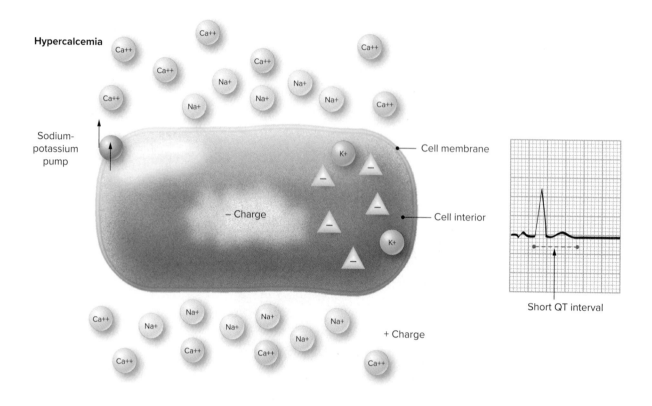

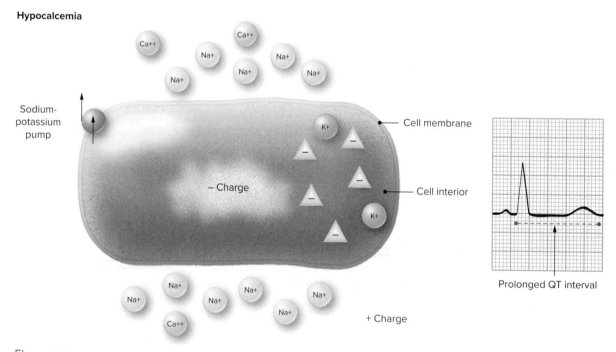

Figure 22-7
With hypercalcemia, the QT interval shortens, and with hypocalcemia, the QT interval is prolonged.

Digoxin

Digoxin is in a class of drugs referred to as *cardiac glycosides.* Digitalis preparations have been used to treat people since the thirteenth century. In certain circumstances or in high concentrations, digitalis can induce deadly dysrhythmias. These

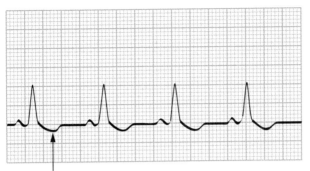

Figure 22-8
Effects of digoxin on the ECG.

Gradual downward curve of the ST segment

drugs slow the influx of sodium while allowing a greater influx of calcium during both depolarization and repolarization. This increases myocardial contractility and improves the heart's pumping ability. Digoxin can be used in heart failure to improve cardiac output.

Two types of ECG changes are seen with digoxin: those associated with therapeutic blood levels of the drug and those associated with toxic blood levels.

In therapeutic doses, digoxin has a parasympathetic effect that slows the pacing rate of the SA node and conduction through the AV node. The slowing of the heart rate is called a *negative chronotropic effect*, while the slowing of the conduction rate is called a *negative dromotropic effect*. These effects can slow the rates associated with fast dysrhythmias. It also inhibits the receptiveness of the AV node to multiple stimuli, allowing fewer stimuli to reach the ventricles. This can be particularly helpful in the treatment of dysrhythmias such as atrial fibrillation and flutter. However, if the heart rate slows too much, decreased cardiac output can result.

Therapeutic Blood Levels—ECG Changes

Digoxin produces a characteristic gradual downward curve of the ST segment (it looks like a ladle) (Figure 22-8). The R wave slurs into the ST segment. Sometimes the T-wave is lost in this scooping effect. The lowest portion of the ST segment is depressed below the baseline. When seen, the T-waves have shorter amplitude and can be biphasic. The QT interval is usually shorter than anticipated, and the U waves are more visible. Also, the PR interval may be prolonged.

Digoxin has a very narrow therapeutic margin of 0.5 to 1.5 mg/dL, and it is excreted from the body slowly. Toxic levels can occur and cause life-threatening dysrhythmias. The chance of developing toxicity is increased in hypokalemia, hypercalcemia, renal impairment, advanced age, acute hypoxia, and hypothyroidism.

Toxic Blood Levels—ECG Changes

Digoxin levels greater than the therapeutic range can cause any dysrhythmia because they increase the automatic behavior of all cardiac conducting cells, causing them to act more like pacemakers. The most common dysrhythmia seen with digoxin toxicity is paroxysmal atrial tachycardia (PAT) with 2nd-degree AV heart block. The conduction block is usually 2:1 but may vary unpredictably. Other abnormal heart rhythms that may be seen include those with slower rates (junctional rhythms, heart blocks), those with faster rates (atrial, junctional, and ventricular tachycardia), and premature complexes (PACs, PJCs, and PVCs). Atrial flutter and fibrillation are least commonly seen with digoxin toxicity.

Other Medications

Class I (specifically type IA) antidysrhythmics are a group of drugs that include procainamide, quinidine, and disopyramide. These antidysrhythmic drugs slow both atrial and ventricular rates. They do not affect rhythm but prolong AV conduction. When patients receive these antidysrhythmic drugs, the ECG may show a slight widening of the QRS complex and a lengthening of the QT interval. The increase of the QT interval is concerning as it can increase the risk of ventricular tachydysrhythmias. The QT interval must be carefully monitored in all patients taking these medications and the drug withheld if significant—usually more than 25%—prolongation occurs. Class I antidysrhythmics can also produce AV blocks, a slowed or completely blocked SA node, and other dysrhythmias.

Quinidine

ECG indicators of quinidine use include a wide, notched P wave; widening of the QRS complex; and, often, ST depression with a prolonged QT interval. The presence of U waves is typical as well. U waves represent delayed repolarization of the ventricular conduction system. These do not require any adjustment in drug dosage.

Hypotension; tachycardia; widening of the QRS complex, QT interval, and PR interval; heart block; and heart failure can develop with quinidine toxicity. Episodes of torsades de pointes can also result.

Procainamide

Adverse cardiovascular effects of procainamide include bradycardia, tachycardia, hypotension, worsening heart failure, AV heart block, torsades de pointes, ventricular fibrillation, and asystole.

Calcium Channel and Beta Blockers

Calcium channel blockers primarily block the AV node, but the extent of block varies significantly within different drugs in this class. Beta blockers slow automaticity of the SA node and the Purkinje system and block the AV node. They can have a number of effects on the cardiovascular system, including bradycardia, hypotension, heart failure, chest pain, and palpitations.

Additional drugs that can increase the QT interval are other antidysrhythmic agents (e.g., amiodarone and dofetilide), the tricyclic antidepressants, the phenothiazines, erythromycin, and various antifungal medications.

Practice Makes Perfect

You can practice identifying the various conditions detailed in this chapter in the following *Practice Makes Perfect* exercises. Identify if pulmonary embolism, low-amplitude waveforms, electrical alternans, electrolyte imbalance, or digitalis use are present. Answers to these exercises can be found in Appendix A.

1.

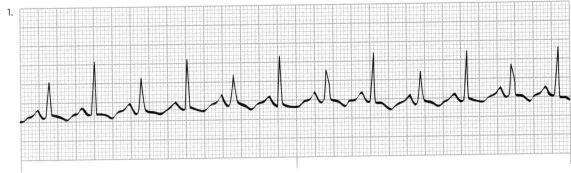

ECG Findings:

2.

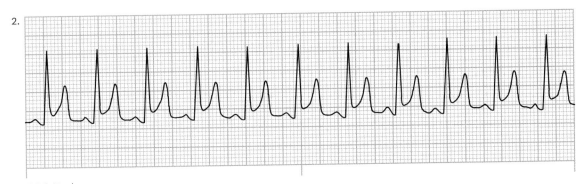

ECG Findings:

3.

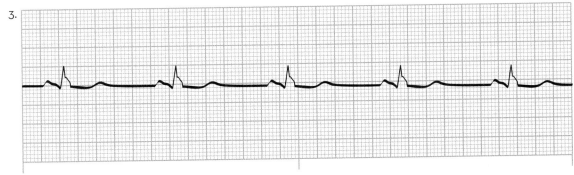

ECG Findings:

4.

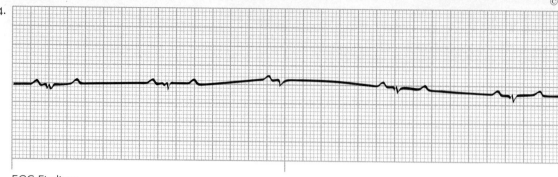

ECG Findings:

5.

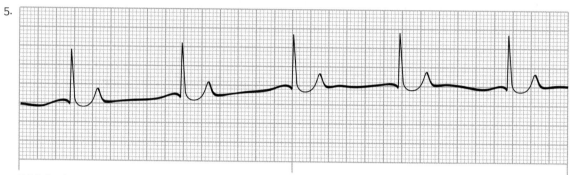

ECG Findings:

6.

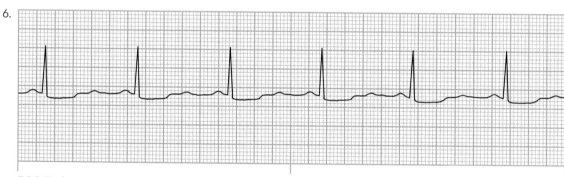

ECG Findings:

7.

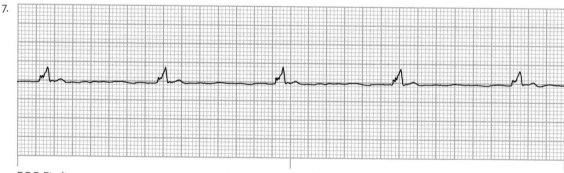

ECG Findings:

8.

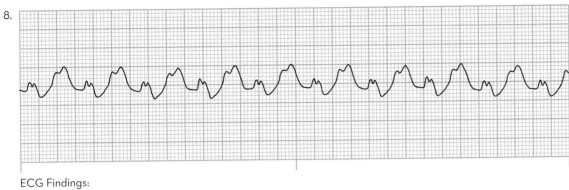

ECG Findings:

9.

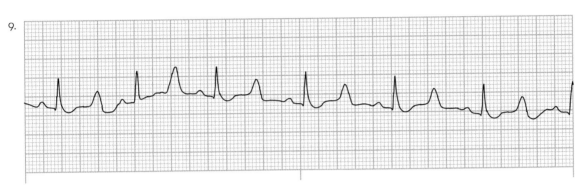

ECG Findings:

10.

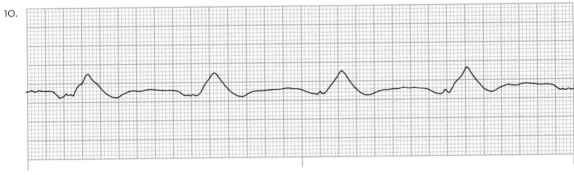

ECG Findings:

11.

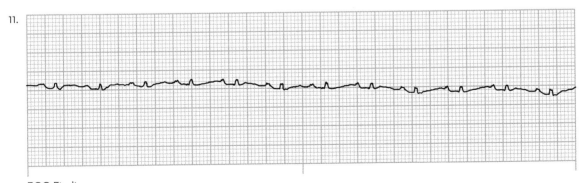

ECG Findings:

Key Points

LO 22.1	• Inflammation of the pericardium is called *pericarditis*. It can produce characteristic changes on the ECG. • In pericarditis, the T-wave is initially upright and may be elevated. During the recovery phase, it inverts. Also in pericarditis, the ST segment is elevated and usually flat or concave. If it is elevated, it appears off the baseline, gradually sloping back down to the next QRS complex.
LO 22.2	• A substantial pericardial effusion can occur with pericarditis. This can cause ECG changes, including low-voltage QRS complexes in all leads and electrical alternans, a condition in which the QRS complexes change in height with each successive beat.
LO 22.3	• An acute blockage of one of the pulmonary arteries that leads to obstruction of blood flow to the lung segment supplied by the artery is called a *pulmonary embolism*. • The characteristic ECG changes seen with massive pulmonary embolus include a large S wave in lead I, a deep Q wave in lead III, and an inverted T-wave in lead III. This is called the *S1 Q3 T3 pattern*.
LO 22.4	• Increases or decreases in the potassium and calcium serum levels can have a profound effect on the ECG. • Key characteristics of hyperkalemia include T-wave peaking, flattened P waves, 1st-degree AV heart block, widened QRS complexes, deepened S waves, merging of S and T-waves, and sine waves. • Key ECG characteristics of hypokalemia include ST segment depression, flattening of the T-wave, and appearance of U waves. • In hypocalcemia, the QT interval is slightly prolonged.
LO 22.5	• Digoxin slows the influx of sodium while allowing a greater influx of calcium. This increases myocardial contractility and improves the heart's pumping ability. It also slows the heart rate and AV conduction, making it useful in the treatment of fast atrial dysrhythmias. • A characteristic gradual downward curve of the ST segment is seen with digoxin. Further, the downward slope of the R wave gradually slurs as it curves downward into the ST segment. • Digoxin has a very narrow therapeutic margin and is excreted from the body slowly. Excessive digitalis levels can cause slower heart rates, faster heart rates, and PVCs. • When patients receive Class I antidysrhythmic drugs such as quinidine or procainamide, the ECG may show a slight widening of the QRS complex and a lengthening of the QT interval. The increase of the QT interval is concerning as it can increase the risk of ventricular tachydysrhythmias. • Calcium channel blockers primarily block the AV node, while beta blockers slow automaticity of the SA node and the Purkinje system and block the AV node.

Assess Your Understanding

The following questions give you a chance to assess your understanding of the material discussed in this chapter. The answers can be found in Appendix A.

1. Define the term "pericarditis." (LO 22.1)

2. ECG changes seen with pericarditis include (LO 22.1)
   a. a large Q wave.
   b. an elevated ST segment, usually flat or concave.
   c. right bundle branch block.
   d. a large S wave in lead I.

3. In comparison to MI, the ECG changes seen with pericarditis include (LO 22.1)
   a. diffuse ST segment and T-wave changes.
   b. T-wave inversion that usually occurs at the onset of the chest pain.
   c. development of a Q wave.
   d. the S1 Q3 T3 pattern.

4. Electrical alternans (LO 22.2)
   a. occurs with a small pericardial effusion.
   b. is a condition in which the electrical axis of the heart varies with each beat.
   c. is recognized on the ECG by QRS complexes, which change height every fifth or sixth beat.
   d. is diagnosed by analyzing just the width of the QRS complexes.

5. ECG changes seen with pulmonary embolism include (LO 22.3)
   a. right bundle branch block.
   b. the S1 Q3 T3 pattern.
   c. tall, symmetrically peaked P waves in leads II, III, and aV_F.
   d. All of the above.

6. Match the following ECG changes with the cardiac condition with which they are seen. (LO 22.3)

ECG Change	Condition
Peaked T-waves, widened QRS complexes, and sine waves _____	a. Hypocalcemia
Appearance of U waves, prolongation of the QT interval _____	b. Hyperkalemia
Prolonged QT interval _____	c. Hypokalemia
Shortened QT interval _____	d. Hypercalcemia

7. Describe how increases or decreases in the potassium and calcium serum levels effect the ECG. (LO 22.3)

8. Prolonged QT intervals are associated with   (LO 22.4)
    a.   atrial fibrillation.
    b.   complete AV heart block.
    c.   sinus bradycardia.
    d.   torsades de pointes.

9. ECG changes associated with digoxin use include a(n)   (LO 22.5)
    a.   longer QT interval.
    b.   elevated ST segment.
    c.   gradual downward curve of the ST segment.
    d.   shorter than normal PR interval.

10. The most common dysrhythmia seen with digoxin toxicity is   (LO 22.5)
    a.   3rd-degree AV heart block.
    b.   paroxysmal atrial tachycardia with 2nd-degree AV heart block.
    c.   junctional tachycardia.
    d.   torsades de pointes.

11. On the following illustrations draw an example of   (LO 22.4)
    a.   tall T-waves.
    b.   sine waves.
    c.   U waves.

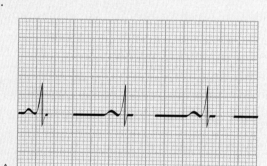

A

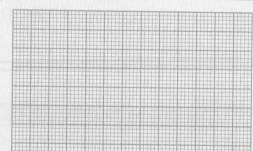

B

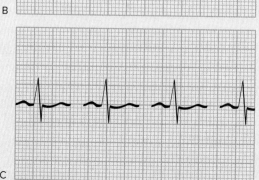

C

Referring to the scenario at the beginning of this chapter, answer the following questions.

12. What electrolyte abnormality is responsible for the patient's cardiac irregularity?   (LO 22.4)
    a. Hyponatremia
    b. Hypercalcemia
    c. Hyperkalemia
    d. Hypoglycemia

13. What is the characteristic T-wave configuration in this condition? (LO 22.4)
    a. Shaped like a tent
    b. Shaped like a mound
    c. Notched
    d. Round and wide

14. What other ECG changes can be expected with this condition?   (LO 22.4)
    a. Prolonged PR interval and widening of the QRS complex
    b. Shortened PR interval and widening of the QRS complex
    c. Prolonged P wave and inverted T-waves
    d. Prolonged PR interval and notching of the T-waves

section 5

Review and Assessment

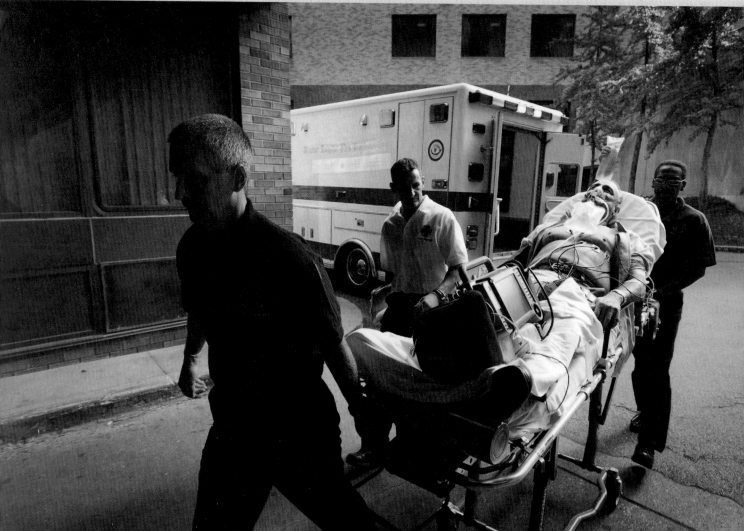

23 Putting It All Together

©rivetti/Getty Images

Courtesy Physio-Control

Learning Outcomes

LO 23.1 Recall how to use the Nine-Step Process and deductive analysis process to interpret ECG tracings and 12 leads.

LO 23.2 List those dysrhythmias that have a slow, normal, or fast heart rate.

LO 23.3 List those dysrhythmias that are regular or irregular. For those that are irregular, identify the types of irregularity with which they are associated.

LO 23.4 Identify the presence or absence of P waves, as well as how they appear with each dysrhythmia.

LO 23.5 Identify the presence or absence of QRS complexes, as well as how they appear with each dysrhythmia.

LO 23.6 Identify the presence or absence of PR intervals, as well as how they appear with each dysrhythmia.

LO 23.7 Recall the normal appearance of the ST segments and T waves and identify the conditions in which they can look different.

LO 23.8 Recall the normal duration of the QT intervals and identify the conditions in which they may be different.

LO 23.9 Identify when U waves may be present.

LO 23.10 Describe the benefits of analyzing ECG tracings using the deductive analysis process.

Case History

Paramedics are dispatched to a nursing home where a middle-aged man is complaining of chest pain. The man has a history of heart disease and hypertension. Upon their arrival, the EMS team is met by a nurse who has been talking with the patient. She tells you that the man was visiting his mother when he became pale and diaphoretic. When she asked if he was okay, he grasped his chest and said "I think I am having a heart attack." The nurse then checked for a pulse, which she says was of a normal rate but irregular. She reports the patient never stopped breathing or became cyanotic.

The paramedics introduce themselves and begin their assessment. Oxygen is applied and vital signs are taken, showing a blood pressure of 170/96, pulse of 120, irregular respirations of 16 per minute, and oxygen saturation of 94% on room air. They apply the cardiac monitor to the patient and see a rhythm that has a faster than normal rate, is irregular, and has narrow QRS complexes, which are each preceded by different looking P wave. One of the paramedics comments, "This looks like an interesting rhythm." His partner remarks, "It won't take long to figure this out; let's just rule out those things it can't be and that will tell us what it is."

The paramedics nod in agreement and begin analyzing the rhythm.

23.1 Using the Nine-Step Process and Deductive Analysis

As described throughout this book, each dysrhythmia and/or condition has certain characteristics. If you remember the characteristics that are associated with each and can identify which characteristics are present, you can determine what dysrhythmia and/or condition (if any) is present. That's why learning about normal and abnormal characteristics is so important for ECG analysis and interpretation.

In this chapter, following the Nine-Step Process, we group dysrhythmias and conditions with similar characteristics together (Figure 23-1). Then we show how to apply the principles of deductive analysis to eliminate those that do not fit the characteristics we see. Then last, we provide you with some examples of how to "put it all together."

23.2 Step 1: Heart Rate

Based on their heart rate, dysrhythmias can be grouped into one of three categories (Figure 23-2): they are either slow, normal, or fast. Some dysrhythmias may produce more than one heart rate or even all three.

Slow Rate

Slow heart rates are those that are less than 60 beats per minute. Dysrhythmias with slow heart rate include sinus bradycardia, junctional escape, and idioventricular rhythm. Several others may also have a slow heart rate including sinus arrest, sick sinus syndrome, 2nd degree AV heart block, and 3rd-degree AV heart block. Whereas the atrial heart in atrial flutter and fibrillation is faster than normal, the ventricular rate can be slow.

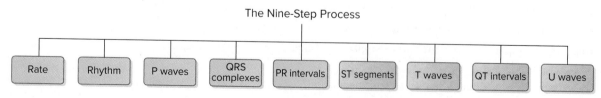

Figure 23-1
The Nine-Step Process.

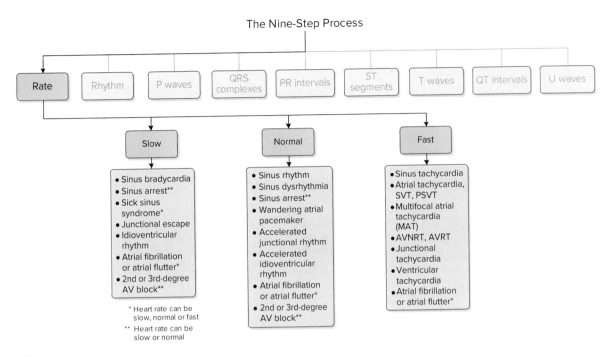

Figure 23-2
Heart rate algorithm.

Normal Rate

Dysrhythmias having a normal heart rate (between 60 and 100 beats per minute) include normal sinus rhythm, sinus dysrhythmia, wandering atrial pacemaker, accelerated junctional rhythm, and accelerated idioventricular rhythm. Note that accelerated idioventricular rhythm spans both slow and normal heart rates, with a rate of 40 to 100 beats per minute. Also, 2nd- and 3rd-degree AV heart blocks, sick sinus syndrome, and sinus arrest can be slow, but they can also have a normal heart rate. Atrial flutter and atrial fibrillation can also have a normal ventricular rate.

Fast Rate

Dysrhythmias having a fast heart rate (greater than 100 beats per minute) include sinus tachycardia, atrial tachycardia, multifocal atrial tachycardia, junctional tachycardia, AVNRT, AVRT, and ventricular tachycardia. Atrial flutter or fibrillation can also be fast.

23.3 Step 2: Regularity

Dysrhythmias will either be regular or irregular (Figure 23-3). As we have discussed, you can further characterize irregularity into the following groups: occasional or very, slight, sudden acceleration or slowing, patterned, total, or variable conduction ratio irregularity.

Regular Rhythms

Dysrhythmias having a regular rhythm include sinus rhythm, sinus bradycardia, sinus tachycardia, junctional escape, accelerated junctional rhythm, junctional tachycardia, idioventricular rhythm and accelerated idioventricular rhythm. Atrial flutter and 2nd degree AV heart block, type II, can also be regular. This occurs if

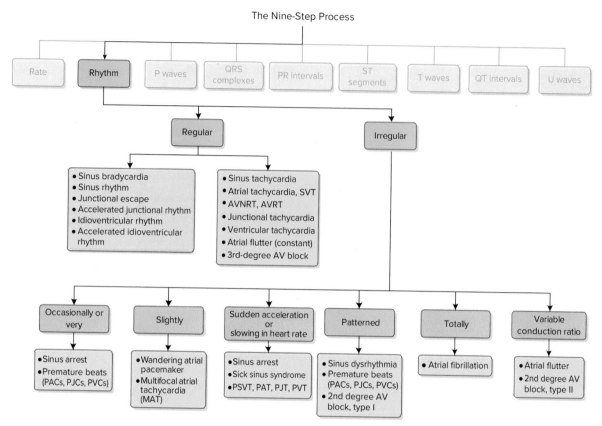

Figure 23-3
Irregularity algorithm.

the conduction ratio remains constant. In 3rd-degree AV heart block, the atrial rhythm is regular, as is the ventricular rhythm; however, the two are not related.

Irregular Rhythms

Irregular rhythms are rhythms in which the distance between the P waves and/or the QRS complexes change. They can be identified by how they appear and as described as follows.

Occasional or Very Irregularity

Sinus arrest and premature beats arising from the atria, AV junction, or ventricles (such as premature atrial complexes, premature junctional complexes, and premature ventricular complexes) can produce an occasional irregularity in the rhythm. Frequently occurring early beats produce a very irregular rhythm.

Slight Irregularity

Wandering atrial pacemaker is characteristically slightly irregular. Multifocal atrial tachycardia, a fast form of wandering atrial pacemaker, can be slightly irregular, but it can also be very irregular.

Sudden Acceleration or Slowing Irregularity

This type of irregularity is characteristically seen with either sinus arrest or paroxysmal tachycardia. You don't always see the sudden slowing or acceleration in the heart rate. When you do see a sudden onset of tachycardia, it is referred to as *paroxysmal*.

Patterned Irregularity

Sinus dysrhythmia, premature beats that occur in a bigeminal, trigeminal, or quadrigeminal pattern, and 2nd degree AV block, type I all produce a patterned irregularity. Sometimes it takes a longer tracing in order to effectively identify this type of irregularity.

Total Irregularity

Only one dysrhythmia, atrial fibrillation, is characterized as totally irregular. It is considered one of the key features of this dysrhythmia.

Variable Conduction Ratio Irregularity

Atrial flutter and 2nd degree AV heart block, type II can produce irregularity due to variable conduction ratio. It occurs when the number of impulses arising from the atria and reaching the ventricles changes.

23.4 Step 3: P Waves

Depending on where each dysrhythmia originates, P waves are either present or not present (Figure 23-4). If P waves are present, they are either normal or abnormal. Sometimes there are more P waves than QRS complexes. Also, with certain dysrhythmias, flutter or fibrillatory waveforms are present instead of P waves.

Normal P Waves

Normal P waves seen preceding each QRS complex occur with normal sinus rhythm, sinus bradycardia, sinus tachycardia, sinus dysrhythmia, and sinus arrest.

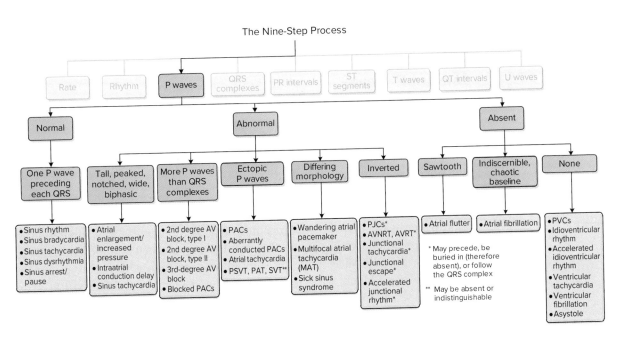

Figure 23-4
P wave algorithm.

Abnormal P Waves

Abnormal P waves are those that look different; are inverted, absent, or follow the QRS complex; or are not followed by a QRS complex. Also, there may be a pacemaker spike that precedes or appears instead of P waves.

Different-Looking Sinus P Waves

P waves that are tall and symmetrically peaked may be seen with increased right atrial pressure and right atrial dilation as well as with sinus tachycardia. Notched or wide (prolonged) P waves may be seen in increased left atrial pressure and left atrial dilation, as well as when there is a conduction delay or block in the interatrial conduction tract. While biphasic P waves may be normal in some ECG leads, they can also occur with both right and left atrial dilation. However, one or both portions of the biphasic P wave should be enlarged or wider than normal if they are the result of atrial enlargement.

Ectopic P′ Waves

The P wave also appears different when the pacemaker site originates from a site other than the SA node. Ectopic P′ waves are seen with premature atrial complexes, aberrantly conducted beats, and atrial tachycardia. With atrial ectopy, the P′ waves may be obscured or buried in the T wave of the preceding beat (resulting in a short P′-P interval). This causes the T wave to appear different than those following the other beats. The T wave may look peaked, notched, or larger than normal.

You may see an atrial P′ wave that is not followed by a QRS complex. This occurs with a nonconducted or blocked premature atrial complex and with atrial tachycardia with block. The appearance of either type of ectopy can be confused with AV heart block.

P′ Waves of Differing Morphology

P′ waves of differing morphology occur with wandering atrial pacemaker and multifocal atrial tachycardia. In these dysrhythmias, the pacemaker site shifts transiently from beat to beat from the SA node to other latent (hidden) pacemaker sites in the atria and AV junction.

Inverted P′ Waves

Inverted P′ waves occur when the electrical impulse travels upward through the AV junction into the atria, causing retrograde (backward) atrial depolarization. The four dysrhythmias in which inverted P′ waves are seen are premature junctional complexes (PJCs), junctional escape rhythm, accelerated junctional rhythm, and junctional tachycardia.

More P Waves than QRS Complexes

More P waves than QRS complexes indicate that the impulse was initiated in the SA node or other ectopic sites in the atria but was blocked and did not reach the ventricles. More P waves than QRS complexes are seen with both types of 2nd degree AV heart block, with 3rd-degree AV heart block, with blocked PACs, and with atrial tachycardia with block.

Sawtooth Waveforms

Sawtooth waveforms or flutter waves (F waves) that appear instead of P waves are seen with atrial flutter.

Indiscernible P Waves

A chaotic-looking baseline with no discernable P waves (in other words an absence of P waves) is seen with atrial fibrillation. Instead of P waves, we see uneven baseline of f waves.

Absent P Waves

With ventricular dysrhythmias, P waves are absent. They can also be absent with junctional rhythms (when the atrial and ventricular depolarization occur simultaneously). Absent P waves may also occur with supraventricular tachycardia (a fast heart rate with narrow QRS complexes).

23.5 Step 4: QRS Complexes

Depending on where each dysrhythmia originates, QRS complexes are either normal or abnormal (Figure 23-5). In certain conditions, they are absent.

Normal QRS Complexes

Depending the lead viewed, QRS complexes should appear normal (upright and narrow) in sinus rhythm, sinus bradycardia, sinus tachycardia, sinus dysrhythmia, sinus arrest/block, wandering atrial pacemaker, premature atrial complexes, atrial

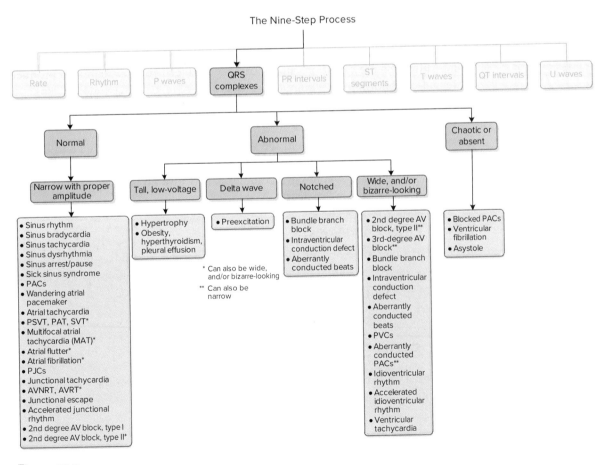

Figure 23-5
QRS complex algorithm.

tachycardia, atrial fibrillation, atrial flutter, junctional escape, accelerated junctional rhythm, premature atrial complexes, and junctional tachycardia.

Abnormal QRS Complexes

The morphology of the abnormal QRS complex can vary from being only slightly abnormal to being extremely wide, notched, or slurred. We refer to these extremely abnormal configurations as bizarre-looking because they deviate so far from normal. Abnormal QRS complexes can occur when the pacemaker is initiated from above the ventricles as well as when it is initiated from within the ventricles.

Tall and Low Amplitude QRS Complexes

Very tall QRS complexes are usually caused by hypertrophy of one or both ventricles or by an abnormal pacemaker or aberrantly conducted beat. Low-voltage or abnormally small QRS complexes may be seen in obese patients, hyperthyroid patients, and patients with pleural effusion.

Wide and/or Abnormal QRS Complexes of Supraventricular Origin

Wide, bizarre QRS complexes seen in supraventricular rhythms are often the result of an intraventricular conduction disturbance usually caused by RBBB or LBBB. With 2nd degree AV block, the QRS complexes may be wide if the block is infranodal.

Other causes of wide QRS complexes include MI; fibrosis; hypertrophy; electrolyte imbalance, such as hypokalemia and hyperkalemia; and excessive administration of cardiac drugs such as quinidine, procainamide, and flecainide. Aberrant conduction, ventricular preexcitation, and pacemaker-induced rhythms can also produce wide and/or abnormal QRS complexes.

Wide, Bizarre QRS Complexes of Ventricular Origin

Dysrhythmias that originate from the ventricular tissue produce wide and bizarre QRS complexes. This includes PVCs, idioventricular rhythm, and ventricular tachycardia. Wide, bizarre-looking QRS complexes may also be produced by an escape ventricular pacemaker in 3rd-degree AV block.

Absent QRS Complexes

Absent QRS complexes result from ventricular standstill and asystole. A chaotic wavy line, rising and falling without any logic, instead of the normal ECG waveforms is seen with ventricular fibrillation.

23.6 Step 5: PR Intervals

The PR interval is either present or absent (Figure 23-6). When the PR interval is present, it can be either normal or abnormal.

Normal PR Intervals

The PR interval is usually normal and constant with normal sinus rhythm, sinus bradycardia, sinus tachycardia, sinus dysrhythmia, sinus arrest, and 2nd degree

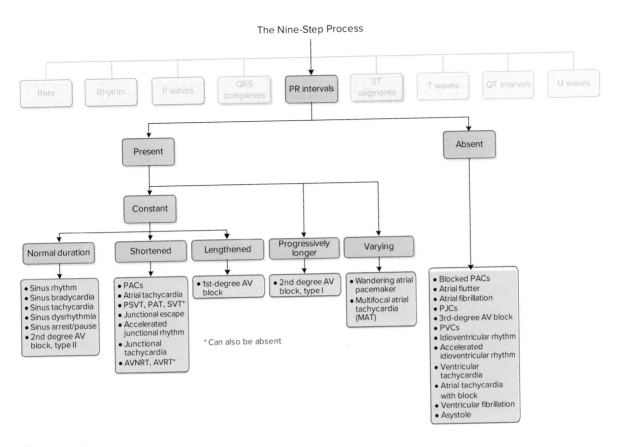

Figure 23-6
PR interval algorithm.

AV block. A PR interval within 0.12 to 0.20 seconds can also be present with some dysrhythmias that arise from the atria, provided the ectopic site is located in the upper-atrial wall.

Abnormal PR Intervals

Abnormal PR intervals are those that are shorter than 0.12 seconds, are longer than 0.20 seconds, are absent, or vary in duration.

Shorter PR Intervals

Shorter-than-normal PR intervals occur with premature beats that arise from the atria (PACs) or junction (PJCs), atrial or junctional tachycardia, and preexcitation.

Longer PR Intervals

Longer-than-normal PR intervals occur in 1st-degree AV heart block. The PR intervals seen in 1st-degree AV block are constant in duration.

Varying PR Intervals

The PR interval varies or changes in two dysrhythmias: wandering atrial pacemaker and 2nd degree AV heart block, type I.

Absent or Not Measurable PR Intervals

The PR interval is absent or not measurable in blocked PACs, atrial flutter, atrial fibrillation, PVCs, idioventricular rhythm, ventricular tachycardia, atrial tachycardia with blocks, and 3rd-degree AV block. It may also be absent or not measurable in PJCs, accelerated junctional rhythm, and junctional tachycardia.

23.7 Step 6: ST Segments

ST segments are either present or not present (Figure 23-7). If they are present, they are either normal or abnormal.

Normal ST Segments

Usually, normal ST segments are seen in sinus bradycardia, sinus rhythm, sinus tachycardia, sinus dysrhythmia, sinus arrest (for the conducted beats), wandering atrial pacemaker, PACs, PJCs, atrial tachycardia, multifocal atrial tachycardia PSVT, PAT, SVT, junctional escape, accelerated junctional rhythm, junctional tachycardia, 2nd degree AV heart block, 3rd-degree AV heart block, AVNRT, and AVRT.

Abnormal ST Segments

Abnormal ST segments are categorized as either elevated or depressed ST segments. Also, the ST segments may be indistinguishable in certain dysrhythmias. Lastly, in bundle branch block, the ST segments are discordant.

Elevated ST Segments

Elevated ST segments are seen in marked myocardial injury and may be an early indicator of MI. Other causes of elevated ST segment include bundle branch block, pericarditis, benigh early repolarization ventricular hypertrophy, ventricular fibrosis or aneurysm, hyperkalemia, and hypothermia.

Depressed ST Segments

A depressed ST segment is seen with myocardial ischemia or reciprocal changes opposite the area of myocardial injury. ST segment depression may also occur with right and left ventricular hypertrophy (strain pattern), right and left bundle branch block, pulmonary embolism, digitalis use, hypokalemia, hyperventilation, hypothermia, and stroke. ST segment depression that appears flat is seen in subendocardial infarction.

Indistinguishable ST Segments

The ST segment is typically indistinguishable in blocked PACs, atrial fibrillation, atrial flutter, AVNRT, AVRT, PVCs, idioventricular rhythm, accelerated idioventricular rhythm, ventricular tachycardia, ventricular fibrillation, and asystole. It may also be indistinguishable in PSVT.

23.8 Step 7: T Waves

T waves are either present or not present (Figure 23-8). If they are present, they are either normal or abnormal.

Normal T Waves

Usually, normal T waves are seen in sinus bradycardia, sinus rhythm, sinus tachycardia, sinus dysrhythmia, sinus arrest (for the conducted beats), wandering atrial

The Nine-Step Process

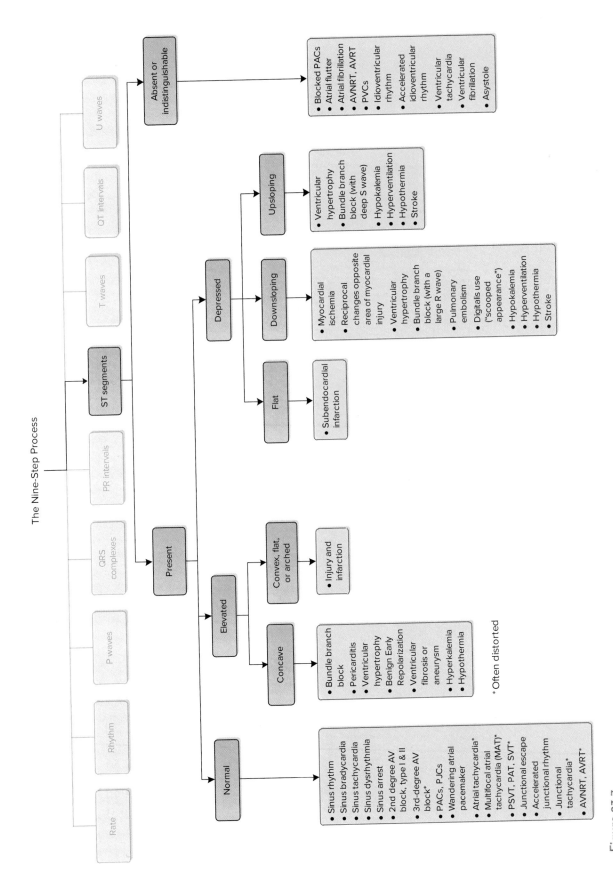

Figure 23-7
ST segment algorithm.

The Nine-Step Process

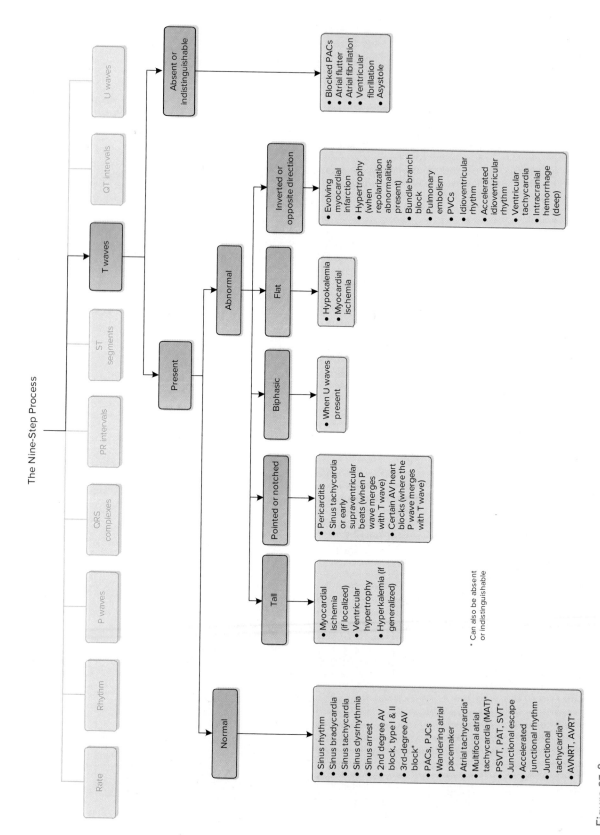

Figure 23-8
T wave algorithm.

pacemaker, PACs, PJCs, atrial tachycardia, multifocal atrial tachycardia, PSVT, PAT, SVT, junctional escape, accelerated junctional rhythm, junctional tachycardia, 2nd degree AV heart block, and 3rd-degree AV heart block. With atrial dysrhythmias, junctional dysrhythmias, and 3rd-degree AV block, the T wave may be distorted if the P wave is buried in the T wave.

Abnormal T Waves

Tall or peaked T waves, also known as *tented T waves*, are seen in myocardial ischemia, ventricular hypertrophy, and hyperkalemia. Heavily notched or pointed T waves in an adult may indicate pericarditis. Inverted T waves in leads where they are characteristically upright may be seen with myocardial ischemia. Discordant T waves are seen with bundle branch block. Deeply inverted T waves are seen with significant cerebral disease such as subarachnoid hemorrhage. Flat T waves are seen with certain electrolyte imbalances. T waves with bumps may be seen when P waves are buried in the T waves. This occurs with early beats that arise from the atria or AV junctional tissue or in certain types of AV heart block.

Also, T waves seen with impulses that arise from the ventricles take a direction opposite of their associated QRS complexes.

23.9 Step 8: QT Intervals

Unless there are no Q waves and/or T waves, the QT interval is present (Figure 23-9). It may be normal or abnormal.

Normal QT Intervals

The normal QT interval is 0.36 to 0.44 seconds in duration. Normal QT intervals are seen in normal sinus rhythm, sinus dysrhythmia, sinus bradycardia (which can also be prolonged), sinus tachycardia (which can also be shortened), sinus arrest (during conducted beats), PACs, atrial tachycardia, wandering atrial pacemaker, PJCs, junctional escape rhythm, accelerated junctional rhythm, junctional tachycardia, 1st-degree AV block, types I & II, 2nd degree AV block, and 3rd-degree AV block.

Abnormal QT Intervals

Abnormal QT intervals are either prolonged, shortened, or indiscernible.

Prolonged QT Intervals

One cause of prolonged QT intervals is congenital conduction system defect. Drugs such as haloperidol and methadone and Class IA antidysrhythmic drugs like amiodarone or sotalol can also produce QT interval prolongation. Moreover, prolonged QT intervals can be present in idioventricular rhythm, accelerated idioventricular rhythm, torsades de points, hypokalemia, and hypocalcemia.

Shortened QT Intervals

A shortened QT interval can be caused by hypercalcemia or digoxin toxicity.

Indiscernible or Not Measurable QT Intervals

Indiscernible QT intervals occur in multifocal atrial tachycardia, paroxysmal atrial tachycardia, atrial flutter, atrial fibrillation, PVCs, ventricular tachycardia, ventricular fibrillation, and asystole.

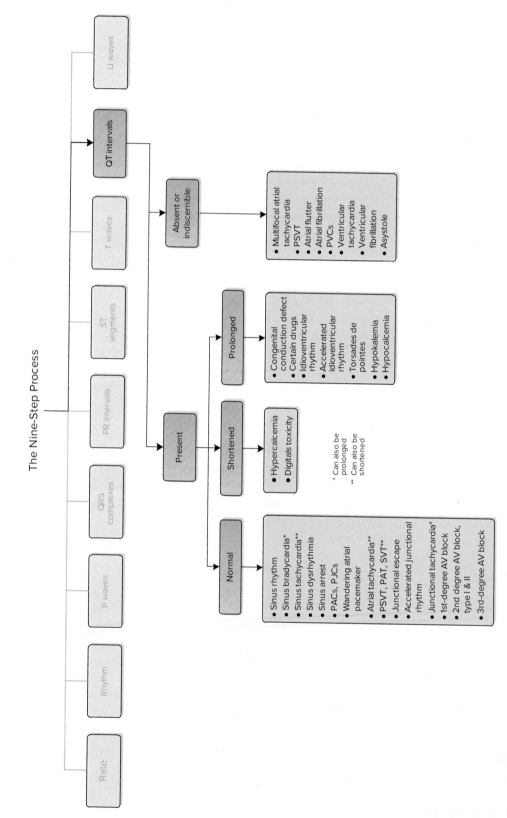

The Nine-Step Process

Rate | Rhythm | P waves | QRS complexes | PR intervals | ST segments | T waves | QT intervals | U waves

QT intervals

Present

Normal
- Sinus rhythm
- Sinus bradycardia*
- Sinus tachycardia**
- Sinus dysrhythmia
- Sinus arrest
- PACs, PJCs
- Wandering atrial pacemaker
- Atrial tachycardia**
- PSVT, PAT, SVT**
- Junctional escape
- Accelerated junctional rhythm
- Junctional tachycardia*
- 1st-degree AV block
- 2nd degree AV block, type I & II
- 3rd-degree AV block

Shortened
- Hypercalcemia
- Digitalis toxicity

Prolonged
- Congenital conduction defect
- Certain drugs
- Idioventricular rhythm
- Accelerated idioventricular rhythm
- Torsades de pointes
- Hypokalemia
- Hypocalcemia

Absent or indiscernible
- Multifocal atrial tachycardia
- PSVT
- Atrial flutter
- Atrial fibrillation
- PVCs
- Ventricular tachycardia
- Ventricular fibrillation
- Asystole

* Can also be prolonged
** Can also be shortened

Figure 23-9
QT interval algorithm.

23.10 Step 9: U Waves

U waves are either present or not present (Figure 23-10).

Prominent U waves are most often seen in hypokalemia but may be present in hypercalcemia, thyrotoxicosis, digitalis toxicity, use of epinephrine and Class 1A and Class 3 antidysrhythmics, mitral valve prolapse, and left ventricular hypertrophy, as well as in congenital long QT syndrome and with intracranial hemorrhage. An inverted U wave may represent myocardial ischemia or left ventricular volume overload.

23.11 So Why Do It?

From the practice tracings that follow, you can see how we are able to narrow down the possibilities when we identify what each problem might be. We start with the heart rate. Just putting it into a slow, normal, or fast group narrows down the field considerably. Then we are able to eliminate other possibilities when we identify the regularity of the tracing. This continues as we move through each step of the analysis process. Sometimes we can quickly identify the dysrhythmia or cardiac condition; at other times it requires us to go through all nine steps to reach a conclusion. Although it is not required that you use this process, you will find that it actually makes it easier to identify cardiac dysrhythmias and conditions.

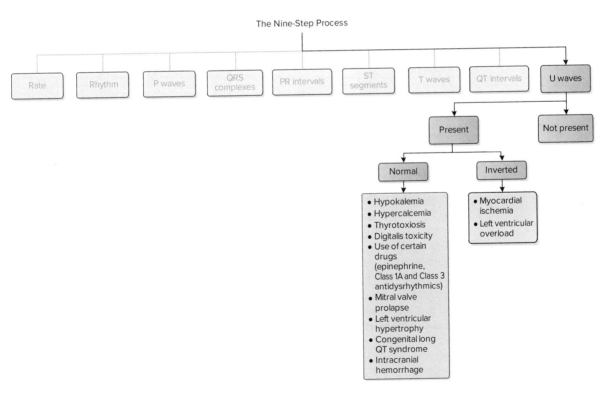

Figure 23-10
U wave algorithm.

Okay, let's analyze 10 ECG tracings. Be sure to apply the principles we learned throughout this book as well as what is described earlier in this chapter. For the dysrhythmias, the Five-Step Process is often sufficient. However, for the 12-lead ECGs, the Nine-Step Process as listed below is needed. Following each practice tracing is an analysis to which you can compare your findings.

1. Is the rate of this dysrhythmia slow, normal, or fast?

2. Is this rhythm regular or irregular? If it is irregular, what type of irregularity is it?

3. Are there P waves? What do they look like? Is each followed by a QRS complex?

4. Are the QRS complexes present? If so, what do they look like?

5. Are the PR intervals present? If so, what is their duration and is it constant?

6. Are the ST segments present? If so, are they normal or abnormal?

7. Are the T waves present? If so, are they normal or abnormal?

8. Are the QT intervals measurable? If so, are they normal or abnormal?

9. Are U waves present?

1.

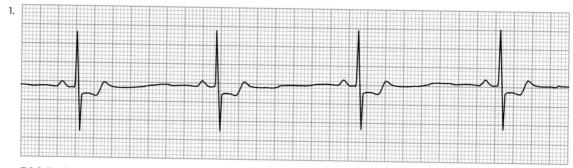

ECG Findings:

Analysis

If you said this rhythm is slow, you are correct. As such, it must be sinus bradycardia, junctional escape, idioventricular rhythm, sinus arrest, sick sinus syndrome, 2nd degree AV heart block, 3rd-degree AV heart block, atrial flutter, or atrial fibrillation. Next, if you said this rhythm is regular, you are correct again. Of the dysrhythmias listed above, only sinus bradycardia, junctional escape, idioventricular rhythm, and 3rd-degree AV block are regular. Atrial flutter can be regular if the conduction ratio remains constant. Because the rhythm is regular, it cannot be sinus arrest or atrial fibrillation. Next, if you said the dysrhythmia has a normal P wave preceding each QRS complex, you are correct. Of the remaining dysrhythmias, only sinus bradycardia has a normal P wave preceding each QRS complex. Thus, we can say with reasonable certainty that this is sinus bradycardia. To make sure, we evaluate the QRS complexes, which are normal, and the PR intervals, which are normal. Further, the ST segments are normal, the T waves are biphasic, and the QT intervals are normal. Lastly, there are no U waves. Therefore, we can conclude this is sinus bradycardia.

2.

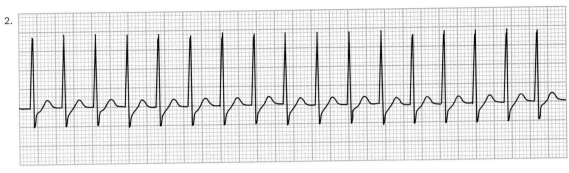

ECG Findings:

Analysis

If you said this dysrhythmia is fast, you are correct. As such, it must be sinus tachycardia, sick sinus syndrome, atrial tachycardia, multifocal atrial tachycardia, junctional tachycardia, SVT, AVRT, AVNRT, ventricular tachycardia, atrial flutter, or atrial fibrillation. Next, if you said the rhythm is regular, you are correct. Sinus tachycardia, junctional tachycardia, and ventricular tachycardia are regular. Atrial flutter can also be regular. Next, we look at the P waves; none are identifiable. Only junctional tachycardia and ventricular tachycardia have no P wave preceding each QRS complex, so this must be one of the two. Next, we evaluate the QRS complexes, which are normal. Of the two dysrhythmias it could be, only junctional tachycardia has narrow QRS complexes. Therefore, it is most likely junctional tachycardia. To further review our findings, we evaluate the PR interval and find none. Also, the ST segments, T waves, and QT intervals are normal, and there are no U waves. This confirms this rhythm is junctional tachycardia. Now, we also recognize that atrial tachycardia can display all the same characteristics, so we can also call this *supraventricular tachycardia.*

3.

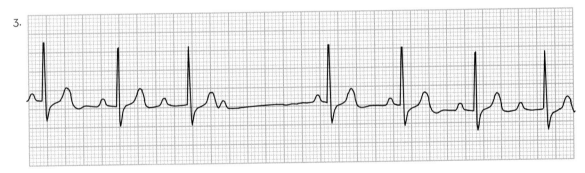

ECG Findings:

Analysis

Your answer is right if you said that this rhythm is slow. As such, it can only be sinus bradycardia, junctional escape, idioventricular rhythm, sinus arrest, sick sinus syndrome, 2nd degree AV heart block, 3rd-degree AV heart block, atrial flutter, or atrial fibrillation. Next, if you identified this rhythm as irregular, you are correct. Of the remaining dysrhythmias, we need to rule out those that are regular. Sinus node arrest, 2nd degree AV heart block, and atrial fibrillation are irregular. Atrial flutter can also be irregular. However, this rhythm has patterned irregularity. Only one of the remaining dysrhythmias, 2nd degree AV block, type I has patterned irregularity, so on the basis of that information, it is most likely that dysrhythmia. Next, we look at the P waves. Only one of the remaining dysrhythmias has upright, normal P waves that are not all followed by a QRS complex. Next, we evaluate the QRS complexes, which are normal. Last, we review the PR intervals. Here we see the characteristic that occurs with just one dysrhythmia, a progressively longer PR interval that leads to a dropped QRS complex. This cycle repeats itself over and over again. Further, the ST segments, T waves, and QT intervals are normal, and there are no U waves present. Therefore, this dysrhythmia can only be 2nd degree AV block, type I.

©rivetti/Getty Images

4.

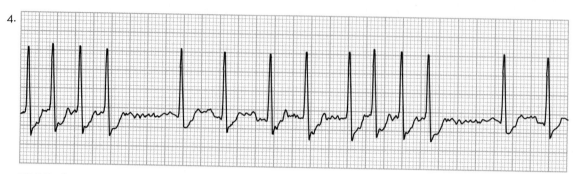

ECG Findings:

Analysis

If you said this dysrhythmia is fast, you are correct. As such, it must be sinus tachycardia, atrial tachycardia, multifocal atrial tachycardia, junctional tachycardia, ventricular tachycardia, atrial flutter, or atrial fibrillation. This rhythm is irregular, specifically a totally irregular rhythm. Of the dysrhythmias listed earlier, only atrial fibrillation is totally irregular. Next, look for the P waves to find that none are identifiable. Instead there is a chaotic baseline preceding each QRS complex. This is a characteristic of only one dysrhythmia, atrial fibrillation. Next, evaluate the QRS complexes. These are normal. Again, this is more evidence that this dysrhythmia is atrial fibrillation. Lastly, we look for PR intervals; we find they are absent. Also, the ST segments, T waves, and QT intervals are indiscernible, and there are no U waves. There is sufficient evidence to call this dysrhythmia *atrial fibrillation* (with a rapid ventricular response).

5.

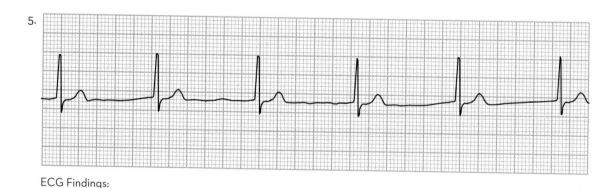

ECG Findings:

Analysis

You are correct if you said this rhythm is slow. As such, it can only be one of the following dysrhythmias: sinus bradycardia, junctional escape, idioventricular rhythm, sinus arrest, sick sinus syndrome, 2nd degree AV heart block, 3rd-degree AV heart block, atrial flutter, or atrial fibrillation. If you said this rhythm is regular, you are also correct. Of the listed dysrhythmias, those that are regular include sinus bradycardia, junctional escape, idioventricular rhythm, and 3rd-degree AV block (although, there is no correlation between the P waves and QRS complexes). Atrial flutter can also be regular. Next, if you said there is an absence of P waves, you are correct. Only two of the listed dysrhythmias have an absence of P wave preceding each QRS complex: junctional escape and idioventricular rhythm. Next, you are correct if you identified this dysrhythmia as having narrow QRS complexes. Of the remaining two dysrhythmias, only junctional escape has narrow QRS complexes. Thus, this dysrhythmia must be junctional escape. To make sure, we evaluate the PR intervals, which are absent. Further, the ST segments, T waves, and QT intervals are normal, and no U waves are present. Therefore, we can conclude this is junctional escape.

6.

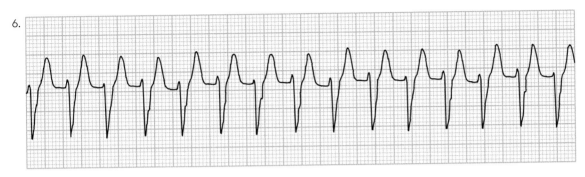

ECG Findings:

Analysis

If you said this rhythm is fast, you are correct. As such, it can only be sinus tachycardia, sick sinus syndrome, atrial tachycardia, multifocal atrial tachycardia, junctional tachycardia, ventricular tachycardia, atrial flutter, or atrial fibrillation. Also, this rhythm is regular. Of the listed dysrhythmias, only sinus tachycardia, junctional tachycardia, and idioventricular tachycardia are regular. Atrial flutter can also be regular. Next, we can see there are no identifiable P waves. Only junctional tachycardia and ventricular tachycardia have no identifiable P waves. Next, we evaluate the QRS complexes, which are wide and bizarre in appearance. Of the two dysrhythmias listed earlier, only ventricular tachycardia has wide, bizarre-looking QRS complexes. Therefore, this dysrhythmia is most likely ventricular tachycardia. A further review of our findings shows that there is no PR interval. Further, the ST segments and QT intervals are indistinguishable, the T waves take the opposite direction of the QRS complexes, and there are no U waves present. This rhythm then is ventricular tachycardia.

7.

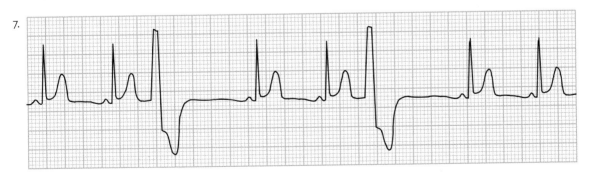

ECG Findings:

Analysis

If you said that this rhythm has a normal rate, you are correct. As such, it can only be sinus rhythm, sinus dysrhythmia, wandering atrial pacemaker, accelerated junctional rhythm, accelerated ventricular rhythm, sinus arrest, 2nd degree AV heart block, 3rd-degree AV heart block, atrial flutter, atrial fibrillation, or a paced rhythm. (These rhythms may or may not include ectopic beats.) Next, we can see this rhythm is irregular. More specifically, it is occasionally irregular. In looking at the irregularity, it is apparent there are two premature beats. Therefore, we need to identify the underlying rhythm and then the origin of the premature beats, in others words, dissect it a bit to see what we have. First, let's look closely at the underlying rhythm. There is one normal-looking P wave preceding each of the normal beats, and the QRS complexes and PR intervals are normal. Further, the distance between the normal beats is the same. Only one of the rhythms listed above has these characteristics. Our underlying rhythm is sinus rhythm. Next, let's look at our early beats. If you said there is an absence of P waves, the QRS complexes are wide and bizarre-looking, and there are no PR intervals with the early beats, you are correct. Remember what we learned earlier, if the QRS complexes are wide, it typically means they have arisen from the ventricles. The exception to this rule is if the impulse arises from the SA node, atria, or AV junction but conduction through the ventricles is abnormal (preexcitation, aberrant conduction, bundle branch block, etc.). Then the QRS complexes can appear wide and bizarre. In looking closer at the early beats, we see there are no P waves or PR intervals, so they can only be PVC. Thus, our dysrhythmia is sinus rhythm with two PVCs. Further, the ST segments, T waves, and QT intervals in the underlying rhythm are normal, and no U waves are present.

8.

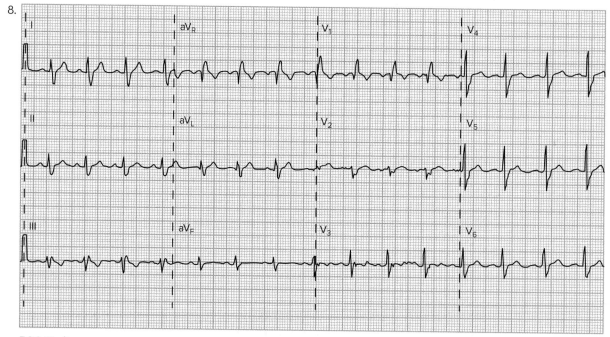

ECG Findings:

Analysis

The rate is 100 beats per minute, so it can be considered fast. Dysrhythmias having a fast rate include sinus tachycardia, sick sinus syndrome, atrial tachycardia, multifocal junctional tachycardia, ventricular tachycardia, atrial flutter, and atrial fibrillation. Whereas atrial tachycardia is also fast, we rule it out as atrial tachycardia normally has a rate of at least 160 beats per minute. The rhythm is regular. Of the listed dysrhythmias, those having a regular rhythm include sinus tachycardia, atrial tachycardia, junctional tachycardia, ventricular tachycardia, and atrial flutter (if the conduction ratio remains constant). There are sinus P waves preceding each QRS complex. However, the P waves are wide and notched in leads III, V_4, V_5, and V_6 and normal in the others. Of the dysrhythmias listed earlier, only one has a P wave: sinus tachycardia. However, the morphology of the P waves indicates left atrial enlargement. The QRS complexes are wider than normal with an rsR′ pattern in lead V_1 and a broad slurred S wave in leads I, aV_L, and V_6. This is characteristic of just one condition, RBBB. The PR intervals are within normal duration (0.14 seconds). The ST segments are normal, and the T waves are inverted in III and V_1 and flattened in lead aV_F. This patient has sinus tachycardia with left atrial enlargement and RBBB.

9.

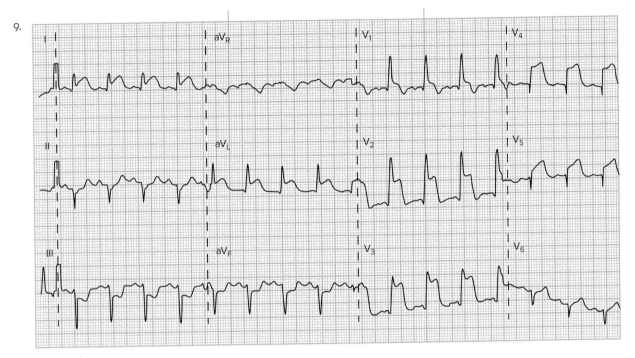

ECG Findings:

Analysis

At 135 beats per minute, this heart rate is fast. Dysrhythmias having a fast rate include sinus tachycardia, sick sinus syndrome, atrial tachycardia, multifocal junctional tachycardia, ventricular tachycardia, atrial flutter, and atrial fibrillation. Whereas atrial tachycardia is also fast, we rule it out because atrial tachycardia normally has a rate of at least 160 beats per minute. This rhythm is regular. Of the listed dysrhythmias, those having a regular rhythm include sinus tachycardia, atrial tachycardia, junctional tachycardia, ventricular tachycardia, and atrial flutter (if the conduction ratio remains constant). There are sinus P waves preceding each QRS complex. They are normal in most leads but biphasic in lead I. Of the dysrhythmias listed earlier, just sinus tachycardia has a sinus P wave preceding each QRS complex. The QRS complexes are narrow with elevated R waves in leads V_2 and V_3. This further supports this dysrhythmia being sinus tachycardia. The PR intervals are 0.18 seconds in duration. This allows us to rule out the presence of 1st-degree AV heart block. The ST segments are elevated (upsloping) in leads I, aV_L, and V_1 through V_6 and depressed in leads III and aV_F. There are Q waves forming in leads V_4 and V_5. The extensive anterior lateral ST segment elevation with reciprocal changes in inferior leads is characteristic of just one cardiac condition, acute MI. Further, no U waves are present. This patient has sinus tachycardia with acute anterolateral STEMI (myocardial injury).

10.

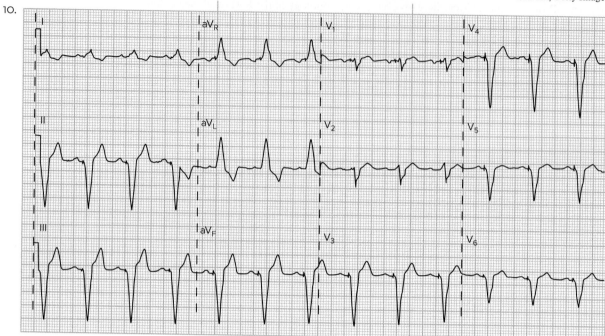

ECG Findings:

Analysis

The heart rate is 88 beats per minute, so it is normal. As such, it can only be sinus rhythm, sinus dysrhythmia, wandering atrial pacemaker, accelerated junctional rhythm, accelerated ventricular rhythm, sinus arrest, 2nd degree AV heart block, 3rd-degree AV heart block, atrial flutter, atrial fibrillation, or a paced rhythm. The rhythm is regular, so we can eliminate a couple of dysrhythmias from consideration. Whereas there are P waves, a more telling characteristic is the small narrow spike preceding each wide QRS complex. The only rhythm having this characteristic is a paced rhythm. Evaluating this rhythm further, we see the PR intervals are within normal duration (0.16 seconds). However, the PR interval is inconsequential because the ventricles are being stimulated to beat by the pacemaker. Along with the wide and bizarre-looking QRS complexes are T waves that take the opposite direction of the QRS complex. The ST segments and QT intervals cannot be identified, and there are no U waves. This patient has sinus rhythm with an AV sequential pacemaker with atrial sensing and ventricular pacing.

Key Points

LO 23.1	• If you remember the characteristics that are associated with each dysrhythmia and condition and can identify which characteristics are present, you can identify which dysrhythmia and/or condition (if any) is present.
LO 23.2	• Slow heart rates are seen with sinus bradycardia, junctional escape rhythm, idioventricular rhythm, AV heart block, and atrial flutter, or fibrillation with slow ventricular response. Sinus arrest and sick sinus syndrome may also have a slow rate.
	• Normal heart rate are seen in normal sinus rhythm, sinus dysrhythmia, wandering atrial pacemaker, accelerated junctional rhythm, and accelerated idioventricular rhythm (which spans across both slow and normal heart rates). Also, 2nd- and 3rd-degree AV heart blocks, sinus arrest, sick sinus syndrome, atrial flutter, and atrial fibrillation can have a normal (ventricular) heart rate.
	• Fast heart rates are seen with sinus tachycardia, atrial tachycardia, SVT, PSVT, junctional tachycardia, AVNRT, AVRT, ventricular tachycardia, and atrial flutter, or fibrillation with rapid ventricular response.
LO 23.3	• Regular rhythms are seen with sinus rhythm, sinus bradycardia, sinus tachycardia, junctional escape, accelerated junctional rhythm, junctional tachycardia, idioventricular rhythm, and accelerated idioventricular rhythm. Atrial flutter, 2nd degree AV heart block, type II, and 3rd-degree AV block can also be regular.
	• The only dysrhythmia characterized as being totally irregular is atrial fibrillation.
LO 23.4	• Normal P waves preceding each QRS complex are seen in normal sinus rhythm, sinus bradycardia, sinus tachycardia, sinus dysrhythmia, and sinus arrest.
	• In atrial fibrillation, there are no discernable P waves. Instead, there is a chaotic-looking baseline of f waves preceding the QRS complexes.
	• Impulses that arise from the AV junction produce an inverted P′ wave that may immediately precede, or occur during or following, the QRS complex.
	• Tall and symmetrically peaked P waves may be seen with increased right atrial pressure and right atrial dilation.
	• Notched or wide (enlarged) P waves may be seen in increased left atrial pressure and left atrial dilation.
	• In wandering atrial pacemaker, the P′ waves appear different, and the P′R intervals appear to vary.
	• More P waves than QRS complexes are characteristic of 2nd degree AV heart block (types I and II), 3rd-degree AV heart block, and blocked PACs.
LO 23.5	• Normal sinus rhythm and dysrhythmias that arise from above the ventricles will usually (unless there is a conduction delay through the ventricles or other type of abnormality as described earlier in the text) have normal QRS complexes.
	• Very tall QRS complexes are usually caused by hypertrophy of one or both ventricles or by an abnormal pacemaker or aberrantly conducted beat.
	• Low-voltage or abnormally small QRS complexes may be seen in obese patients, hyperthyroid patients, and patients with pleural effusion.

	• Wide, bizarre-looking QRS complexes with T waves that deflect in an opposite direction to the R waves are key characteristics seen with PVCs, idioventricular rhythm, accelerated idioventricular rhythm, and ventricular tachycardia.
	• 3rd-degree AV heart block is another dysrhythmia where there may be abnormal QRS complexes.
LO 23.6	• With type I, 2nd degree AV heart block, PR intervals are progressively longer until a QRS complex is dropped and then the cycle starts over.
	• There will be an absence of PR intervals in atrial flutter and fibrillation and in ventricular dysrhythmias.
	• In 3rd-degree AV heart block, the PR interval is absent.
	• In type II, 2nd degree AV heart block, the PR intervals associated with the P waves conducted through to the ventricles are constant.
LO 23.7	• The ST segment is normal in most dysrhythmias but typically indistinguishable in blocked PACs, atrial fibrillation, atrial flutter, AVNRT, AVRT, PVCs, idioventricular rhythm, accelerated idioventricular rhythm, ventricular tachycardia, ventricular fibrillation, and asystole. It may also be indistinguishable in PSVT.
	• Elevated ST segments are seen in myocardial injury and may be an early indicator of MI.
	• Depressed ST segments are seen with myocardial ischemia or reciprocal changes opposite the area of myocardial injury.
	• The ST segment is typically indistinguishable in blocked PACs, atrial fibrillation, atrial flutter, AVNRT, AVRT, PVCs, idioventricular rhythm, accelerated idioventricular rhythm, and ventricular tachycardia
	• In most dysrhythmias, the T waves are normal. With atrial dysrhythmias, junctional dysrhythmias, and 3rd-degree AV block, the T wave may be distorted if the P wave is buried in the T wave.
	• Tall or peaked T waves, also known as tented T waves, are seen in myocardial ischemia, ventricular hypertrophy, and hyperkalemia.
	• T waves seen with impulses that arise from the ventricles take a direction opposite of their associated QRS complexes.
LO 23.8	• The QT extends from the beginning of the QRS complex to the end of the T wave. Its duration is usually between 0.36 and 0.44 seconds.
	• In certain conditions, such as myocardial ischemia or infarction, a prolonged QT interval can predispose the patient to life-threatening ventricular dysrhythmias.
LO 23.9	• Prominent U waves are most often seen in hypokalemia.
LO 23.10	• Using an organized, systematic process to analyze ECG tracings allows us to narrow down the possibilities until we are left with a reasonable conclusion as to what is the dysrhythmia or condition.

Assess Your Understanding

The following questions give you a chance to assess your understanding of the material discussed in this chapter. The answers can be found in Appendix A.

1. A rhythm that arises from the SA node and has a heart rate of less than 60 beats per minute is called   (LO 23.2)
   a. junctional escape.
   b. idioventricular rhythm.
   c. sinus bradycardia.
   d. normal sinus rhythm.

2. Which of the following dysrhythmias has a normal heart rate?   (LO 23.2)
   a. Accelerated junctional rhythm
   b. Sinus tachycardia
   c. Idioventricular rhythm
   d. Junctional tachycardia

3. Sinus bradycardia has a/an _____ rhythm.   (LO 23.2)
   a. patterned irregular
   b. regular
   c. occasionally irregular
   d. variable conduction ratio

4. Which of the following produces a totally irregular rhythm?   (LO 23.3)
   a. Paroxysmal atrial tachycardia
   b. Wenckebach
   c. Atrial fibrillation
   d. Atrial flutter with varying conduction ratio

5. A patterned irregular rhythm is seen when   (LO 23.3)
   a. the heart rate suddenly accelerates.
   b. there are many early beats.
   c. initiation of the heartbeat changes from site to site with each heartbeat.
   d. premature complexes occur every other (second) complex, every third complex, or every fourth complex.

6. Which of the following has more P waves than QRS complexes?   (LO 23.4)
   a. A dysrhythmia in which the pacemaker changes from site to site with each beat
   b. Premature complexes that arise from the atria
   c. AV heart block
   d. Dysrhythmias that arise from the AV junction

7. Which of the following will produce P waves that look different than sinus P waves?   (LO 23.4)
   a. Enlarged or damaged atria
   b. 2nd degree AV heart block

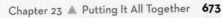

  c. Sinus dysrhythmia

  d. 3rd-degree AV heart block

8. Which of the following dysrhythmias will likely have narrow QRS complexes? (LO 23.5)

  a. Bundle branch block

  b. Idioventricular rhythm

  c. Paced rhythm

  d. Sinus tachycardia

9. Low-voltage or abnormally small QRS complexes are seen in (LO 23.5)

  a. hypertrophy.

  b. obese patients.

  c. premature ventricular complexes.

  d. sinus bradycardia.

10. Premature ventricular complexes (LO 23.5)

  a. arise from the atria.

  b. always have an inverted QRS complex.

  c. produce QRS complexes that look different than those that arise above or at the AV junction.

  d. have narrow, upright QRS complexes.

11. With 3rd-degree AV block, the QRS complexes (LO 23.5)

  a. are always wide and bizarre-looking.

  b. are associated with irregular and frequent PVCs.

  c. follow each P wave but at a slower rate.

  d. are slower than the P wave rate because there is complete blockage of the AV node.

12. Which of the following is characteristic of 1st-degree AV heart block? (LO 23.6)

  a. Varying PR intervals

  b. PR intervals of less than 0.12 seconds

  c. More P waves than QRS complexes

  d. Constant PR intervals of greater than 0.20 seconds

13. With wandering atrial pacemaker, the P′R intervals (LO 23.6)

  a. are constant.

  b. vary.

  c. are longer in duration than 0.20 seconds.

  d. become progressively longer.

14. The PR intervals with 2nd degree AV heart block, type I (LO 23.6)

  a. are constant.

  b. are absent.

  c. are shorter in duration than 0.12 seconds.

  d. become progressively longer.

15. There will be an absence of PR intervals in   (LO 23.6)
    a. premature atrial complexes.
    b. 1st-degree AV heart block.
    c. 3rd-degree AV heart block.
    d. wandering atrial pacemaker.

16. Type II, 2nd degree AV heart block has PR intervals that   (LO 23.6)
    a. continually change in duration.
    b. are constant.
    c. have no relationship.
    d. are less than 0.12 seconds in duration.

17. With atrial dysrhythmias, the   (LO 23.4)
    a. atrial waveforms differ in appearance from normal sinus P waves.
    b. P'R intervals are almost always prolonged.
    c. QRS complexes are wider than normal.
    d. site of origin is in the bundle of His.

18. Your patient is a 67-year-old female with a history of cardiac problems. After attaching her to the monitor, you see a slightly irregular rhythm with normal QRS complexes, but each P' wave is different. This rhythm is   (LO 23.3)
    a. frequent PACs.
    b. sinus arrest.
    c. sinus dysrhythmia.
    d. wandering atrial pacemaker

19. List the conditions that lead to ST segment depression.   (LO 23.7)

20. In which of the following dysrhythmias will the ST segment be indistinguishable?   (LO 23.7)
    a. Sinus bradycardia
    b. 3rd-degree AV block
    c. PVCs
    d. Junctional escape

21. Describe when ST elevation occurs during a myocardial infarction. (LO 23.7)

22. Describe when tall T waves are seen on the ECG.   (LO 23.7)

23. In which of the following dysrhythmias does the T wave deflect opposite of the QRS complex?   (LO 23.7)
    a. Atrial tachycardia
    b. Idioventricular rhythm
    c. 2nd degree AV block, type I
    d. Atrial fibrillation

24. List the conditions that cause a longer than normal QT interval.   (LO 23.8)

25. List the conditions in which a U wave is seen.   (LO 23.9)

26. Describe how to perform a systematic analysis of an ECG tracing. (LO 23.1)

27. Describe why a systematic analysis is used to analyze and interpret ECG tracings.   (LO 23.10)

Referring to the scenario at the beginning of this chapter, answer the following questions.

28. What is the origin of the narrow QRS complexes?   (LO 23.5)
    a. Supraventricular
    b. Ventricular

29. The underlying rhythm is most likely   (LO 23.4)
    a. multifocal atrial tachycardia.
    b. atrial flutter.
    c. idioventricular rhythm.
    d. sinus rhythm.

For each of the tracings on the following pages, practice the Nine-Step Process for analyzing ECGs. To achieve the greatest learning, you should practice assessing and interpretating the ECGs immediately after reading Chapter 23. Below are questions you should consider as you assess each tracing. Your answers can be written into the area below each ECG marked "ECG Findings." Your findings can be compared to the answers provided in Appendix A. All dysrhythmia tracings are 6 seconds in length.

1. Determine the heart rate. Is it slow? Normal? Fast? What is the ventricular rate? What is the atrial rate?

2. Determine if the rhythm is regular or irregular. If it is irregular, what type of irregularity is it? Occasional or frequent? Slight? Sudden acceleration or slowing in heart rate? Total? Patterned? Does it have a variable conduction ratio?

3. Determine if P waves are present. If so, how do they appear? Do they have normal height and duration? Are they tall? Notched? Wide? Biphasic? Of differing morphology? Inverted? One for each QRS complex? More than one preceding some or all the QRS complexes? Do they have a sawtooth appearance? An indiscernible chaotic baseline?

4. Determine if QRS complexes are present. If so, how do they appear? Narrow with proper amplitude? Tall? Low amplitude? Delta wave? Notched? Wide? Bizarre-looking? With chaotic waveforms?

5. Determine the presence of PR intervals. If present, how do they appear? Constant? Of normal duration? Shortened? Lengthened? Progressively longer? Varying?

6. Evaluate the ST segments. Do they have normal duration and position? Are they elevated? (If so, are they flat, concave, convex, arched?) Depressed? (If so, are they normal, flat, downsloping, or upsloping?)

7. Determine if T waves are present. If so, how do they appear? Of normal height and duration? Tall? Wide? Notched? Inverted?

8. Determine the presence of QT intervals. If present, what is their duration? Normal? Shortened? Prolonged?

9. Determine if U waves are present. If present, how do they appear? Of normal height and duration? Inverted?

10. Identify the rhythm, dysrhythmia, or condition.

1.

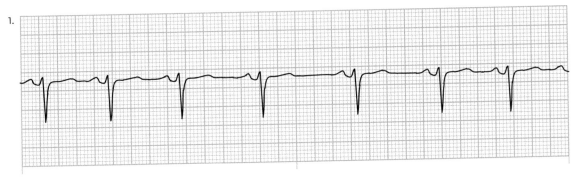

ECG Findings:

2.

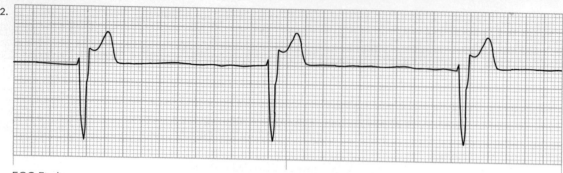

ECG Findings: _____

3.

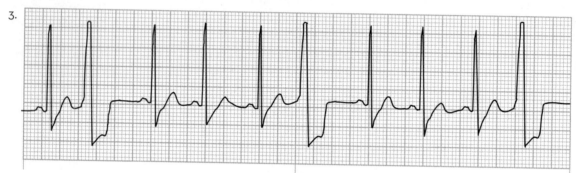

ECG Findings: _____

4.

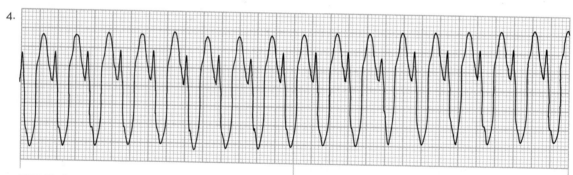

ECG Findings: _____

5.

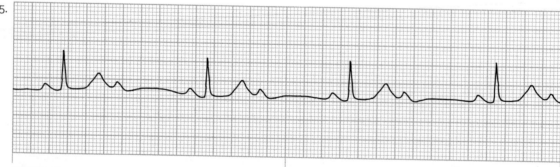

ECG Findings: _____

6.

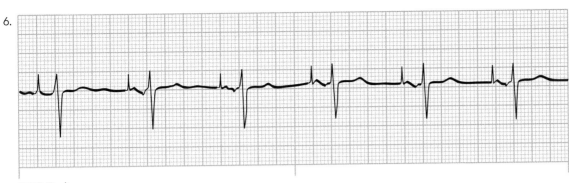

ECG Findings:

7.

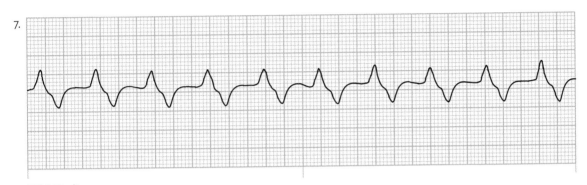

ECG Findings:

8.

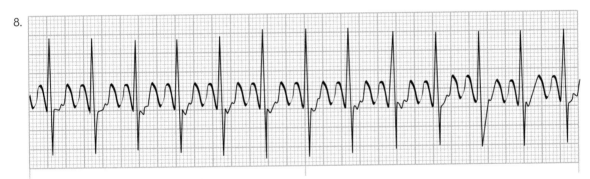

ECG Findings:

9.

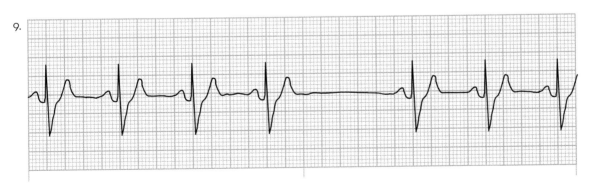

ECG Findings:

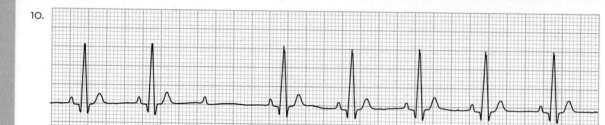

10.

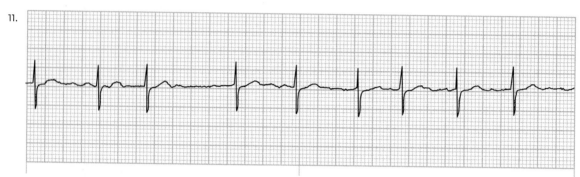

ECG Findings:

11.

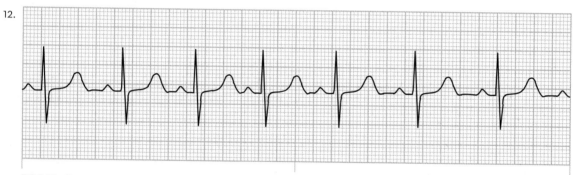

ECG Findings:

12.

ECG Findings:

13.

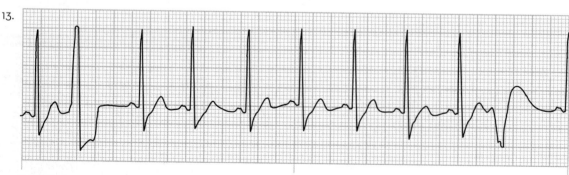

ECG Findings:

14.

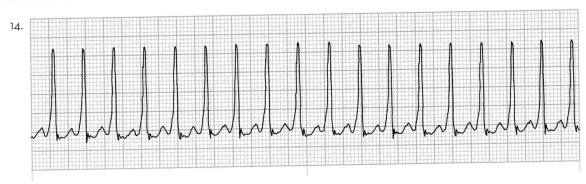

ECG Findings:

15.

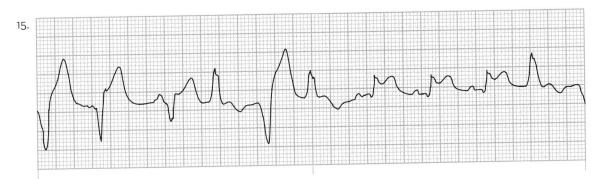

ECG Findings:

16.

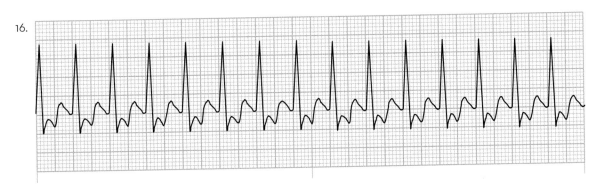

ECG Findings:

17.

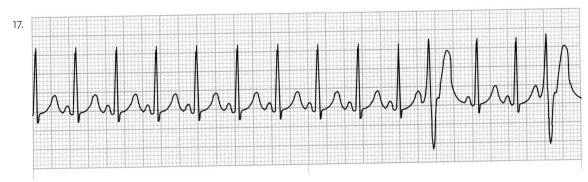

ECG Findings:

18.

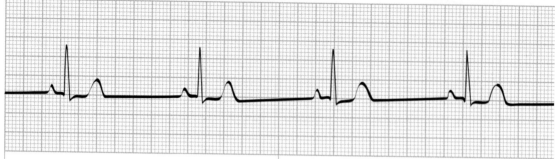

ECG Findings:

19.

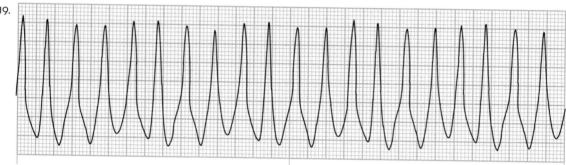

ECG Findings:

20.

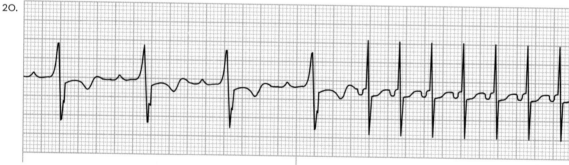

ECG Findings:

21.

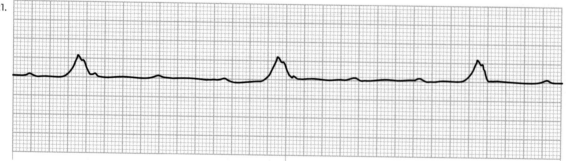

ECG Findings:

22.

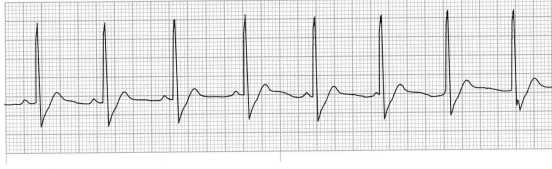

ECG Findings:

23.

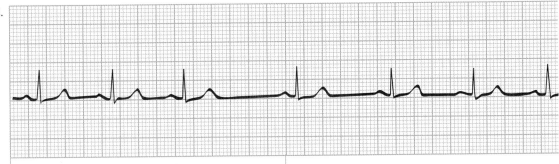

ECG Findings:

24.

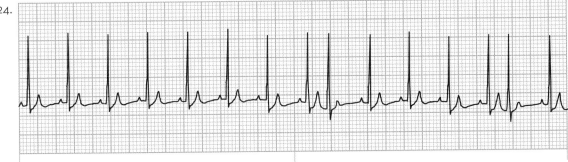

ECG Findings:

25.

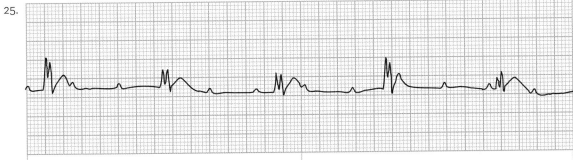

ECG Findings:

26.

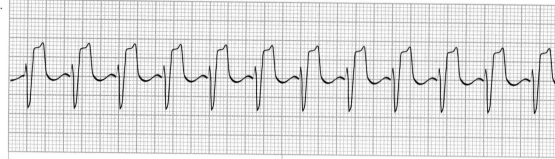

ECG Findings:

27.

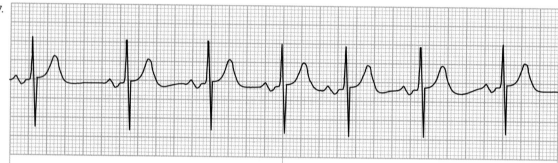

ECG Findings:

28.

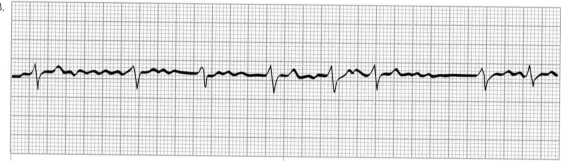

ECG Findings:

29.

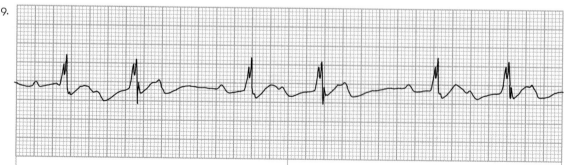

ECG Findings:

30.

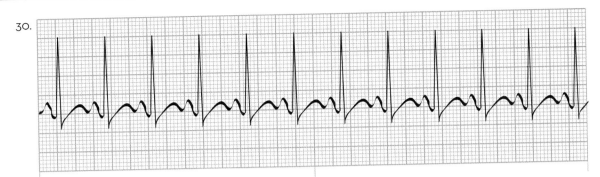

ECG Findings:

31.

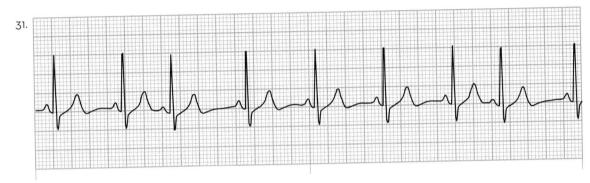

ECG Findings:

32.

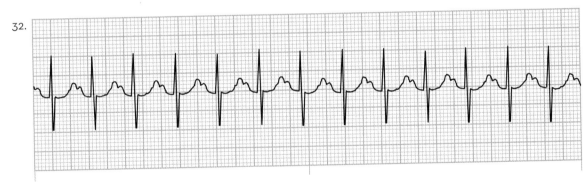

ECG Findings:

33.

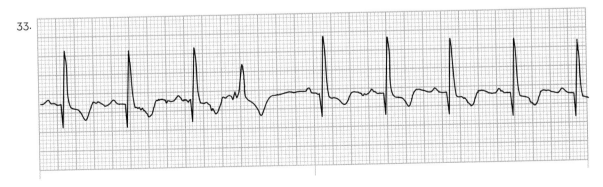

ECG Findings:

34.

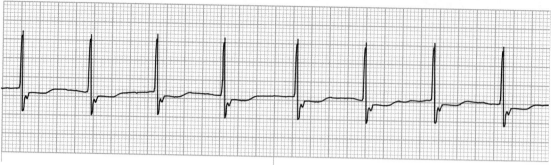

ECG Findings:

35.

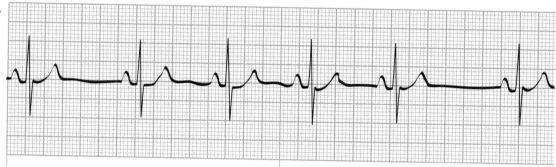

ECG Findings:

36.

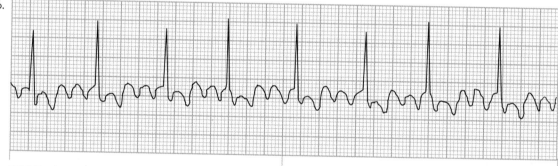

ECG Findings:

37.

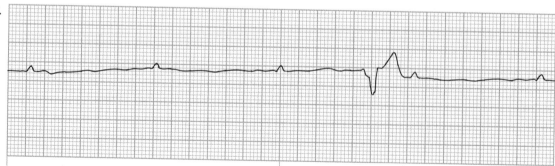

ECG Findings:

38.

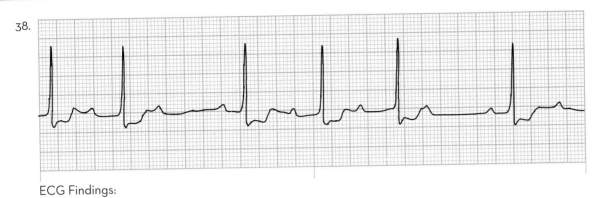

ECG Findings:

39.

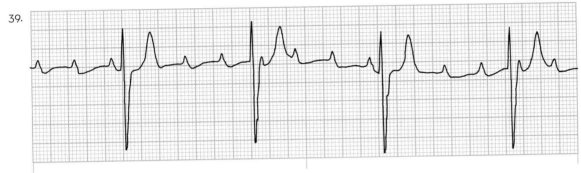

ECG Findings:

40.

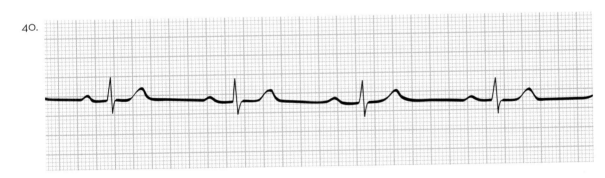

ECG Findings:

41.

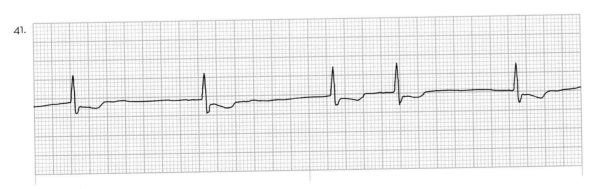

ECG Findings:

42.

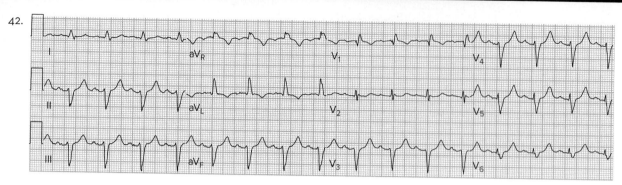

ECG Findings: _____

43.

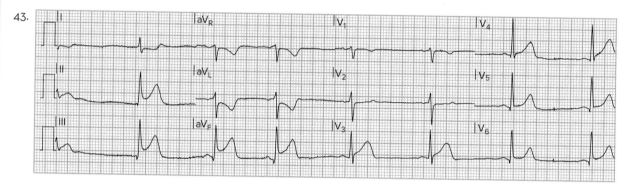

ECG Findings: _____

44.

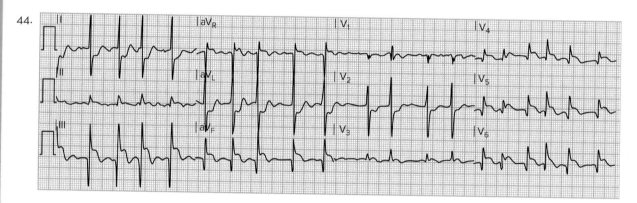

ECG Findings: _____

45.

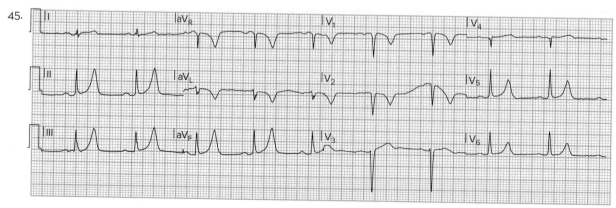

ECG Findings: _____

46.

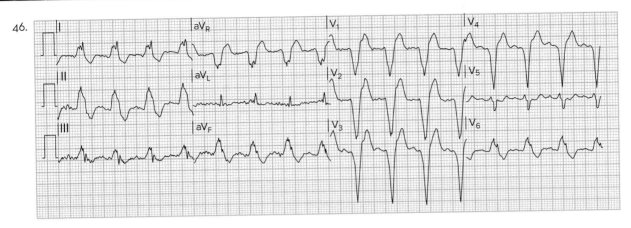

ECG Findings:

47.

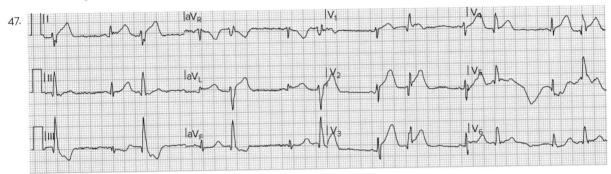

ECG Findings:

48.

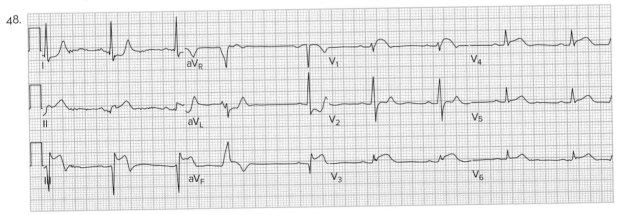

ECG Findings:

49.

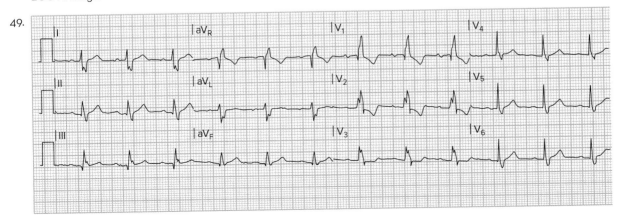

ECG Findings:

50.

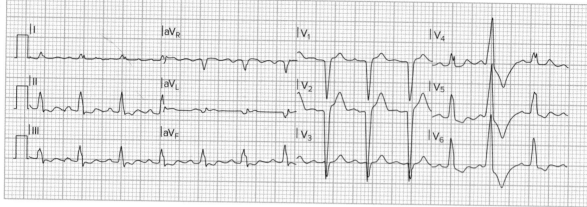

ECG Findings:

51.

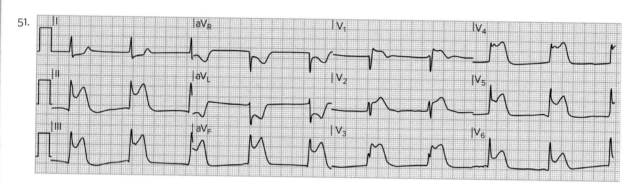

ECG Findings:

52.

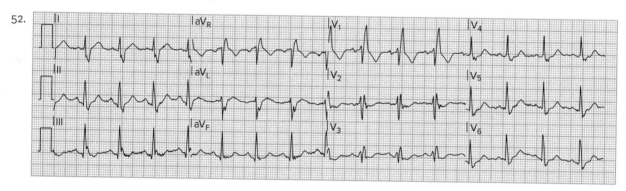

ECG Findings:

53.

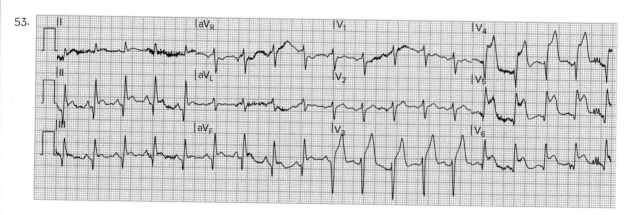

ECG Findings:

54.

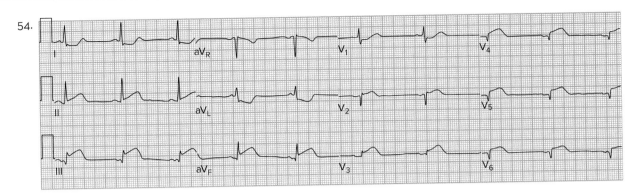

ECG Findings:

55.

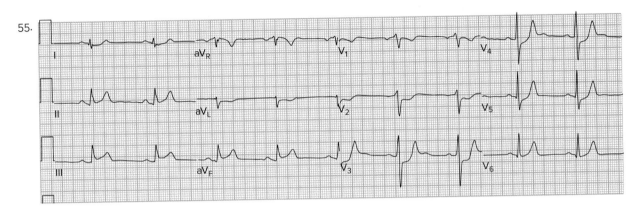

ECG Findings:

56.

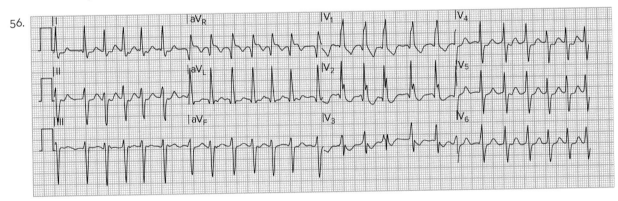

ECG Findings:

57.

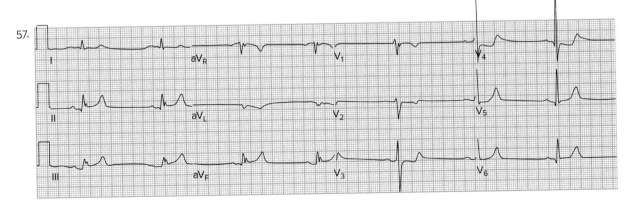

ECG Findings:

58.

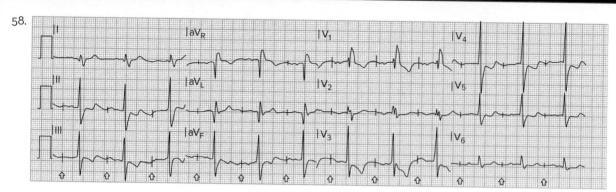

ECG Findings:

59.

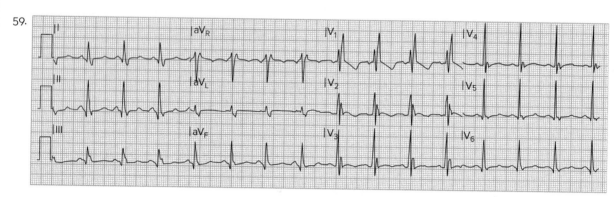

ECG Findings:

60.

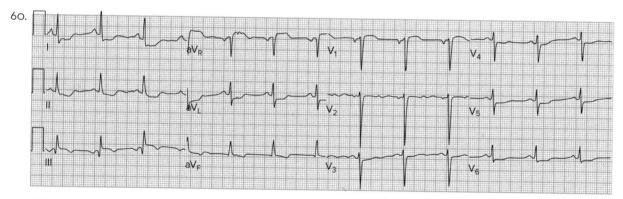

ECG Findings:

61.

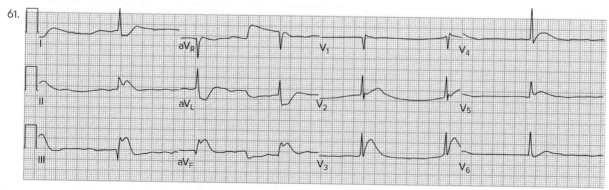

ECG Findings:

62.

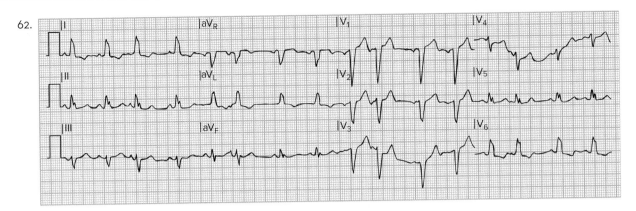

ECG Findings:

63.

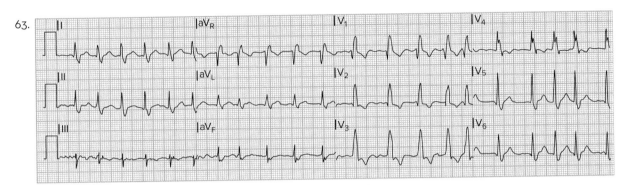

ECG Findings:

64.

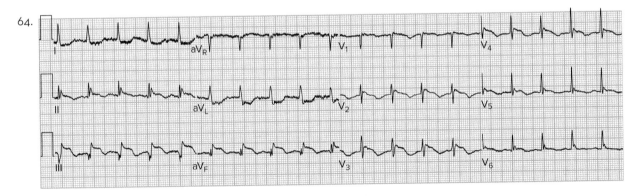

ECG Findings:

65.

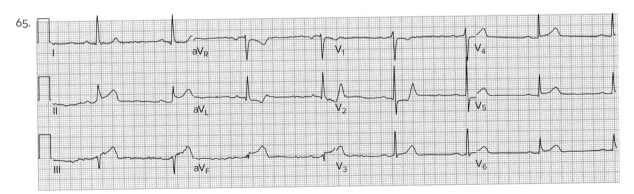

ECG Findings:

66.

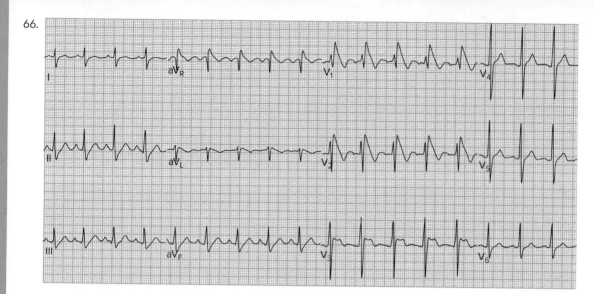

ECG Findings:

67.

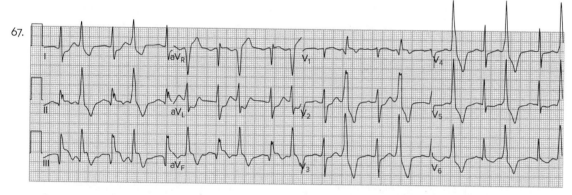

ECG Findings:

68.

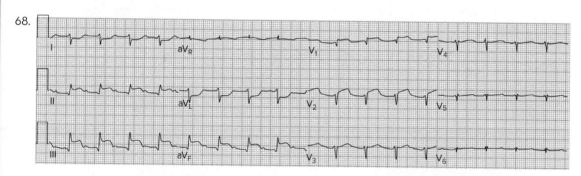

ECG Findings:

69.

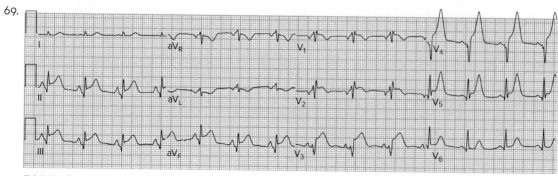

ECG Findings:

70.

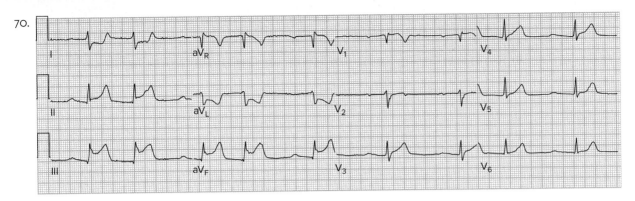

ECG Findings:

71.

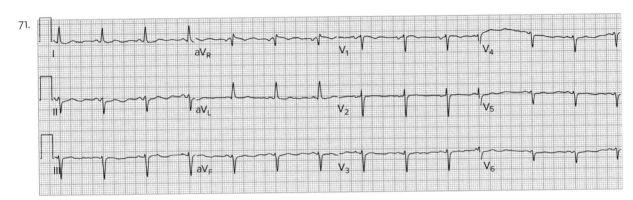

ECG Findings:

72.

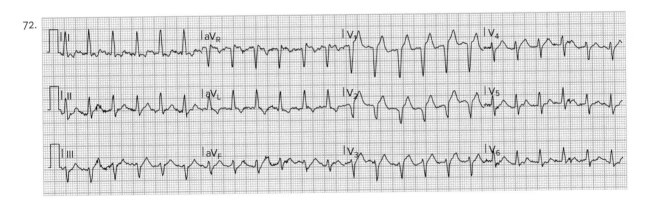

ECG Findings:

73.

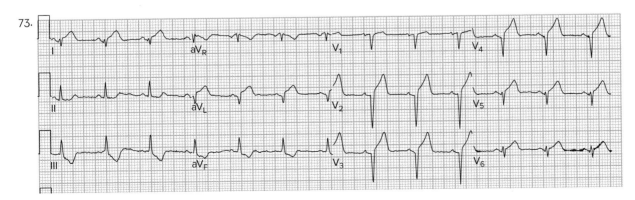

ECG Findings:

Answers

Assess Your Understanding

The following are answers to the **Assess Your Understanding** sections at the end of each chapter.

Chapter 1

1. d
2. b
3. d
4. b
5. c
6. The heart is shaped like an inverted blunt cone. Its top is the larger, flat part while its inferior end tapers to a blunt, rounded point.
7. d
8. a
9. b
10. a
11. The working cells contract to propel blood out of the heart's chambers. The pacemaker cells and electrical conducting cells generate and carry impulses throughout the heart. These cells differ from the working cells as they lack myofibrils and cannot contract.
12. The six structures of the heart's conduction system are the SA node, intraatrial conductive pathway and internodal pathways, AV node, bundle of His, right and left bundle branches, and Purkinje fibers.
13. a
14. a

15. b
16. The right coronary artery supplies the right ventricle with blood.
17. The left coronary artery, which branches into the left anterior descending and circumflex arteries, supplies the left ventricle.
18. d
19. d
20. d
21. b
22. c
23. Once the cell has returned to its polarized state, it can be stimulated again, but during the relative refractory period, a sufficiently strong stimulus will depolarize the myocardium. This aspect of depolarization will become important later when we discuss ventricular ectopy.
24. b
25. b
26. c
27. c
28. b
29. b
30. They are called the coronary arteries because they encircle the heart like a crown.

31. a
32. 60 to 100 beats per minute
33. b
34. d

Chapter 2

1. a
2. b
3. a
4. In the first phase, ventricular activation begins in the septum as it is depolarized from left to right. Early depolarization of the right ventricle also occurs. In the second phase, the right and left ventricular apex are depolarized, and the depolarization of the right ventricle is completed. In the third phase, the remainder of the left ventricle is depolarized toward the lateral wall.
5. d
6. a
7. Views of the heart obtained via ECG electrodes placed at specific locations on the body can be thought of as ECG leads.
8. a
9. Leads I, II, and III are referred to as the standard limb leads.
10. d

11. a

12. The leads that view the inferior portion of the left ventricle are II, III, and aV$_F$.

13. Lead V$_4$ is placed at the 5th intercostal space (between the fifth and sixth ribs) in the midclavicular line (the imaginary line that extends down from the midpoint of the clavicle).

14. Lead V$_1$ faces and is close to the right ventricle. It also has a view of the ventricular septum.

15. Lead V$_4$R is placed in the 5th intercostal space even with right midclavicular line.

16. d

17. b

18. c

19. b

20. a

21. a

22. b

23. d

24. c

Chapter 3

1. c

2. d

3. c

4. a

5. a

6. b

7. c

8. c

9. a

10. various dysrhythmias and cardiac conditions.

11. c

12. a

13. b

14. c

15. b

16. a registration or calibration mark.

17. c

18. With a regular rhythm, the distance between the consecutive P waves should be the same, just as the distance between the consecutive QRS complexes should be the same throughout the tracing.

19. A normal P wave arises from the isoelectric line, appearing as an upright and slightly asymmetrical waveform. There is one P wave preceding each QRS complex. Its amplitude is 0.5 to 2.5 mm and its duration is 0.06 to 0.10 seconds.

20. The T wave represents ventricular repolarization.

Chapter 4

1. d

2. a abnormal ECG rhythms.

3. c

4. a

5. d

6. a

7. c

8. b

9. d

10. d

11. 6

12. Cardiac output is a product of heart rate × stroke volume. Extremely fast heart rates reduce stroke volume by not allowing enough time between heartbeats for the ventricles (and coronary arteries) to refill with blood. A reduction in stroke volume can lead to decreased cardiac output.

13. d

14. a

Chapter 5

1. d

2. c

3. a

4. d

5. c

6. d

7. a

8. c

9. a

10. b

11. d

12. a

13. d

14. a

15. The six second × 10 method of calculating the heart rate can be used with irregular rhythms.

16. You know a heart rhythm is irregular when using either the calipers method or paper and pen method when the distances between the intervals differs; in other words, the distance from the location you identify as the starting point and the location where the second caliper point or pen mark falls is either before or after the second P wave or R wave in one or more P-P intervals or R-R intervals.

17. a

Chapter 6

1. b

2. b

3. d

4. d

5. c

6. b

7. a

8. c

9. c

10. b

11. a

12. a

13. b

14. d

15. c

16. A sinus P wave represents that the impulse originated in the SA node.

17. Impulses that originate outside the SA node, produce P waves (called P prime or P′) that look different than the sinus P waves.

18. Tall and symmetrically peaked P waves suggest increased right atrial pressure and right atrial enlargement. Notched or wide (prolonged) P waves indicate increased left atrial pressure and left atrial enlargement.

19. d

20. d

Chapter 7

1. QRS complex

2. b

3. c

4. The QRS is normally 0.08 to 0.12 seconds in duration.

5. d

6. a

7. c

8. c

9. c

10. b

11. c

12. a

13. d

14. c

15. a

16. b

17. d

18. b

19. d

20. a

21. c

Chapter 8

1. b

2. b

3. b

4. d

5. a

6. c

7. d

8. b

9. c

10. b

11. d

12. a

13. c

14. d

Chapter 9

1. The ST segment yields important information because it represents the end of ventricular depolarization and the beginning of ventricular repolarization. Elevation or depression of the ST segment is a hallmark feature of myocardial ischemia and injury. Assessing T waves is important as it represents the completion of ventricular repolarization (recovery). Abnormal T waves indicate the presence of abnormal ventricular repolarization. As an example tall, peaked, or inverted T waves can indicate myocardial ischemia. A prolonged QT interval indicates prolonged ventricular repolarization, which means the relative refractory period is longer. In certain conditions, such as myocardial ischemia or infarction, a prolonged QT interval can predispose the patient to life-threatening ventricular dysrhythmias such as torsades de pointes. A shortened QT interval can be caused by hypercalcemia or digoxin toxicity.

2. c

3. a

4. ST segment depression or elevation can be identified by comparing the position of the ST segment to baseline. The baseline is even horizontally with the TP or the PR segment. It is considered elevated if it is above the baseline and depressed if it is below it. From the baseline, draw a vertical line to the J point of the QRS complex. Then move over one small box (0.04 seconds) to the right, Then, count the number of small boxes between that point and the baseline. That is how much ST segment elevation or depression is present (if any).

5. d

6. a

7. c

8. b

9. During repolarization, positively charged ions, such as potassium, leave the cell, causing the positive charge to lower. Then the other positively charged ions, such as sodium, are removed by special transport systems, such as the sodium-potassium pumps, until the electrical potential inside the cell reaches its original negative charge.

10. c

11. Measure from where the T wave starts to leave or return to the baseline to the tip of the T wave.

12. d

13. The QT interval represents the time needed for ventricular depolarization and repolarization.

14. c

15. a

16. d

17. c

18. a

19. a

20. c

21. d

22. c

Chapter 10

1. Heart disease is defined as any medical condition of the heart or the blood vessels supplying it, or of the muscles, valves, or internal electrical pathways, that impairs cardiac functioning. Learning about these diseases will help you understand how the dysrhythmias and cardiac conditions that can be detected by the ECG occur.

2. b

3. **Hypertension** can lead to hardening and thickening of the arteries, which narrows the vessels through which blood flows.

4. a

5. Unrelieved stress may damage the arteries and worsen other risk factors for heart disease.

6. d

7. **With heart failure, while the heart is weak and unable to pump enough blood, it is more of a progressive condition. With cardiogenic shock, a suddenly weakened heart cannot pump enough blood to meet the body's needs.**

8. c

9. A pulmonary embolism is an acute blockage of one of the pulmonary arteries by a blood clot or other foreign matter. It leads to obstruction of blood flow to the lung segment supplied by the artery.

10. d

11. c

12. Atherosclerosis begins with small deposits of fatty material, particularly cholesterol, invading the intima, the inner lining of the arteries. The body has an inflammatory response to protect itself and sends white blood cells called macrophages to engulf the invading cholesterol in the arterial wall. This produces a non-obstructive lesion called a fatty streak, which is situated between the intima and media of the artery. Later, these areas of deposit become invaded by fibrous tissue, including lipoprotein-filled smooth muscle cells and collagen that become calcified and harden into plaque called a fibrous cap—hard on the outside but soft and mushy or sticky on the inside. This leads to narrowing of the affected vessels and a reduction of blood flow through them.

13. d

14. b

15. c

16. c

17. The inflammatory process associated with pericarditis can stimulate the body's immune response, resulting in white cells or serous, fibrous, purulent, and hemorrhagic exudates being sent to the injured area. This results in a buildup of an abnormal amount of fluid and/or a change in the character of the fluid in the pericardial space.

18. b

19. Aortic stenosis is a narrowing or blockage of the aortic valve that results from calcification and degeneration of the aortic leaflets. The changed shape of the leaflets reduces blood flow through the valve.

20. c

Chapter 11

1. b

2. Decreased cardiac output can decrease the blood pressure and perfusion of the body's cells, leaving the patient in a perilous condition; it may even become life threatening.

3. b

4. A dysrhythmia is an ECG rhythm that differs from normal sinus rhythm. More specifically, a dysrhythmia is a condition in which there is abnormal electrical activity in the heart. The heartbeat may be slower or faster than normal, it may be irregular, or conduction through the heart may be delayed or blocked.

5. d

6. b

7. b, c, d, a

8. b

9. d, c, a, b

10. b

11. a

12. b

13. d

14. b

15. a

16. c, d, b, a

17. c

18. a

19. b

Chapter 12

1. b

2. b

3. b

4. b

5. c

6. a

7. c

8. c

9. b

10. a

11. d

12. d

13. d

14. d, c, b, a

15. a

16. c

17. a

18. b

19. a

Chapter 13

1. a

2. It is important to check for a pulse with premature beats as they don't always produce effective contraction of the heart. In some cases, there is no pulse with a premature beat.

3. d

4. A PAC causes the ECG rhythm to be irregular.

5. c

6. d

7. c

8. a

9. Treatments for stable SVT include, aside from placing an IV, administering oxygen and attaching the ECG monitor, employing vagal manuevers, and administering 6mg of adenosine in the first dose and 12 mg in a second dose (if indicated). You should also consider the use of beta blockers or calcium channel blockers.

10. Treatments for unstable SVT include IV placement, oxygen administration, considering sedation and delivering synchronized cardioversion starting at 50 to 100 joules (for atrial tachycardia and atrial flutter), and using escalating energy levels until termination of the SVT. A trial use of vagal manuevers and/or a dose of adenosine may also be employed.

11. b

12. c

13. c

14. c

15. b

16. d

17. d

18. c

19. b

20. c

Chapter 14

1. c

2. d

3. c

4. b

5. b, a, c

6. b

7. b

8. c, a, b

9. c

10. d

11. d

12. b

13. c

14. b

15. b

Chapter 15

1. c

2. d

3. d

4. c, b, a, d

5. c

6. b

7. a

8. b

9. d

10. c

11. a

12. b

13. c

14. a

15. c

Chapter 16

1. a

2. a

3. c

4. b

5. a

6. c

7. a

8. d

9. c

10. b

11. b, d, c, a

12. c

13. b

14. a

Chapter 17

1. a

2. d

3. b

4. d

5. a

6. d

7. b

8. c

9. a

10. c

11. b

12. With a dual-chamber pacemaker, there will be an initial pacing spike followed by a P wave and a second pacing spike followed by a broad QRS complex.

13. a

14. d

15. a

16. d

Chapter 18

1. The 12-lead ECG can be used to identify various cardiac conditions as well as to differentiate between dysrhythmias where the origin is uncertain

(i.e., supraventricular vs. ventricular).

2. c

3. b

4. a

5. right arm, left arm, left leg, and right leg

6. a

7. d

8. c, e, a, b, d

9. b

10. To analyze a 12-lead ECG printout, we begin by looking at the left column from top to bottom, the middle from top to bottom, and the right from top to bottom.

11. b

12. d

13. a

14. c

15. a

16. The left ventricular vectors are larger and persist longer than those of the smaller right ventricle, primarily because of the greater thickness of the left ventricular wall.

17. b

18. a

19. c

20. d

21. c

22. b, c, a

23. a. normal; b. left axis deviation; c. right axis deviation; d. extreme axis deviation

24. Positive

25. Right is negative, left is positive

26. Top is negative, bottom is positive

27. a

28. b

29. Infarcted tissue cannot depolarize and therefore has no vectors. The vectors from the other side are unopposed vectors because of this, so the mean QRS vector tends to point away from the infarct.

Chapter 19

1. a

2. The heart constantly pumps blood to the body. Because of this, its oxygen consumption is proportionately greater than that of any other single organ. With the oxygen demand in the myocardial cells being so high, the heart must have its own blood supply.

3. b

4. c

5. d

6. b

7. c

8. b

9. 1. Changes in the T wave (peaking or inversion)

   2. Changes in the ST segment (depression or elevation)

   3. Enlarged Q waves or appearance of new Q waves

   4. A new or presumably new bundle branch block

   5. Reciprocal changes in leads facing the affected area from opposing angles

10. b

11. Myocardial injury produces ST segment elevation in the leads facing the affected area. The other conditions produce specific ECG changes as well.

12. b, c, a

13. a

14.

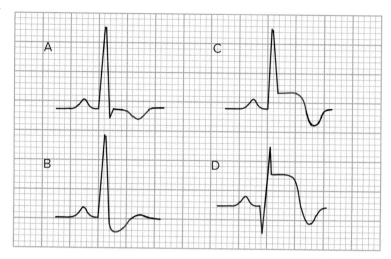

15. a

16. c

17. Tall, peaked T waves, depressed ST segments, and inverted T waves are signs of ischemia.

18. d

19. The presence of pathologic Q waves indicates myocardial infarction has occurred. For the Q waves to be considered pathologic, they must:

- Be greater than 0.04 seconds in duration or;
- Have a depth at least one-third the height of the R wave in the same QRS complex and;
- Be present in two or more contiguous leads (leads that look at the same area of the heart).

20. a

21. d, b, a, c

22. d

23. a

24. d

Chapter 20

1. Ventricular conduction disturbances involve delays in electrical conduction through either the AV node, the bundle branches (right or left), the fascicles (anterior, posterior), or a combination of these.

2. a

3. b

4. b

5. c

6. c

7. a

8. In bundle branch block the ST segment and T waves take an opposite direction to the QRS complex.

9. a

10. Appearance of QRS complexes and T waves in each of the leads listed below.

	I	aV_L	V_1	V_5	V_6
RBBB	Usually, wide, positively deflecting QRS complexes with broad (wide) terminal S waves	Usually, wide, positively deflecting QRS complexes with broad (wide) terminal S waves	Wide, positively deflecting, triphasic QRS complexes (with some variation of an rSR' configuration) with discordant ST segments and T waves	Usually, wide, positively deflecting QRS complexes with broad (wide) terminal S waves	Usually, wide, positively deflecting QRS complexes with broad (wide) terminal S waves
LBBB	Positively deflecting, wide and slurred or notched QRS complexes with discordant ST segments and T waves	Positively deflecting, wide and slurred or notched QRS complexes with discordant ST segments and T waves	Wide, slurred or notched negatively deflecting QRS complexes and discordant ST segments and T waves	Positively deflecting, wide and slurred or notched QRS complexes with discordant ST segments and T waves	Positively deflecting, wide and slurred or notched QRS complexes with discordant ST segments and T waves

11. Appearance of q waves, R waves, and S waves in each of the leads listed below.

	I	aV$_L$	II	III	aV$_F$
Left Anterior Fascicular Block	Small q waves and tall R waves	Small q waves and tall R waves	Small r waves and deep S waves	Small r waves and deep S waves	Small r waves and deep S waves
Left Posterior Fascicular Block	Small r waves and deep S waves	Small r waves and deep S waves	Small q waves and tall R waves	Small q waves and tall R waves	Small q waves and tall R waves

12. Brugada syndrome has a similar appearance to RBBB and is characterized by persistent ST segment elevation in leads V, V$_2$, and V$_3$. However, unlike RBBB, the rSR′ is not more than 0.12 seconds wide and there are no broad S waves in leads I and V$_6$.

13. The most recognizable feature of ARVD is that it looks like an incomplete RBBB pattern with a post-depolarization epsilon wave at the end of the QRS complex. The epsilon wave is described as a terminal notch in the QRS complex and is seen in approximately 40% of the cases. It is due to slowed intraventricular conduction and is best seen in leads V$_1$ and V$_2$. There is also T wave inversion in leads V$_1$, V$_2$, and V$_3$ and prolonged S-wave upstroke. There may also be localized QRS widening (0.08 to 0.11 seconds) in those same leads.

14. Bundle branch block generally requires no treatment other than monitoring the patient for progression to complete heart block. If the patient experiences complete heart block, then pacing is generally required to maintain adequate heart rate and cardiac output.

15. b

16. d

Chapter 21

1. ECG changes seen with hypertrophy include an increase in the amplitude of the R waves or S waves in given leads and axis deviation.

2. b

3. d

4. b

5. d

6. b

7. c

8. c

9. c

10. a

11. d

12. a

13. d

14. d

15. a

16. c

17. b

18. d

19. a

20. a

21. d, b, a, c

22. d

23. a

24. d

Chapter 22

1. Pericarditis is defined as inflammation of the pericardium.

2. b

3. a

4. b

5. d

6. b, c, a

7. Those involving potassium are the most immediately life threatening. Levels that are too high (hyperkalemia) or too low (hypokalemia) can quickly result in serious cardiac dysrhythmias. Hyperkalemia can generate a rapid progression of changes in the ECG that can end in ventricular fibrillation and death. Severe hypokalemia can lead to dysrhythmias and pulseless electrical activity (PEA) or asystole. Hypocalcemia prolongs the QT interval while hypercalcemia shortens it. Torsades de pointes, a variant of ventricular tachycardia, is seen in patients with prolonged QT intervals. Hypocalcemia also results in decreased cardiac contraction.

8. d

9. c

10. b

11.

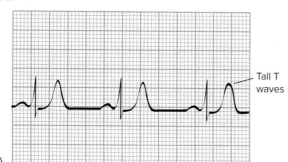

A — Tall T waves

— Sine wave pattern

B

— U wave

C

12. c

13. a

14. a

Chapter 23

1. c

2. a

3. b

4. c

5. d

6. c

7. a

8. d

9. b

10. c

11. d

12. d

13. b

14. d

15. c

16. b

17. a

18. d

19. Conditions that lead to ST segment depression include: myocardial ischemia or reciprocal changes opposite the area of myocardial injury, right and left ventricular hypertrophy (strain pattern), right and left bundle branch block, pulmonary embolism, digitalis use, hypokalemia, hyperventilation, hypothermia, and stroke.

20. c

21. ST elevation occurs during a myocardial infarction when injury to the heart muscle actually occurs.

22. Tall or peaked T waves, also known as *tented T waves*, are seen in myocardial ischemia, ventricular hypertrophy, and hyperkalemia.

23. b

24. A longer than normal QT interval is seen in congenital conduction system defect. Also, drugs such as haloperidol and methadone and Class IA antidysrhythmic drugs like amiodarone or sotalol can also produce QT interval prolongation. Lastly, prolonged QT intervals can be present in idioventricular rhythm, accelerated idioventricular rhythm, torsades de points, hypokalemia, and hypocalcemia.

25. Prominent U waves are most often seen in hypokalemia but may be present in hypercalcemia, thyrotoxicosis, digitalis toxicity, use of epinephrine and Class 1A and Class 3 antidysrhythmics, mitral valve prolapse, and left ventricular hypertrophy, as well as in congenital long QT syndrome and with intracranial hemorrhage.

26. It is useful to use an organized approach such as the Nine-Step Process described in this book. That way you are gathering

all the information needed to properly analyze and interpret each ECG tracing.

27. It is easy to leave out essential information or to miss key characteristics when not using a systematic approach. Whether it is the Nine-Step Process or some other assessment process, the key is to employ a systematic approach.

28. a

29. a

Practice Makes Perfect

Answers to the **Practice Makes Perfect** exercises. More in-depth answers for each can be found at the Fast & Easy ECGs Website at www.mhhe.com/shadeECG2e.

Section 2

1. 39 BPM, regular, normal P waves, normal QRS complexes, normal PR intervals, ST segment depression, biphasic T waves, prolonged QT interval.

2. Underlying rhythm, 94 BPM, patterned irregularity, normal P waves, normal QRS complexes, normal PR intervals, ST segment depression, normal T waves, and normal QT interval. Early beats wide and bizarre-looking QRS complexes not preceded by a P wave and T waves deflect opposite of QRS complexes.

3. Underlying rhythm, 60 BPM, irregular due to sudden onset of tachycardic rhythm, P waves present in underlying rhythm but not the tachycardia, QRS complexes normal throughout, normal PR intervals in underlying ST segment, depression in tachycardia, T waves normal in underlying but indiscernible in tachycardia. QT intervals normal in underlying, indeterminable in tachycardia.

4. 120 BPM, regular, normal P waves, normal QRS complexes, borderline PR intervals (at 0.20 seconds in duration), normal ST segment, normal T waves, short QT interval.

5. 140 BPM, totally irregular, no discernible P waves, narrow but abnormal QRS complexes, absent PR intervals, depressed ST segments and biphasic T waves where they can be seen, indiscernible QT intervals.

6. 48 BPM, regular, normal P waves marching through wide bizarre-looking QRS complexes, absent PR intervals, elevated ST segment and T wave, normal QT intervals.

7. 160 BPM, regular, absent P waves, narrow but abnormal QRS complexes, absent PR intervals, normal ST segment, normal T waves, short QT interval.

8. Underlying rhythm, 56 BPM, occasionally irregular, normal P waves, normal QRS complexes, normal PR intervals, ST segment depression, biphasic T waves, and normal QT interval. Two early beats have wide and bizarre-looking QRS complexes not preceded by a P wave and T waves deflect opposite of QRS complexes.

9. 130 BPM, regular, absent P waves, wide bizarre-looking QRS complexes, absent PR intervals, indiscernible ST segments, T waves deflect opposite QRS complexes, indiscernible QT intervals.

10. 70 BPM, patterned irregularity, upright normal P waves but not all are followed by a QRS complex, normal QRS complexes but some are dropped, cyclical progressively longer PR intervals, depressed ST segments, biphasic T waves, normal QT intervals.

11. 90 BPM, patterned irregularity, normal P waves, normal QRS complexes, normal PR intervals, normal ST segments, normal T waves, normal QT intervals.

12. 130 BPM, regular, tall P waves, QRS complexes have deep Q wave, normal PR intervals, normal ST segments, normal T waves, and short QT intervals.

13. 45 BPM, regular, normal P waves, biphasic QRS complexes, normal PR intervals, normal ST segment, inverted T waves, short QT intervals.

Chapter 12

1. Sinus rhythm 86 BPM, regular rhythm, normal P waves, QRS complexes have

notched R wave, deep S wave, normal PR intervals, normal ST segments, normal T waves, and normal QT interval

2. Sinus arrest with sinus escape Underlying rhythm is 79 BPM (pause slow heart rate), irregular (due to pause), normal P waves, tall QRS complexes, normal PR intervals, depressed ST segments, inverted T waves, and normal QT interval

3. Sinus bradycardia 39 BPM, regular rhythm, normal P waves, normal QRS complexes, normal PR intervals, slightly depressed - upward sloping ST segments, normal, biphasic T waves, and normal QT interval

4. Sinus dysrhythmia 80 BPM, irregular rhythm (patterned), normal P waves, normal QRS complexes, normal PR intervals, slightly depressed - upward sloping ST segments, normal T waves, and normal QT interval (although when it speeds up the QTI shortens)

5. Sinus tachycardia 115 BPM, regular rhythm, normal P waves, normal QRS complexes, normal PR intervals, normal ST segments, normal T waves, and shortened QT interval (0.28 seconds)

6. Sinus rhythm with intraventricular conduction defect (IVD) 88 BPM, regular rhythm, biphasic P waves, wide rS complexes, normal PR intervals, indistinguishable ST segments, tall wide T waves, and normal QT intervals

7. Sinus dysrhythmia 60 BPM, irregular rhythm (patterned),

tall P waves, RS complexes, normal PR intervals, slightly depressed - normal ST segments, normal T waves, and normal QT interval

8. Sinus bradycardia
9. Sinus rhythm
10. Sinus rhythm with IVD
11. Sinus dysrhythmia
12. Sinus bradycardia
13. Sinus rhythm/borderline sinus tachycardia
14. Sinus tachycardia
15. Sinus dysrhythmia
16. Sinus tachycardia
17. Sinus bradycardia (severe)
18. Sinus tachycardia with IVD
19. Sinus bradycardia with IVD
20. Sinus rhythm/borderline sinus bradycardia
21. Sinus dysrhythmia
22. Sinus rhythm
23. Sinus bradycardia
24. Sinus tachycardia
25. Sinus tachycardia with IVD
26. Sinus pause with sinus escape
27. Sick sinus syndrome
28. Sinus tachycardia with IVD
29. Sinus rhythm with IVD
30. Sinus tachycardia
31. Sinus pause with sinus escape
32. Sinus tachycardia with elevated ST segments
33. Sinus rhythm/borderline sinus tachycardia
34. Sinus bradycardia
35. Sinus rhythm
36. Sinus bradycardia
37. Sinus tachycardia
38. Sinus bradycardia

39. Sinus bradycardia
40. Sinus arrest with sinus escape
41. Sinus dysrhythmia
42. Sinus rhythm with ST elevation
43. Sinus tachycardia
44. Sinus arrest with junctional escape
45. Sinus tachycardia with IVD
46. Sinus dysrhythmia
47. Sinus bradycardia
48. Sinus tachycardia
49. Sinus arrest with junctional escape
50. Sinus rhythm
51. Sick sinus syndrome
52. Sick sinus syndrome
53. Sinus tachycardia
54. Sinus pause (2 episodes)
55. Sinus bradycardia
56. Sick sinus syndrome
57. Sinus tachycardia with artifact

Chapter 13

1. Sinus rhythm with one PAC
2. Wandering atrial pacemaker
3. Atrial fibrillation
4. Atrial flutter (2 to 1 conduction ratio)
5. Atrial tachycardia
6. Sinus rhythm with two PACs
7. Atrial fibrillation with rapid ventricular response (uncontrolled)
8. Atrial flutter
9. Sinus rhythm with two PACs
10. Sinus rhythm with PSVT (sustained)
11. Sinus rhythm with two PACs
12. Supraventricular tachycardia

13. Atrial fibrillation with rapid ventricular response (uncontrolled)

14. Sinus tachycardia with one PAC

15. Atrial tachycardia

16. Atrial flutter (with variable conduction ratio)

17. Sinus rhythm with bigeminal PACs

18. Atrial fibrillation/flutter

19. Wandering atrial pacemaker

20. Sinus rhythm with trigeminal aberrantly conducted PACs

21. Multifocal atrial tachycardia

22. Atrial fibrillation

23. Sinus tachycardia with two PACs

24. Atrial fibrillation

25. Sinus rhythm with one PAC

26. Bradycardic wandering atrial pacemaker

27. Atrial fibrillation with rapid ventricular response (uncontrolled)

28. Sinus rhythm with bigeminal PACs

29. Atrial fibrillation

30. Sinus rhythm with one PAC

31. Atrial flutter

32. Atrial tachycardia

33. Sinus rhythm with two PACs (it is bigeminal if it continues in this pattern)

34. Atrial flutter (with variable conduction ratio)

35. Sinus dysrhythmia with two PACs

36. Multifocal atrial tachycardia

37. Atrial flutter (with variable conduction ratio)

38. Multifocal atrial tachycardia

39. Sinus rhythm with bigeminal PACs

40. Sinus rhythm with two PACs (one blocked and one conducted)

41. Atrial fibrillation with rapid ventricular response (uncontrolled)

42. Atrial flutter (with variable conduction ratio)

43. Sinus rhythm with two blocked PACs

44. Wandering atrial pacemaker

45. Multifocal atrial tachycardia

46. Atrial flutter (with variable conduction ratio)

47. Atrial fibrillation with rapid ventricular response (uncontrolled)

48. Sinus tachycardia with two PACs

49. Atrial fibrillation

50. Atrial flutter (with variable conduction ratio)

51. Atrial fibrillation with rapid ventricular response (uncontrolled)

52. Sinus rhythm with two PACs

53. Multifocal atrial tachycardia

54. Sinus rhythm with paroxysmal atrial tachycardia

55. Sinus rhythm with trigeminal aberrantly conducted PACs

56. Sinus rhythm with one blocked PAC

57. Atrial tachycardia with areas of 2 to 1 block

Chapter 14

1. Sinus rhythm with bigeminal PJCs

2. Supraventricular tachycardia

3. Junctional escape rhythm

4. Accelerated junctional rhythm

5. Junctional tachycardia

6. Sinus rhythm with two PJCs

7. Bradycardic junctional escape

8. Junctional tachycardia

9. Junctional escape rhythm

10. Sinus rhythm with one PJC

11. Junctional tachycardia

12. Accelerated junctional rhythm

13. Accelerated junctional rhythm

14. Junctional tachycardia

15. Accelerated junctional rhythm

16. Sinus rhythm with trigeminal PJCs

17. Accelerated junctional rhythm

18. Junctional escape rhythm

19. Sinus rhythm with two PJCs

20. Junctional escape rhythm

21. Junctional tachycardia

22. Accelerated junctional rhythm

23. Junctional tachycardia

24. Sinus tachycardia with two PJCs

25. Accelerated junctional rhythm

26. Sinus rhythm with trigeminal PJCs

27. Accelerated junctional rhythm

28. Sinus rhythm with bigeminal PJCs

29. Junctional tachycardia

30. Sinus tachycardia with two PJCs

31. Sinus rhythm with trigeminal PJCs

32. Junctional escape rhythm

33. Sinus rhythm with a couplet of PJCs

34. Sinus rhythm with WPW and run of paroxysmal junctional tachycardia

35. Junctional escape rhythm

36. Junctional tachycardia

37. Sinus rhythm with two PJCs (may be trigeminal PJCs if it continues in this pattern)

38. Accelerated junctional rhythm

39. Sinus rhythm with one PJC

40. Antedromic AV reentrant tachycardia

41. Sinus tachycardia with quadrigeminal PJCs

42. Sinus pause with junctional escape rhythm

43. Sinus rhythm with bigeminal PJCs

44. Junctional escape rhythm

45. Junctional tachycardia

46. Accelerated junctional rhythm

47. Sinus rhythm with two PJCs

48. Junctional escape rhythm

49. Antedromic AV reentrant tachycardia

50. Sinus rhythm with bigeminal PJCs

51. Accelerated junctional rhythm

52. Sinus rhythm with two PJCs

53. Accelerated junctional rhythm

Chapter 15

1. Sinus rhythm with two PVCs

2. Ventricular tachycardia

3. Idioventricular rhythm

4. Sinus rhythm with bigeminal PVCs

5. Sinus tachycardia with multiform PVCs

6. Ventricular fibrillation

7. Sinus tachycardia with trigeminal PVCs

8. Accelerated idioventricular rhythm

9. Sinus rhythm with trigeminal PVCs

10. Ventricular tachycardia

11. Sinus rhythm with bigeminal PVCs

12. Ventricular tachycardia

13. Sinus rhythm with one PVC

14. Idioventricular rhythm

15. Sinus tachycardia with trigeminal PVCs

16. Sinus bradycardia with trigeminal PVCs

17. Asystole

18. Sinus bradycardia with a couplet of PVCs

19. Ventricular fibrillation

20. Sinus rhythm with one PVC

21. Ventricular tachycardia

22. Idioventricular rhythm

23. Sinus tachycardia with quadrigeminal PVCs

24. Sinus rhythm with two interpolated PVCs

25. Sinus rhythm with two multiform PVCs

26. Sinus rhythm with four PVCs (there is a couplet of PVCs among the four)

27. Ventricular tachycardia

28. Sinus rhythm with trigeminal PVCs

29. Ventricular tachycardia

30. Sinus rhythm with multiform PVCs

31. Idioventricular rhythm

32. Sinus rhythm with quadrigeminal PVCs

33. Sinus rhythm with quadrigeminal PVCs

34. Sinus rhythm with bigeminal PVCs

35. Sinus tachycardia with two PVCs

36. Idioventricular rhythm

37. Accelerated idioventricular rhythm

38. Sinus rhythm with four multiform PVCs (two of which occur as a couplet)

39. Idioventricular rhythm

40. Sinus rhythm with ventricular escape

41. Sinus rhythm with two couplets of PVCs

42. Atrial fibrillation with 2 multiform PVCs

43. Sinus rhythm with one PVC and a run of ventricular tachycardia

44. Accelerated idioventricular rhythm

45. Sinus dysrhythmia with one PVC

46. Sinus tachycardia with two runs of ventricular tachycardia

47. Junctional tachycardia with three multiform PVCs

48. Sinus rhythm with trigeminal PVCs

49. Ventricular tachycardia

50. Sinus bradycardia with one PVC

51. Sinus tachycardia with two uniform PVCs

52. Ventricular tachycardia

53. Sinus tachycardia with three multiform PVCs

54. Accelerated idioventricular rhythm

55. Sinus rhythm (borderline sinus bradycardia) with trigeminal PVCs

56. Multiform ventricular tachycardia—Torsades de pointes

57. Sinus pause with ventricular escape rhythm

Chapter 16

1. 2nd-degree AV heart block, type I

2. Sinus bradycardia with 1st-degree AV block

3. 2nd-degree AV heart block, type II

4. 3rd-degree AV block

5. Sinus bradycardia with 1st-degree AV block

6. 2nd-degree AV heart block, type I

7. 3rd-degree AV heart block

8. 2nd-degree AV heart block, type II

9. 3rd-degree AV heart block

10. Sinus tachycardia with 1st-degree AV block

11. 2nd-degree AV block, type I

12. 3rd-degree AV block

13. 2nd-degree AV block, type I

14. Sinus bradycardia with 1st-degree AV block (and IVD)

15. 2nd-degree AV block, type I

16. 2nd-degree AV block, type I

17. 2nd-degree AV block, type II

18. AV dissociation

19. Sinus bradycardia with 1st-degree AV block

20. 2nd-degree AV block, type II

21. 3rd-degree AV block

22. 2nd-degree AV block, type I

23. 2nd-degree AV block, type II

24. Sinus bradycardia with 1st-degree AV block

25. 2nd-degree AV block, type I

26. 3rd-degree AV block

27. 2nd-degree AV block, type II (with variable conduction ratio)

28. 2nd-degree AV block, type I

29. 3rd-degree AV block

30. Sinus bradycardia with 1st-degree AV block

31. 2nd-degree AV block, type II

32. Sinus tachycardia with 1st-degree AV block

33. 2nd-degree AV block, type I

34. 2nd-degree AV block, type II

35. 2nd-degree AV block, type I

36. 2nd-degree AV block, type I

37. 2nd-degree AV block, type II

38. 3rd-degree AV block

39. 2nd-degree AV block, type I

40. 3rd-degree AV block

41. Sinus tachycardia with 1st-degree AV block

42. 2nd-degree AV block, type II

43. Sinus bradycardia with 1st-degree AV block

44. 3rd-degree AV block

45. 2nd-degree AV block, type II

46. 3rd-degree AV block

47. Sinus bradycardia with 1st-degree AV block

48. Sinus bradycardia with IVD and 1st-degree AV block

49. 3rd-degree AV block

50. 2nd-degree AV block, type II

51. Sinus rhythm with 1st AV block and 2nd-degree AV block, type II

52. AV dissociation

53. Sinus rhythm with 1st AV degree and 2nd-degree AV block, type II

Chapter 17

1. Pacemaker failure to pace

2. Ventricular pacemaker

3. Atrioventricular pacemaker

4. Ventricular pacemaker

5. Demand pacemaker

6. Pacemaker failure to capture

7. Pacemaker failure to sense

8. Pacemaker failure to pace

9. Pacemaker failure to capture

10. Pacemaker mediated tachycardia

11. Pacemaker failure to pace

12. Pacemaker failure to capture

13. Pacemaker failure to pace

14. Sinus rhythm with demand pacemaker

15. Ventricular pacemaker

16. Pacemaker failure to capture

17. Accelerated junctional rhythm with demand pacemaker

18. Pacemaker mediated tachycardia

19. Pacemaker failure to pace with ventricular escape and resumption of sinus rhythm

Chapter 18

1. Left axis deviation

2. Normal axis

3. Left axis deviation

4. Extreme left axis deviation

Chapter 19

1. Junctional escape rhythm; POSSIBLE STEMI (inferior and anterior acute injury or infarction): ST segment elevation in inferior (II, III, and aV_F) with reciprocal ST segment depression in lead aV_L. ST segment elevation in anterior leads V_3 and V_4. The culprit vessel is likely the RCA, but it could also be the LAD or both the RCA and the LAD.

2. Sinus bradycardia; POSSIBLE STEMI (inferior acute injury or infarction): ST segment elevation in inferior (II, III, and aV$_F$). The culprit vessel is the RCA.

3. Atrial fibrillation with rapid ventricular response; septal infarct (age undetermined), ST segment and T wave abnormality–likely lateral ischemia.

4. Sinus rhythm with PVCs; POSSIBLE STEMI (anteroseptal acute injury or infarction): ST segment elevation in anteroseptal leads (V$_1$-V$_3$).

5. Sinus rhythm; left axis deviation, flat ST depression in lateral leads, possible lateral subendocardial injury.

6. Sinus bradycardia with 1st-degree AV block; possible left atrial enlargement (enlarged P waves); POSSIBLE STEMI (inferoposterior acute injury or infarction): ST segment elevation in inferior (II, III, and aV$_F$), reciprocal ST segment depression in leads I and aV$_L$; pronounced ST segment depression in lead V$_2$.

7. Atrial fibrillation with rapid ventricular response, left axis deviation, low voltage QRS complexes, cannot rule out anterior infarct (age undetermined), inferior infarct (age undetermined), marked ST depression in leads I, II, III, aV$_F$, V$_3$, V$_4$, V$_5$, and V$_6$ likely due to myocardial ischemia.

8. Sinus rhythm; POSSIBLE STEMI (anterolateral acute injury or infarction): ST segment elevation in anterior leads (V$_2$ through V$_4$) and in

high la F. The likely culprit vessel is the LAD or the LCx.

9. Sinus bradycardia with PVC; POSSIBLE STEMI (inferolateral acute injury or infarction): ST segment elevation in inferior (II, III, and aV$_F$) and low lateral leads (V$_5$, V$_6$) and reciprocal ST segment depression in lead aV$_L$.

10. Sinus bradycardia; POSSIBLE STEMI (inferoposterior acute injury or infarction): ST segment elevation in inferior (II, III, and aV$_F$) (acute injury or infarction) and reciprocal ST segment depression in leads I and aV$_L$. Pronounced ST segment depression in leads V$_1$-V$_3$.

11. Sinus bradycardia with 1st-degree AV block; POSSIBLE STEMI (inferolateral acute injury or infarction): ST segment elevation in inferior (II, III, and aV$_F$) and low lateral leads (V$_5$, V$_6$), developing Q wave in lead III and reciprocal ST segment depression in lead aV$_L$. There also appears to be ST segment elevation in lead V$_4$. The culprit vessel is likely the RCA, but it could also be the LAD or both the RCA and the LAD as there is inferolateral wall involvement.

12. Sinus rhythm with PAC; possible left atrial enlargement (enlarged P waves); POSSIBLE STEMI (anteroseptal acute injury or infarction): ST segment elevation in anteroseptal leads (V$_1$-V$_4$). There also appears to be ST segment elevation in lead V$_5$ which could be related to the anterior injury but it could be, instead,

repolarization abnormality. LAD is likely culprit vessel. Lastly, there appears to be a pathologic Q wave in lead III and possibly lead aV$_F$ which could reflect "inferior MI– Age undetermined."

13. Sinus rhythm; possible left atrial enlargement (enlarged P waves); low voltage QRS complexes; POSSIBLE STEMI (inferior and anterolateral acute injury or infarction): ST segment elevation in inferior leads and Q wave development in leads III and aV$_F$. There is also reciprocal ST segment depression in lead aV$_L$. There is also ST segment elevation in the anterior leads and low lateral leads. The likely culprit vessel is the RCA but the LAD may also be occluded or significantly narrowed.

14. Sinus rhythm; POSSIBLE STEMI (inferolateral acute injury or infarction): ST segment elevation in inferior leads. There is also ST segment depression in leads I and aV$_L$ which may reciprocal to the inferior ST segment elevation or may indicate involvement of other coronary arteries. RCA is the likely culprit vessel.

15. Sinus rhythm; POSSIBLE STEMI (inferior acute injury or infarction): ST segment elevation in inferior leads (II, III and aV$_F$) with reciprocal ST segment depression in lead aV$_L$. RCA is the likely culprit vessel. ST segment depression in leads V$_1$ and V$_2$ may indicate posterior injury. A further consideration is that ST segment elevation in the low lateral leads (V$_5$ and V$_6$) may indicate lateral

injury. If so, it would be referred to as an inferolateral STEMI. The elevation in the low lateral leads may also be repolarization abnormality.

16. Sinus tachycardia; POSSIBLE STEMI (anterolateral acute injury or infarction): ST elevation in V_2-V_6 (acute injury or infarct). Also, there appears to be a pathologic Q wave in lead III and possibly lead aV_F which could reflect "inferior infarct—age undetermined." Low QRS voltages in precordial leads.

17. Sinus bradycardia; POSSIBLE STEMI (inferoposterior acute injury or infarction): ST segment elevation in inferior leads (II, III, and aV_F) (acute injury) and ST segment depression in leads V_1 and V_2. RCA or circumflex artery is the likely culprit vessel (perhaps both vessels are occluded or narrowed). There is also widespread ST segment depression. The ST segment depression in leads I and aV_L may be reciprocal.

18. Sinus rhythm; right bundle branch block; left axis deviation; POSSIBLE STEMI (anteroseptal acute injury or infarction): ST segment elevation in anteroseptal leads (V_1-V_4) along with concordant ST segments and T waves. Also, there is reciprocal ST segment depression in leads III and aV_F. RCA is likely culprit vessel. There may also be lateral ST segment elevation as well which would suggest a lateral injury pattern.

19. Sinus rhythm with borderline 1st-degree AV block; POSSIBLE STEMI (inferoposterior acute injury or infarction): ST segment elevation in inferior leads (leads II, III, and aV_F) and ST segment depression in V_1 and V_2 which may indicate posterior STEMI (acute injury or infarction). There is also reciprocal ST segment depression in lead aV_L. Additionally, there may be elevation in the low lateral leads. The likely culprit vessel is the RCA.

20. Sinus rhythm (borderline sinus bradycardia); widespread T wave inversion and ST segment depression in lead I. This could be due to myocardial ischemia. In cases where the patient is experiencing chest pain, this may represent an NSTEMI.

21. Sinus rhythm; POSSIBLE STEMI (anterolateral acute injury or infarction): ST segment elevation in anterior leads (V_2 through V_4) and in lateral leads (I, aV_L, V_5, and V_6). There is also reciprocal ST segment depression in lead III and aV_F. The likely culprit vessel is the LAD or the LCx.

22. 3rd-degree AV Heart Block; POSSIBLE STEMI (inferior acute injury or infarction): ST segment elevation in inferior leads (II, III, and aV_F). There is also reciprocal ST segment depression in lead aV_L. Additionally, there is ST segment elevation in the anterior leads. The likely culprit vessel is the RCA but the LAD may also be occluded or significantly narrowed.

23. Sinus rhythm; right bundle branch block; right axis deviation; POSSIBLE STEMI (anterolateral acute injury or infarction): ST segment elevation in anterior leads (V_1-V_4) along with concordant ST segments and T waves. There is also ST segment elevation in the lateral leads (I, aV_L, V_5, and V_6). Also, there is reciprocal ST segment depression in leads III and aV_F. The LCA or LCx is the likely culprit vessel.

24. Sinus bradycardia; possible left atrial enlargement (enlarged P waves); POSSIBLE STEMI (anterolateral acute injury or infarction): ST segment elevation in anterior leads (V_2-V_4) and in the lateral leads (I, aV_L, V_5, and V_6). Also, there is reciprocal ST segment depression in leads III and aV_F. The LCA or LCx is the likely culprit vessel.

25. Sinus rhythm with borderline 1st-degree AV block; POSSIBLE STEMI (inferior acute injury or infarction): ST segment elevation in inferior leads (II, III, and aV_F) along with pathologic Q waves in leads III and aV_F (acute injury or infarction). There is also reciprocal ST segment depression in lead aV_L. Additionally, there is ST segment depression in leads V_2 and V_3. The likely culprit vessel is the RCA.

Chapter 20

1. Atrial fibrillation, right bundle branch block

2. Sinus tachycardia, left atrial enlargement, right axis deviation, incomplete right bundle branch block

3. Sinus rhythm, left bundle branch block

4. Sinus rhythm, left axis deviation indicating left anterior (fascicular) hemiblock, voltage criteria for left ventricular hypertrophy, cannot rule out Septal infarct (age undetermined)

5. Sinus tachycardia with occasional PVCs, incomplete left bundle branch block, ST & T wave abnormality suggestive of inferolateral ischemia

6. Sinus rhythm, right bundle branch block, left axis deviation indicating left anterior (fascicular) hemiblock

7. Sinus bradycardia with 1st-degree AV block; right bundle branch block and right axis deviation.

8. Sinus rhythm; right bundle branch block and right axis deviation.

9. Sinus rhythm; right bundle branch block.

10. Atrial fibrillation with rapid ventricular response; left bundle branch block and left ventricular hypertrophy.

11. Sinus rhythm with 1st-degree AV block; anterior fascicular block.

12. Sinus bradycardia; left atrial enlargement (enlarged P waves); left bundle branch block with left ventricular hypertrophy.

13. AV dissociation; right bundle branch block and anterior fascicular block.

14. 3rd-degree AV block; left bundle branch block; left axis deviation; possible inferior infarct—age undetermined.

15. Sinus rhythm; left atrial enlargement (enlarged P

waves); left bundle branch block; repolarization abnormality; left axis deviation.

16. Sinus dysrhythmia; right bundle branch block; left axis deviation.

17. Sinus rhythm with PVCs and PAC; right bundle branch block.

18. Sinus rhythm; right bundle branch block; left anterior fascicular block.

19. Atrial fibrillation with rapid ventricular response; left bundle branch block and left axis deviation.

20. Sinus tachycardia; Brugada Syndrome.

21. Sinus rhythm with 1st-degree AV block; left bundle block.

22. Sinus rhythm; left atrial enlargement (enlarged P waves); right bundle branch block; right axis deviation.

23. Sinus rhythm; right bundle branch block; right axis deviation.

24. Sinus tachycardia; left atrial enlargement (enlarged P waves); Brugada Syndrome.

25. Sinus rhythm with 1st-degree AV block; left posterior fascicular block; widespread inverted T waves that may be due to ischemia or repolarization abnormality.

Chapter 21

1. Sinus rhythm, left atrial enlargement, left ventricular hypertrophy

2. Sinus tachycardia, right atrial enlargement, right ventricular hypertrophy

3. 2nd-degree AV block, type I, right ventricular hypertrophy

4. Sinus rhythm, left ventricular hypertrophy

5. Sinus rhythm, right atrial enlargement, right ventricular hypertrophy

6. Atrial fibrillation, rSR′ pattern in V_1 suggests right ventricular conduction delay, voltage criteria for left ventricular hypertrophy, marked ST abnormality suggestive of possible inferolateral subendocardial injury

Chapter 22

1. Sinus tachycardia, electrical alternans

2. Sinus tachycardia, pericarditis

3. Sinus bradycardia, hypocalcemia

4. Sinus bradycardia, low amplitude QRS complexes

5. Sinus bradycardia, digitalis effect

6. Sinus rhythm, hypokalemia

7. Junctional escape rhythm, hypercalcemia

8. Sinus tachycardia, hyperkalemia

9. Sinus rhythm, hypocalcemia

10. Sinus bradycardia, hyperkalemia

11. Sinus tachycardia, low amplitude QRS complexes

Section 5

1. Sinus dysrhythmia

2. Idioventricular rhythm

3. Sinus tachycardia with quadrigeminal PVCs

4. Ventricular tachycardia

5. 2nd-degree AV block, type II

6. Pacemaker rhythm

7. Ventricular tachycardia

8. Atrial flutter (2 to 1 conduction)

9. Sinus pause with sinus escape

10. Sinus rhythm with 2nd-degree AV block, type II.

11. Atrial fibrillation

12. Sinus dysrhythmia

13. Sinus tachycardia with two multiform PVCs

14. Antedromic atrioventricular reentrant tachcyardia

15. Sinus rhythm with a run of ventricular tachycardia and multiform PVCs

16. Atrial flutter (with 2 to 1 conduction ratio)

17. Sinus tachycardia with two PVCs

18. Sinus bradycardia

19. Ventricular tachycardia

20. Sinus rhythm with 1st degree AV block and paroxysmal junctional tachycardia

21. 3rd-degree AV heart block

22. Atrioventricular dissociation

23. Sinus dysrhythmia

24. Sinus tachycardia with two PACs

25. 3rd-degree AV block

26. Sinus tachycardia

27. Sinus dysrhythmia

28. Atrial fibrillation

29. 2nd-degree AV heart block, type I

30. Sinus tachycardia

31. Sinus rhythm with two PACs

32. Sinus tachycardia with 1st-degree AV block

33. Sinus rhythm with one aberrantly conducted PAC

34. Accelerated junctional rhythm

35. Sinus dysrhythmia

36. Atrial flutter

37. Ventricular standstill (asystole) with a ventricular escape beat

38. 2nd-degree AV block, type I

39. 3rd-degree AV heart block (with sinus tachycardia)

40. Sinus bradycardia

41. Junctional escape with one PJC

42. Sinus tachycardia with 1st-degree AV block; left atrial enlargement (enlarged P waves); right bundle branch block with left anterior fascicular block.

43. Sinus dysrhythmia; POSSIBLE STEMI (inferior acute injury or infarction): ST segment elevation in inferior leads (II, III, and aV_F). There is also reciprocal ST segment depression in lead aV_L. The likely culprit vessel is the RCA.

44. Atrial fibrillation with rapid ventricular response; POSSIBLE STEMI (inferolateral acute injury or infarction): ST segment elevation in inferior leads (II, III, and aV_F), pathological Q waves in leads III and aV_F and reciprocal ST segment depression in leads I and aV_L; ST segment elevation in low lateral leads (V_3-V_6). The RCA or LCx is the likely culprit artery.

45. Sinus bradycardia with borderline 1st-degree AV block; tall peaked T waves that may be due to myocardial ischemia or hyperkalemia. There are also inverted T waves in aV_L, V_1, and V_2. Lastly, there is right axis deviation.

46. Sinus tachycardia; left atrial enlargement (enlarged P waves); left bundle branch block.

47. Sinus rhythm with bigeminal PVCs or aberrantly conducted PACs; POSSIBLE STEMI (anterolateral): ST segment elevation in anterior leads (V_2, V_3, and V_4) and high lateral leads (I and aV_L) (acute injury or infarction).

48. Sinus bradycardia with 1st-degree AV block and occasional PVC; POSSIBLE STEMI (inferoanterior acute injury or infarction): ST segment elevation in inferior leads (II, III, and aV_F), pathologic Q waves in leads III and aV_F and reciprocal ST segment depression in leads I, aV_L; ST segment elevation in anterior leads (V_1, V_3, and V_4). The RCA and/or LAD is the likely culprit vessel but it could also be the LCx. An alternative explanation for the tall, wide QRS complexes in leads I, aV_L, and V_2 is left ventricular hypertrophy with repolarization abnormality.

49. Sinus rhythm with 1st-degree AV block; left atrial enlargement (enlarged P waves); right bundle branch block with right axis deviation or left posterior fascicular block.

50. Sinus rhythm with occasional PVC; right atrial enlargement (enlarged P waves); left bundle branch block.

51. Junctional escape; POSSIBLE STEMI (inferior, anteroseptal and lateral acute injury): ST segment elevation in inferior leads (II, III, and aV_F) and reciprocal ST segment depression in leads I and

aV_L; ST segment elevation in septal and anterior leads (V_1-V_4); ST segment elevation in V_5 and V_6. The RCA and LAD are the likely culprit vessels.

52. Sinus rhythm (borderline sinus tachycardia); right bundle branch block with right axis deviation or left posterior fascicular block.

53. Sinus rhythm; POSSIBLE STEMI (inferior and anterolateral acute injury or infarction): ST segment elevation in inferior leads (II, III, and aV_F) and pathologic Q waves; ST segment elevation in anterior and lateral leads (V_3-V_6). The RCA and LAD or LCx is the likely culprit vessel.

54. Sinus dysrhythmia; right atrial enlargement (enlarged P waves); low voltage QRS complexes; POSSIBLE STEMI (inferior and anterolateral acute injury or infarction): ST segment elevation in inferior leads (II, III, and aV_F) and reciprocal ST segment depression in leads I and aV_L; ST segment elevation in anterior and lateral leads (V_3-V_6). There are also pathologic Q waves in V_4-V_6. The RCA and LAD are the likely culprit vessels.

55. Sinus bradycardia; POSSIBLE STEMI (inferior acute injury or infarction): ST segment elevation in inferior leads (II, III, and aV_F). The likely culprit vessel is the RCA. There is also ST segment depression in lead aV_L which could be reciprocal to the inferior STEMI or it could be related to the ST Segment depression in leads V_2 through V_5. The ST segment

depression may indicate anterior and high lateral wall myocardial ischemia which may be caused by narrowing of the LCx.

56. Atrial fibrillation with rapid ventricular response; right bundle branch block with left anterior fascicular block.

57. Sinus bradycardia; POSSIBLE STEMI (inferior): ST segment elevation in inferior leads (II, III and aV_F) and reciprocal ST segment depression in lead aV_L. The RCA is the likely culprit vessel.

58. Atrial paced rhythm; right bundle branch block, left ventricular hypertrophy with repolarization abnormality.

59. Sinus rhythm; right bundle branch block with right axis deviation.

60. Sinus rhythm; POSSIBLE SUBENDOCARDIAL INFARCTION (anterolateral): flat ST segment depression in anterior leads (V_3 and V_4) and lateral leads (I, aV_L, V_5, and V_6). There is also ST segment elevation in III, aV_R, and V_1. Alternatively, this may be POSSIBLE ACUTE LEFT MAIN CORONARY ARTERY OCCLUSION: As mentioned in Chapter 19, ST segment elevation in lead aV_R is now given considered when assessing for STEMI. One rule that can be used is that if ST segment elevation in lead aV_R is greater than any ST segment elevation in lead V_1 plus ST segment depression in 7 or more other leads it is suggestive of acute left main coronary occlusion. This patient is a candidate

for an urgent cardiac catheterization.

61. Sinus bradycardia; POSSIBLE STEMI (inferolateral acute injury or infarction): ST segment elevation in inferior leads (II, III, and aV_F) with Q wave formation in lead III. The ST segment depression in leads I and aV_L could be reciprocal to the inferior STEMI but could also be related to the ST segment elevation the low lateral lead, V_5. The ST segment elevation in the anterior leads V_2-V_4 could be repolarization abnormality but also could be due to myocardial injury. The RCA or the LCx is the likely culprit vessel.

62. Sinus tachycardia with premature atrial complexes; left atrial enlargement (enlarged P waves); left bundle branch block.

63. Atrial fibrillation with rapid ventricular response; right bundle branch block.

64. Sinus tachycardia; POSSIBLE STEMI (inferoanterior acute injury or infarction): ST segment elevation in inferior leads (II, III, and aV_F) and reciprocal ST segment depression in lead aV_L. There is ST segment elevation in the anterior leads, V_2-V_5. The likely culprit vessels could be the RCA, LAD and/or the LCx.

65. Sinus bradycardia; POSSIBLE STEMI (inferior acute injury or infarction): ST segment elevation in inferior leads (II, III, and aV_F) and reciprocal ST segment depression in leads and aV_L. The RCA is the likely culprit vessel.

66. Sinus tachycardia; Brugada Syndrome.

67. Sinus rhythm with bigeminal PVC; POSSIBLE STEMI (inferior acute injury or infarction): ST segment elevation in inferior leads (II, III, and aV_F) and reciprocal ST segment depression in leads I and aV_L. The RCA is the likely culprit vessel.

68. Sinus rhythm with 1st-degree AV block; POSSIBLE STEMI (inferoseptal acute injury or infarction): ST segment elevation in inferior leads (II, III, and aV_F) and septal leads (V_1 and V_2). There is

also reciprocal ST segment depression in leads and aV_L. The RCA is the likely culprit vessel.

69. Sinus rhythm; right atrial enlargement (enlarged P waves); incomplete right bundle branch block; POSSIBLE STEMI (inferior and anterior acute injury or infarction): ST segment elevation in inferior leads (II, III, and aV_F); ST segment elevation in anterior leads (V_2-V_5). and reciprocal ST segment depression in leads and aV_L. The RCA is the likely culprit vessel.

70. Sinus rhythm with 1st-degree AV block and premature junctional complex (PJC); POSSIBLE STEMI (inferior acute injury or infarction): ST segment elevation in inferior leads (II, III and aV_F) and reciprocal ST segment depression in leads I and aV_L. There is also 1 mm ST segment elevation in leads V_3 and V_4 that is likely due to early repolarization. The RCA is the likely culprit vessel.

71. Sinus rhythm; left anterior fascicular block.

Glossary

1st-degree AV heart block A delay in conduction of the impulse arising from the SA node as it passes through the AV node to the ventricles. This delay produces a longer-than-normal PR interval. Because all the impulses are conducted, each P wave will be followed by a QRS complex.

2nd-degree AV heart block, type I A dysrhythmia in which there is a progressively longer delay in conduction of the impulse from the SA node through the AV node until finally the impulse fails to conduct and a ventricular beat is dropped. This results in progressively longer PR intervals until a QRS complex is absent. This occurs in a cyclical manner. It is considered an incomplete block. It is also referred to as *Wenckebach* and sometimes Mobitz Type I.

2nd-degree AV heart block, type II A dysrhythmia in which not all the impulses arising from the SA node are conducted through the AV node to the ventricles, resulting in more P waves than QRS complexes. It is considered an incomplete block.

3rd-degree AV heart block A dysrhythmia in which none of the impulses arising from the SA node are conducted through the AV node to the ventricles. The ventricles are stimulated to beat by an escape pacemaker that arises below the level of the AV node. This results in the P waves having no relationship to the QRS complexes because each is beating independently of each other. This is a complete block.

A

Absolute refractory period The portion of repolarization during which no stimulus, no matter how strong, will depolarize the cell.

Accelerated junctional rhythm A dysrhythmia that arises from the AV junction and produces a heart rate of 60 to 100 BPM

Action potential The measure over time of the change in electrical charge of the cell. The action potential reflects the ability of the cell to depolarize.

Actin One of the two components of the contractile filaments involved in muscular contraction.

Accessory pathway A conduction pathway from the SA node to the ventricles that does not pass through the normal AV node pathway.

Acute Sudden and recent onset of a condition, sign, or symptom.

Acute coronary syndrome The condition of ischemic chest pain consisting of the following conditions: unstable angina, non-ST segment elevation MI, and ST segment elevation MI.

Acute cor pulmonale enlargement of the right ventricle due to disease of the lungs or of the pulmonary blood vessels

Adrenalin A natural catecholamine that causes bronchodilation and increased AV conduction, muscular contractility, heart rate, and vasoconstriction; also called *epinephrine*.

Adrenergic Nerves that release epinephrine or epinephrine-like substances.

Afterload The pressure against which the left ventricle must pump blood during contraction.

Alpha adrenergic receptors Any of the adrenergic receptor tissues that respond to norepinephrine.

Amplitude The height and depth, or waveform size, of an ECG complex.

Anemia A deficiency in the oxygen-carrying hemoglobin molecule contained in blood.

Aneurysm A pathological blood-filled dilation of a blood vessel.

Angina See angina pectoris.

Angina pectoris Chest pain resulting when the supply of oxygen and coronary perfusion is insufficient to that demanded by the heart muscle.

Antegrade Occurring or performed in the normal or forward direction of conduction or flow.

Artery A vessel that carries blood away from the heart. Arteries generally carry oxygenated blood. The exception is the pulmonary artery, which carries deoxygenated blood to the lungs.

Aorta The major vessel that carries oxygenated blood from the heart to the body.

Aortic valve The valve that separates the left ventricle from the aorta.

Apex The tip of a structure that may be either at the top or bottom. The apex of the heart is the bottom of the heart.

Arrhythmia Technically speaking, the absence of a rhythm, but this term is used synonymously with the term *dysrhythmia*.

Artifact Extraneous spikes and waves produced by anything other than the physiological activity of the heart. Common causes include electrical or mechanical factors such as muscle tremor, alternating current interference, loose leads, chest compressions, and patient movement. Also called *electrical interference* or *noise*.

Asymmetrical Not identical on both sides of a central line; lacking symmetry.

Asystole Absence of cardiac electrical activity; seen on the ECG as an absence of upward or downward deflection on the isoelectric line of the ECG.

Atherosclerosis An accumulation of fat-containing deposits within the arterial wall.

Atria The chamber of the heart that collects blood returning from the rest of the body. The right atrium collects deoxygenated blood from the body and passes it to the right ventricle. The left atrium collects oxygenated blood from the lungs and passes it to the left ventricle.

Atrial kick The additional volume of blood pushed into the ventricle by the contraction of the atria.

Atrial tachycardia with block Tachycardia that arises from the atria at a rate of between 150 and 250 BPM. However, the AV node fails to carry some of the impulses through to the ventricles. This results in a faster-than-normal rate with more P' waves that QRS complexes.

Atrioventricular dissociation (AV dissociation) Any rhythm characterized by the atria and ventricles beating independently but at the same rate per minute.

Atrioventricular junction (AV junction) The area of conductive tissue that includes the AV node, its atrial pathways, and the bundle of His.

Atrioventricular node (AV node) Specialized tissue at the base of the wall between the two upper heart chambers (atria). Electrical impulses pass from the SA node to the AV node, then on to the bundle of His.

Augmented limb leads The limb leads aV_R, aV_L, and aV_F. The ECG waveforms produced by these leads is so small that the ECG machine enhances, or augments, them by 50% so their amplitude is comparable to other leads.

Automaticity The ability of cardiac cells to initiate spontaneous electrical impulses.

Autonomic nervous system The involuntary portion of the peripheral nervous system that regulates vital body functions. It is separated into the parasympathetic and sympathetic divisions. The autonomic nervous system controls cardiac, smooth muscle, and glandular activity.

Axis The sum of all the vectors of electrical activity associated with conduction of the action potential through the heart.

Axis deviation An abnormal shift of the normal axis of the cardiac muscle action potential, often associated with ventricular hypertrophy and certain conduction defects.

B

Baroreceptors Pressure-sensitive nerve endings in the walls of the atria of the heart and in some larger blood vessels such as the carotid artery. Baroreceptors stimulate reflex mechanisms that allow the body to adapt to changes in blood pressure by dilating or constricting blood vessels.

Beta receptors Adrenergic receptors of the nervous system that respond to adrenaline. Activation causes relaxation of smooth muscles and increases in cardiac rate and contractility.

Bifascicular block Conduction abnormality in the heart where two of the three main fascicles of the His-Purkinje system are blocked.

Bicuspid valve See mitral valve.

Bigeminy Cardiac electrical activity characterized by a pattern of one ectopic beat followed by one normal beat. These ectopic beats can be either atrial or ventricular in origin and indicate cardiac irritability.

Bipolar limb lead An ECG lead with both a positive and negative electrode. Leads I, II, and III are bipolar limb leads.

Biphasic A waveform having both a positive and negative deflection.

Blocked PACs Premature beats that arise from the atria but are not conducted through the AV node to the ventricles. This lack of conduction results in a P wave that is seen earlier than normal in the cardiac cycle and is not followed by a QRS complex.

Blood pressure The pressure exerted by circulating blood upon the walls of blood vessels. It is one of the key patient vital signs.

Bradycardia A heart rate below 60 beats per minute, when related to atrial or sinus rhythms.

Bradycardic junctional escape rhythm A dysrhythmia that arises from the AV junction as an escape mechanism but is slower than the inherent rate of the AV junction, which is normally 40 to 60 BPM. Also known as junctional escape rhythm with bradycardia.

Bundle branch Either of the parts of the bundle of His passing respectively to the right and left ventricles.

Bundle branch block Disorder that leads to one or both of the bundle branches failing to conduct impulses. This produces a delay in depolarization of the ventricle it supplies.

Bundle of His Specialized muscle fibers in the intraventricular septum that carry the electric impulses to the ventricles.

Burst See Salvo.

C

Calibration Is the process of determining that the settings on the ECG match the proper standards. The internal regulation of the ECG machine is such that a 1-mV electrical signal results in a 10-mm deflection.

Cardiac cycle The sequence of events that occur when the heart beats. The cycle has two main phases: diastole, when the heart ventricles are relaxed, and systole, when the ventricles contract. One cardiac cycle is defined as the contraction of the two atria followed by contraction of the two ventricles.

Cardiac output The volume of blood pumped by the heart, usually measured in milliliters per minute. Cardiac output is calculated by multiplying the heart rate by the stroke volume.

Cardiac tamponade See pericardial tamponade.

Cardiogenic shock is a life-threatening condition resulting from inadequate circulation of blood to the organs of the body due to primary failure of the heart.

Catecholamine Sympathetic nervous system neurotransmitter such as epinephrine, norepinephrine, and dopamine.

Cardiomyopathy Any structural or functional disease of heart muscle that is marked especially by hypertrophy of cardiac muscle, enlargement of the heart, rigidity and loss of flexibility of the heart walls, or narrowing of the ventricles.

Chemoreceptor A sensory nerve cell that is activated by chemicals such as the carbon dioxide detectors in the brainstem.

Cholesterol is a major component of the blood. Higher than normal amounts of cholesterol in the blood, which can occur from eating too many fatty foods, may lead to diseases of the arteries such as atherosclerosis.

Cholinergic Nerve fibers that release acetylcholine.

Chordae tendineae Thin, strong strings of connective tissue that anchor the papillary muscles to the floor of the heart.

Chronotropic Affecting the rate of the heartbeat. A positive chronotropic effect increases the heart rate whereas a negative chronotropic effect decreases the heart rate.

Circadian rhythm The biological clock in humans based on a 24-hour cycle. At regular intervals each day, the body becomes active or tired. Some medications affect the body more at certain times during the day than at others.

Compensatory pause The pause following a premature complex that allows the original rhythm to begin again at its normal rate. The compensatory pause will be exactly twice the R-R interval of the normally conducted beat, as seen following a PVC.

Complete AV heart block See 3rd-degree AV heart block.

Congenital heart disease is a problem in the structure of the heart that is present at birth.

Contractility Ability of the cardiac muscle to contract in response to electrical stimulation.

Conductive pathway Route followed by nerve impulses as they pass through the heart. A pathway is made of two nodes (special conduction cells) and a series of conduction fibers or bundles.

Conductivity The ability of the cardiac muscle to transmit an electrical stimulus from cell to cell.

Coronary angioplasty also known as percutaneous coronary intervention (PCI), is a nonsurgical procedure used to open occluded or narrowed coronary arteries. By doing this, it improves blood flow to the heart muscle. The procedure involves threading a catheter through a small puncture in a leg or arm artery to the heart, then inserting a thin, expandable balloon into the clogged artery and inflating it. This opens the artery by pushing the plaque against the artery wall. The balloon is then removed and blood flows more easily through the artery. A stent, a thin hollow tube, may be placed inside the artery at the site to keep the artery open.

Coronary artery bypass graft (CABG) is a procedure used to improve blood flow to the heart muscle. It involves taking a healthy artery or vein from the body and grafting it to a blocked coronary artery. The grafted artery or vein bypasses the blocked portion of the coronary artery, creating a new path for oxygen-rich blood to flow to the heart muscle.

Congenital heart defects are problems with the structure of the heart. They are present at birth and can involve the heart walls, valves and the arteries and veins near the heart.

Coronary arteries A pair of arteries that branch from the ascending aorta and supply oxygenated blood to the myocardium.

Coronary artery disease The gradual narrowing and hardening of the coronary arteries. Coronary artery disease is usually the result of atherosclerosis.

Coronary veins Vessels that transport deoxygenated blood from the capillaries of the heart to the right atrium.

Couplet Two consecutive complexes.

Cyanosis A condition in which a person's skin is discolored to a bluish hue because of inadequate oxygenation of the blood.

D

Delta wave A slurred or widened upstroke at the beginning of the cardiac QRS complex causing prolongation of the complex. It indicates anomalous impulse conduction and is diagnostic of Wolff-Parkinson-White syndrome.

Depolarization Loss of the difference in charge between the inside and outside of the plasma membrane of a muscle or nerve cell due to a change in permeability and migration of positively charged ions into the cell.

Desemones A specialized local thickening of the plasma membrane of myocytes that serves to anchor contiguous cells together and prevent them from pulling apart during contraction.

Diastole The period of relaxation during the cardiac cycle when the cardiac muscle fibers lengthen, causing the heart to dilate and fill with blood. Coronary perfusion occurs during this phase.

Dilation Expansion or widening of an organ, opening, or vessel.

Dynamic ECGs ECGs which are displayed on an ECG oscilloscope. They reflect what is currently occurring in the heart.

Dysrhythmia Any abnormality in the otherwise normal rhythmic pattern of the heartbeat.

Dyspnea Difficulty in breathing.

E

Early beats Premature beats that arise from somewhere in the heart other the SA node before the SA node has a chance to initiate the heartbeat.

Ectopic A cardiac complex originating from somewhere other than the SA node. It may be early (premature) or late (escape) in nature.

ECG tracings Electrocardiogram printouts. Also referred to as static ECGs.

Einthovan's Triangle An electrical triangle formed by the patients' right arm, left arm, and left leg, used to position electrodes for ECG monitoring. This positioning is used to determine the direction of cardiac vectors to various leads.

Electrical axis See axis.

Electrocardiogram The tracing made by an electrocardiograph; also known as *ECG* or *EKG*.

Electrocardiograph Device used to record electrical variation within cardiac tissue.

Electrode The electrical sensor applied to the body to record the ECG.

Electrolyte Element or compound that, when melted or dissolved in water or other solvent, dissociates into ions (atoms able to carry an electric charge). Sodium, potassium, and calcium are the primary electrolytes involved in myocardial activity.

Electromechanical dissociation (EMD) A condition characterized by seemingly normal electrical heart activity without discernible mechanical activity. This term is no longer used and has been replaced by the term *Pulseless Electrical Activity* (PEA).

Endocarditis is an infection of the inner lining of the heart chambers and valves or the inner lining of the blood vessels attached to the heart. When it is caused by an infection, it is called infective endocarditis (IE).

Embolism Is undissolved material such as thrombi or fragments of thrombi, pieces of fatty deposit, tissue fragments, clumps of bacteria, protozoan parasites, fat globules, or air bubbles carried by the blood and lodging in a blood vessel to obstruct blood flow.

Endocardium The serous membrane that lines the inner aspect of the four chambers of the heart and its valves.

Enlargement An increase in size.

Epinephrine See adrenalin.

Epicardium A thick serous membrane that constitutes the smooth outer surface of the heart.

Escape complex A beat that occurs after the normal pacemaker fails to fire.

Excitability The degree to which a myocardial cell is reactive to external stimuli.

Extrasystole A premature complex.

F

f waves Waves associated with atrial fibrillation. They occur with an atrial rate of greater than 350 times per minute.

F waves Saw-toothed waves associated with atrial flutter. They occur with atrial rate of 250 to 350 times per minute.

Fascicle A small bundle of conducting fibers.

Fibrillation Quivering of the muscle fibers of the heart, resulting in a lack of heartbeat and pulse.

Fibrillatory waves Chaotic baseline that lacks any semblance of organized electrical waveforms. Also know as f waves.

Fibrinolytic agent is any drug capable of stimulating the dissolution of a blood clot (thrombus).

Fibrous cap is a layer of fibrous connective tissue which forms on atherosclerotic plaques. It is thicker and less cellular than the normal intima, Rupturing of the cap is the cause of acute arterial obstruction during myocardial infarction.

Flutter An abnormal rapid spasmodic and usually rhythmic motion or contraction of the atria or ventricles.

Flutter waves Waveforms seen in atrial flutter where the atria are firing at a rate of 250–350 BPM. They are sometimes described as having a *picket fence* or *sawtooth* appearance.

Fusion beat A beat that occurs when two complexes originating from different pacemakers fuse to form one complex as seen with some types of PVCs.

G

Ganglia A group of nerve cells forming a nerve center along the thoracic and sacral spine associated with the parasympathetic nervous system.

Gap junctions An area of the conducting cell that improves the rate of electrical conduction.

H

Heart failure is the inability of the heart to keep up with the demands on it and, specifically, failure of the heart to pump blood with normal efficiency

Hemiblock Failure of conduction of the muscular excitatory impulse in either of the two fascicles of the left branch bundle branch.

Hyperkalemia An abnormally high level of potassium in the blood.

Hypertrophy Increase in bulk (as by thickening of muscle fibers) without multiplication of parts.

Hypertension A condition in which a person's blood pressure is abnormally high. For normal adults, the blood pressure should be less than 130 mm Hg systolic and less than 85 mm Hg diastolic. Pressures above 140/90 indicate a mild form of hypertension; above 180/110 is considered severe.

Hypokalemia An abnormally low level of potassium in the blood.

I

Idioventricular rhythm Relating to, or arising in the ventricles of the heart independent of the atria.

Incomplete bundle branch block A partial block of the bundle branch in which the QRS complexes are less than 0.12 seconds in duration

Infarction Irreversible cell death caused by prolonged obstruction of arterial blood supply to an area of the body.

Innervated A part of the body supplied with nerves.

Inotropic Something that increases the force of contraction of the heart.

Intercalated disks The specialized regions of the cardiac muscle cells that comprise the longitudinal and end-to-end junctions between adjacent cells and that function to connect them mechanically and electrically.

Interpolated A beat that occurs between normal heartbeats without disturbing the succeeding beat or the basic rhythm of the heart.

Intraatrial pathway Electrical pathways that travel through the atria from the SA node to the AV node.

Intranodal pathway The conduction pathways through the AV node.

Ion An atom or group of atoms that carries a positive or negative electric charge as a result of having lost or gained one or more electrons.

Ischemia Local tissue hypoxia caused by a reduction in arterial blood supply and/or oxygen to an area of the body.

Isoelectric A condition of the cell that is neither negatively nor positively charged in relation to its neighboring cells. This also refers to the baseline on the ECG that indicates neither a positive or negative deflection.

J

J point The point at which the QRS complex meets the ST segment; also called the *junction.*

Junctional A cardiac rhythm resulting from impulses coming from a locus of tissue in the area of the AV junction.

Junctional escape rhythm with bradycardia See Bradycardic junctional escape rhythm.

L

Latent Present but not visible, apparent, or actualized; existing as potential.

Left bundle branch The portion of the heart's conduction system that supplies the left ventricle. It is composed of two fascicles: the left anterior fascicle and the left posterior fascicle.

Left bundle branch block In this condition, activation of the left ventricle is delayed, which results in the left ventricle contracting later than the right ventricle.

Left main coronary artery Abbreviated as LCA and also known as the left coronary artery. It arises from the aorta above the left cusp of the aortic valve. It then divides into the left anterior descending artery and the circumflex branch both of which supply blood to the left ventricle and left atrium.

Lipoproteins are biochemical compounds that contain both protein and lipid. Most lipids in plasma are present in the form of lipoproteins.

M

Mediastinum The space in the chest between the pleural sacs of the lungs that contains all the viscera of the chest except the lungs and includes the heart, aorta, and vena cava.

Mitral valve A valve in the heart between the left atrium and the left ventricle. It prevents the blood in the ventricle from returning to the atrium during systole and consists of two triangular flaps; also called *bicuspid valve* or *left atrioventricular valve.*

Mobitz Type I See 2nd-degree AV heart block, type I

Morphology The configuration or shape of a wave or complex on the ECG.

Multifocal Impulses that arise from different sites within the heart.

Myocardial infarction Irreversible cell death caused by prolonged obstruction of arterial blood supply to an area of the myocardium.

Myocardial injury A degree of cellular damage beyond ischemia.

Myocardial ischemia Local tissue hypoxia caused by a reduction in arterial blood supply and/or oxygen to an area of the myocardium.

Myocarditis is an inflammation of the heart muscle.

Myocardium The muscular layer of the heart.

Myocytes The individual muscle cells of the heart.

Myosin One of the two components of the contractile filaments involved in the muscular contraction.

N

Neurotransmitter A substance, such as norepinephrine or acetylcholine, used to transmit a nerve impulse across the synapse between nerve cells or to the organ being stimulated (innervated).

Noise See Artifact.

Noncompensatory pause The pause following a premature complex that resets the rate of the original rhythm. The R-R interval following a noncompensatory pause is less than two preceding R-R intervals; seen following a PAC.

Nonconducted PACs See blocked PAC.

Nonperfusing PVC A premature ventricular complex that fails to produce a pulse.

Norepinephrine The chemical compound used to transmit impulses by the sympathetic nervous system. A precursor to epinephrine and referred to as *noradrenaline.*

Normal sinus rhythm (NSR) Normal beating of the heart, as measured by an ECG. It has certain generic features that serve as hallmarks for comparison with normal ECGs. The rate is between 60 and 100 beats per minute, and the beats are synchronized.

O

Osborn wave A narrow, positive deflection wave at the junction of the QRS complex and the ST segment associated with hypothermia.

Oscilloscope An instrument that displays the fluctuating electrical quantity as a visible waveform on the fluorescent screen of a cathode-ray tube.

P

P wave The electrical representation of atrial contraction. It precedes the QRS complex.

P mitrale Right atrial enlargement associated with severe lung disease such as COPD and pulmonary hypertension.

P pulmonale Hypertrophy of the right ventricle associated with severe lung disease such as emphysema.

Palpitations Heartbeat sensations that feel as if the heart is pounding or racing. It may present as an unpleasant awareness of a person's own heartbeat or feel like a skipped beat. Palpitations can be felt in the chest, throat, or neck.

Papillary muscles The thin muscles that connect to the cusps of the mitral and tricuspid valves and contract to facilitate their ability to resist prolapsing during systole.

Parasympathetic The part of the autonomic nervous system that chiefly contains cholinergic fibers. The nerves of this system tend to induce secretion, to increase the tone and contractility of smooth muscle, and to slow the heart rate.

Pericardial effusion TK.

Parietal pericardium The tough thickened membranous outer layer of the pericardium that is attached to the central part of the diaphragm and the posterior part of the sternum.

Paroxysmal An abrupt beginning and ending of an event such as a different cardiac rhythm.

PEA See Pulseless Electrical Activity.

Pericardial tamponade A condition characterized by a collection of fluid within the pericardial sac that constricts the heart and inhibits its ability to fill during diastole; also called *cardiac tamponade.*

Pericarditis Inflammation of the pericardium.

Pericardium The conical sac of serous membranes that encloses the heart and the roots of the aorta, pulmonary artery, and vena cava. It is composed of one layer that is closely adherent to the heart while the other lines the inner surface of the outer coat, with the intervening space being filled with pericardial fluid.

Peripheral artery disease is a condition in which plaque builds up in and narrows the arteries that carry blood to the head, organs, and limbs.

Plaque is a lesion of atherosclerosis, a pearly white area on the inner lining of an arterial wall that causes the intimal surface to bulge into the lumen; it is composed of fatty material, cell debris, smooth muscle cells, and collagen.

Polarization Refers to the cell returning to its resting state during which no electrical activity occurs and its intracellular fluid is negatively charged relative to the extracellular fluid.

Polarized In a resting state during which no electrical activity takes place.

Polymorphic ventricular tachycardia A type of ventricular tachycardia that is irregular in rate and rhythm and has varying shapes or morphologies on the ECG.

PR intervals Is the distance from the beginning of the P wave to the beginning of the Q wave (or R wave if the Q wave is absent).

PR segment The flat isoelectric line that represents the electrical impulse traveling through the His-Purkinje system.

Precordial Situated in front of the heart.

Preload The stretched condition of the heart muscle at the end of diastole just before contraction.

Premature ventricular complexes (PVCs) Early beats that arise from the ventricles before the SA node has a chance to initiate a heartbeat.

Prinzmetal's angina is chest pain caused by coronary artery spasms. Coronary artery spasm is a temporary, abrupt, and focal (restricted to one location) contraction of the muscles in the wall of a coronary artery that constricts the artery, slowing or stopping blood flow.

Pulmonic valve The heart valve that separates the right ventricle from the pulmonary artery; also called the *semilunar valve*.

Pulmonary embolism is a sudden blockage in a lung artery.

Pulseless Electrical Activity (PEA) A condition characterized by seemingly normal electrical heart activity without discernible mechanical activity.

Purkinje fibers Specialized heart cells capable of conducting electrical impulses and directly stimulating myocardial cells to contract.

Q

Q wave The first negative deflection of the QRS complex not following an R wave.

QRS complex The waveform that represents electrical conduction through the ventricles. It is associated with ventricular contraction.

QT interval Period of time measured from the beginning of the Q wave to the end of the T wave, representing the time required to depolarize and repolarize the ventricles.

Quadrigeminal Cardiac electrical activity characterized by a pattern of one ectopic beat followed by three normal beats. These ectopic beats can be either atrial or ventricular in origin and indicate cardiac instability.

R

R wave The positive deflection in the QRS complex.

Reentry A cardiac mechanism that explains certain abnormal heart actions (such as tachycardia) and that involves the transmission of a wave of depolarization along an alternate pathway when the original pathway is blocked, with return of the impulse along the blocked pathway resulting in a reinitiation of the impulse.

Refractory period The phase during repolarization during which a stimulus may or may not cause the cell to depolarize; also see relative and absolute refractory period.

Regular sinus rhythm See normal sinus rhythm.

Relative refractory period A later phase of repolarization during which a sufficiently strong stimulus will depolarize the cell.

Renin-angiotensin-aldosterone system is a hormone system that regulates blood pressure and fluid balance.

Repolarization Restoration of the difference in charge between the inside and outside of the plasma membrane of a muscle fiber or cell following depolarization.

Resting membrane potential (RMP) The relative difference between the electrical charge of the inside as compared with the outside of the cell.

Retrograde Occurring or performed in a direction opposite to the normal or forward direction of conduction or flow.

Right bundle branch The portion of the heart's conduction system that supplies the right ventricle.

Right bundle branch block Occurs when transmission of the electrical impulse is delayed or not conducted along the right bundle branch. This leads to the right ventricle depolarizing by means of cell-to-cell conduction that spreads from the interventricular septum and left ventricle to the right ventricle.

Right main coronary artery (RCA) One of the two main coronary arteries. It divides into the right posterior descending and acute marginal arteries and supplies blood to the right ventricle, right atrium, sinoatrial node, and atrioventricular node.

R-R interval The interval of time between consecutive R waves.

Run Three or more premature beats presenting in a row.

S

SA node See sinoatrial node.

S wave First negative deflection that *extends below* the baseline in the QRS complex following the R wave

Salvo A series of three or more consecutive and identical beats, as in a salvo of ventricular tachycardia.

Sarcolemma The plasma membrane surrounding the myocytes of the heart.

Sarcoplasmic reticulum The specialized area of the myocyte for storage of calcium required to trigger contraction of the actin and myosin filaments.

Septal necrosis Infarction of an area of the septum.

Semilunar valve See pulmonic valve.

Septum The curved slanting wall that separates the right and left ventricles of the heart and is composed of a muscular lower part and a thinner more membranous upper part.

Sick sinus syndrome A collection of heart rhythm disorders that include persistent sinus bradycardia, tachycardias, bradycardia-tachycardia (alternating slow and fast heart rhythms), atrial fibrillation and atrial flutter.

Sinoatrial node A small mass of tissue made up of nerve fibers that is embedded in the musculature of the right atrium. It initiates the impulses stimulating the heartbeat and is the primary pacemaker of the heart; also called *SA node, sinus node*.

Sinus arrest A pause in the normal cardiac rhythm, due to a momentary failure of the sinus node to initiate an impulse, that lasts three or more intervals. The pause is not an exact multiple of the normal cardiac cycle.

Sinus pause A pause in the normal cardiac rhythm, due to a momentary failure of the sinus node to initiate an impulse, that lasts for one to two intervals. The pause is not an exact multiple of the normal cardiac cycle.

Sodium-potassium pump The energy-requiring process by which the cellular membrane actively moves sodium out of the cell and potassium into the cell.

ST segment The phase following ventricular depolarization before repolarization begins. It is helpful in assessing the presence of cardiac ischemia.

Static ECGs An ECG tracing that is printed.

Stroke is injury or death to an area of the brain caused by an interruption in blood flow. It is often due to a blood clot but may also be caused by hemorrhage into the brain.

Stroke volume The volume of blood pumped from a ventricle of the heart in one beat.

Superimposed (waveform) An ECG waveform that forms over another waveform, distorting the appearance of both. It is commonly seen as a P wave superimposed on the T wave of the preceding beat.

Supraventricular Relating to or being a rhythmic abnormality of the heart caused by impulses originating above the ventricles.

Supraventricular tachycardia (SVT) Tachycardia that arises from a site above the ventricles.

Sustained Continuous; failing to cease spontaneously.

Sympathetic The part of the autonomic nervous system that contains chiefly adrenergic fibers. It tends to decrease glandular secretion and to speed the heart rate.

Syncytium The network of anastomoses (connections) between adjoining myocytes.

Systole The contraction of the ventricles during which blood is ejected from the ventricles into the aorta and pulmonary artery.

T

T wave A wave representing ventricular repolarization.

Tachycardia A heart rate greater than 100 when referring to sinus rhythms or a rate greater than is normally expected with any rhythm.

Thrombus is a fibrinous clot that forms within a blood vessel or inside the heart and remains at the site of its formation, often hindering blood flow

Thrombolytics Medications that attack and accelerate the body's ability to break up or lyse a blood clot (thrombus).

Torsades de pointes Ventricular tachycardia characterized by fluctuation of the QRS complexes around the electrocardiographic baseline. It is typically caused by a long QT interval.

Tricuspid valve Heart valve situated between the right atrium and the right ventricle. It resembles the mitral valve in structure but consists of three triangular membranous flaps; also called the *right atrioventricular valve.*

Trifascicular block Is a combination of bifascicular block and prolongation of the PR interval.

Trigeminal Cardiac electrical activity characterized by a pattern of one ectopic beat followed by two normal beats. These ectopic beats can be either atrial or ventricular in origin and indicate cardiac instability.

Triglyceride is an ester formed from glycerol and three fatty acid groups. Triglycerides are the main constituents of natural fats and oils, and high concentrations in the blood indicate an elevated risk of heart attack and stroke.

U

U wave A positive wave that may follow the T wave, representing the final repolarization phase of the ventricles.

Unifocal Complexes originating from the same location (focus).

V

Vagus nerve Either of the pair of 10th cranial nerves that arise from the medulla and chiefly supply the heart with autonomic sensory and motor fibers.

Valvular diseases is any disease process involving one or more of the four heart valves. These conditions often occur as a result of aging, but may also be due to congenital abnormalities or specific disease or physiologic processes including rheumatic heart disease and pregnancy.

Vasospasm is the sudden constriction of an artery, reducing its diameter and amount of blood it can deliver.

Vector The geometric direction of travel of an electrical impulse in the heart.

Vein A vessel that carries blood toward the heart.

Ventricles The chambers of the heart that receive blood from a corresponding atrium and from which blood is forced into the arteries.

Ventricular conduction disturbances Delayed intraventricular conduction that results in abnormal-appearing QRS complexes.

Ventricular standstill A condition in which the atria continue to beat but the ventricles have stopped.

Visceral pleura The fibrous connective tissue adhering to the outer surface of an organ.

W

Wenckebach Mobitz type I See 2nd-degree AV block, type I.

Wide complex tachycardia of unknown origin Tachycardia with wide and bizarre-looking QRS complexes but with an unknown site of origin.

Wide complex regular tachycardia Tachycardia with wide and bizarre-looking QRS complexes but with a regular rhythm. Indicates rhythm is arising from one site.

Wide complex irregular tachycardia Tachycardia with wide and bizarre-looking QRS complexes but with an irregular rhythm. Indicates rhythm is arising from more than one site.

Index